Springer Tracts in Modern Physics

Volume 248

For further volumes:
http://www.springer.com/series/426

Springer Tracts in Modern Physics

Springer Tracts in Modern Physics provides comprehensive and critical reviews of topics of current interest in physics. The following fields are emphasized: elementary particle physics, solid-state physics, complex systems, and fundamental astrophysics.

Suitable reviews of other fields can also be accepted. The editors encourage prospective authors to correspond with them in advance of submitting an article. For reviews of topics belonging to the above mentioned fields, they should address the responsible editor, otherwise the managing editor.

See also springer.com

Springer Tracts in Modern Physics

Volume 248

For further volumes:
http://www.springer.com/series/426

Springer Tracts in Modern Physics

Springer Tracts in Modern Physics provides comprehensive and critical reviews of topics of current interest in physics. The following fields are emphasized: elementary particle physics, solid-state physics, complex systems, and fundamental astrophysics.

Suitable reviews of other fields can also be accepted. The editors encourage prospective authors to correspond with them in advance of submitting an article. For reviews of topics belonging to the above mentioned fields, they should address the responsible editor, otherwise the managing editor.

See also springer.com

Stefan Yoshi Buhmann

Dispersion Forces II

Many-Body Effects, Excited Atoms, Finite Temperature and Quantum Friction

Springer

Stefan Yoshi Buhmann
Quantum Optics and Laser Science
Imperial College London
London
UK

ISSN 0081-3869 ISSN 1615-0430 (electronic)
ISBN 978-3-642-32465-9 ISBN 978-3-642-32466-6 (eBook)
DOI 10.1007/978-3-642-32466-6
Springer Heidelberg New York Dordrecht London

Library of Congress Control Number: 2012944385

Printed on acid-free paper

Springer is part of Springer Science+Business Media (www.springer.com)

For Daniela, Ole and Jona

Preface

This book is volume I in a two-piece study of dispersion forces as described within the context of macroscopic quantum electrodynamics (QED) in dispersing and absorbing media. Its purpose is threefold: To provide insights and intuitions into macroscopic QED and dispersion forces; to enable the reader to perform his/her own calculations of such forces; and to serve as a reference for dispersion forces in concrete geometries and scenarios. For these purposes, calculations and derivations are laid out in detail and broken down into small steps. Common tricks and approximations are explicitly shown. The results are linked to the pioneering historic works as well as recent research in the field and made plausible by simple physical models.

The book is mainly aimed at three groups of readers. First, it shall provide graduate and postgraduate students with a practical introduction to the field of dispersion forces. While mainly intended for self-study, it can also serve as the basis for a graduate lecture course where many of the worked examples can be used as exercises. Second, this book shall provide researchers from various fields with an overview on macroscopic QED and dispersion forces, providing them with both qualitative results and the theoretical tools for quantitative calculations. Finally, it should serve experimentalists as a means to numerically evaluate dispersion forces and potentials for relevant practical scenarios.

While the basics of macroscopic QED as well as dispersion forces between ground-state objects have been covered in detail in Vol. I, this Vol. II addresses more advanced topics most of which are subject to the current research. These include relations between dispersion forces, Casimir–Polder (CP) potentials of excited or moving atoms, and the impact of finite temperature. To ensure that Vol. II can be read independently, the material of Vol. I is briefly reviewed in the beginning. Occasionally, references to the more detailed material in Vol. I are given.

The content of this volume is laid out as follows. Chapter 1 contains a review of the main results of Vol. I. It summarizes the formalism of macroscopic QED as well as the calculations of ground-state dispersion potentials. For CP potentials, these calculations will be generalized and presented in more detail in Chap. 4.

Readers of Vol. I can skip this Chap. 1, although it might give a new, more condensed, and unified view on macroscopic QED and ground-state dispersion forces.

In Vol. I, ground-state dispersion forces have been calculated explicitly for highly symmetric geometries. In Chap. 2 of this volume, we develop methods for approximating CP potentials for bodies of arbitrary shapes. Based on a Born expansion, it is shown that the potential can be alternatively obtained from a series of volume integrals over the bodies or by summing over appropriately chosen body parts. These approximations are illustrated for the examples of a ring and an inhomogeneous half space.

In Chap. 3, we compare Casimir forces between bodies, CP forces between atoms and bodies, and van der Waals (vdW) forces between atoms and draw connections between them. Reviewing asymptotic power laws for various geometries in the long and short-distance limits, we show that they are special cases of the general scaling behavior of dispersion forces. Using the methods of Chap. 2, we show that forces on bodies are simple sums over the forces on the atoms contained therein in the dilute-gas limit. For more dense bodies, many-atom contributions need to be taken into account. This is explicitly demonstrated for the CP potential, leading to general expressions for many-atom vdW potentials.

The CP potential of a ground-state atom is studied in detail in Chap. 4 of Vol. I, as reviewed in Chap. 1 of this volume. These results are extended to excited atoms in Chap. 4, where we derive the CP potential of an excited atom by means of perturbation theory. The alternative minimal and multipolar coupling schemes are seen to lead to equivalent results. Invoking the Green's tensor given in App. A, we discuss the examples of an excited atom in front of a perfectly conducting plate or a magnetodielectric half space. The more advanced scenario of an atom in front of a meta-material superlens is also considered.

The results are further generalized in Chap. 5 where the dynamics of the excited-state force is considered. As shown, the time-dependent force can be found from the quantum averaged Lorentz force. It is governed by the spontaneous decay of the initially excited atom. As illustrated by the example of an atom near a plate, the strength of the excited state force sensitively depends on the environment-induced shifts and broadenings of the atomic transition frequencies.

Chapter 6 focusses of the resonant force on an excited atom under strong-coupling conditions in cavity QED. Using the Jaynes–Cummings model together with a dressed-state approach, we generalize the approach of Chap. 4 beyond perturbation theory. The CP potential follows from the eigenenergies of the strongly coupled atom–field system. In close similarity to Chap. 5, we also address the dynamics of the strong-coupling force.

The impact of finite temperature on the CP force is addressed in Chap. 7. We first use a perturbative approach to calculate the CP potential of a ground-state or excited atom in a finite-temperature environment. Using the examples of an atom in front of a perfectly conducting plate or a metal half space, we illustrate the intertwined dependence of the thermal CP potential on distance, temperature, and atomic transition frequencies. Following the Lorentz-force approach of Chap. 5, we then consider the dynamics of the force for non-equilibrium scenarios.

The final Chap. 8 is devoted to the effect of motion on the CP force. Using the Lorentz-force approach, the leading non-relativistic velocity dependence of the force is derived. The results are applied to the quantum friction on an atom moving parallel to a plate. The differences of quantum friction for excited versus ground-state atoms near metal or dielectric plates are discussed.

Two appendices provide technical background and reference material. Appendix A collects information about the classical Green's tensor for the electromagnetic field. In addition to reviewing the general properties and specific examples contained in App. B of volume I, the scaling behavior and Born expansion of the Green's tensor are given. Appendix B is a brief review of atomic physics as needed for the examples studied in Chaps. 7 and 8.

The content of this book has originated in my research at Friedrich-Schiller-University of Jena during my Ph.D. and subsequently at Imperial College London. It was supported by Thuringian Ministry of Science, the E.–W. Kuhlmann-Foundation, the German Research Foundation, the Alexander von Humboldt Foundation, and the Engineering and Physical Sciences Research Council, UK. I am deeply indebted to my Ph.D. supervisor D.-G. Welsch who has introduced to research in general and macroscopic QED in particular. This work would not have been possible without the support of my hosts at Imperial College, S. Scheel and E. A. Hinds. Some of the results contained in this book were obtained in collaboration with Ho Trung Dung and A. Sambale at Friedrich-Schiller-University of Jena; J. A. Crosse and M. R. Tarbutt at Imperial College; and S. Å. Ellingsen from the Norwegian University of Science and Technology in Trondheim. I am grateful to G. Barton, D. Bloch, M. Ducloy, H. Haakh and M. S. Kim for discussions. In addition, I would like to thank L. Arntzen, A. V. Chizhov, D. A. R. Dalvit, C. Farina, M. DeKieviet, F. Haake, D. Meschede, P. Milonni, F. S. S. da Rosa, and H. Ulbricht for their kind hospitality and I. V. Bondarev, D. A. R. Dalvit, S. Å. Ellingsen, F. Intraviaia, A. Jacob, V. N. Marachevsky, F. S. S. da Rosa, A. Sambale, Y. Sherkunov, and M. S. Tomaš for their visits. Some of these visits were made possible by the network 'New Trends and Applications of the Casimir Effect' for which I am grateful to both the organisers of the network and the European Science Foundation. I would like to thank D. Baumgärtel, S. Å. Ellingsen, A. Sambale and M. R. Tarbutt for valuable feedback on various parts of the manuscript and P. Hertel for encouragement and advice. I am grateful to C. Ascheron of Springer for his enthusiasm for this project and to P. Wölfle for his useful comments. Finally, I thank my wife, children, parents, and sister for their encouragement, support and distractions.

London, June 2012 Stefan Yoshi Buhmann

Contents

Symbols

$n! = \prod_{k=1}^{n} k$	Factorial
$(2n+1)!! = \prod_{k=1}^{n}(2k+1)$	Double factorial
$\cdot$	Scalar product
$\times$	Vector product
$\partial/\partial x$	Partial derivative with respect to a variable x
∇	Gradient
$\overleftarrow{\nabla}$	Gradient acting to the left
$\dot{} = \partial/\partial t$	Time derivative
$\bar{}$	Fourier transform
* or C.c.	Complex conjugate
† or H.c.	Hermitian conjugate
Re	Real part
Im	Imaginary part
$\parallel$	Longitudinal part
$\perp$	Transverse part
^{T}	Transpose
$\circledast$	Duality transform
$\hat{}$	Operator
$'$	Power–Zienau–Woolley transform
$[\,,]$	Commutator
$\lvert\psi\rangle$	Quantum state
$\langle\hat{f}\rangle = \langle\psi\lvert\hat{f}\rvert\psi\rangle$	Quantum average of an observable $\hat{f}$
$\langle\hat{f}\rangle_T = \mathrm{tr}(\hat{f}\hat{\rho}_T)$	Thermal average of an observable $\hat{f}$
$\Delta\hat{f} = \hat{f} - \langle\hat{f}\rangle$	Quantum fluctuation of an observable $\hat{f}$
$\lvert\{0\}\rangle$	Ground state of the medium-assisted electromagnetic field

$\lvert \mathbf{1}_\lambda(\boldsymbol{r},\omega)\rangle$, $\lvert \mathbf{1}_\lambda(\boldsymbol{r},\omega)\mathbf{1}_{\lambda'}(\boldsymbol{r}',\omega')\rangle$ $\lvert \mathbf{1}_{\lambda_1}(\mathbf{r}_1,\omega_1)\cdots\mathbf{1}_{\lambda_n}(\boldsymbol{r}_n,\omega_n)\rangle$	Single-, two- and n-quantum Fock states state of the medium-assisted electromagnetic field
$1_V(\boldsymbol{r})$	Characteristic function of volume V
$\hat{a}$, $\hat{a}^\dagger$	Photon annihilation and creation operators
$\boldsymbol{A}$	Vector potential for the electromagnetic field
$\hat{A}_{mn} = \lvert m\rangle\langle n\rvert$	Atomic flip operators
$\boldsymbol{\alpha}$, $\boldsymbol{\alpha}_A$, α, α_A	Polarisability of atom A
$\boldsymbol{B}$, $\mathcal{B}$	Magnetic field
$\boldsymbol{\beta}$, $\boldsymbol{\beta}_A$, β, β_A	Magnetisability of atom A
$\boldsymbol{\beta}_d$, $\boldsymbol{\beta}_{d,A}$, β_d, $\beta_{d,A}$	Diamagnetic magnetisability of atom A
$\boldsymbol{\beta}_p$, $\boldsymbol{\beta}_{p,A}$, β_p, $\beta_{p,A}$	Paramagnetic magnetisability of atom A
$c = 2.99792458 \times 10^8$ m/s	Speed of light
C_n	Coefficients for asymptotic power laws of dispersion potentials
χ	Electric susceptibility
χ_λ with $\lambda = e, m$	*e*lectric or *m*agnetic susceptibility
$\boldsymbol{d}$, $\boldsymbol{d}_A$	Electric dipole moment of atom A
$\boldsymbol{d}_{mn} = \langle m\lvert\hat{\boldsymbol{d}}\rvert n\rangle$ $\boldsymbol{d}^A_{mn} = \langle m_A\lvert\hat{\boldsymbol{d}}_A\rvert n_A\rangle$	Electric-dipole matrix elements of atom A
$D_{(n)}$	Frequency denominators for perturbative calculations of van der Waals potentials
$\boldsymbol{D}$, $\mathcal{D}$	Electric excitation
$\delta(\boldsymbol{r})$	Delta function
$\boldsymbol{\delta}(\boldsymbol{r})$	Delta tensor
$\boldsymbol{\delta}^{\parallel}(\boldsymbol{r})$	Longitudinal delta function
$\boldsymbol{\delta}^{\perp}(\boldsymbol{r})$	Transverse delta function
$\delta\omega_n = \sum_k \delta\omega_n^k$	Frequency shift of level n
$\delta\omega_n^k$	Frequency shift of level n due to the influence of level k
$e = 1.60217733 \times 10^{-19}$ C	electron charge
$\boldsymbol{e}_i = (\partial\boldsymbol{r}/\partial i)/\lvert\partial\boldsymbol{r}/\partial i\rvert$ with $i = x, y, z, r, \varphi, \theta$	Unit vectors in cartesian or spherical coordinates
$\boldsymbol{e}_\sigma$ with $\sigma = s, p$	polarisation unit vectors for s- and p-polarised waves
$\boldsymbol{e}_{ij} = (\boldsymbol{r}_i - \boldsymbol{r}_j)/\lvert\boldsymbol{r}_i - \boldsymbol{r}_j\rvert$	Unit vector pointing from $\boldsymbol{r}_j$ to $\boldsymbol{r}_i$
E_n, E_n^A	Eigenenergy of atom A
$\boldsymbol{E}$, $\mathcal{E}$	Electric field
ε	Relative permittivity
$\varepsilon_0 = 8.854187871 \times 10^{-12}$ As/(Vm)	Vacuum permittivity
ϵ_{ijk}	Levi–Civita symbol
η	Number density
$\boldsymbol{f}_\lambda$, $\boldsymbol{f}_\lambda^\dagger$ with $\lambda = e, m$	Fundamental fields

$\boldsymbol{F}$	Force
$\boldsymbol{F}_A$	Force acting on atom A
$\boldsymbol{F}_\alpha$	Force acting on particle α
$\boldsymbol{F}^{(0)}$	Van der Waals force in free space
$\boldsymbol{F}^{(1)}$	Body-induced van der Waals force
$\boldsymbol{F}^{\mathrm{res}}$	Resonant Casimir–Polder force
$\boldsymbol{F}^{\mathrm{nres}}$	Non-resonant Casimir–Polder force
$\boldsymbol{G}$	Green's tensor
$\boldsymbol{G}^{(0)}$	Bulk part of the Green's tensor
$\boldsymbol{G}^{(1)}$	Scattering part of the Green's tensor
$\boldsymbol{G}_{\lambda\lambda'}$ with $\lambda, \lambda' = e, m$	Green's tensor for the *e*lectric or *m*agnetic field ($\lambda = e, m$) created by an *e*lectric or *m*agnetic source ($\lambda' = e, m$)
γ	Damping constant
γ_α	Gyromagnetic ratio of particle α
$\Gamma_n = \sum_{k<n} \Gamma_n^k$	Decay or heating rate of level n
Γ_n^k	Transition rate from level n to level k
H	Hamiltonian
H_A	Hamiltonian of atom A
H_{AF}	Hamiltonian coupling atom A to the medium-assisted electromagnetic field
H_{F}	Hamiltonian of the medium-assisted electromagnetic field
$\boldsymbol{H}$, $\mathcal{H}$	Magnetic excitation
$\hbar = 1.05457266 \times 10^{-34}\,\mathrm{Js}$	Planck constant
$\hat{I}$	Unit operator
$\boldsymbol{I}$	Unit tensor
$\boldsymbol{j}$	Current density
$\boldsymbol{j}_A$	Current density of atom A
$\boldsymbol{j}_{\mathrm{in}}$	Internal current density
$\boldsymbol{j}_{\mathrm{N}}$	Noise current density
$\boldsymbol{j}_{\mathrm{R}}$	Röntgen current density
$\boldsymbol{k}$	Wave vector
$k_{\mathrm{B}} = 1.380658 \times 10^{-23}\,\mathrm{J/K}$	Boltzmann constant
m_A	Mass of atom A
m_α	Mass of particle α
$m_e = 9.1093897 \times 10^{-31}\,\mathrm{kg}$	Electron mass
$\boldsymbol{m}$, $\boldsymbol{m}_A$	Magnetic dipole moment of atom A
$\hat{\boldsymbol{m}}_\alpha = \gamma_\alpha \hat{\boldsymbol{s}}_\alpha$	Spin-induced magnetic dipole moment of particle α
$\boldsymbol{M}$	Magnetisation
$\boldsymbol{M}_A$	Magnetisation of atom A
$\boldsymbol{M}_{\mathrm{N}}$	Noise magnetisation

μ	Relative permeability
$\mu_0 = 4\pi \times 10^{-7}\,\mathrm{Vs/(Am)}$	Vacuum permeability
$n = \sqrt{\varepsilon\mu}$	Refractive index
$\lvert n\rangle$, $\lvert n_A\rangle$	Eigenstate of atom A
$n(\omega)$	Average thermal photon number
$\omega_\mathrm{L} = \sqrt{\omega_\mathrm{T}^2 + \omega_\mathrm{P}^2}$	Longitudinal resonance frequency
ω_{mn}, ω_{mn}^A	Bare transition frequency of atom A
$\tilde{\omega}_{mn} = \omega_{mn} + \delta\omega_m - \delta\omega_n$	Shifted atomic transition frequency
ω_P	Plasma frequency
$\omega_\mathrm{S} = \sqrt{\omega_\mathrm{T}^2 + \omega_\mathrm{P}^2/2}$	Surface plasmon frequency
ω_T	Transverse resonance frequency
ω_-	Minimum of all relevant medium and atomic transition frequencies
ω_+	Maximum of all relevant medium and atomic transition frequencies
$\Omega_{nk} = \tilde{\omega}_{nk} + \frac{\mathrm{i}}{2}(\Gamma_n + \Gamma_k)$	Complex atomic transition frequency
Ω_R	Vacuum Rabi frequency
p_n	Probability of an atom to be in internal state $\lvert n\rangle$
$\boldsymbol{p}_A$	Canonical center-of-mass momentum of atom A
$\boldsymbol{p}_\alpha$	Canonical momentum of particle α
$\bar{\boldsymbol{p}}_\alpha$	Canonical relative momentum of particle α
$\boldsymbol{P}$	Polarisation
$\boldsymbol{P}_A$	Polarisation of atom A
$\boldsymbol{P}_\mathrm{N}$	Noise polarisation
$\mathcal{P}$	Principal value
$\boldsymbol{\Pi}$	Canonically conjugate momenta of the electromagnetic field
ϕ	Scalar potential for the electromagnetic field
ϕ_A	Coulomb potential of atom A
q_α	Charge of particle α
$Q = \omega_\nu/\gamma_\nu$	Q-factor of a cavity with mode frequency ω_ν and width γ_ν
$\boldsymbol{r}_A$	Center-of-mass position of atom A
$\boldsymbol{r}_\alpha$	Position of particle α
$\bar{\boldsymbol{r}}_\alpha$	Relative position of particle α
ρ	Charge density
ρ_A	Charge density of atom A
ρ_in	Internal charge density
ρ_N	Noise charge density
ρ_T	Energy density of the thermal electromagnetic field

$\hat{\rho}$	Density matrix of the electromagnetic field
$\hat{\rho}_T$	Thermal density matrix of the electromagnetic field
$\hat{\boldsymbol{s}}_\alpha$	Spin of particle α
$\mathcal{S}$	Symmetrisation operator
$\mathrm{sgn}(x)$	Sign function
$\hat{\sigma}$	Internal density matrix of an atom
$\hat{\sigma}$, $\hat{\sigma}^\dagger$, $\hat{\sigma}_z$	Pauli lowering, raising and z-operators for a two-level atom
σ_{mn}	Atomic density matrix elements
T	Temperature
$T_\omega = \hbar\omega_\pm/k_\mathrm{B}$	Spectroscopic temperature
$T_z = \hbar c/(z_A k_\mathrm{B})$	Geometric temperature
$\boldsymbol{T}$	Maxwell stress tensor
θ_c	Coupling angle
$\Theta(z)$	Unit step function
U	Potential
$U^{(0)}$	Van der Waals potential in free space
$U^{(1)}$	Body-induced van der Waals potential
U^res	Resonant Casimir–Polder potential
U^nres	Non-resonant Casimir–Polder potential
V	Volume
∂V	Surface of volume V
$z_\omega = c/\omega_\pm$	Spectroscopic length
$z_T = \hbar c/(k_\mathrm{B} T)$	Thermal length
Z	Partition function
ζ	Magnetic susceptibility

Acronyms

CP	Casimir–Polder
QED	Quantum electrodynamics
vdW	van der Waals

Chapter 1
Introduction

As a foundation for the remainder of this book, we first need to review the basic principles of macroscopic quantum electrodynamics (QED) and its application to the calculation of dispersion forces between ground-state atoms and/or bodies. We focus on the main results while also giving a brief idea of how they can be obtained.

1.1 Macroscopic Quantum Electrodynamics

Macroscopic QED is the theory of the quantised electromagnetic field in the presence of macroscopic media. We begin by discussing the properties of the free field and then proceed by introducing the coupling of this field to atoms.

1.1.1 Medium-Assisted Electromagnetic Field

The quantum electromagnetic fields are operator-valued functions of position $\boldsymbol{r}$ and, in the Heisenberg picture, time t. In the presence of magnetoelectric media, we have to distinguish between the electric and magnetic fields $\hat{\boldsymbol{E}}$ and $\hat{\boldsymbol{B}}$ which act on free charges and currents via the Lorentz force, and the electric and magnetic excitations $\hat{\boldsymbol{D}}$ and $\hat{\boldsymbol{H}}$ which are generated by such free charges and currents.[1] An exact quantum theory of the macroscopic electromagnetic field can be constructed by expressing the fields in terms of fundamental degrees of freedom with specified commutation relations and giving the Hamiltonian of the system. This can be uniquely done on the basis of the fluctuation–dissipation theorem for the medium response functions while requiring the macroscopic Maxwell equations and constitutive relations to hold.

[1] Note that alternatively, $\hat{\boldsymbol{B}}$ is commonly referred to as the induction field while $\hat{\boldsymbol{H}}$ is called the magnetic field; and $\hat{\boldsymbol{D}}$ is known as the displacement field.

S. Y. Buhmann, *Dispersion Forces II*, Springer Tracts in Modern Physics 248,
DOI: 10.1007/978-3-642-32466-6_1, © Springer-Verlag Berlin Heidelberg 2012

The constitutive relations describing the dependence of $\hat{\boldsymbol{D}}$ and $\hat{\boldsymbol{H}}$ on $\hat{\boldsymbol{E}}$ and $\hat{\boldsymbol{B}}$ are best formulated in frequency space. Introducing frequency components of the fields according to

$$\hat{f} = \int_0^\infty \mathrm{d}\omega\, \underline{\hat{f}}(\omega) + \mathrm{H.\,c.}\,, \tag{1.1}$$

the constitutive relations may be written as

$$\underline{\hat{\boldsymbol{D}}} = \varepsilon_0 \varepsilon \underline{\hat{\boldsymbol{E}}} + \underline{\hat{\boldsymbol{P}}}_{\mathrm{N}}\,, \tag{1.2}$$

$$\underline{\hat{\boldsymbol{H}}} = \frac{1}{\mu_0 \mu} \underline{\hat{\boldsymbol{B}}} - \underline{\hat{\boldsymbol{M}}}_{\mathrm{N}}\,, \tag{1.3}$$

The present dispersing and absorbing media are hence represented by linear, causal response functions ε and μ as well as random noise fields $\hat{\boldsymbol{P}}_{\mathrm{N}}$ and $\hat{\boldsymbol{M}}_{\mathrm{N}}$. We assume the medium response to be local and isotropic, so that the electric permittivity ε and the magnetic permeability μ are complex-valued scalar functions of position and frequency. Their imaginary parts must be positive for absorbing media, $\mathrm{Im}\,\varepsilon(\boldsymbol{r},\omega) > 0$ and $\mathrm{Im}\,\mu(\boldsymbol{r},\omega) > 0$ for real frequencies. The reactive and random contributions of the medium are closely related via the fluctuation–dissipation theorem [1]. When applied to our case, it states that the quantum ground-state fluctuations of the noise polarisation and magnetisation are related to the imaginary parts of the permittivity and permeability, respectively:

$$\langle \mathcal{S}[\Delta\hat{\boldsymbol{P}}_{\mathrm{N}}(\boldsymbol{r},\omega)\Delta\hat{\boldsymbol{P}}_{\mathrm{N}}^{\dagger}(\boldsymbol{r}',\omega')]\rangle = \frac{\hbar\varepsilon_0}{2\pi}\,\mathrm{Im}\,\varepsilon(\boldsymbol{r},\omega)\boldsymbol{\delta}(\boldsymbol{r}-\boldsymbol{r}')\delta(\omega-\omega')\,, \tag{1.4}$$

$$\langle \mathcal{S}[\Delta\hat{\boldsymbol{M}}_{\mathrm{N}}(\boldsymbol{r},\omega)\Delta\hat{\boldsymbol{M}}_{\mathrm{N}}^{\dagger}(\boldsymbol{r}',\omega')]\rangle = \frac{\hbar}{2\pi\mu_0}\,\frac{\mathrm{Im}\mu(\boldsymbol{r},\omega)}{|\mu(\boldsymbol{r},\omega)|^2}\,\boldsymbol{\delta}(\boldsymbol{r}-\boldsymbol{r}')\delta(\omega-\omega')\,. \tag{1.5}$$

Here, $\mathcal{S}(\hat{a}\hat{b}) = \frac{1}{2}(\hat{a}\hat{b}+\hat{b}\hat{a})$ denotes a symmetrised operator product.

As mentioned, we require the electromagnetic fields to obey the macroscopic Maxwell equations. In the absence of free charges or currents, they read

$$\boldsymbol{\nabla}\cdot\underline{\hat{\boldsymbol{D}}} = 0\,, \tag{1.6}$$

$$\boldsymbol{\nabla}\cdot\underline{\hat{\boldsymbol{B}}} = 0\,, \tag{1.7}$$

$$\boldsymbol{\nabla}\times\underline{\hat{\boldsymbol{E}}} - \mathrm{i}\omega\underline{\hat{\boldsymbol{B}}} = \boldsymbol{0}\,, \tag{1.8}$$

$$\boldsymbol{\nabla}\times\underline{\hat{\boldsymbol{H}}} + \mathrm{i}\omega\underline{\hat{\boldsymbol{D}}} = \boldsymbol{0}\,. \tag{1.9}$$

Substituting the constitutive relations into (1.9) and making use of (1.8), one finds that the electric field obeys an inhomogeneous Helmholtz equation

$$\left[\nabla \times \frac{1}{\mu}\nabla \times \; - \frac{\omega^2}{c^2}\,\varepsilon\right]\underline{\hat{\boldsymbol{E}}} = \mathrm{i}\mu_0\omega\,\underline{\hat{\boldsymbol{j}}}_{\mathrm{N}} \tag{1.10}$$

with the source being given by the noise current density

$$\underline{\hat{\boldsymbol{j}}}_{\mathrm{N}} = -\mathrm{i}\omega\underline{\hat{\boldsymbol{P}}}_{\mathrm{N}} + \nabla \times \underline{\hat{\boldsymbol{M}}}_{\mathrm{N}}\,. \tag{1.11}$$

Together with the noise charge density

$$\underline{\hat{\rho}}_{\mathrm{N}} = -\nabla \cdot \underline{\hat{\boldsymbol{P}}}_{\mathrm{N}}\,, \tag{1.12}$$

it fulfils the equation of continuity

$$-\,\mathrm{i}\omega\underline{\hat{\rho}}_{\mathrm{N}} + \nabla \cdot \underline{\hat{\boldsymbol{j}}}_{\mathrm{N}} = 0\,. \tag{1.13}$$

By introducing the classical Green's tensor $\boldsymbol{G}$ as the unique solution to the differential equation

$$\left[\nabla \times \frac{1}{\mu(\boldsymbol{r},\omega)}\nabla \times \; - \frac{\omega^2}{c^2}\,\varepsilon(\boldsymbol{r},\omega)\right]\boldsymbol{G}(\boldsymbol{r},\boldsymbol{r}',\omega) = \boldsymbol{\delta}(\boldsymbol{r}-\boldsymbol{r}') \tag{1.14}$$

with the boundary condition $\boldsymbol{G}(\boldsymbol{r},\boldsymbol{r}',\omega) \to \boldsymbol{0}$ for $|\boldsymbol{r}-\boldsymbol{r}'| \to \infty$ for absorbing media, the electric field can be given as

$$\underline{\hat{\boldsymbol{E}}}(\boldsymbol{r},\omega) = \mathrm{i}\mu_0\omega \int \mathrm{d}^3r'\,\boldsymbol{G}(\boldsymbol{r},\boldsymbol{r}',\omega)\cdot\underline{\hat{\boldsymbol{j}}}_{\mathrm{N}}(\boldsymbol{r}',\omega)\,. \tag{1.15}$$

Recall that

$$\boldsymbol{\delta}(\boldsymbol{r}) = \delta(\boldsymbol{r})\boldsymbol{I} = -\Delta\,\frac{1}{4\pi r}\,\boldsymbol{I} \tag{1.16}$$

($\boldsymbol{I}$: unit tensor) is the delta tensor. The Green's tensor is an analytic function of frequency in the upper half of the complex plane. Further useful properties such as the Schwarz reflection principle, Onsager reciprocity and an integral relation are laid out in App. A.1.

Having solved the Maxwell equations in frequency space, an explicit field quantisation can be constructed by expressing noise polarisation and magnetisation in terms of fundamental variables with well-defined commutation relations. We require that the ground-state averages of noise polarisation and magnetisation vanish and that their fluctuations satisfy the fluctuation–dissipation theorem. This can be achieved by introducing bosonic creation and annihilation operators $\hat{\boldsymbol{f}}^{\dagger}_{\lambda}(\boldsymbol{r},\omega)$ and $\hat{\boldsymbol{f}}_{\lambda}(\boldsymbol{r},\omega)$ ($\lambda = e, m$) for the elementary electric and magnetic excitations of the system with commutation relations

$$\left[\hat{\boldsymbol{f}}_{\lambda}(\boldsymbol{r},\omega), \hat{\boldsymbol{f}}_{\lambda'}(\boldsymbol{r}',\omega')\right] = \left[\hat{\boldsymbol{f}}^{\dagger}_{\lambda}(\boldsymbol{r},\omega), \hat{\boldsymbol{f}}^{\dagger}_{\lambda'}(\boldsymbol{r}',\omega')\right] = \boldsymbol{0}\,, \tag{1.17}$$

$$\left[\hat{\boldsymbol{f}}_{\lambda}(\boldsymbol{r},\omega), \hat{\boldsymbol{f}}^{\dagger}_{\lambda'}(\boldsymbol{r}',\omega')\right] = \delta_{\lambda\lambda'}\boldsymbol{\delta}(\boldsymbol{r}-\boldsymbol{r}')\delta(\omega-\omega') \tag{1.18}$$

and a ground-state $|\{0\}\rangle$ defined by

$$\hat{\boldsymbol{f}}_{\lambda}(\boldsymbol{r},\omega)|\{0\}\rangle = \boldsymbol{0} \quad \forall\, \lambda, \boldsymbol{r}, \omega\,. \tag{1.19}$$

Relating them to noise polarisation and magnetisation via [2–5],

$$\underline{\hat{\boldsymbol{P}}}_{\mathrm{N}}(\boldsymbol{r},\omega) = \mathrm{i}\sqrt{\frac{\hbar\varepsilon_0}{\pi}\,\mathrm{Im}\,\varepsilon(\boldsymbol{r},\omega)}\,\hat{\boldsymbol{f}}_{e}(\boldsymbol{r},\omega)\,, \tag{1.20}$$

$$\underline{\hat{\boldsymbol{M}}}_{\mathrm{N}}(\boldsymbol{r},\omega) = \sqrt{\frac{\hbar}{\pi\mu_0}\frac{\mathrm{Im}\,\mu(\boldsymbol{r},\omega)}{|\mu(\boldsymbol{r},\omega)|^2}}\,\hat{\boldsymbol{f}}_{m}(\boldsymbol{r},\omega)\,, \tag{1.21}$$

one can easily verify that the fluctuation–dissipation theorem holds in the form of (1.4) and (1.5).

The electric field can now be expressed in terms of the fundamental variables $\hat{\boldsymbol{f}}_{\lambda}$ and $\hat{\boldsymbol{f}}^{\dagger}_{\lambda}$ by substituting (1.11), (1.20) and (1.21) into (1.15) and recalling (1.1). One finds

$$\begin{aligned}\hat{\boldsymbol{E}}(\boldsymbol{r}) &= \int_0^\infty \mathrm{d}\omega\,\underline{\hat{\boldsymbol{E}}}(\boldsymbol{r},\omega) + \mathrm{H.c.} \\ &= \int_0^\infty \mathrm{d}\omega \sum_{\lambda=e,m}\int \mathrm{d}^3r'\,\boldsymbol{G}_{\lambda}(\boldsymbol{r},\boldsymbol{r}',\omega)\cdot\hat{\boldsymbol{f}}_{\lambda}(\boldsymbol{r}',\omega) + \mathrm{H.c.}\end{aligned} \tag{1.22}$$

with coefficients

$$\boldsymbol{G}_{e}(\boldsymbol{r},\boldsymbol{r}',\omega) = \mathrm{i}\,\frac{\omega^2}{c^2}\sqrt{\frac{\hbar}{\pi\varepsilon_0}\,\mathrm{Im}\,\varepsilon(\boldsymbol{r}',\omega)}\,\boldsymbol{G}(\boldsymbol{r},\boldsymbol{r}',\omega)\,, \tag{1.23}$$

$$\boldsymbol{G}_{m}(\boldsymbol{r},\boldsymbol{r}',\omega) = \mathrm{i}\,\frac{\omega}{c}\sqrt{\frac{\hbar}{\pi\varepsilon_0}\frac{\mathrm{Im}\,\mu(\boldsymbol{r}',\omega)}{|\mu(\boldsymbol{r}',\omega)|^2}}\left[\nabla'\times\boldsymbol{G}(\boldsymbol{r}',\boldsymbol{r},\omega)\right]^{\mathrm{T}}. \tag{1.24}$$

They obey the useful integral relation

$$\sum_{\lambda=e,m} \int \mathrm{d}^3 s\, \boldsymbol{G}_\lambda(\boldsymbol{r},\boldsymbol{s},\omega) \cdot \boldsymbol{G}_\lambda^{*\mathrm{T}}(\boldsymbol{r}',\boldsymbol{s},\omega) = \frac{\hbar\mu_0}{\pi}\,\omega^2\,\mathrm{Im}\,\boldsymbol{G}(\boldsymbol{r},\boldsymbol{r}',\omega) \tag{1.25}$$

which follows directly from the respective integral relation (A.5) of the Green's tensor as given in App. A.1. Using Maxwell equation (1.8), the magnetic field can be expanded in a similar way:

$$\begin{aligned}\hat{\boldsymbol{B}}(\boldsymbol{r}) &= \int_0^\infty \mathrm{d}\omega\, \underline{\hat{\boldsymbol{B}}}(\boldsymbol{r},\omega) + \mathrm{H.\,c.} \qquad (1.26)\\ &= \int_0^\infty \frac{\mathrm{d}\omega}{\mathrm{i}\omega} \sum_{\lambda=e,m} \int \mathrm{d}^3 r'\, \nabla\times\boldsymbol{G}_\lambda(\boldsymbol{r},\boldsymbol{r}',\omega)\cdot\hat{\boldsymbol{f}}_\lambda(\boldsymbol{r}',\omega) + \mathrm{H.\,c.}\end{aligned}$$

It can then be shown that the electric and magnetic fields obey the equal-time commutation relations [2–5]

$$\big[\hat{\boldsymbol{E}}(\boldsymbol{r}),\hat{\boldsymbol{B}}(\boldsymbol{r}')\big] = \frac{\mathrm{i}\hbar}{\varepsilon_0}\nabla\times\boldsymbol{\delta}(\boldsymbol{r}-\boldsymbol{r}') \tag{1.27}$$

as known from free-space QED, cf. App. A of Vol. I. Furthermore, the ground-state fluctuations of the electric field are given by [2, 5]

$$\big\langle \mathcal{S}\big[\Delta\hat{\boldsymbol{E}}(\boldsymbol{r},\omega)\Delta\hat{\boldsymbol{E}}^\dagger(\boldsymbol{r}',\omega')\big]\big\rangle = \frac{\hbar}{2\pi}\,\mu_0\omega^2\,\mathrm{Im}\,\boldsymbol{G}(\boldsymbol{r},\boldsymbol{r}',\omega)\delta(\omega-\omega')\,, \tag{1.28}$$

as required by the fluctuation–dissipation theorem.

Having explicitly quantised the electromagnetic field, we next need to specify the Hamiltonian which governs its dynamics. We require the time-dependent fields in the Heisenberg picture to obey the Maxwell equations

$$\nabla\cdot\hat{\boldsymbol{D}} = 0\,, \tag{1.29}$$

$$\nabla\cdot\hat{\boldsymbol{B}} = 0\,, \tag{1.30}$$

$$\nabla\times\hat{\boldsymbol{E}} + \dot{\hat{\boldsymbol{B}}} = \boldsymbol{0}\,, \tag{1.31}$$

$$\nabla\times\hat{\boldsymbol{H}} - \dot{\hat{\boldsymbol{D}}} = \boldsymbol{0}\,. \tag{1.32}$$

Recall that the first of these equations is the Gauss law, the second one states the non-existence of magnetic monopoles, the third one is the Faraday law of induction and the last one is the Ampère law. With the Maxwell equations in the frequency domain being valid by construction, this requires the time-dependent frequency components of the fields to be Fourier components, in particular, $\hat{\boldsymbol{f}}_\lambda(\boldsymbol{r},\omega,t)=\hat{\boldsymbol{f}}_\lambda(\boldsymbol{r},\omega)\mathrm{e}^{-\mathrm{i}\omega t}$. The Hamiltonian of the medium-assisted electromagnetic field must hence be given by [2–5]

$$\hat{H}_{\mathrm{F}}=\sum_{\lambda=e,m}\int \mathrm{d}^3r\int_0^\infty \mathrm{d}\omega\,\hbar\omega\,\hat{\boldsymbol{f}}_\lambda^\dagger(\boldsymbol{r},\omega)\cdot\hat{\boldsymbol{f}}_\lambda(\boldsymbol{r},\omega)\,; \tag{1.33}$$

it generates the correct Heisenberg equations

$$\dot{\hat{\boldsymbol{f}}}_\lambda(\boldsymbol{r},\omega)=\frac{1}{\mathrm{i}\hbar}\big[\hat{\boldsymbol{f}}_\lambda(\boldsymbol{r},\omega),\hat{H}_{\mathrm{F}}\big]=-\mathrm{i}\omega\hat{\boldsymbol{f}}_\lambda(\boldsymbol{r},\omega)\,, \tag{1.34}$$

recall the commutation relations (1.17) and (1.18).

Eigenstates of the Hamiltonian $\hat{H}_{\mathrm{F}}$ are the Fock states which can be generated by repeated action of the creation operators on the ground state $|\{0\}\rangle$. Single- and two-quantum Fock states are given by

$$|\mathbf{1}_\lambda(\boldsymbol{r},\omega)\rangle=\hat{\boldsymbol{f}}_\lambda^\dagger(\boldsymbol{r},\omega)|\{0\}\rangle\,, \tag{1.35}$$

$$|\mathbf{1}_\lambda(\boldsymbol{r},\omega)\mathbf{1}_{\lambda'}(\boldsymbol{r}',\omega')\rangle=\frac{1}{\sqrt{2}}\hat{\boldsymbol{f}}_{\lambda'}^\dagger(\boldsymbol{r}',\omega')\hat{\boldsymbol{f}}_\lambda^\dagger(\boldsymbol{r},\omega)|\{0\}\rangle\,. \tag{1.36}$$

More generally, an n-quantum Fock state reads

$$|\mathbf{1}_{\lambda_1}(\boldsymbol{r}_1,\omega_1)\dots\mathbf{1}_{\lambda_n}(\boldsymbol{r}_n,\omega_n)\rangle=\frac{1}{\sqrt{n!}}\hat{\boldsymbol{f}}_{\lambda_n}^\dagger(\boldsymbol{r}_n,\omega_n)\cdots\hat{\boldsymbol{f}}_{\lambda_1}^\dagger(\boldsymbol{r}_1,\omega_1)|\{0\}\rangle\,. \tag{1.37}$$

Another important quantum state of the electromagnetic field is the thermal state of uniform temperature T, described by the density matrix

$$\hat{\rho}_T=\frac{\mathrm{e}^{-\hat{H}_{\mathrm{F}}/(k_{\mathrm{B}}T)}}{\mathrm{tr}\big[\mathrm{e}^{-\hat{H}_{\mathrm{F}}/(k_{\mathrm{B}}T)}\big]} \tag{1.38}$$

(k_{B}: Boltzmann constant). Non-vanishing thermal averages $\langle\dots\rangle_T=\mathrm{tr}(\dots\hat{\rho}_T)$ of the fundamental fields are given by [6]

$$\big\langle\hat{\boldsymbol{f}}_\lambda^\dagger(\boldsymbol{r},\omega)\hat{\boldsymbol{f}}_{\lambda'}(\boldsymbol{r}',\omega')\big\rangle_T=n(\omega)\delta_{\lambda\lambda'}\boldsymbol{\delta}(\boldsymbol{r}-\boldsymbol{r}')\delta(\omega-\omega')\,, \tag{1.39}$$

$$\big\langle\hat{\boldsymbol{f}}_\lambda(\boldsymbol{r},\omega)\hat{\boldsymbol{f}}_{\lambda'}^\dagger(\boldsymbol{r}',\omega')\big\rangle_T=[n(\omega)+1]\delta_{\lambda\lambda'}\boldsymbol{\delta}(\boldsymbol{r}-\boldsymbol{r}')\delta(\omega-\omega')\,, \tag{1.40}$$

with

$$n(\omega)=\frac{1}{\mathrm{e}^{\hbar\omega/(k_{\mathrm{B}}T)}-1} \tag{1.41}$$

being the average thermal photon number as governed by Bose–Einstein statistics. Using the field expansion (1.22), the fluctuation–dissipation theorem takes the form

$$\left\langle \mathcal{S}\left[\Delta\hat{\boldsymbol{E}}(\boldsymbol{r},\omega)\Delta\hat{\boldsymbol{E}}^{\dagger}(\boldsymbol{r}',\omega')\right]\right\rangle_T = \frac{\hbar}{\pi}\left[n(\omega)+\tfrac{1}{2}\right]\mu_0\omega^2\,\mathrm{Im}\,\boldsymbol{G}(\boldsymbol{r},\boldsymbol{r}',\omega)\delta(\omega-\omega') \tag{1.42}$$

for finite temperature T. It reduces to the ground-state result (1.28) in the zero-temperature limit. For large temperatures, one finds

$$\left\langle \mathcal{S}\left[\Delta\hat{\boldsymbol{E}}(\boldsymbol{r},\omega)\Delta\hat{\boldsymbol{E}}^{\dagger}(\boldsymbol{r}',\omega')\right]\right\rangle_T = \frac{k_\mathrm{B}T}{\pi\omega}\,\mu_0\omega^2\,\mathrm{Im}\,\boldsymbol{G}(\boldsymbol{r},\boldsymbol{r}',\omega)\delta(\omega-\omega')\,, \tag{1.43}$$

in agreement with classical physics, cf. (2.174) in Vol. I.

Finally, we introduce scalar and vector potentials for the electromagnetic field

$$\hat{\boldsymbol{E}} = -\boldsymbol{\nabla}\hat{\phi} - \dot{\hat{\boldsymbol{A}}}\,, \tag{1.44}$$

$$\hat{\boldsymbol{B}} = \boldsymbol{\nabla}\times\hat{\boldsymbol{A}}\,, \tag{1.45}$$

which are useful for the introduction of atom–field interactions in the next section. We will work in Coulomb gauge $\boldsymbol{\nabla}\cdot\hat{\boldsymbol{A}} = 0$ throughout this book, so that the potentials are uniquely related to the longitudinal and transverse parts of the electric field:

$$\hat{\boldsymbol{E}}^{\parallel} = -\boldsymbol{\nabla}\hat{\phi}\,, \qquad \hat{\boldsymbol{E}}^{\perp} = -\dot{\hat{\boldsymbol{A}}}\,. \tag{1.46}$$

The longitudinal ($\parallel$) and transverse ($\perp$) parts of a vector field $\boldsymbol{f}$ are given by

$$\boldsymbol{f}^{\parallel(\perp)}(\boldsymbol{r}) = \int \mathrm{d}^3r'\,\boldsymbol{\delta}^{\parallel(\perp)}(\boldsymbol{r}-\boldsymbol{r}')\cdot\boldsymbol{f}(\boldsymbol{r}')\,, \tag{1.47}$$

with

$$\boldsymbol{\delta}^{\parallel}(\boldsymbol{r}) = -\boldsymbol{\nabla}\boldsymbol{\nabla}\frac{1}{4\pi r}\,, \qquad \boldsymbol{\delta}^{\perp}(\boldsymbol{r}) = \boldsymbol{\nabla}\times(\boldsymbol{\nabla}\times\boldsymbol{I})\frac{1}{4\pi r} \tag{1.48}$$

being the longitudinal and transverse delta functions. Recalling (1.22) for the electric field, we can easily express the potentials in terms of the fundamental fields:

$$\begin{aligned}\boldsymbol{\nabla}\hat{\phi}(\boldsymbol{r}) &= \int_0^\infty \mathrm{d}\omega\,\boldsymbol{\nabla}\underline{\hat{\phi}}(\boldsymbol{r},\omega) + \mathrm{H.\,c.}\\ &= -\sum_{\lambda=e,m}\int \mathrm{d}^3r'\int_0^\infty \mathrm{d}\omega\,{}^{\parallel}\boldsymbol{G}_\lambda(\boldsymbol{r},\boldsymbol{r}',\omega)\cdot\hat{\boldsymbol{f}}_\lambda(\boldsymbol{r}',\omega) + \mathrm{H.\,c.}\,,\end{aligned} \tag{1.49}$$

$$\hat{\boldsymbol{A}}(\boldsymbol{r}) = \int_0^\infty \mathrm{d}\omega\, \underline{\hat{\boldsymbol{A}}}(\boldsymbol{r},\omega) + \mathrm{H.\,c.}$$

$$= \sum_{\lambda=e,m} \int \mathrm{d}^3 r' \int_0^\infty \frac{\mathrm{d}\omega}{\mathrm{i}\omega}\, {}^{\perp}\boldsymbol{G}_\lambda(\boldsymbol{r},\boldsymbol{r}',\omega)\cdot \hat{\boldsymbol{f}}_\lambda(\boldsymbol{r}',\omega) + \mathrm{H.\,c.} \tag{1.50}$$

where left/right longitudinal or transverse components of a tensor field $\boldsymbol{T}(\boldsymbol{r},\boldsymbol{r}')$ have been denoted as

$${}^{\|/\perp}\boldsymbol{T}^{\|/\perp}(\boldsymbol{r},\boldsymbol{r}') = \int \mathrm{d}^3 s \int \mathrm{d}^3 s'\, \boldsymbol{\delta}^{\|/\perp}(\boldsymbol{r}-\boldsymbol{s})\cdot \boldsymbol{T}(\boldsymbol{s},\boldsymbol{s}')\cdot \boldsymbol{\delta}^{\|/\perp}(\boldsymbol{s}'-\boldsymbol{r}')\,. \tag{1.51}$$

The vector potential and its canonically conjugate momentum

$$\hat{\boldsymbol{\Pi}} = -\varepsilon_0 \hat{\boldsymbol{E}}^{\perp} \tag{1.52}$$

obey the canonical equal-time commutation relations

$$\left[\hat{\boldsymbol{A}}(\boldsymbol{r}), \hat{\boldsymbol{\Pi}}(\boldsymbol{r}')\right] = \mathrm{i}\hbar \boldsymbol{\delta}^{\perp}(\boldsymbol{r}-\boldsymbol{r}')\,. \tag{1.53}$$

1.1.2 Atom–Field Interactions

A neutral atom or molecule A (briefly referred to as atom in the following) is a bound system of particles $\alpha \in A$ with charges q_α ($\sum_{\alpha\in A} q_\alpha = 0$), masses m_α, positions $\hat{\boldsymbol{r}}_\alpha$, canonically conjugate momenta $\hat{\boldsymbol{p}}_\alpha$, spins $\hat{\boldsymbol{s}}_\alpha$ and associated magnetic moments $\hat{\boldsymbol{m}}_\alpha = \gamma_\alpha \hat{\boldsymbol{s}}_\alpha$ (γ_α: gyromagnetic ratio). The position and spin variables satisfy the canonical equal-time commutation relations

$$\left[\hat{\boldsymbol{r}}_\alpha, \hat{\boldsymbol{p}}_\beta\right] = \mathrm{i}\hbar\delta_{\alpha\beta}\boldsymbol{I}\,, \tag{1.54}$$

$$\left[\hat{\boldsymbol{s}}_\alpha, \hat{\boldsymbol{s}}_\beta\right] = -\mathrm{i}\hbar\delta_{\alpha\beta}\boldsymbol{I}\times\hat{\boldsymbol{s}}_\alpha\,. \tag{1.55}$$

The particles contained in the atom give rise to charge and and current densities

$$\hat{\rho}_A(\boldsymbol{r}) = \sum_{\alpha\in A} q_\alpha \delta(\boldsymbol{r}-\hat{\boldsymbol{r}}_\alpha)\,, \tag{1.56}$$

$$\hat{\boldsymbol{j}}_A(\boldsymbol{r}) = \sum_{\alpha\in A} q_\alpha \mathcal{S}\left[\dot{\hat{\boldsymbol{r}}}_\alpha \delta(\boldsymbol{r}-\hat{\boldsymbol{r}}_\alpha)\right] - \sum_{\alpha\in A} \hat{\boldsymbol{m}}_\alpha \times \nabla\delta(\boldsymbol{r}-\hat{\boldsymbol{r}}_\alpha) \tag{1.57}$$

and a Coulomb potential

$$\hat{\phi}_A(\boldsymbol{r}) = \int \mathrm{d}^3r' \, \frac{\hat{\rho}_A(\boldsymbol{r}')}{4\pi\varepsilon_0|\boldsymbol{r}-\boldsymbol{r}'|} = \sum_{\alpha\in A} \frac{q_\alpha}{4\pi\varepsilon_0|\boldsymbol{r}-\hat{\boldsymbol{r}}_\alpha|} \,. \tag{1.58}$$

The symmetrisation operator $\mathcal{S}$ indicates full symmetrisation of any function of $\hat{\boldsymbol{r}}_\alpha$ and $\hat{\boldsymbol{p}}_\alpha$ after integrating out delta functions, see Sect. 2.4 in Vol. I for details. Note that the charge and current densities obey the continuity equation

$$\dot{\hat{\rho}}_A(\boldsymbol{r}) + \nabla\cdot\hat{\boldsymbol{j}}_A(\boldsymbol{r}) = 0 \,. \tag{1.59}$$

Introducing centre-of-mass and relative coordinates

$$\hat{\boldsymbol{r}}_A = \sum_{\alpha\in A} \frac{m_\alpha}{m_A}\hat{\boldsymbol{r}}_\alpha \,, \qquad \hat{\overline{\boldsymbol{r}}}_\alpha = \hat{\boldsymbol{r}}_\alpha - \hat{\boldsymbol{r}}_A \tag{1.60}$$

($m_A = \sum_{\alpha\in A} m_\alpha$) with associated momenta [7]

$$\hat{\boldsymbol{p}}_A = \sum_{\alpha\in A} \hat{\boldsymbol{p}}_\alpha \,, \qquad \hat{\overline{\boldsymbol{p}}}_\alpha = \hat{\boldsymbol{p}}_\alpha - \frac{m_\alpha}{m_A}\hat{\boldsymbol{p}}_A \tag{1.61}$$

and commutation relations

$$\left[\hat{\boldsymbol{r}}_A, \hat{\boldsymbol{p}}_A\right] = \mathrm{i}\hbar\boldsymbol{I} \,, \tag{1.62}$$

we can define the atomic polarisation and magnetisation as

$$\hat{\boldsymbol{P}}_A(\boldsymbol{r}) = \sum_{\alpha\in A} q_\alpha \hat{\overline{\boldsymbol{r}}}_\alpha \int_0^1 \mathrm{d}\sigma \, \delta\big(\boldsymbol{r} - \hat{\boldsymbol{r}}_A - \sigma\hat{\overline{\boldsymbol{r}}}_\alpha\big) \,, \tag{1.63}$$

$$\hat{\boldsymbol{M}}_A(\boldsymbol{r}) = \sum_{\alpha\in A} q_\alpha \int_0^1 \mathrm{d}\sigma \, \sigma\mathcal{S}\Big[\hat{\overline{\boldsymbol{r}}}_\alpha \times \dot{\hat{\overline{\boldsymbol{r}}}}_\alpha \delta\big(\boldsymbol{r} - \hat{\boldsymbol{r}}_A - \sigma\hat{\overline{\boldsymbol{r}}}_\alpha\big)\Big] + \sum_{\alpha\in A} \hat{\boldsymbol{m}}_\alpha \delta(\boldsymbol{r} - \hat{\boldsymbol{r}}_\alpha) \,. \tag{1.64}$$

They are related to the atomic charge and current densities via

$$\hat{\rho}_A = -\nabla\cdot\hat{\boldsymbol{P}}_A \,, \tag{1.65}$$

$$\hat{\boldsymbol{j}}_A = \dot{\hat{\boldsymbol{P}}}_A + \nabla\times\hat{\boldsymbol{M}}_A + \hat{\boldsymbol{j}}_\mathrm{R} \tag{1.66}$$

where

$$\hat{\boldsymbol{j}}_\mathrm{R}(\boldsymbol{r}) = \nabla\times\mathcal{S}\Big[\hat{\boldsymbol{P}}_A(\boldsymbol{r})\times\dot{\hat{\boldsymbol{r}}}_A\Big] \tag{1.67}$$

is the Röntgen current density [7, 8] associated with the centre-of-mass motion of the atom. In addition, the polarisation is related to the Coulomb potential via

$$\nabla\hat{\phi}_A = \frac{1}{\varepsilon_0}\hat{\boldsymbol{P}}_A^{\parallel}\,. \tag{1.68}$$

Closely related to polarisation and magnetisation are the atomic electric and magnetic dipole moments,

$$\hat{\boldsymbol{d}}_A = \sum_{\alpha\in A} q_\alpha \hat{\overline{\boldsymbol{r}}}_\alpha = \sum_{\alpha\in A} q_\alpha \hat{\boldsymbol{r}}_\alpha\,, \tag{1.69}$$

$$\hat{\boldsymbol{m}}_A = \sum_{\alpha\in A}\left(\frac{q_\alpha}{2}\hat{\overline{\boldsymbol{r}}}_\alpha \times \dot{\hat{\overline{\boldsymbol{r}}}}_\alpha + \hat{\boldsymbol{m}}_\alpha\right). \tag{1.70}$$

The dynamics of a non-relativistic free atom is governed by the Hamiltonian

$$\hat{H}_A = \sum_{\alpha\in A}\frac{\hat{\boldsymbol{p}}_\alpha^2}{2m_\alpha} + \sum_{\substack{\alpha,\beta\in A\\ \alpha\neq\beta}}\frac{q_\alpha q_\beta}{8\pi\varepsilon_0|\hat{\boldsymbol{r}}_\alpha - \hat{\boldsymbol{r}}_\beta|}\,. \tag{1.71}$$

With the aid of the centre-of-mass and relative coordinates (1.60) and (1.61), it can be cast into the alternative form

$$\begin{aligned}\hat{H}_A &= \frac{\hat{\boldsymbol{p}}_A^2}{2m_A} + \sum_{\alpha\in A}\frac{\hat{\overline{\boldsymbol{p}}}_\alpha^2}{2m_\alpha} + \sum_{\substack{\alpha,\beta\in A\\ \alpha\neq\beta}}\frac{q_\alpha q_\beta}{8\pi\varepsilon_0|\hat{\boldsymbol{r}}_\alpha - \hat{\boldsymbol{r}}_\beta|}\\ &= \frac{\hat{\boldsymbol{p}}_A^2}{2m_A} + \sum_n E_n^A|n_A\rangle\langle n_A|\end{aligned} \tag{1.72}$$

where E_n^A and $|n_A\rangle$ denote the eigenenergies and eigenstates of the atom's internal Hamiltonian. By virtue of the canonical commutation relations (1.54) and definition (1.61) of the relative momenta, the atomic Hamiltonian implies the relation

$$\sum_{\alpha\in A}\frac{q_\alpha}{m_\alpha}\langle m_A|\hat{\overline{\boldsymbol{p}}}_\alpha|n_A\rangle = \mathrm{i}\omega_{mn}^A\boldsymbol{d}_{mn}^A \tag{1.73}$$

[$\omega_{mn}^A = (E_m^A - E_n^A)/\hbar$; $\boldsymbol{d}_{mn}^A = \langle m_A|\hat{\boldsymbol{d}}_A|n_A\rangle$]. It leads to the Thomas–Reiche–Kuhn sum rule [9–11]

$$\frac{1}{2\hbar}\sum_k \omega_{kn}^A(\boldsymbol{d}_{nk}^A\boldsymbol{d}_{kn}^A + \boldsymbol{d}_{kn}^A\boldsymbol{d}_{nk}^A) = \sum_{\alpha\in A}\frac{q_\alpha^2}{2m_\alpha}\boldsymbol{I}\,. \tag{1.74}$$

1.1.2.1 Minimal Coupling

The coupling of the charged particles contained in one or several atom A to the electromagnetic field may be implemented by means of the minimal coupling scheme (cf., e.g., Ref. [8]). We combine the Hamiltonians of the free field (1.33) and atoms (1.71), make the replacement $\hat{\boldsymbol{p}}_\alpha \mapsto \hat{\boldsymbol{p}}_\alpha - q_\alpha \hat{\boldsymbol{A}}(\hat{\boldsymbol{r}}_\alpha)$ in the atomic Hamiltonians and add the Coulomb interactions of the atoms with each other and with the body-assisted field. For a complete description of magnetic effects, we furthermore include a Pauli interaction terms coupling the particle spins to the magnetic field. The resulting total Hamiltonian of the system [2–5, 12–14]

$$\begin{aligned}\hat{H} =& \sum_A \sum_{\alpha\in A} \frac{\left[\hat{\boldsymbol{p}}_\alpha - q_\alpha \hat{\boldsymbol{A}}(\hat{\boldsymbol{r}}_\alpha)\right]^2}{2m_\alpha} + \sum_{A,B} \sum_{\substack{\alpha\in A,\beta\in B\\ \alpha\neq\beta}} \frac{q_\alpha q_\beta}{8\pi\varepsilon_0 |\hat{\boldsymbol{r}}_\alpha - \hat{\boldsymbol{r}}_\beta|} \\ &+ \sum_{\lambda=e,m} \int \mathrm{d}^3r \int_0^\infty \mathrm{d}\omega\, \hbar\omega\, \hat{\boldsymbol{f}}_\lambda^\dagger(\boldsymbol{r},\omega)\cdot\hat{\boldsymbol{f}}_\lambda(\boldsymbol{r},\omega) + \sum_A \sum_{\alpha\in A} q_\alpha \hat{\phi}(\hat{\boldsymbol{r}}_\alpha) \\ &- \sum_A \sum_{\alpha\in A} \gamma_\alpha \hat{\boldsymbol{s}}_\alpha \cdot \hat{\boldsymbol{B}}(\hat{\boldsymbol{r}}_\alpha) \\ =& \sum_A \hat{H}_A + \sum_{A\neq B} \hat{H}_{AB} + \hat{H}_{\mathrm{F}} + \sum_A \hat{H}_{A\mathrm{F}} \end{aligned} \tag{1.75}$$

can be separated into field (1.33) and atomic Hamiltonians (1.72), interatomic Coulombs interactions

$$\hat{H}_{AB} = \sum_{\alpha\in A,\beta\in B} \frac{q_\alpha q_\beta}{4\pi\varepsilon_0 |\hat{\boldsymbol{r}}_\alpha - \hat{\boldsymbol{r}}_\beta|} \tag{1.76}$$

and atom–field couplings

$$\begin{aligned}\hat{H}_{A\mathrm{F}} =& \sum_{\alpha\in A} q_\alpha \hat{\phi}(\hat{\boldsymbol{r}}_\alpha) - \sum_{\alpha\in A} \frac{q_\alpha}{m_\alpha} \hat{\boldsymbol{p}}_\alpha \cdot \hat{\boldsymbol{A}}(\hat{\boldsymbol{r}}_\alpha) \\ &+ \sum_{\alpha\in A} \frac{q_\alpha^2}{2m_\alpha} \hat{\boldsymbol{A}}^2(\hat{\boldsymbol{r}}_\alpha) - \sum_{\alpha\in A} \gamma_\alpha \hat{\boldsymbol{s}}_\alpha \cdot \hat{\boldsymbol{B}}(\hat{\boldsymbol{r}}_\alpha)\,. \end{aligned} \tag{1.77}$$

The total electric/magnetic fields and excitations in the presence of the atoms are given by

$$\hat{\boldsymbol{\mathcal{E}}} = \hat{\boldsymbol{E}} - \nabla\hat{\phi}_A\,, \qquad \hat{\boldsymbol{\mathcal{B}}} = \hat{\boldsymbol{B}}\,, \tag{1.78}$$

$$\hat{\boldsymbol{\mathcal{D}}} = \hat{\boldsymbol{D}} - \varepsilon_0\nabla\hat{\phi}_A\,, \qquad \hat{\boldsymbol{\mathcal{H}}} = \hat{\boldsymbol{H}}\,. \tag{1.79}$$

Using the above Hamiltonian, it can be shown that they obey the Maxwell equations [2–5, 12, 14]

$$\nabla \cdot \hat{\mathcal{B}} = 0\,, \tag{1.80}$$

$$\nabla \cdot \hat{\mathcal{D}} = \sum_A \hat{\rho}_A\,, \tag{1.81}$$

$$\nabla \times \hat{\mathcal{E}} + \dot{\hat{\mathcal{B}}} = \boldsymbol{0}\,, \tag{1.82}$$

$$\nabla \times \hat{\mathcal{H}} - \dot{\hat{\mathcal{D}}} = \sum_A \hat{\boldsymbol{j}}_A\,, \tag{1.83}$$

with the atomic charge and current densities acting as sources in the Gauss, and Ampère laws (cf. Sect. 2.4.1 in Vol. I). Conversely, the motion of the charged particles under the influence of the electromagnetic field is governed by the Newton equations

$$m_\alpha \ddot{\hat{\boldsymbol{r}}}_\alpha = q_\alpha \hat{\mathcal{E}}(\hat{\boldsymbol{r}}_\alpha) + q_\alpha \mathcal{S}\Big[\dot{\hat{\boldsymbol{r}}}_\alpha \times \hat{\mathcal{B}}(\hat{\boldsymbol{r}}_\alpha)\Big] + \nabla_\alpha\big[\hat{\boldsymbol{m}}_\alpha \cdot \hat{\mathcal{B}}(\hat{\boldsymbol{r}}_\alpha)\big] \tag{1.84}$$

with particle velocities

$$\dot{\hat{\boldsymbol{r}}}_\alpha = \frac{1}{m_\alpha}\left[\hat{\boldsymbol{p}}_\alpha - q_\alpha \hat{\mathbf{A}}(\hat{\boldsymbol{r}}_\alpha)\right]. \tag{1.85}$$

Here, the first two terms describe the Lorentz force while the last term is the Zeeman force.

When the atom is small compared to the wavelength of the relevant electromagnetic field, we may employ the long-wavelength approximation, where the atoms interact with each other via a dipole–dipole term

$$\hat{H}_{AB} = \frac{\hat{\boldsymbol{d}}_A \cdot \hat{\boldsymbol{d}}_B - 3(\hat{\boldsymbol{d}}_A \cdot \hat{\boldsymbol{e}}_{AB})(\hat{\boldsymbol{e}}_{AB} \cdot \hat{\boldsymbol{d}}_B)}{4\pi\varepsilon_0 \hat{r}^3_{AB}} \tag{1.86}$$

$[\hat{r}_{AB} = |\hat{\boldsymbol{r}}_A - \hat{\boldsymbol{r}}_B|, \hat{\boldsymbol{e}}_{AB} = (\hat{\boldsymbol{r}}_A - \hat{\boldsymbol{r}}_B)/\hat{r}_{AB}]$ and the atom–field interaction simplifies to

$$\hat{H}_{\mathrm{AF}} = -\hat{\boldsymbol{d}}_A \cdot \hat{\mathbf{E}}^{\parallel}(\hat{\boldsymbol{r}}_A) - \sum_{\alpha\in A} \frac{q_\alpha}{m_\alpha} \hat{\overline{\boldsymbol{p}}}_\alpha \cdot \hat{\mathbf{A}}(\hat{\boldsymbol{r}}_A) + \sum_{\alpha\in A} \frac{q_\alpha^2}{2m_\alpha} \hat{\mathbf{A}}^2(\hat{\boldsymbol{r}}_A) - \sum_{\alpha\in A} \gamma_\alpha \hat{\boldsymbol{s}}_\alpha \cdot \hat{\boldsymbol{B}}(\hat{\boldsymbol{r}}_A)\,. \tag{1.87}$$

For non-magnetic atoms, the Pauli term can be neglected and we may employ the electric-dipole approximation

$$\hat{H}_{\mathrm{AF}} = -\hat{\boldsymbol{d}}_A \cdot \hat{\boldsymbol{E}}^{\parallel}(\hat{\boldsymbol{r}}_A) - \sum_{\alpha\in A} \frac{q_\alpha}{m_\alpha} \hat{\bar{\boldsymbol{p}}}_\alpha \cdot \hat{\boldsymbol{A}}(\hat{\boldsymbol{r}}_A) + \sum_{\alpha\in A} \frac{q_\alpha^2}{2m_\alpha} \hat{\boldsymbol{A}}^2(\hat{\boldsymbol{r}}_A) \,. \quad (1.88)$$

1.1.2.2 Multipolar Coupling

The multipolar coupling scheme is an alternative, equivalent description of the interacting atom–field system. It can be obtained by subjecting all variables to a Power–Zienau–Woolley transformation [15–17]

$$\hat{f}' = \hat{U} \hat{f} \hat{U}^\dagger \quad \text{with } \hat{U} = \exp\left[\frac{\mathrm{i}}{\hbar} \int \mathrm{d}^3 r \sum_A \hat{\boldsymbol{P}}_A \cdot \hat{\boldsymbol{A}}\right]. \quad (1.89)$$

Expressing the Hamiltonian (1.75) in terms of these new, primed variables, we obtain the multipolar Hamiltonian [2–5, 12–14, 18]

$$\begin{aligned}
\hat{H} &= \sum_{\lambda=e,m} \int \mathrm{d}^3 r \int_0^\infty \mathrm{d}\omega\, \hbar\omega\, \hat{\boldsymbol{f}}_\lambda'^{\dagger}(\boldsymbol{r},\omega) \cdot \hat{\boldsymbol{f}}_\lambda'(\boldsymbol{r},\omega) \\
&\quad + \sum_A \sum_{\alpha\in A} \frac{1}{2m_\alpha} \mathcal{S}\left[\hat{\boldsymbol{p}}_\alpha' + \int \mathrm{d}^3 r\, \hat{\boldsymbol{\Xi}}_\alpha' \times \hat{\boldsymbol{B}}'\right]^2 + \frac{1}{2\varepsilon_0} \int \mathrm{d}^3 r \sum_A \hat{\boldsymbol{P}}_A'^2 \\
&\quad - \int \mathrm{d}^3 r \sum_A \hat{\boldsymbol{P}}_A' \cdot \hat{\boldsymbol{E}}' - \sum_A \sum_{\alpha\in A} \gamma_\alpha \hat{\boldsymbol{s}}_\alpha' \cdot \hat{\boldsymbol{B}}'(\hat{\boldsymbol{r}}_\alpha') \\
&= \sum_A \hat{H}_A' + \hat{H}_{\mathrm{F}}' + \sum_A \hat{H}_{\mathrm{AF}}'
\end{aligned} \quad (1.90)$$

with

$$\begin{aligned}
\hat{\boldsymbol{\Xi}}_\alpha'(\boldsymbol{r}) &= q_\alpha \hat{\bar{\boldsymbol{r}}}_\alpha' \int_0^1 \mathrm{d}\sigma\, \sigma \delta(\boldsymbol{r} - \hat{\boldsymbol{r}}_A' - \sigma \hat{\bar{\boldsymbol{r}}}_\alpha') \\
&\quad - \frac{m_\alpha}{m_A} \sum_{\beta\in A} q_\beta \hat{\bar{\boldsymbol{r}}}_\beta' \int_0^1 \mathrm{d}\sigma\, \sigma \delta(\boldsymbol{r} - \hat{\boldsymbol{r}}_A' - \sigma \hat{\bar{\boldsymbol{r}}}_\beta') + \frac{m_\alpha}{m_A} \hat{\boldsymbol{P}}_A'(\boldsymbol{r}) \,.
\end{aligned} \quad (1.91)$$

While the total Hamiltonian is the same as the minimal coupling one, its separation into atom, field and interaction Hamiltonians is different. They now read

$$\hat{H}'_A = \sum_{\alpha\in A} \frac{\hat{\boldsymbol{p}}'^2_\alpha}{2m_\alpha} + \frac{1}{2\varepsilon_0}\int \mathrm{d}^3r\, \hat{\boldsymbol{P}}'^2_A$$
$$= \frac{\hat{\boldsymbol{p}}'^2_A}{2m_A} + \sum_{\alpha\in A} \frac{\hat{\overline{\boldsymbol{p}}}'^2_\alpha}{2m_\alpha} + \frac{1}{2\varepsilon_0}\int \mathrm{d}^3r\, \hat{\boldsymbol{P}}'^2_A$$
$$= \frac{\hat{\boldsymbol{p}}'^2_A}{2m_A} + \sum_n E'_n |n'\rangle\langle n'| \,, \tag{1.92}$$

$$\hat{H}'_\mathrm{F} = \sum_{\lambda=e,m} \int \mathrm{d}^3r \int_0^\infty \mathrm{d}\omega\, \hbar\omega\, \hat{\boldsymbol{f}}'^\dagger_\lambda(\boldsymbol{r},\omega)\cdot\hat{\boldsymbol{f}}'_\lambda(\boldsymbol{r},\omega)\,, \tag{1.93}$$

$$\hat{H}'_{A\mathrm{F}} = -\int \mathrm{d}^3r\, \hat{\boldsymbol{P}}'_A\cdot\hat{\boldsymbol{E}}' - \int \mathrm{d}^3r\, \hat{\boldsymbol{M}}'_A\cdot\hat{\boldsymbol{B}}' + \sum_{\alpha\in A}\frac{1}{2m_\alpha}\left[\int \mathrm{d}^3r\, \hat{\boldsymbol{\Xi}}'_\alpha \times \hat{\boldsymbol{B}}'\right]^2$$
$$- \frac{1}{m_A}\int \mathrm{d}^3r\, \hat{\boldsymbol{P}}'_A \times \hat{\boldsymbol{p}}'_A \cdot \hat{\boldsymbol{B}}'\,. \tag{1.94}$$

where the canonical magnetisation

$$\hat{\boldsymbol{M}}'_A(\boldsymbol{r}) = \sum_{\alpha\in A}\frac{q_\alpha}{2m_\alpha}\int_0^1 \mathrm{d}\sigma\,\sigma\mathcal{S}\big[\hat{\overline{\boldsymbol{r}}}'_\alpha \times \hat{\overline{\boldsymbol{p}}}'_\alpha \delta(\boldsymbol{r}-\hat{\boldsymbol{r}}'_A-\sigma\hat{\overline{\boldsymbol{r}}}'_\alpha)\big] + \sum_{\alpha\in A}\gamma_\alpha\hat{\boldsymbol{s}}'_\alpha\delta(\boldsymbol{r}-\hat{\boldsymbol{r}}'_\alpha) \tag{1.95}$$

differs from the physical one (1.64) given above. In long-wavelength approximation, the multipolar coupling Hamiltonian reduces to

$$\hat{H}'_{A\mathrm{F}} = -\hat{\boldsymbol{d}}'_A\cdot\hat{\boldsymbol{E}}'(\hat{\boldsymbol{r}}'_A) - \hat{\boldsymbol{m}}'_A\cdot\hat{\boldsymbol{B}}'(\hat{\boldsymbol{r}}'_A) + \sum_{\alpha\in A}\frac{q_\alpha^2}{8m_\alpha}[\hat{\overline{\boldsymbol{r}}}'_\alpha\times\hat{\boldsymbol{B}}'(\hat{\boldsymbol{r}}'_A)]^2$$
$$+\frac{3}{8m_A}[\hat{\boldsymbol{d}}'_A\times\hat{\boldsymbol{B}}'(\hat{\boldsymbol{r}}'_A)]^2 - \frac{1}{m_A}\hat{\boldsymbol{d}}'_A\times\hat{\boldsymbol{p}}'_A\cdot\hat{\boldsymbol{B}}'(\hat{\boldsymbol{r}}'_A) \tag{1.96}$$

with canonical magnetic dipole moment

$$\hat{\boldsymbol{m}}'_A = \sum_{\alpha\in A}\left(\frac{q_\alpha}{2m_\alpha}\hat{\overline{\boldsymbol{r}}}'_\alpha\times\hat{\overline{\boldsymbol{p}}}'_\alpha + \gamma_\alpha\hat{\boldsymbol{s}}'_\alpha\right). \tag{1.97}$$

Here, the first two terms represent electric and magnetic dipole interactions; the next two terms are diamagnetic interactions; and the last term is the Röntgen interaction associated with the centre-of-mass motion. For non-magnetic atoms, we may employ the electric-dipole approximation

$$\hat{H}'_{\mathrm{AF}} = -\hat{\boldsymbol{d}}'_A \cdot \hat{\boldsymbol{E}}'(\hat{\boldsymbol{r}}'_A) - \frac{1}{m_A}\hat{\boldsymbol{d}}'_A \times \hat{\boldsymbol{p}}'_A \cdot \hat{\boldsymbol{B}}'(\hat{\boldsymbol{r}}'_A)\,. \tag{1.98}$$

The main advantages of the multipolar Hamiltonian are the fact that it allows for a systematic multipole expansion and its great simplicity. For well-separated atoms with $\hat{\boldsymbol{P}}'_A \cdot \hat{\boldsymbol{P}}'_B = 0$ for $A \neq B$, there is no direct interaction between the atoms. Furthermore, when neglecting the effect of atomic centre-of-mass motion, the atom–field interaction in electric-dipole approximation consists of a single term only.

Due to the unitarity of the Power–Zienau–Woolley transformation (1.89), the variables of the multipolar coupling scheme obey the same commutation relations as the original ones. The expansions (1.22), (1.26), (1.49), and (1.50) remain valid with primed fundamental fields

$$\hat{\boldsymbol{f}}'_\lambda(\boldsymbol{r},\omega) = \hat{\boldsymbol{f}}_\lambda(\boldsymbol{r},\omega) + \frac{1}{\hbar\omega}\int \mathrm{d}^3r'\,\hat{\boldsymbol{P}}^\perp_A(\boldsymbol{r}')\cdot \boldsymbol{G}^*_\lambda(\boldsymbol{r}',\boldsymbol{r},\omega) \tag{1.99}$$

instead of the original ones. The transformation leads to a change in canonical particle and field momenta,

$$\hat{\boldsymbol{p}}'_\alpha = \hat{\boldsymbol{p}}_\alpha - q_\alpha \hat{\boldsymbol{A}}(\hat{\boldsymbol{r}}_\alpha) - \int \mathrm{d}^3r\,\hat{\boldsymbol{\Xi}}_\alpha \times \hat{\boldsymbol{B}}\,, \tag{1.100}$$

$$\hat{\boldsymbol{\Pi}}' = \hat{\boldsymbol{\Pi}} - \hat{\boldsymbol{P}}^\perp_A \tag{1.101}$$

while leaving the particle positions and the vector potential of the electromagnetic field invariant. As a consequence, the canonical momenta in the multipolar coupling scheme

$$m_\alpha \dot{\hat{\boldsymbol{r}}}_\alpha = \hat{\boldsymbol{p}}'_\alpha + \int \mathrm{d}^3r\,\hat{\boldsymbol{\Xi}}_\alpha \times \hat{\boldsymbol{B}} \tag{1.102}$$

agree with the physical ones in the long-wavelength approximation, $\hat{\boldsymbol{p}}'_\alpha = m_\alpha \dot{\hat{\boldsymbol{r}}}_\alpha$. Recall that this is not the case in the minimal coupling scheme (1.85). On the other hand, the canonical field momentum (1.52) is identical with the transverse part of the physical electric field in the minimal coupling scheme, whereas these quantities differ in the multipolar scheme,

$$\hat{\boldsymbol{\Pi}}' = -\varepsilon_0\,\hat{\boldsymbol{\mathcal{E}}}^\perp - \hat{\boldsymbol{P}}^\perp_A\,. \tag{1.103}$$

The total Hamiltonians for minimal or multipolar coupling being identical, the eigenenergies of the total system and the equations of motion for the physical variables are the same in both schemes. This is not true in general for the eigenenergies and states of the separate atom and field Hamiltonians: The ground state $|\{0'\}\rangle$ of $\hat{H}'_{\mathrm{F}}$,

$$\hat{\boldsymbol{f}}'_\lambda(\boldsymbol{r},\omega)|\{0'\}\rangle = \boldsymbol{0}\quad \forall\,\lambda,\boldsymbol{r},\omega\,, \tag{1.104}$$

is different from that of $\hat{H}_{\mathrm{F}}$; and similarly the eigenstates $|n'_A\rangle$ of $\hat{H}'_A$ have to be distinguished from the eigenstates $|n_A\rangle$ of $\hat{H}_A$. Perturbative treatments in the two schemes will lead to two alternative approximations to the exact eigenenergies of the total Hamiltonian.

1.2 Dispersion Forces

Dispersion forces are effective electromagnetic forces between electrically neutral and unpolarised objects in the absence of external electromagnetic fields. Their existence is one of the surprising consequences of correlated quantum fluctuations: In a classical theory, objects with vanishing charge and current densities are not subject to any electromagnetic force; in particular, when the electromagnetic field also vanishes. In quantum electrodynamics, the closest analogue to the above situation is the ground state where charge and current densities as well as the electromagnetic field vanish on average, but still exhibit non-zero fluctuations. The fluctuations are mutually correlated; fluctuating charge and current densities induce a fluctuating electromagnetic field and vice versa. As a consequence, the Lorentz force

$$\hat{\boldsymbol{F}} = \int_V \mathrm{d}^3r\, \left(\hat{\rho}\hat{\boldsymbol{E}} + \hat{\boldsymbol{j}} \times \hat{\boldsymbol{B}}\right) \tag{1.105}$$

on an object containing charged matter acquires a non-zero ground-state average $\boldsymbol{F} = \langle\hat{\boldsymbol{F}}\rangle$ which is commonly known as the dispersion force. Throughout this book, we will distinguish three types of dispersion forces: Casimir forces between bodies [19, 20]; Casimir–Polder (CP) forces between atoms and bodies [21, 22]; and van der Waals (vdW) forces between atoms [22, 23].

1.2.1 *Casimir Forces*

The Casimir force acting on a magnetoelectric body of volume V in the presence of other bodies is the average Lorentz force

$$\boldsymbol{F} = \int_V \mathrm{d}^3r\, \left\langle \hat{\rho}_{\mathrm{in}}(\boldsymbol{r})\hat{\boldsymbol{E}}(\boldsymbol{r}') + \hat{\boldsymbol{j}}_{\mathrm{in}}(\boldsymbol{r}) \times \hat{\boldsymbol{B}}(\boldsymbol{r}')\right\rangle_{\boldsymbol{r}'\to\boldsymbol{r}} \tag{1.106}$$

acting on its internal charge and current densities when bodies and electromagnetic field are in their ground state $|\{0\}\rangle$. The charge and current densities consist of a random, fluctuating part, and a part that is induced by the electromagnetic field:

$$\underline{\hat{\rho}}_{\mathrm{in}}(\boldsymbol{r},\omega)=\underline{\hat{\rho}}_{\mathrm{N}}(\boldsymbol{r},\omega)-\varepsilon_0\nabla\cdot\left[\chi(\boldsymbol{r},\omega)\underline{\hat{\boldsymbol{E}}}(\boldsymbol{r},\omega)\right], \tag{1.107}$$

$$\underline{\hat{\boldsymbol{j}}}_{\mathrm{in}}(\boldsymbol{r},\omega)=\underline{\hat{\boldsymbol{j}}}_{\mathrm{N}}(\boldsymbol{r},\omega)-\mathrm{i}\varepsilon_0\omega\chi(\boldsymbol{r},\omega)\underline{\hat{\boldsymbol{E}}}(\boldsymbol{r},\omega)+\frac{1}{\mu_0}\nabla\times\left[\zeta(\boldsymbol{r},\omega)\underline{\boldsymbol{B}}(\boldsymbol{r},\omega)\right] \tag{1.108}$$

where

$$\chi(\boldsymbol{r},\omega)=\varepsilon(\boldsymbol{r},\omega)-1\,,\qquad \zeta(\boldsymbol{r},\omega)=1-\frac{1}{\mu(\boldsymbol{r},\omega)} \tag{1.109}$$

are the electric and magnetic susceptibilities.

As shown in detail in Chap. 3 of Vol. I, the ground-state average in (1.106) can be easily evaluated by expressing all quantities in terms of the fundamental fields via (1.11), (1.12), (1.20), (1.21), (1.22) and (1.26) and making use of the commutation relations (1.17) and (1.18). The emerging spatial integrals over products of Green's tensors can be performed by means of the integral relation (1.25) and one finds

$$\boldsymbol{F}=-\frac{\hbar}{\pi}\int_V \mathrm{d}^3r\int_0^\infty \mathrm{d}\omega\,\mathrm{Im}\biggl(\frac{\omega^2}{c^2}\nabla\cdot\left[\chi(\boldsymbol{r},\omega)\boldsymbol{G}(\boldsymbol{r},\boldsymbol{r}',\omega)\right]-\mathrm{tr}\left\{\boldsymbol{I}\times\left[\nabla\times\zeta(\boldsymbol{r},\omega)\nabla\times+\frac{\omega^2}{c^2}\chi(\boldsymbol{r},\omega)\right]\boldsymbol{G}(\boldsymbol{r},\boldsymbol{r}',\omega)\times\overleftarrow{\nabla}'\right\}\biggr)_{\boldsymbol{r}'\to\boldsymbol{r}}, \tag{1.110}$$

note that $[\boldsymbol{G}\times\overleftarrow{\nabla}']_{ij}=G_{ik}\epsilon_{jkl}\overleftarrow{\partial}/\partial x'_l$. The coincidence limit $\boldsymbol{r}'\to\boldsymbol{r}$ has to be performed such that divergent self-forces due to current and charge densities being acted on by their own fields are discarded. To this end, we decompose the body into regions of approximately constant magnetoelectric properties and note that in each of these regions the Green's tensor can then be separated into bulk and scattering parts according to

$$\boldsymbol{G}(\boldsymbol{r},\boldsymbol{r}',\omega)=\boldsymbol{G}^{(0)}(\boldsymbol{r},\boldsymbol{r}',\omega)+\boldsymbol{G}^{(1)}(\boldsymbol{r},\boldsymbol{r}',\omega)\,. \tag{1.111}$$

The bulk part is the Green's tensor of an infinite homogeneous medium and gives rise to self-forces. We discard them by retaining only the scattering Green's tensor which describes the effect of the inhomogeneous environment. After rotating the ω-integral to the purely imaginary axis by means of contour-integral techniques, the Casimir force on a magnetoelectric body of arbitrary shape and material in an arbitrary environment can be given as [5, 14, 24]

$$\begin{aligned}\boldsymbol{F} = &-\frac{\hbar}{\pi}\int\limits_V \mathrm{d}^3r \int\limits_0^\infty \mathrm{d}\omega\, \mathrm{Im}\biggl(\frac{\omega^2}{c^2}\,\boldsymbol{\nabla}\cdot\bigl[\chi(\boldsymbol{r},\omega)\boldsymbol{G}^{(1)}(\boldsymbol{r},\boldsymbol{r}',\omega)\bigr] \\ &-\mathrm{tr}\biggl\{\boldsymbol{I}\times\biggl[\boldsymbol{\nabla}\times\zeta(\boldsymbol{r},\omega)\boldsymbol{\nabla}\times+\frac{\omega^2}{c^2}\chi(\boldsymbol{r},\omega)\biggr]\boldsymbol{G}^{(1)}(\boldsymbol{r},\boldsymbol{r}',\omega)\times\overleftarrow{\boldsymbol{\nabla}}'\biggr\}\biggr)_{\boldsymbol{r}'=\boldsymbol{r}} \\ = &\frac{\hbar}{\pi}\int\limits_V \mathrm{d}^3r \int\limits_0^\infty \mathrm{d}\xi\,\biggl(\frac{\xi^2}{c^2}\,\boldsymbol{\nabla}\cdot\bigl[\chi(\boldsymbol{r},\mathrm{i}\xi)\boldsymbol{G}^{(1)}(\boldsymbol{r},\boldsymbol{r}',\mathrm{i}\xi)\bigr] \\ &+\mathrm{tr}\biggl\{\boldsymbol{I}\times\biggl[\boldsymbol{\nabla}\times\zeta(\boldsymbol{r},\mathrm{i}\xi)\boldsymbol{\nabla}\times-\frac{\xi^2}{c^2}\chi(\boldsymbol{r},\mathrm{i}\xi)\biggr]\boldsymbol{G}^{(1)}(\boldsymbol{r},\boldsymbol{r}',\mathrm{i}\xi)\times\overleftarrow{\boldsymbol{\nabla}}'\biggr\}\biggr)_{\boldsymbol{r}'=\boldsymbol{r}}.\end{aligned} \tag{1.112}$$

For a body that is not immersed in any medium, such that $\chi(\boldsymbol{r},\mathrm{i}\xi)\equiv\zeta(\boldsymbol{r},\mathrm{i}\xi)\equiv 0$ on its surface, the Casimir force can be given in the alternative form

$$\begin{aligned}\boldsymbol{F} = &-\frac{\hbar}{2\pi}\int\limits_V \mathrm{d}^3r \int\limits_0^\infty \mathrm{d}\xi\,\biggl(\frac{\xi^2}{c^2}\chi(\boldsymbol{r},\mathrm{i}\xi)\boldsymbol{\nabla}\mathrm{tr}\,\boldsymbol{G}^{(1)}(\boldsymbol{r},\boldsymbol{r},\mathrm{i}\xi) \\ &-\boldsymbol{\nabla}\Bigl\{\zeta(\boldsymbol{r},\mathrm{i}\xi)\mathrm{tr}\bigl[\boldsymbol{\nabla}\times\boldsymbol{G}^{(1)}(\boldsymbol{r},\boldsymbol{r}',\mathrm{i}\xi)\times\overleftarrow{\boldsymbol{\nabla}}'\bigr]_{\boldsymbol{r}'=\boldsymbol{r}}\Bigr\}\biggr).\end{aligned} \tag{1.113}$$

We will employ it in Sect. 3 to illuminate the microscopic origin of Casimir forces.

When evaluating the Casimir force in concrete examples, it is often useful to give it in terms of a surface rather than a volume integral. To this end, we employ the inhomogeneous Maxwell equations (1.29) and (1.32) in the form

$$\varepsilon_0\boldsymbol{\nabla}\cdot\hat{\boldsymbol{E}} = \hat{\rho}_{\mathrm{in}}\,,\qquad \frac{1}{\mu_0}\boldsymbol{\nabla}\times\hat{\boldsymbol{B}} - \varepsilon_0\dot{\hat{\boldsymbol{E}}} = \hat{\boldsymbol{j}}_{\mathrm{in}} \tag{1.114}$$

together with the homogeneous ones (1.30) and (1.31) to rewrite the Lorentz force as

$$\hat{\boldsymbol{F}} = \int\limits_V \mathrm{d}^3r\,\hat{\rho}_{\mathrm{in}}\hat{\boldsymbol{E}} + \hat{\boldsymbol{j}}_{\mathrm{in}}\times\hat{\boldsymbol{B}} = \int\limits_{\partial V}\mathrm{d}\boldsymbol{A}\cdot\hat{\boldsymbol{T}} - \varepsilon_0\int\limits_V \mathrm{d}^3r\,\frac{\partial}{\partial t}\bigl(\hat{\boldsymbol{E}}\times\hat{\boldsymbol{B}}\bigr)\,. \tag{1.115}$$

The Lorentz force on the internal charges and currents in a body of volume V is thus seen to be given by a surface integral over the Maxwell stress tensor

$$\hat{\boldsymbol{T}} = \varepsilon_0\hat{\boldsymbol{E}}\hat{\boldsymbol{E}} + \frac{1}{\mu_0}\hat{\boldsymbol{B}}\hat{\boldsymbol{B}} - \tfrac{1}{2}\Bigl(\varepsilon_0\,\hat{\boldsymbol{E}}^2 - \frac{1}{\mu_0}\,\hat{\boldsymbol{B}}^2\Bigr)\boldsymbol{I} \tag{1.116}$$

plus a volume integral over a term containing a total time derivative. When the body-assisted field is prepared in its stationary ground-state, the quantum average of the latter term vanishes. The Casimir force is hence determined by the average stress tensor which can readily be evaluated by again using the field expansions (1.22) and (1.26). After removing unphysical self-forces by dropping the contributions from the bulk Green's tensor, one finds

$$\begin{aligned}
\boldsymbol{F} = \frac{\hbar}{\pi} \int_0^\infty \mathrm{d}\omega \int_{\partial V} \mathrm{d}\boldsymbol{A} \cdot \Big\{ & \frac{\omega^2}{c^2} \operatorname{Im} \boldsymbol{G}^{(1)}(\boldsymbol{r}, \boldsymbol{r}, \omega) \\
& - \nabla \times \operatorname{Im} \boldsymbol{G}^{(1)}(\boldsymbol{r}, \boldsymbol{r}', \omega) \times \overleftarrow{\nabla}' |_{\boldsymbol{r}'=\boldsymbol{r}} \\
& - \tfrac{1}{2} \operatorname{tr}\Big[\frac{\omega^2}{c^2} \operatorname{Im} \boldsymbol{G}^{(1)}(\boldsymbol{r}, \boldsymbol{r}, \omega) - \nabla \times \operatorname{Im} \boldsymbol{G}^{(1)}(\boldsymbol{r}, \boldsymbol{r}', \omega) \times \overleftarrow{\nabla}' |_{\boldsymbol{r}'=\boldsymbol{r}} \Big] \boldsymbol{I} \Big\} \\
= -\frac{\hbar}{\pi} \int_0^\infty \mathrm{d}\xi \int_{\partial V} \mathrm{d}\boldsymbol{A} \cdot \Big\{ & \frac{\xi^2}{c^2} \boldsymbol{G}^{(1)}(\boldsymbol{r}, \boldsymbol{r}, \mathrm{i}\xi) + \nabla \times \boldsymbol{G}^{(1)}(\boldsymbol{r}, \boldsymbol{r}', \mathrm{i}\xi) \times \overleftarrow{\nabla}' |_{\boldsymbol{r}'=\boldsymbol{r}} \\
& - \tfrac{1}{2} \operatorname{tr}\Big[\frac{\xi^2}{c^2} \boldsymbol{G}^{(1)}(\boldsymbol{r}, \boldsymbol{r}, \mathrm{i}\xi) + \nabla \times \boldsymbol{G}^{(1)}(\boldsymbol{r}, \boldsymbol{r}', \mathrm{i}\xi) \times \overleftarrow{\nabla}' |_{\boldsymbol{r}'=\boldsymbol{r}} \Big] \boldsymbol{I} \Big\} .
\end{aligned} \tag{1.117}$$

Using the Maxwell stress tensor, the Casimir force on an arbitrary body can thus be given as an integral over its surface.

1.2.2 Casimir–Polder Forces

The CP force on an atom in the presence of magnetoelectric bodies is the quantum average of the Lorentz force (1.105) on the atomic charge and current densities $\hat{\rho}_A$ and $\hat{\boldsymbol{j}}_A$, with the body-assisted field being in its ground state. As discussed in detail in Chap. 5, the average Lorentz force is time-dependent in general, and its evaluation requires solving the coupled atom–field dynamics. For ground-state atoms, the CP force may be alternatively obtained from a time-independent calculation of the CP potential, as shown by Casimir and Polder [22]: Assuming both atom and field to be prepared in their ground-state, they considered the shift of the ground-state energy due to the atom–field coupling. The position-dependent part of this energy shift is the CP potential

$$U(\boldsymbol{r}_A) = \Delta E(\boldsymbol{r}_A) \tag{1.118}$$

from which the CP force can be obtained,

$$F(r_A) = -\nabla U(r_A)\,. \tag{1.119}$$

The approach relies on the Born–Oppenheimer approximation, assuming that the fast internal, electronic motion effectively decouples from the slow centre-of-mass motion. It neglects the effect of this latter motion on the CP force, which is treated in Chap. 8.

Following Casimir and Polder's method, we start from the uncoupled ground state $|0\rangle = |0_A\rangle|\{0\}\rangle$ of $\hat{H}_A + \hat{H}_F$ and calculate the energy shift using perturbation theory:

$$\Delta E = \Delta_1 E + \Delta_2 E + \cdots \tag{1.120}$$

with first and second-order shifts

$$\Delta_1 E = \langle 0|\hat{H}_{AF}|0\rangle\,, \tag{1.121}$$

$$\Delta_2 E = \sum_{I\neq 0} \frac{\langle 0|\hat{H}_{AF}|I\rangle\langle I|\hat{H}_{AF}|0\rangle}{E_0 - E_I}\,. \tag{1.122}$$

The second-order shift has intermediate states $|I\rangle = |k_A\rangle|\mathbf{1}_\lambda(\boldsymbol{r},\omega)\rangle$ and denominators $E_0 - E_I = E_0^A - (E_k^A + \hbar\omega) = -\hbar(\omega_k + \omega)$ where $\omega_k = (E_k^A - E_0^A)/\hbar$ (labels A for atomic operators, transition frequencies and dipole matrix elements will be omitted whenever a single atom is present only). Reading from right to left, it may be envisaged to consist of two processes: Interacting with the electromagnetic field, the atom makes a transition to a higher-energy state while emitting a photon, $\langle I|\hat{H}_{AF}|0\rangle$, followed by a transition back to the ground state accompanied by the reabsorption of the photon, $\langle 0|\hat{H}_{AF}|I\rangle$. As the emission process violates energy conservation, the transition and the emitted photon are purely virtual.

According to the Born–Oppenheimer approximation, we need to consider an atom at given position $\boldsymbol{r}_A$. In the minimal coupling scheme, the required interaction Hamiltonian (1.88) is given by

$$\hat{H}_{AF} = -\hat{\boldsymbol{d}}\cdot\hat{\boldsymbol{E}}^{\parallel}(\boldsymbol{r}_A) - \sum_{\alpha\in A}\frac{q_\alpha}{m_\alpha}\hat{\boldsymbol{p}}_\alpha\cdot\hat{\boldsymbol{A}}(\boldsymbol{r}_A) + \sum_{\alpha\in A}\frac{q_\alpha^2}{2m_\alpha}\hat{\boldsymbol{A}}^2(\boldsymbol{r}_A)\,, \tag{1.123}$$

in long-wavelength approximation for a non-magnetic atom. To leading order in the particle charges, the energy-shift hence consists of the first-order contribution due to the $\hat{\boldsymbol{A}}^2$ coupling plus the second-order corrections due to the $\hat{\boldsymbol{d}}\cdot\hat{\boldsymbol{E}}^{\parallel}$ and $\hat{\boldsymbol{p}}\cdot\hat{\boldsymbol{A}}$ terms. As shown in detail in Chap. 4 of Vol. I (cf. also Chap. 4 of this volume where we will perform a similar calculation for an excited atom), these contributions can be calculated by using the field expansions (1.22) and (1.50) together with the commutation relations (1.17) and (1.18). Using the integral relation (1.25) and the Thomas–Reiche–Kuhn sum rule (1.74), the first-order energy shift is found to be

$$\Delta_1 E = \frac{\mu_0}{\pi} \sum_k \omega_k \int_0^\infty \mathrm{d}\omega\, \boldsymbol{d}_{0k} \cdot \mathrm{Im}\, {}^\perp\boldsymbol{G}^\perp(\boldsymbol{r}_A, \boldsymbol{r}_A, \omega) \cdot \boldsymbol{d}_{k0}\,. \tag{1.124}$$

The second-order shift reads

$$\Delta_2 E = -\frac{\mu_0}{\pi} \sum_k \int_0^\infty \frac{\mathrm{d}\omega}{\omega_k + \omega} \boldsymbol{d}_{0k} \cdot \mathrm{Im} \Big\{ \omega^{2\,\parallel}\boldsymbol{G}^\parallel(\boldsymbol{r}_A, \boldsymbol{r}_A, \omega) + \omega_k^{2\,\perp}\boldsymbol{G}^\perp(\boldsymbol{r}_A, \boldsymbol{r}_A, \omega)$$
$$- \omega_k \omega \Big[{}^\parallel\boldsymbol{G}^\perp(\boldsymbol{r}_A, \boldsymbol{r}_A, \omega) + {}^\perp\boldsymbol{G}^\parallel(\boldsymbol{r}_A, \boldsymbol{r}_A, \omega) \Big] \Big\} \cdot \boldsymbol{d}_{k0} \tag{1.125}$$

where the identity $\boldsymbol{G} = {}^\perp\boldsymbol{G}^\perp + {}^\perp\boldsymbol{G}^\parallel + {}^\parallel\boldsymbol{G}^\perp + {}^\parallel\boldsymbol{G}^\parallel$ has been used. In order to obtain the CP potential, we add the two results and discard the position-independent infinite self-energy associated with the bulk Green's tensor, which is a part of the free-space Lamb-shift (as discussed in Sect. 4.1 of Vol. I).

After rotating the frequency integral to the positive imaginary axis via contour-integral techniques, the CP potential of an electric ground-state atom in an arbitrary environment of magnetoelectric bodies reads $U(\boldsymbol{r}_A) = U_e(\boldsymbol{r}_A)$ with [5, 12, 25]

$$U_e(\boldsymbol{r}_A) = \frac{\hbar\mu_0}{2\pi} \int_0^\infty \mathrm{d}\xi\, \xi^2 \mathrm{tr}\big[\boldsymbol{\alpha}(\mathrm{i}\xi) \cdot \boldsymbol{G}^{(1)}(\boldsymbol{r}_A, \boldsymbol{r}_A, \mathrm{i}\xi)\big] \tag{1.126}$$

where $\boldsymbol{\alpha}(\omega)$ is the atomic ground-state polarisability as given by the dispersion formula [26]

$$\boldsymbol{\alpha}(\omega) = \lim_{\epsilon\to 0+} \frac{1}{\hbar} \sum_k \left(\frac{\boldsymbol{d}_{k0}\boldsymbol{d}_{0k}}{\omega + \omega_k + \mathrm{i}\epsilon} - \frac{\boldsymbol{d}_{0k}\boldsymbol{d}_{k0}}{\omega - \omega_k + \mathrm{i}\epsilon} \right). \tag{1.127}$$

For an isotropic atom, the potential simplifies to

$$U(\boldsymbol{r}_A) = U_e(\boldsymbol{r}_A) = \frac{\hbar\mu_0}{2\pi} \int_0^\infty \mathrm{d}\xi\, \xi^2 \alpha(\mathrm{i}\xi)\, \mathrm{tr}\, \boldsymbol{G}^{(1)}(\boldsymbol{r}_A, \boldsymbol{r}_A, \mathrm{i}\xi)\,. \tag{1.128}$$

with

$$\alpha(\omega) = \lim_{\epsilon\to 0+} \frac{2}{3\hbar} \sum_k \frac{\omega_k |\boldsymbol{d}_{0k}|^2}{\omega_k^2 - \omega^2 - \mathrm{i}\omega\epsilon}\,. \tag{1.129}$$

The calculation of the CP potential is much simpler in the multipolar coupling scheme, because the interaction Hamiltonian (1.98) for a non-magnetic atom consists of a single term when neglecting the effect of atomic centre-of-mass motion:

$$\hat{H}'_{\mathrm{AF}} = -\hat{\boldsymbol{d}}' \cdot \hat{\boldsymbol{E}}'(\boldsymbol{r}_A) . \tag{1.130}$$

Within this scheme, we start from the uncoupled ground state $|0\rangle = |0'_A\rangle|\{0'\}\rangle$ of $\hat{H}'_A + \hat{H}'_F$ and calculate the leading, second-order energy shift with intermediate states $|I\rangle = |n'_A\rangle|\mathbf{1}'_\lambda(\boldsymbol{r},\omega)\rangle$. Using the field expansion (1.22) and the commutation relations (1.17) and (1.18) for the transformed fields, we obtain

$$\Delta E' = \Delta_2 E' = -\frac{\mu_0}{\pi}\sum_k \int_0^\infty \frac{\mathrm{d}\omega}{\omega'_k + \omega}\,\omega^2 \boldsymbol{d}'_{0k}\cdot \operatorname{Im}\boldsymbol{G}(\boldsymbol{r}_A,\boldsymbol{r}_A,\omega)\boldsymbol{d}'_{k0} \tag{1.131}$$

where $\omega'_k = (E_k^{A\prime} - E_0^{A\prime})/\hbar$, $\boldsymbol{d}'_{mn} = \langle m'|\hat{\boldsymbol{d}}'|n'\rangle$. The CP potential can again be obtained by discarding the Lamb shift contribution associated with the bulk Green's tensor (see Sect. 4.2 of Vol. I) and rotating the frequency integral to the imaginary axis. One finds $U'(\boldsymbol{r}_A) = U'_e(\boldsymbol{r}_A)$ with

$$\begin{aligned} U'_e(\boldsymbol{r}_A) &= \frac{\hbar\mu_0}{2\pi}\int_0^\infty \mathrm{d}\xi\,\xi^2 \operatorname{tr}\left[\boldsymbol{\alpha}'(\mathrm{i}\xi)\cdot\boldsymbol{G}^{(1)}(\boldsymbol{r}_A,\boldsymbol{r}_A,\mathrm{i}\xi)\right] \\ &= \frac{\hbar\mu_0}{2\pi}\int_0^\infty \mathrm{d}\xi\,\xi^2 \alpha'(\mathrm{i}\xi)\operatorname{tr}\boldsymbol{G}^{(1)}(\boldsymbol{r}_A,\boldsymbol{r}_A,\mathrm{i}\xi) \end{aligned} \tag{1.132}$$

with

$$\boldsymbol{\alpha}'(\omega) = \lim_{\epsilon\to 0+}\frac{1}{\hbar}\sum_k\left(\frac{\boldsymbol{d}'_{k0}\boldsymbol{d}'_{0k}}{\omega+\omega'_k+\mathrm{i}\epsilon} - \frac{\boldsymbol{d}'_{0k}\boldsymbol{d}'_{k0}}{\omega-\omega'_k+\mathrm{i}\epsilon}\right), \tag{1.133}$$

$$\alpha'(\omega) = \lim_{\epsilon\to 0+}\frac{2}{3\hbar}\sum_k \frac{\omega'_k|\boldsymbol{d}'_{0k}|^2}{\omega'^2_k - \omega^2 - \mathrm{i}\omega\epsilon} \tag{1.134}$$

where the second lines are valid for isotropic atoms.

The leading-order results in the minimal and multipolar coupling schemes have exactly the same form, but the unperturbed energy levels and wave functions are different in general, and so are the transitions frequencies, dipole matrix elements and polarisabilities. As we will see in Chap. 4, the same is true for an atom in an excited energy eigenstate. The calculation in the multipolar coupling scheme is considerably simpler, so we will almost exclusively use it throughout this book, in particular for more complex calculations. We will omit the primes distinguishing the two coupling schemes when either our discussion applies to both schemes or when we are exclusively working in one of the schemes, as will be the case in the following.

The multipolar coupling scheme allows for studying the CP potential of an atom with magnetic properties in a systematic way. For such an atom, we employ the full multipolar interaction Hamiltonian (1.96) in its long-wavelength form, once more neglecting velocity-dependent effects ($m_A \to \infty$)

$$\hat{H}_{\mathrm{AF}} = -\hat{\boldsymbol{d}} \cdot \hat{\boldsymbol{E}}(\hat{\boldsymbol{r}}_A) - \hat{\boldsymbol{m}} \cdot \hat{\boldsymbol{B}}(\hat{\boldsymbol{r}}_A) + \sum_{\alpha \in A} \frac{q_\alpha^2}{8m_\alpha} \big[\hat{\bar{\boldsymbol{r}}}_\alpha \times \hat{\boldsymbol{B}}(\hat{\boldsymbol{r}}_A)\big]^2 . \qquad (1.135)$$

The magnetic properties of the atom manifest themselves in a second-order energy shift due to the magnetic-dipole or paramagnetic interaction $\hat{\boldsymbol{m}} \cdot \hat{\boldsymbol{B}}$ plus a first-order shift due to the diamagnetic $(\hat{\bar{\boldsymbol{r}}} \times \hat{\boldsymbol{B}})^2$ term. Note that throughout this book we restrict our attention to non-chiral atoms whose ground state is an eigenstate of the parity operator. Mixed contributions to the second-order energy shift from the electric-dipole interaction $\hat{\boldsymbol{d}} \cdot \hat{\boldsymbol{E}}$ and the paramagnetic $\hat{\boldsymbol{m}} \cdot \hat{\boldsymbol{B}}$ coupling can then be excluded by a symmetry argument. For a discussion of the CP potential of chiral molecules, see Sect. 4.5 of Vol. I.

The second-order energy shift due to the paramagnetic interaction can be found by using the magnetic-field expansion (1.26),

$$\Delta_2 E = \frac{\mu_0}{\pi} \sum_k \int_0^\infty \frac{\mathrm{d}\omega}{\omega_k + \omega} \boldsymbol{m}_{0k} \cdot \nabla \times \operatorname{Im} \boldsymbol{G}(\boldsymbol{r}_A, \boldsymbol{r}_A, \omega) \times \overleftarrow{\nabla}' \cdot \boldsymbol{m}_{k0} \qquad (1.136)$$

($\boldsymbol{m}_{mn} = \langle m | \hat{\boldsymbol{m}} | n \rangle$) where ∇ and $\overleftarrow{\nabla}'$ are thought to act on the first and second position arguments of the Green's tensor only. Following our usual steps, one can then derive the CP potential of a paramagnetic ground-state atom [13, 14, 27]

$$\begin{aligned} U_p(\boldsymbol{r}_A) &= \frac{\hbar\mu_0}{2\pi} \int_0^\infty \mathrm{d}\xi \, \mathrm{tr}\big[\boldsymbol{\beta}_p(\mathrm{i}\xi) \cdot \nabla \times \boldsymbol{G}^{(1)}(\boldsymbol{r}_A, \boldsymbol{r}_A, \mathrm{i}\xi) \times \overleftarrow{\nabla}'\big] \\ &= \frac{\hbar\mu_0}{2\pi} \int_0^\infty \mathrm{d}\xi \, \beta_p(\mathrm{i}\xi) \mathrm{tr}\big[\nabla \times \boldsymbol{G}^{(1)}(\boldsymbol{r}_A, \boldsymbol{r}_A, \mathrm{i}\xi) \times \overleftarrow{\nabla}'\big] \end{aligned} \qquad (1.137)$$

where the paramagnetic magnetisability

$$\boldsymbol{\beta}_p(\omega) = \lim_{\epsilon\to 0+} \frac{1}{\hbar}\sum_k \left(\frac{\boldsymbol{m}_{k0}\boldsymbol{m}_{0k}}{\omega+\omega_k+\mathrm{i}\epsilon} - \frac{\boldsymbol{m}_{0k}\boldsymbol{m}_{k0}}{\omega-\omega_k+\mathrm{i}\epsilon}\right), \tag{1.138}$$

$$\beta_p(\omega) = \lim_{\epsilon\to 0+} \frac{2}{3\hbar}\sum_k \frac{\omega_k|\boldsymbol{m}_{0k}|^2}{\omega_k^2-\omega^2-\mathrm{i}\omega\epsilon} \tag{1.139}$$

has been introduced.

Following similar steps, the first-order energy shift

$$\Delta_1 E = \frac{\hbar\mu_0}{2\pi}\sum_{\alpha\in A}\frac{q_\alpha^2}{4m_\alpha}\int_0^\infty \mathrm{d}\omega\, \mathrm{tr}\langle 0|\hat{\overline{\boldsymbol{r}}}_\alpha \times \left[\boldsymbol{\nabla}\times \mathrm{Im}\,\boldsymbol{G}(\boldsymbol{r}_A,\boldsymbol{r}_A,\omega)\times\overleftarrow{\boldsymbol{\nabla}}'\right]\times\hat{\overline{\boldsymbol{r}}}_\alpha|0\rangle \tag{1.140}$$

is found to lead to the CP potential [28]

$$\begin{aligned} U_d(\boldsymbol{r}_A) &= \frac{\hbar\mu_0}{2\pi}\int_0^\infty \mathrm{d}\xi\, \mathrm{tr}\left[\boldsymbol{\beta}_d\cdot\boldsymbol{\nabla}\times\boldsymbol{G}^{(1)}(\boldsymbol{r}_A,\boldsymbol{r}_A,\mathrm{i}\xi)\times\overleftarrow{\boldsymbol{\nabla}}'\right] \\ &= \frac{\hbar\mu_0}{2\pi}\int_0^\infty \mathrm{d}\xi\, \beta_d \mathrm{tr}\left[\boldsymbol{\nabla}\times\boldsymbol{G}^{(1)}(\boldsymbol{r}_A,\boldsymbol{r}_A,\mathrm{i}\xi)\times\overleftarrow{\boldsymbol{\nabla}}'\right] \end{aligned} \tag{1.141}$$

of a diamagnetic atom, with the respective diamagnetic magnetisability being given by

$$\boldsymbol{\beta}_d = -\sum_{\alpha\in A}\frac{q_\alpha^2}{4m_\alpha}\langle\hat{\overline{\boldsymbol{r}}}_\alpha^2\boldsymbol{I}-\hat{\overline{\boldsymbol{r}}}_\alpha\hat{\overline{\boldsymbol{r}}}_\alpha\rangle, \tag{1.142}$$

$$\beta_d = -\sum_{\alpha\in A}\frac{q_\alpha^2\langle\hat{\overline{\boldsymbol{r}}}_\alpha^2\rangle}{6m_\alpha}. \tag{1.143}$$

Adding the results (1.137) and (1.141) and introducing the total magnetisability

$$\boldsymbol{\beta}(\omega) = \boldsymbol{\beta}_p(\omega)+\boldsymbol{\beta}_d, \qquad \beta(\omega) = \beta_p(\omega)+\beta_d, \tag{1.144}$$

the total potential of a magnetic atom can be given as

$$
\begin{aligned}
U_m(\boldsymbol{r}_A) &= U_p(\boldsymbol{r}_A) + U_d(\boldsymbol{r}_A) \\
&= \frac{\hbar\mu_0}{2\pi}\int_0^\infty \mathrm{d}\xi\,\mathrm{tr}\big[\boldsymbol{\beta}(\mathrm{i}\xi)\cdot\boldsymbol{\nabla}\times\boldsymbol{G}^{(1)}(\boldsymbol{r}_A,\boldsymbol{r}_A,\mathrm{i}\xi)\times\overleftarrow{\boldsymbol{\nabla}}'\big] \\
&= \frac{\hbar\mu_0}{2\pi}\int_0^\infty \mathrm{d}\xi\,\beta(\mathrm{i}\xi)\mathrm{tr}\big[\boldsymbol{\nabla}\times\boldsymbol{G}^{(1)}(\boldsymbol{r}_A,\boldsymbol{r}_A,\mathrm{i}\xi)\times\overleftarrow{\boldsymbol{\nabla}}'\big]\,. \qquad (1.145)
\end{aligned}
$$

By combining it with our result (1.132) for an electric atom, the full CP potential of an electromagnetic ground-state atom reads

$$
U(\boldsymbol{r}_A) = U_e(\boldsymbol{r}_A) + U_m(\boldsymbol{r}_A)\,. \qquad (1.146)
$$

1.2.3 Van der Waals Forces

The vdW force between two atoms in the presence of magnetoelectric bodies can be derived in close analogy to the CP force: For two atoms A and B, the energy shift induced by the atom–field interactions will have a component that depends on the position of both atoms. According to Casimir and Polder [22], it can be identified as the vdW potential,

$$
U(\boldsymbol{r}_A,\boldsymbol{r}_B) = \Delta E(\boldsymbol{r}_A,\boldsymbol{r}_B) \qquad (1.147)
$$

The body-assisted vdW force on atom A due to atom B is then given by

$$
\boldsymbol{F}(\boldsymbol{r}_A,\boldsymbol{r}_B) = -\boldsymbol{\nabla}_A U(\boldsymbol{r}_A,\boldsymbol{r}_B)\,. \qquad (1.148)
$$

Note that the total force on atom A in the presence of atom B and a body is the sum of the CP force (1.119) between the atom and the body and the body-assisted vdW force between the two atoms,

$$
\boldsymbol{F}_A = \boldsymbol{F}(\boldsymbol{r}_A) + \boldsymbol{F}(\boldsymbol{r}_A,\boldsymbol{r}_B)\,. \qquad (1.149)
$$

To calculate the vdW potential, we work within the multipolar coupling scheme for convenience. Starting from the uncoupled ground state $|0\rangle = |0_A\rangle|0_B\rangle|\{0\}\rangle$ of $\hat{H}_A+\hat{H}_B+\hat{H}_\mathrm{F}$, let us first consider the energy shift due to the interaction Hamiltonian

$$
\hat{H}_{\mathrm{AF}} + \hat{H}_{\mathrm{BF}} = -\hat{\boldsymbol{d}}_A\cdot\hat{\boldsymbol{E}}(\boldsymbol{r}_A) - \hat{\boldsymbol{d}}_B\cdot\hat{\boldsymbol{E}}(\boldsymbol{r}_B) \qquad (1.150)
$$

of two non-magnetic atoms in long-wavelength approximation. Each atom must undergo at least two transitions in order to return to its ground state, so the leading two-atom contributions are contained in the fourth-order shift

$$\Delta_4 E = \sum_{I,II,III\neq 0} \frac{\langle 0|\hat{H}_{\mathrm{AF}}+\hat{H}_{B\mathrm{F}}|III\rangle\langle III|\hat{H}_{\mathrm{AF}}+\hat{H}_{B\mathrm{F}}|II\rangle}{(E_0-E_{III})(E_0-E_{II})} \times \frac{\langle II|\hat{H}_{\mathrm{AF}}+\hat{H}_{B\mathrm{F}}|I\rangle\langle I|\hat{H}_{\mathrm{AF}}+\hat{H}_{B\mathrm{F}}|0\rangle}{(E_0-E_I)}\,. \tag{1.151}$$

Expanding the fourfold product of $\hat{H}_{\mathrm{AF}}+\hat{H}_{B\mathrm{F}}$, we retain those terms where both $\hat{H}_{\mathrm{AF}}$ and $\hat{H}_{B\mathrm{F}}$ appear exactly twice, i.e., each atom emits/absorbs two photons. There is a total of $\binom{4}{2}=6$ possibilities as to which two of the four subsequent interactions atom A is involved in. For each of these, there are two choices for the intermediate photon states, i.e., for the order in which the photons emitted/absorbed by atom A are emitted/absorbed by atom B. There is hence a total of $\binom{4}{2}\times 2=12$ distinct contributions to the vdW potential.

Let us give an example: Atom A undergoes a transition to a higher energy eigenstate while emitting a virtual photon $[|I\rangle=|k_A\rangle|0_B\rangle|\mathbf{1}_\lambda(\boldsymbol{r},\omega)\rangle]$, immediately followed by a downward transition of the same atom accompanied by the emission of a second virtual photon $[|II\rangle=|0_A\rangle|0_B\rangle|\mathbf{1}_\lambda(\boldsymbol{r},\omega)\mathbf{1}_{\lambda'}(\boldsymbol{r}',\omega')\rangle]$. Atom B subsequently absorbs the two emitted photons in the order of their emission while making a transition to a higher energy eigenstate and back $[|III\rangle=|0_A\rangle|l_B\rangle|\mathbf{1}_{\lambda'}(\boldsymbol{r}',\omega')\rangle]$. The corresponding contribution to the energy shift can be evaluated by using the field expansion (1.22), the commutation relations (1.17), (1.18) and the integral relation (1.25):

$$\Delta E_{(1)} = -\frac{\mu_0^2}{\hbar\pi^2}\sum_{k,l}\int_0^\infty \mathrm{d}\omega\,\omega^2\int_0^\infty \mathrm{d}\omega'\,\omega'^2 \times \frac{\boldsymbol{d}_{0l}^B\cdot\operatorname{Im}\mathbf{G}(\boldsymbol{r}_B,\boldsymbol{r}_A,\omega')\cdot\boldsymbol{d}_{0k}^A\boldsymbol{d}_{l0}^B\cdot\operatorname{Im}\mathbf{G}(\boldsymbol{r}_B,\boldsymbol{r}_A,\omega)\cdot\boldsymbol{d}_{k0}^A}{(\omega_l^B+\omega')(\omega+\omega')(\omega_k^A+\omega)}\,. \tag{1.152}$$

The other eleven contributions lead to similar results with different intermediate-state frequency denominators (see Table 1.1). In addition, depending on which of the atoms emits or absorbs the photons, $\boldsymbol{d}_{k0}^A$ and $\boldsymbol{d}_{l0}^B$ (emission) have to be exchanged for $\boldsymbol{d}_{0k}^A$ and $\boldsymbol{d}_{0l}^B$ (absorption) and vice versa. We restrict our attention to atoms with a time-reversal invariant internal Hamiltonian, where the dipole-matrix elements are always real, $\boldsymbol{d}_{k0}^A=\boldsymbol{d}_{0k}^A$, $\boldsymbol{d}_{l0}^B=\boldsymbol{d}_{0l}^B$. The sum over the different contributions to the energy shift is then simply a sum over the frequency denominators. After rearranging this sum, one may employ contour-integral techniques to perform the ω'-integral and rotate the remaining ω-integral to the positive imaginary axis.

Table 1.1 Frequency denominators for the different contributions to the vdW potential

Denominator
$D_{(1)} = (\omega_l^B + \omega')(\omega + \omega')(\omega_k^A + \omega)$
$D_{(2)} = (\omega_l^B + \omega)(\omega + \omega')(\omega_k^A + \omega)$
$D_{(3)} = (\omega_k^A + \omega')(\omega_k^A + \omega_l^B)(\omega_k^A + \omega)$
$D_{(4)} = (\omega_l^B + \omega')(\omega_k^A + \omega_l^B)(\omega_k^A + \omega)$
$D_{(5)} = (\omega_k^A + \omega')(\omega_k^A + \omega_l^B + \omega + \omega')(\omega_k^A + \omega)$
$D_{(6)} = (\omega_l^B + \omega)(\omega_k^A + \omega_l^B + \omega + \omega')(\omega_k^A + \omega)$
$D_{(7)} = (\omega_k^A + \omega')(\omega + \omega')(\omega_l^B + \omega)$
$D_{(8)} = (\omega_k^A + \omega)(\omega + \omega')(\omega_l^B + \omega)$
$D_{(9)} = (\omega_k^A + \omega')(\omega_k^A + \omega_l^B)(\omega_l^B + \omega)$
$D_{(10)} = (\omega_l^B + \omega')(\omega_k^A + \omega_l^B)(\omega_l^B + \omega)$
$D_{(11)} = (\omega_k^A + \omega)(\omega_k^A + \omega_l^B + \omega + \omega')(\omega_l^B + \omega)$
$D_{(12)} = (\omega_l^B + \omega')(\omega_k^A + \omega_l^B + \omega + \omega')(\omega_l^B + \omega)$

The resulting vdW potential of two electric isotropic ground-state atoms in the possible presence of magnetoelectric bodies is given by $U(\boldsymbol{r}_A, \boldsymbol{r}_B) = U_{ee}(\boldsymbol{r}_A, \boldsymbol{r}_B)$ with [14, 18, 29, 30]

$$U_{ee}(\boldsymbol{r}_A, \boldsymbol{r}_B) = -\frac{\hbar\mu_0^2}{2\pi}\int_0^\infty \mathrm{d}\xi\, \xi^4 \alpha_A(\mathrm{i}\xi)\alpha_B(\mathrm{i}\xi) \times \mathrm{tr}\big[\boldsymbol{G}(\boldsymbol{r}_A, \boldsymbol{r}_B, \mathrm{i}\xi)\cdot\boldsymbol{G}(\boldsymbol{r}_B, \boldsymbol{r}_A, \mathrm{i}\xi)\big] \tag{1.153}$$

for isotropic atoms. Note that the atomic polarisabilities (1.134) may be different for the two atoms as they depend on the atomic species via the transition frequencies and dipole-matrix elements. The general result for the vdW potential looks very similar to that for the CP potential (1.132) where the polarisabilities of two atoms appear instead of one and two full Green's tensors connecting the positions of the atoms occur rather than a single, equal-position scattering Green's tensor.

By using the decomposition (1.111) of the Green's tensor into its bulk and scattering parts, the vdW potential can be separated according to

$$U(\boldsymbol{r}_A, \boldsymbol{r}_B) = U^{(0)}(r_{AB}) + U^{(1)}(\boldsymbol{r}_A, \boldsymbol{r}_B) \tag{1.154}$$

into a free-space potential

$$U^{(0)}(\boldsymbol{r}_A, \boldsymbol{r}_B) = -\frac{\hbar\mu_0^2}{2\pi}\int_0^\infty \mathrm{d}\xi\, \xi^4 \alpha_A(\mathrm{i}\xi)\alpha_B(\mathrm{i}\xi) \times \mathrm{tr}\big[\boldsymbol{G}^{(0)}(\boldsymbol{r}_A, \boldsymbol{r}_B, \mathrm{i}\xi)\cdot\boldsymbol{G}^{(0)}(\boldsymbol{r}_B, \boldsymbol{r}_A, \mathrm{i}\xi)\big] \tag{1.155}$$

and a body-induced modification

$$U^{(1)}(\boldsymbol{r}_A,\boldsymbol{r}_B)=-\frac{\hbar\mu_0^2}{2\pi}\int\limits_0^\infty \mathrm{d}\xi\,\xi^4\alpha_A(\mathrm{i}\xi)\alpha_B(\mathrm{i}\xi)$$
$$\times\,\mathrm{tr}\big[2\boldsymbol{G}^{(1)}(\boldsymbol{r}_A,\boldsymbol{r}_B,\mathrm{i}\xi)\cdot\boldsymbol{G}^{(0)}(\boldsymbol{r}_B,\boldsymbol{r}_A,\mathrm{i}\xi)$$
$$+\,\boldsymbol{G}^{(1)}(\boldsymbol{r}_A,\boldsymbol{r}_B,\mathrm{i}\xi)\cdot\boldsymbol{G}^{(1)}(\boldsymbol{r}_B,\boldsymbol{r}_A,\mathrm{i}\xi)\big] \qquad (1.156)$$

of this potential. An analogous decomposition holds for the associated vdW force:

$$\boldsymbol{F}(\boldsymbol{r}_A,\boldsymbol{r}_B)=\boldsymbol{F}^{(0)}(\boldsymbol{r}_A,\boldsymbol{r}_B)+\boldsymbol{F}^{(1)}(\boldsymbol{r}_A,\boldsymbol{r}_B)\,. \qquad (1.157)$$

The free-space vdW force is a pure atom–atom force that obeys Newton's third law of mechanics,

$$\boldsymbol{F}^{(0)}(\boldsymbol{r}_B,\boldsymbol{r}_A)=-\boldsymbol{F}^{(0)}(\boldsymbol{r}_A,\boldsymbol{r}_B)\,. \qquad (1.158)$$

The body-assisted correction $\boldsymbol{F}^{(1)}$ is an atom–atom–body force. Here, a relation of the kind (1.158) does not hold, because the bodies also contribute to the momentum balance of the system.

For atoms with magnetic properties, the total interaction Hamiltonian reads

$$\hat{H}_{\mathrm{AF}}+\hat{H}_{\mathrm{BF}}=-\hat{\boldsymbol{d}}_A\cdot\hat{\boldsymbol{E}}(\boldsymbol{r}_A)-\hat{\boldsymbol{m}}_A\cdot\hat{\boldsymbol{B}}(\hat{\boldsymbol{r}}_A)+\sum_{\alpha\in A}\frac{q_\alpha^2}{8m_\alpha}\big[\hat{\bar{\boldsymbol{r}}}_\alpha\times\hat{\boldsymbol{B}}(\hat{\boldsymbol{r}}_A)\big]^2$$
$$-\,\hat{\boldsymbol{d}}_B\cdot\hat{\boldsymbol{E}}(\boldsymbol{r}_B)-\hat{\boldsymbol{m}}_B\cdot\hat{\boldsymbol{B}}(\hat{\boldsymbol{r}}_B)+\sum_{\beta\in B}\frac{q_\beta^2}{8m_\beta}\big[\hat{\bar{\boldsymbol{r}}}_\beta\times\hat{\boldsymbol{B}}(\hat{\boldsymbol{r}}_B)\big]^2\,. \qquad (1.159)$$

The contributions of the paramagnetic $\hat{\boldsymbol{m}}\cdot\hat{\boldsymbol{B}}$ interactions to the vdW potential can be treated in complete analogy with the case of purely electric atoms. Again considering the fourth-order energy shift, but exchanging $\hat{\boldsymbol{d}}\cdot\hat{\boldsymbol{E}}\mapsto\hat{\boldsymbol{m}}\cdot\hat{\boldsymbol{B}}$, we have to use the magnetic-field expansion (1.26). Following the same steps as above, the vdW potential of two isotropic paramagnetic ground-state atoms is found to be [14, 27, 13]

$$U_{pp}(\boldsymbol{r}_A,\boldsymbol{r}_B)=-\frac{\hbar\mu_0^2}{2\pi}\int\limits_0^\infty \mathrm{d}\xi\,\beta_{p,A}(\mathrm{i}\xi)\beta_{p,B}(\mathrm{i}\xi)\mathrm{tr}\big\{\big[\nabla\times\boldsymbol{G}(\boldsymbol{r}_A,\boldsymbol{r}_B,\mathrm{i}\xi)\times\overleftarrow{\nabla}'\big]$$
$$\cdot\,\big[\nabla\times\boldsymbol{G}(\boldsymbol{r}_B,\boldsymbol{r}_A,\mathrm{i}\xi)\times\overleftarrow{\nabla}'\big]\big\}\,, \qquad (1.160)$$

recall (1.139). Similarly, contributions with combinations of $\hat{\boldsymbol{d}}\cdot\hat{\boldsymbol{E}}$ and $\hat{\boldsymbol{m}}\cdot\hat{\boldsymbol{B}}$ interactions lead to the mixed electric–paramagnetic vdW potentials [14, 13, 27, 30]

$$U_{ep}(\boldsymbol{r}_A,\boldsymbol{r}_B) = -\frac{\hbar\mu_0^2}{2\pi}\int_0^\infty d\xi\,\xi^2\alpha_A(i\xi)\beta_{p,B}(i\xi)$$
$$\times\,\mathrm{tr}\big[\boldsymbol{G}(\boldsymbol{r}_A,\boldsymbol{r}_B,i\xi)\times\overleftarrow{\nabla}'\cdot\nabla\times\boldsymbol{G}(\boldsymbol{r}_B,\boldsymbol{r}_A,i\xi)\big]\,, \tag{1.161}$$

$$U_{pe}(\boldsymbol{r}_A,\boldsymbol{r}_B) = -\frac{\hbar\mu_0^2}{2\pi}\int_0^\infty d\xi\,\xi^2\beta_{p,A}(i\xi)\alpha_B(i\xi)$$
$$\times\,\mathrm{tr}\big\{\big[\nabla\times\boldsymbol{G}(\boldsymbol{r}_A,\boldsymbol{r}_B,i\xi)\big]\cdot\big[\boldsymbol{G}(\boldsymbol{r}_B,\boldsymbol{r}_A,i\xi)\times\overleftarrow{\nabla}'\big]\big\}\,. \tag{1.162}$$

We again assume the atoms to be non-chiral, so that cases where one atom undergoes electric and magnetic-dipole interactions do not contribute.

Contributions involving diamagnetic interactions have to be treated separately. Electric and paramagnetic interactions always occur in pairs in order to return each atom to its ground state. On the contrary, diamagnetic interactions do not necessarily change an atom's internal state so that a single interaction per atom is sufficient. The leading two-atom contributions to the vdW interaction of two diamagnetic atoms are hence of second order in the $(\hat{\boldsymbol{r}}\times\hat{\boldsymbol{B}})^2$ term, cf. (1.122). They correspond to processes where one of the atoms simultaneously emits two photons without changing its internal state while the other atom simultaneously absorbs them. Summing the purely diamagnetic contributions with the aid of the field expansion (1.26) and applying contour-integral techniques, one finds the vdW potential [28]

$$U_{dd}(\boldsymbol{r}_A,\boldsymbol{r}_B) = -\frac{\hbar\mu_0^2}{2\pi}\int_0^\infty d\xi\,\beta_{d,A}\beta_{d,B}\mathrm{tr}\big\{\big[\nabla\times\boldsymbol{G}(\boldsymbol{r}_A,\boldsymbol{r}_B,i\xi)\times\overleftarrow{\nabla}'\big]$$
$$\cdot\,\big[\nabla\times\boldsymbol{G}(\boldsymbol{r}_B,\boldsymbol{r}_A,i\xi)\times\overleftarrow{\nabla}'\big]\big\} \tag{1.163}$$

of two isotropic diamagnetic ground-state atoms, recall (1.143).

Finally, we need to consider mixed contributions where one atom undergoes a diamagnetic transition and the other one makes either two electric or two paramagnetic transitions. These third-order contributions lead to the mixed electric–diamagnetic vdW potentials [28]

$$U_{de}(\boldsymbol{r}_A,\boldsymbol{r}_B) = -\frac{\hbar\mu_0^2}{2\pi}\int_0^\infty d\xi\,\xi^2\beta_{d,A}\alpha_B(i\xi)$$
$$\times\,\mathrm{tr}\big\{\big[\nabla\times\boldsymbol{G}(\boldsymbol{r}_A,\boldsymbol{r}_B,i\xi)\big]\cdot\big[\boldsymbol{G}(\boldsymbol{r}_B,\boldsymbol{r}_A,i\xi)\times\overleftarrow{\nabla}'\big]\big\}\,, \tag{1.164}$$

$$U_{ed}(\boldsymbol{r}_A,\boldsymbol{r}_B) = -\frac{\hbar\mu_0^2}{2\pi}\int_0^\infty d\xi\,\xi^2\alpha_A(i\xi)\beta_{d,B}$$

$$\times \operatorname{tr}\left[\boldsymbol{G}(\boldsymbol{r}_A,\boldsymbol{r}_B,\mathrm{i}\xi)\times\overleftarrow{\nabla}'\cdot\nabla\times\boldsymbol{G}(\boldsymbol{r}_B,\boldsymbol{r}_A,\mathrm{i}\xi)\right] \tag{1.165}$$

and the mixed paramagnetic–diamagnetic ones

$$U_{dp}(\boldsymbol{r}_A,\boldsymbol{r}_B) = -\frac{\hbar\mu_0^2}{2\pi}\int_0^\infty \mathrm{d}\xi\,\beta_{d,A}\beta_{p,B}(\mathrm{i}\xi)\operatorname{tr}\left\{\left[\nabla\times\boldsymbol{G}(\boldsymbol{r}_A,\boldsymbol{r}_B,\mathrm{i}\xi)\times\overleftarrow{\nabla}'\right]\right.$$
$$\left.\cdot\left[\nabla\times\boldsymbol{G}(\boldsymbol{r}_B,\boldsymbol{r}_A,\mathrm{i}\xi)\times\overleftarrow{\nabla}'\right]\right\}, \tag{1.166}$$

$$U_{pd}(\boldsymbol{r}_A,\boldsymbol{r}_B) = -\frac{\hbar\mu_0^2}{2\pi}\int_0^\infty \mathrm{d}\xi\,\beta_{p,A}(\mathrm{i}\xi)\beta_{d,B}\operatorname{tr}\left\{\left[\nabla\times\boldsymbol{G}(\boldsymbol{r}_A,\boldsymbol{r}_B,\mathrm{i}\xi)\times\overleftarrow{\nabla}'\right]\right.$$
$$\left.\cdot\left[\nabla\times\boldsymbol{G}(\boldsymbol{r}_B,\boldsymbol{r}_A,\mathrm{i}\xi)\times\overleftarrow{\nabla}'\right]\right\}. \tag{1.167}$$

Summing the individual magnetic potentials (1.160), (1.163), (1.166) and (1.167), the resulting total vdW of two magnetic atoms can be represented in a compact form by introducing the total magnetisability (1.144):

$$U_{mm}(\boldsymbol{r}_A,\boldsymbol{r}_B) = -\frac{\hbar\mu_0^2}{2\pi}\int_0^\infty \mathrm{d}\xi\,\beta_A(\mathrm{i}\xi)\beta_B(\mathrm{i}\xi)\operatorname{tr}\left\{\left[\nabla\times\boldsymbol{G}(\boldsymbol{r}_A,\boldsymbol{r}_B,\mathrm{i}\xi)\times\overleftarrow{\nabla}'\right]\right.$$
$$\left.\cdot\left[\nabla\times\boldsymbol{G}(\boldsymbol{r}_B,\boldsymbol{r}_A,\mathrm{i}\xi)\times\overleftarrow{\nabla}'\right]\right\}. \tag{1.168}$$

Similarly, we may combine (1.161) with (1.165) and (1.162) with (1.164) to obtain the total mixed electric–magnetic potentials

$$U_{em}(\boldsymbol{r}_A,\boldsymbol{r}_B) = -\frac{\hbar\mu_0^2}{2\pi}\int_0^\infty \mathrm{d}\xi\,\xi^2\alpha_A(\mathrm{i}\xi)\beta_B(\mathrm{i}\xi)$$
$$\times \operatorname{tr}\left[\boldsymbol{G}(\boldsymbol{r}_A,\boldsymbol{r}_B,\mathrm{i}\xi)\times\overleftarrow{\nabla}'\cdot\nabla\times\boldsymbol{G}(\boldsymbol{r}_B,\boldsymbol{r}_A,\mathrm{i}\xi)\right], \tag{1.169}$$

$$U_{me}(\boldsymbol{r}_A,\boldsymbol{r}_B) = -\frac{\hbar\mu_0^2}{2\pi}\int_0^\infty \mathrm{d}\xi\,\xi^2\beta_A(\mathrm{i}\xi)\alpha_B(\mathrm{i}\xi)$$
$$\times \operatorname{tr}\left\{\left[\nabla\times\boldsymbol{G}(\boldsymbol{r}_A,\boldsymbol{r}_B,\mathrm{i}\xi)\right]\cdot\left[\boldsymbol{G}(\boldsymbol{r}_B,\boldsymbol{r}_A,\mathrm{i}\xi)\times\overleftarrow{\nabla}'\right]\right\}. \tag{1.170}$$

Summarising the main results (1.153), (1.168), (1.169) and (1.170) of this section, the general vdW potential of an electromagnetic ground-state atom is given by

Table 1.2 Duality transformation of electromagnetic fields, response functions, dispersion forces and potentials

Dual partners	Duality transformation	
$\hat{\boldsymbol{E}}, \hat{\boldsymbol{H}}$:	$\hat{\boldsymbol{E}}^{\circledast} = \sqrt{\mu_0/\varepsilon_0}\,\hat{\boldsymbol{H}}$,	$\hat{\boldsymbol{H}}^{\circledast} = -\sqrt{\varepsilon_0/\mu_0}\,\hat{\boldsymbol{E}}$
$\hat{\boldsymbol{D}}, \hat{\boldsymbol{B}}$:	$\hat{\boldsymbol{D}}^{\circledast} = \sqrt{\varepsilon_0/\mu_0}\,\hat{\boldsymbol{B}}$,	$\hat{\boldsymbol{B}}^{\circledast} = -\sqrt{\mu_0/\varepsilon_0}\hat{\boldsymbol{D}}$
$\hat{\boldsymbol{P}}_{\mathrm{N}}, \hat{\boldsymbol{M}}_{\mathrm{N}}$:	$\hat{\boldsymbol{P}}_{\mathrm{N}}^{\circledast} = \mu\hat{\boldsymbol{M}}_{\mathrm{N}}/c$,	$\hat{\boldsymbol{M}}_{\mathrm{N}}^{\circledast} = -c\hat{\boldsymbol{P}}_{\mathrm{N}}/\varepsilon$
$\hat{\boldsymbol{f}}_e, \hat{\boldsymbol{f}}_m$:	$\hat{\boldsymbol{f}}_e^{\circledast} = -\mathrm{i}(\mu/\lvert\mu\rvert)\hat{\boldsymbol{f}}_m$,	$\hat{\boldsymbol{f}}_m^{\circledast} = -\mathrm{i}(\lvert\varepsilon\rvert/\varepsilon)\hat{\boldsymbol{f}}_e$
ε, μ:	$\varepsilon^{\circledast} = \mu$,	$\mu^{\circledast} = \varepsilon$
α, β:	$\alpha^{\circledast} = c^2\beta$,	$\beta^{\circledast} = \alpha/c^2$
$\boldsymbol{G}_{ee}, \boldsymbol{G}_{mm}$:	$\boldsymbol{G}_{ee}^{\circledast} = (1/\mu)\boldsymbol{G}_{mm}(1/\mu) + (1/\mu)\boldsymbol{\delta}$,	$\boldsymbol{G}_{mm}^{\circledast} = \varepsilon\boldsymbol{G}_{ee}\varepsilon - \varepsilon\boldsymbol{\delta}$
$\boldsymbol{G}_{em}, \boldsymbol{G}_{me}$:	$\boldsymbol{G}_{em}^{\circledast} = -(1/\mu)\boldsymbol{G}_{me}\varepsilon$,	$\boldsymbol{G}_{me}^{\circledast} = -\varepsilon\boldsymbol{G}_{em}(1/\mu)$
$\boldsymbol{F}$:	$\boldsymbol{F}^{\circledast} = \boldsymbol{F}$	
U_e, U_m:	$U_e^{\circledast} = U_m$,	$U_m^{\circledast} = U_e$
U_{ee}, U_{mm}:	$U_{ee}^{\circledast} = U_{mm}$,	$U_{mm}^{\circledast} = U_{ee}$
U_{em}, U_{me}:	$U_{em}^{\circledast} = U_{me}$,	$U_{me}^{\circledast} = U_{em}$

$$U(\boldsymbol{r}_A, \boldsymbol{r}_B) = U_{ee}(\boldsymbol{r}_A, \boldsymbol{r}_B) + U_{em}(\boldsymbol{r}_A, \boldsymbol{r}_B) + U_{me}(\boldsymbol{r}_A, \boldsymbol{r}_B) + U_{mm}(\boldsymbol{r}_A, \boldsymbol{r}_B)\,. \tag{1.171}$$

1.3 Duality

Electric–magnetic duality is a useful symmetry of the electromagnetic field. A duality transformation $\circledast$ consists of a simultaneous global exchange of electric and magnetic field quantities as laid out in Table 1.2, accompanied by a global exchange of electric permittivity and magnetic permeability.

As shown in Sec. 2.1.4 of Vol. I, the Maxwell equations in the absence of free charges and currents (1.28)–(1.32) together with the constitutive relations (1.2) and (1.3) are invariant under such a duality transformation [14, 27, 31]. The Hamiltonian (1.33) of the free electromagnetic field is also duality invariant.

The presence of free charges and currents may lead to a breakdown of duality invariance. Furthermore, a duality symmetry does not hold for the Lorentz force on individual charged particles. The situation is different when only electrically neutral objects are present, as is the case for dispersion forces.

To see how dispersion interactions behave when subject to a duality transformation, we note that they uniquely depend on the atomic response functions α and β as well as the body response functions ε and μ, cf. Table 1.2. The application of a duality transformation hence amounts to simultaneous global exchanges $\alpha \leftrightarrow \beta$ and $\varepsilon \leftrightarrow \mu$. As dispersion interactions depend on the latter quantities only indirectly via the Green's tensor, we first need to determine the transformation behaviour of the

Green's tensor. To that end, it is useful to introduce the tensors

$$\boldsymbol{G}_{ee}(\boldsymbol{r},\boldsymbol{r}',\omega) = \frac{\mathrm{i}\omega}{c}\,\boldsymbol{G}(\boldsymbol{r},\boldsymbol{r}',\omega)\,\frac{\mathrm{i}\omega}{c}\,, \tag{1.172}$$

$$\boldsymbol{G}_{mm}(\boldsymbol{r},\boldsymbol{r}',\omega) = \boldsymbol{\nabla}\times\boldsymbol{G}(\boldsymbol{r},\boldsymbol{r}',\omega)\times\overleftarrow{\boldsymbol{\nabla}}'\,, \tag{1.173}$$

$$\boldsymbol{G}_{em}(\boldsymbol{r},\boldsymbol{r}',\omega) = \frac{\mathrm{i}\omega}{c}\,\boldsymbol{G}(\boldsymbol{r},\boldsymbol{r}',\omega)\times\overleftarrow{\boldsymbol{\nabla}}'\,, \tag{1.174}$$

$$\boldsymbol{G}_{me}(\boldsymbol{r},\boldsymbol{r}',\omega) = \boldsymbol{\nabla}\times\boldsymbol{G}(\boldsymbol{r},\boldsymbol{r}',\omega)\,\frac{\mathrm{i}\omega}{c}\,. \tag{1.175}$$

In terms of these quantities, the Casimir force (1.118) can be given in the compact form

$$\boldsymbol{F} = \frac{\hbar}{\pi}\int_0^\infty \mathrm{d}\xi \int_{\partial V} \mathrm{d}\boldsymbol{A}\cdot\sum_{\lambda=e,m}\left[\boldsymbol{G}^{(1)}_{\lambda\lambda}(\boldsymbol{r},\boldsymbol{r},\mathrm{i}\xi) - \tfrac{1}{2}\mathrm{tr}\,\boldsymbol{G}^{(1)}_{\lambda\lambda}(\boldsymbol{r},\boldsymbol{r},\mathrm{i}\xi)\boldsymbol{I}\right]. \tag{1.176}$$

Defining $\alpha_e = \alpha$, $\alpha_m = \beta/c^2$, the CP potentials (1.132) and (1.145) can be written as

$$U_\lambda(\boldsymbol{r}_A) = \frac{\hbar}{2\pi\varepsilon_0}\int_0^\infty \mathrm{d}\xi\,\alpha_\lambda(\mathrm{i}\xi)\mathrm{tr}\,\boldsymbol{G}^{(1)}_{\lambda\lambda}(\boldsymbol{r}_A,\boldsymbol{r}_A,\mathrm{i}\xi) \quad (\lambda = e,m)\,, \tag{1.177}$$

while the vdW potentials (1.153), (1.168), (1.169) and (1.170) read

$$U_{\lambda\lambda'}(\boldsymbol{r}_A,\boldsymbol{r}_B) = -\frac{\hbar}{2\pi\varepsilon_0^2}\int_0^\infty \mathrm{d}\xi\,\alpha^A_\lambda(\mathrm{i}\xi)\alpha^B_{\lambda'}(\mathrm{i}\xi) \times \mathrm{tr}\left[\boldsymbol{G}_{\lambda\lambda'}(\boldsymbol{r}_A,\boldsymbol{r}_B,\mathrm{i}\xi)\cdot\boldsymbol{G}_{\lambda'\lambda}(\boldsymbol{r}_B,\boldsymbol{r}_A,\mathrm{i}\xi)\right] \quad (\lambda,\lambda' = e,m)\,. \tag{1.178}$$

Under a global exchange $\varepsilon \leftrightarrow \mu$, the tensors (1.172)–(1.175) transform into one another as laid out in Table 1.2 (cf. Appendix A.1). The transformation behaviour of the dispersion interactions follows immediately: The Casimir force an arbitrary magnetoelectric body (1.3) is duality invariant [14, 27, 31],

$$\boldsymbol{F}^{\circledast} = \boldsymbol{F}\,, \tag{1.179}$$

when the body is not immersed in any medium, such that $\varepsilon(\boldsymbol{r},\mathrm{i}\xi) = \mu(\boldsymbol{r},\mathrm{i}\xi) \equiv 1$ on its surface. The electric and magnetic components (1.177) of the CP potential of an atom in a free-space region transform into one another under a duality transformation [14, 27, 31],

$$U_e^{\circledast}(\boldsymbol{r}_A) = U_m(\boldsymbol{r}_A)\,, \tag{1.180}$$

$$U_m^{\circledast}(\boldsymbol{r}_A) = U_e(\boldsymbol{r}_A)\,, \tag{1.181}$$

such that the total potential remains invariant. The same holds for the total vdW potential, where [14, 27, 31]

$$U_{ee}^{\circledast}(\boldsymbol{r}_A, \boldsymbol{r}_B) = U_{mm}(\boldsymbol{r}_A, \boldsymbol{r}_B)\,, \tag{1.182}$$

$$U_{mm}^{\circledast}(\boldsymbol{r}_A, \boldsymbol{r}_B) = U_{ee}(\boldsymbol{r}_A, \boldsymbol{r}_B)\,, \tag{1.183}$$

$$U_{em}^{\circledast}(\boldsymbol{r}_A, \boldsymbol{r}_B) = U_{me}(\boldsymbol{r}_A, \boldsymbol{r}_B)\,, \tag{1.184}$$

$$U_{me}^{\circledast}(\boldsymbol{r}_A, \boldsymbol{r}_B) = U_{em}(\boldsymbol{r}_A, \boldsymbol{r}_B)\,. \tag{1.185}$$

The duality invariance of dispersion interactions is extremely useful when studying concrete examples. For instance, once we know the explicit formula for the CP potential of an electric atom in a specific magnetoelectric environment, the corresponding result for a magnetic atom can be obtained immediately by applying a duality transformation, i.e., by making the replacements $\alpha \mapsto \beta$ and $\varepsilon \leftrightarrow \mu$. In this way, dispersion interactions need to be calculated explicitly only for half the conceivable combinations of electric and magnetic objects, while the dual combinations can be treated via simple replacement rules.

It should be stressed that duality invariance and the associated replacement rules are only valid for dispersion forces on objects that are situated in a free-space region, they do not apply to objects immersed in a medium. However, for dispersion forces on atoms, duality invariance can be restored by including local-field corrections [13, 32].

References

1. R. Kubo, Rep. Prog. Phys. 29(I), 255 (1966).
2. L. Knöll, S. Scheel, D.G. Welsch, in *Coherence and Statistics of Photons and Atoms*, ed. by J. Perina (Wiley, New York, 2001), p. 1
3. D.T. Ho, S.Y. Buhmann, L. Knöll, D.G. Welsch, S. Scheel, J. Kästel, Phys. Rev. A **68**(4), 043816 (2003)
4. D.T. Ho, S.Y. Buhmann, L. Knöll, S. Scheel, D.G. Welsch, in *Proceedings of the 8th International Conference on Squeezed States and Uncertainty Relations*, ed. by H. Moya-Cessa, R. Jáuregui, S. Hacyan, O. Castaños (Rinton Press, Princeton, New Jersey, 2003), p. 117
5. S.Y. Buhmann, D.G. Welsch, Prog. Quantum Electron. **31**(2), 51 (2007)
6. S.Y. Buhmann, M.R. Tarbutt, S. Scheel, E.A. Hinds, Phys. Rev. A **78**(5), 052901 (2008)
7. C. Baxter, M. Babiker, R. Loudon, Phys. Rev. A **47**(2), 1278 (1993)

8. D.P. Craig, T. Thirunamachandran, *Molecular Quantum Electrodynamics* (Dover, New York, 1998)
9. W. Thomas, Naturwissenschaften **13**(28), 627 (1925)
10. W. Kuhn, Z. Phys. **33**(1), 408 (1925)
11. F. Reiche, W. Thomas, Z. Phys. **34**(1), 408 (1925)
12. S.Y. Buhmann, L. Knöll, D.G. Welsch, D.T. Ho, Phys. Rev. A **70**(5), 052117 (2004)
13. H. Safari, D.G. Welsch, S.Y. Buhmann, S. Scheel, Phys. Rev. A **78**(6), 062901 (2008)
14. S. Scheel, S.Y. Buhmann, Acta Phys. Slovaca **58**(5), 675 (2008)
15. E.A. Power, S. Zienau, Philos. Trans. R. Soc. Lond. Ser. A **251**(999), 427 (1959)
16. R.G. Woolley, Proc. R. Soc. Lond. Ser. A **321**(1547), 557 (1971)
17. C. Cohen-Tannoudji, J. Dupont-Roc, G. Grynberg, *Photons and Atoms* (Wiley, New York, 1989)
18. H. Safari, S.Y. Buhmann, D.G. Welsch, D.T. Ho, Phys. Rev. A **74**(4), 042101 (2006)
19. H.B.G. Casimir, Proc. K. Ned. Akad. Wet. **51**, 793 (1948)
20. E.M. Lifshitz, Sov. Phys. JETP **2**(1), 73 (1956)
21. J.E. Lennard-Jones, Trans. Faraday Soc. **28**, 333 (1932)
22. H.B.G. Casimir, D. Polder, Phys. Rev. **73**(4), 360 (1948)
23. F. London, Z. Phys. **63**(3–4), 245 (1930)
24. C. Raabe, D.G. Welsch, Phys. Rev. A **73**(6), 063822 (2006)
25. S.Y. Buhmann, D.T. Ho, D.G. Welsch, J. Opt. B: Quantum Semiclass. Opt. **6**(3), S127 (2004)
26. V.M. Fain, Y.I. Khanin, *Quantum Electronics* (MIT Press, Cambridge, 1969)
27. S.Y. Buhmann, S. Scheel, H. Safari, D.G. Welsch, Int. J. Mod. Phys. A **24**(8–9), 1796 (2009)
28. S.Y. Buhmann, H. Safari, S. Scheel, A. Salam, Phys. Rev. A. **87**(1), 012507 (2013)
29. S.Y. Buhmann, H. Safari, D.G. Welsch, D.T. Ho, Open Syst. Inf. Dyn. **13**(4), 427 (2006)
30. S.Y. Buhmann, H. Safari, D.T. Ho, D.G. Welsch, Opt. Spectrosc. (USSR) **103**(3), 374 (2006)
31. S.Y. Buhmann, S. Scheel, Phys. Rev. Lett. **102**(14), 140404 (2009)
32. A. Sambale, S.Y. Buhmann, D.G. Welsch, M.S. Tomaš, Phys. Rev. A **75**(4), 042109 (2007)

Chapter 2
Approximating Casimir–Polder Potentials

As seen in the previous chapter, dispersion forces can be expressed in terms of the classical Green's tensor for the electromagnetic field and the polarisabilities and magnetisabilities of the atoms. In order to study the position-dependence of a dispersion force for a particular arrangement of bodies, one needs to calculate the respective Green's tensor by solving the inhomogeneous Helmholtz equation (2.149). For many arrangements displaying a high degree of symmetry, e.g., free space, planar, spherical, or cylindrical multilayer systems, the Green's tensor is available in closed form [1]. Exploiting this fact, one can find exact and explicit expressions for, e.g., the Casimir force between two plates (Sect. 3.3 of Vol. I); the CP potential of an atom in various planar multilayer systems (Sect. 4.6 of Vol. I) or next to a sphere (Sect. 4.7 of Vol. I); and the vdW potential of two atoms in free space (Sect. 5.4 of Vol. I), in front of a plate (Sect. 5.5.1 of Vol. I) or next to a sphere (Sect. 5.5.2 of Vol. I). A brief summary of most of these results can be found in Table 3.1 of Sect. 3.1 in this volume.

For configurations displaying less symmetry, approximative methods are required. In this chapter, we consider arrangements which deviate only slightly from a highly symmetrical one. We begin by showing how the Green's tensor can be approximated in this case. We use the approximate Green's tensor to express the CP potential in terms of multiple volume integrals or as a sum over bodies. These two alternative forms are illustrated by considering the CP potential of an atom interacting with a weakly dielectric ring and an inhomogeneous half space. In addition, we discuss the convergence of the Born expansion by studying an atom next to a metal plate or sphere.

2.1 Born Expansions of the Green's Tensor

A powerful tool for obtaining approximate solutions to the Helmholtz equation is the Dyson equation. As shown in the following, it can be used to obtain the Born expansion of the Green's tensor as a systematic power-series expansion. We begin

S. Y. Buhmann, *Dispersion Forces II*, Springer Tracts in Modern Physics 248,
DOI: 10.1007/978-3-642-32466-6_2, © Springer-Verlag Berlin Heidelberg 2012

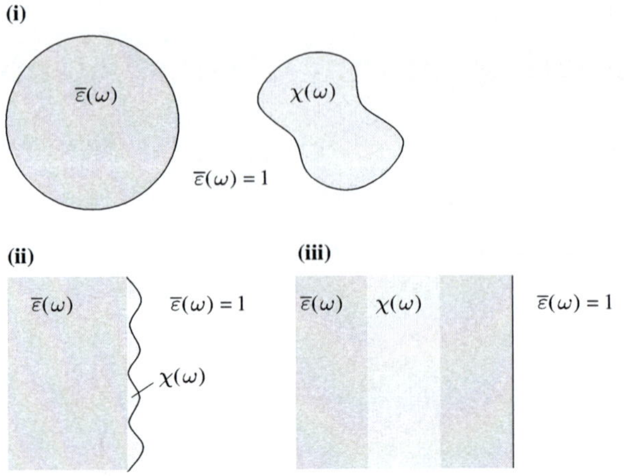

Fig. 2.1 Examples for permittivity decompositions: **(i)** weakly dielectric body next to a sphere; **(ii)** surface roughness of a plate; **(iii)** inhomogeneous half space

with the case of purely electric bodies and then proceed to the purely magnetic and fully magnetoelectric cases.

2.1.1 Electric Bodies

Let us consider an arrangement of purely electric bodies for which the permittivity can be decomposed as

$$\varepsilon(\boldsymbol{r},\omega) = \overline{\varepsilon}(\boldsymbol{r},\omega) + \chi(\boldsymbol{r},\omega)\,. \tag{2.1}$$

Here, $\overline{\varepsilon}(\boldsymbol{r},\omega)$ describes some background bodies with the corresponding Green's tensor being the known solution to

$$\left[\nabla\times\nabla\times\ -\frac{\omega^2}{c^2}\,\overline{\varepsilon}(\boldsymbol{r},\omega)\right]\overline{\boldsymbol{G}}(\boldsymbol{r},\boldsymbol{r}',\omega) = \boldsymbol{\delta}(\boldsymbol{r}-\boldsymbol{r}') \tag{2.2}$$

and $\chi(\boldsymbol{r},\omega)$ describes small corrections to this background. The decomposition (2.1) applies to a large variety of cases. As illustrated in Fig. 2.1, it can be used to study a weakly dielectric body of unusual shape in the possible presence of highly symmetric background bodies (i), surface roughness (ii) or inhomogeneities of a body's permittivity (iii).

Using the decomposition of the permittivity, the differential equation (1.14) for the full Green's tensor can be written as

$$\left[\nabla\times\nabla\times-\frac{\omega^2}{c^2}\overline{\varepsilon}(\boldsymbol{r},\omega)\right]\boldsymbol{G}(\boldsymbol{r},\boldsymbol{r}',\omega)$$
$$=\delta(\boldsymbol{r}-\boldsymbol{r}')+\frac{\omega^2}{c^2}\chi(\boldsymbol{r},\omega)\boldsymbol{G}(\boldsymbol{r},\boldsymbol{r}',\omega)\,. \tag{2.3}$$

Its solution can be written in the form of a Dyson equation [2]

$$\boldsymbol{G}(\boldsymbol{r},\boldsymbol{r}',\omega)=\overline{\boldsymbol{G}}(\boldsymbol{r},\boldsymbol{r}',\omega)+\frac{\omega^2}{c^2}\int \mathrm{d}^3s\,\chi(\boldsymbol{s},\omega)\overline{\boldsymbol{G}}(\boldsymbol{r},\boldsymbol{s},\omega)\cdot\boldsymbol{G}(\boldsymbol{s},\boldsymbol{r}',\omega)\,, \tag{2.4}$$

as can easily be verified by direct substitution upon exploiting the fact that $\overline{\boldsymbol{G}}$ is a solution to (2.3).

By repeated use of the Dyson equation, one can obtain an expansion of $\boldsymbol{G}$ in powers of χ, which is known as the Born expansion. We start the series by using the zero-order approximation $\boldsymbol{G}=\overline{\boldsymbol{G}}$ in the Dyson equation to obtain the solution to linear order in χ

$$\boldsymbol{G}(\boldsymbol{r},\boldsymbol{r}',\omega)=\overline{\boldsymbol{G}}(\boldsymbol{r},\boldsymbol{r}',\omega)+\frac{\omega^2}{c^2}\int \mathrm{d}^3s\,\chi(\boldsymbol{s},\omega)\overline{\boldsymbol{G}}(\boldsymbol{r},\boldsymbol{s},\omega)\cdot\overline{\boldsymbol{G}}(\boldsymbol{s},\boldsymbol{r}',\omega)\,. \tag{2.5}$$

Substituting this solution back into the Dyson equation, we obtain a better approximation which is correct to quadratic order χ,

$$\begin{aligned}\boldsymbol{G}(\boldsymbol{r},\boldsymbol{r}',\omega)=&\,\overline{\boldsymbol{G}}(\boldsymbol{r},\boldsymbol{r}',\omega)+\frac{\omega^2}{c^2}\int \mathrm{d}^3s\,\chi(\boldsymbol{s},\omega)\overline{\boldsymbol{G}}(\boldsymbol{r},\boldsymbol{s},\omega)\cdot\overline{\boldsymbol{G}}(\boldsymbol{s},\boldsymbol{r}',\omega)\\&+\frac{\omega^4}{c^4}\int \mathrm{d}^3s\,\chi(\boldsymbol{s},\omega)\int \mathrm{d}^3s'\,\chi(\boldsymbol{s}',\omega)\\&\times\overline{\boldsymbol{G}}(\boldsymbol{r},\boldsymbol{s},\omega)\cdot\overline{\boldsymbol{G}}(\boldsymbol{s},\boldsymbol{s}',\omega)\cdot\overline{\boldsymbol{G}}(\boldsymbol{s}',\boldsymbol{r}',\omega)\,.\end{aligned} \tag{2.6}$$

Iterating in this way, the full Born expansion of the Green's tensor is found to be [3]

$$\boldsymbol{G}(\boldsymbol{r},\boldsymbol{r}',\omega)=\overline{\boldsymbol{G}}(\boldsymbol{r},\boldsymbol{r}',\omega)+\sum_{K=1}^{\infty}\Delta_K\boldsymbol{G}(\boldsymbol{r},\boldsymbol{r}',\omega) \tag{2.7}$$

with

$$\begin{aligned}\Delta_K\boldsymbol{G}(\boldsymbol{r},\boldsymbol{r}',\omega)=&\,\frac{\omega^{2K}}{c^{2K}}\int \mathrm{d}^3s_1\,\chi(\boldsymbol{s}_1,\omega)\cdots\int \mathrm{d}^3s_K\,\chi(\boldsymbol{s}_K,\omega)\\&\times\overline{\boldsymbol{G}}(\boldsymbol{r},\boldsymbol{s}_1,\omega)\cdot\overline{\boldsymbol{G}}(\boldsymbol{s}_1,\boldsymbol{s}_2,\omega)\cdots\overline{\boldsymbol{G}}(\boldsymbol{s}_K,\boldsymbol{r}',\omega)\end{aligned} \tag{2.8}$$

denoting corrections of order K in χ.

An alternative expansion can be obtained by isolating the singular part of the Green's tensor [4]: According to (A.18), the Green's tensor in an infinite bulk medium can be written as

$$\boldsymbol{G}(\boldsymbol{r},\boldsymbol{r}',\omega) = -\frac{c^2}{3\omega^2\varepsilon(\boldsymbol{r},\omega)}\,\delta(\boldsymbol{r}-\boldsymbol{r}') + \boldsymbol{H}(\boldsymbol{r},\boldsymbol{r}',\omega) \tag{2.9}$$

where $\boldsymbol{H}$ is free of delta-function singularities. This decomposition remains true in the general case of an arbitrary arrangement of bodies. Applying it to $\overline{\boldsymbol{G}}$ and substituting it into the Dyson equation, we find

$$\begin{aligned}\frac{\overline{\varepsilon}(\boldsymbol{r},\omega)+\frac{1}{3}\chi(\boldsymbol{r},\omega)}{\overline{\varepsilon}(\boldsymbol{r},\omega)}\,&\boldsymbol{G}(\boldsymbol{r},\boldsymbol{r}',\omega)\\ &= \overline{\boldsymbol{G}}(\boldsymbol{r},\boldsymbol{r}',\omega) + \frac{\omega^2}{c^2}\int \mathrm{d}^3s\,\chi(\boldsymbol{s},\omega)\overline{\boldsymbol{H}}(\boldsymbol{r},\boldsymbol{s},\omega)\cdot\boldsymbol{G}(\boldsymbol{s},\boldsymbol{r}',\omega)\,.\end{aligned} \tag{2.10}$$

Introducing the auxiliary tensor

$$\boldsymbol{F}(\boldsymbol{r},\boldsymbol{r}',\omega) = \frac{\overline{\varepsilon}(\boldsymbol{r},\omega)+\frac{1}{3}\chi(\boldsymbol{r},\omega)}{\overline{\varepsilon}(\boldsymbol{r},\omega)}\,\boldsymbol{G}(\boldsymbol{r},\boldsymbol{r}',\omega)\,, \tag{2.11}$$

the new Dyson equation takes the more explicit form

$$\begin{aligned}\boldsymbol{F}(\boldsymbol{r},\boldsymbol{r}',\omega) &= \overline{\boldsymbol{G}}(\boldsymbol{r},\boldsymbol{r}',\omega)\\ &\quad+\frac{\omega^2}{c^2}\int \mathrm{d}^3s\,\frac{\chi(\boldsymbol{s},\omega)\overline{\varepsilon}(\boldsymbol{s},\omega)}{\overline{\varepsilon}(\boldsymbol{s},\omega)+\frac{1}{3}\chi(\boldsymbol{s},\omega)}\overline{\boldsymbol{H}}(\boldsymbol{r},\boldsymbol{s},\omega)\cdot\boldsymbol{F}(\boldsymbol{s},\boldsymbol{r}',\omega)\,.\end{aligned} \tag{2.12}$$

It can easily be solved by repeated iteration, leading to a Born series

$$\boldsymbol{F}(\boldsymbol{r},\boldsymbol{r}',\omega) = \overline{\boldsymbol{G}}(\boldsymbol{r},\boldsymbol{r}',\omega) + \sum_{K=1}^{\infty}\Delta_K\boldsymbol{F}(\boldsymbol{r},\boldsymbol{r}',\omega) \tag{2.13}$$

with

$$\begin{aligned}&\Delta_K\boldsymbol{F}(\boldsymbol{r},\boldsymbol{r}',\omega)\\ &= \frac{\omega^{2K}}{c^{2K}}\int \mathrm{d}^3s_1\,\frac{\chi(\boldsymbol{s}_1,\omega)\overline{\varepsilon}(\boldsymbol{s}_1,\omega)}{\overline{\varepsilon}(\boldsymbol{s}_1,\omega)+\frac{1}{3}\chi(\boldsymbol{s}_1,\omega)}\cdots\int \mathrm{d}^3s_K\,\frac{\chi(\boldsymbol{s}_K,\omega)\overline{\varepsilon}(\boldsymbol{s}_K,\omega)}{\overline{\varepsilon}(\boldsymbol{s}_K,\omega)+\frac{1}{3}\chi(\boldsymbol{s}_K,\omega)}\\ &\quad\times\overline{\boldsymbol{H}}(\boldsymbol{r},\boldsymbol{s}_1,\omega)\cdot\overline{\boldsymbol{H}}(\boldsymbol{s}_1,\boldsymbol{s}_2,\omega)\cdots\overline{\boldsymbol{G}}(\boldsymbol{s}_K,\boldsymbol{r}',\omega)\,.\end{aligned} \tag{2.14}$$

When both field point $\boldsymbol{r}$ and source point $\boldsymbol{r}'$ are situated in free space, we have $\chi(\boldsymbol{r},\omega)=0$ and $\overline{\varepsilon}(\boldsymbol{r},\omega)=1$. The Green's tensor then coincides with the auxiliary tensor, $\boldsymbol{F}(\boldsymbol{r},\boldsymbol{r}',\omega) = \boldsymbol{G}(\boldsymbol{r},\boldsymbol{r}',\omega)$, and its alternative Born series is given by (2.7) with

$$\begin{aligned}\Delta_K \boldsymbol{G}(\boldsymbol{r},\boldsymbol{r}',\omega) &= \frac{\omega^{2K}}{c^{2K}} \int \mathrm{d}^3 s_1 \frac{\chi(\boldsymbol{s}_1,\omega)\overline{\varepsilon}(\boldsymbol{s}_1,\omega)}{\overline{\varepsilon}(\boldsymbol{s}_1,\omega)+\frac{1}{3}\chi(\boldsymbol{s}_1,\omega)} \cdots \int \mathrm{d}^3 s_K \frac{\chi(\boldsymbol{s}_K,\omega)\overline{\varepsilon}(\boldsymbol{s}_K,\omega)}{\overline{\varepsilon}(\boldsymbol{s}_K,\omega)+\frac{1}{3}\chi(\boldsymbol{s}_K,\omega)} \\ &\quad \times \overline{\boldsymbol{H}}(\boldsymbol{r},\boldsymbol{s}_1,\omega)\cdot\overline{\boldsymbol{H}}(\boldsymbol{s}_1,\boldsymbol{s}_2,\omega)\cdots\overline{\boldsymbol{H}}(\boldsymbol{s}_K,\boldsymbol{r}',\omega) \end{aligned} \tag{2.15}$$

denoting corrections of order K in $\chi\overline{\varepsilon}/(\overline{\varepsilon}+\frac{1}{3}\chi)$.

The two Born series differ in their expansion parameters. The expansion (2.7) with (2.8) is more intuitive, because the perturbative parameter $\chi = \varepsilon - 1$ on a free-space background is simply the electric susceptibility (1.109). The alternative Born expansion with (2.15) is based on the perturbative parameter $\chi\overline{\varepsilon}/(\overline{\varepsilon}+\frac{1}{3}\chi)$. It is favourable for metals with large χ, ensuring better convergence in this case. In particular, in the perfect conductor limit $\chi \to \infty$, each of the terms in the series (2.8) obviously diverges, whereas the terms (2.15) remain finite with a perturbative parameter $\chi\overline{\varepsilon}/\left(\overline{\varepsilon}+\frac{1}{3}\chi\right) \to 3$.

2.1.2 Magnetic Bodies

In the case of purely magnetic bodies, we decompose the inverse permeability according to

$$\frac{1}{\mu(\boldsymbol{r},\omega)} = \frac{1}{\overline{\mu}(\boldsymbol{r},\omega)} - \zeta(\boldsymbol{r},\omega)\,. \tag{2.16}$$

Note that the correction ζ coincides with the magnetic susceptibility (1.09) in the case of a free-space background, $\zeta = 1 - 1/\mu$. The unperturbed Green's tensor is now given by

$$\left[\nabla\times\frac{1}{\overline{\mu}(\boldsymbol{r},\omega)}\nabla\times - \frac{\omega^2}{c^2}\right]\overline{\boldsymbol{G}}(\boldsymbol{r},\boldsymbol{r}',\omega) = \boldsymbol{\delta}(\boldsymbol{r}-\boldsymbol{r}') \tag{2.17}$$

and the Helmholtz equation (1.14) for the full Green's tensor reads

$$\begin{aligned}&\left[\nabla\times\frac{1}{\overline{\mu}(\boldsymbol{r},\omega)}\nabla\times - \frac{\omega^2}{c^2}\right]\boldsymbol{G}(\boldsymbol{r},\boldsymbol{r}',\omega) \\ &\qquad = \boldsymbol{\delta}(\boldsymbol{r}-\boldsymbol{r}') + \nabla\times\zeta(\boldsymbol{r},\omega)\nabla\times\boldsymbol{G}(\boldsymbol{r},\boldsymbol{r}',\omega)\,. \end{aligned} \tag{2.18}$$

Using the background solution (2.17) and employing partial integration, we can easily verify that the Helmholtz equation is solved by the Dyson equation

$$\begin{aligned}\boldsymbol{G}(\boldsymbol{r},\boldsymbol{r}',\omega) &= \overline{\boldsymbol{G}}(\boldsymbol{r},\boldsymbol{r}',\omega) \\ &\quad - \int \mathrm{d}^3 s\,\zeta(\boldsymbol{s},\omega)\Big[\overline{\boldsymbol{G}}(\boldsymbol{r},\boldsymbol{s},\omega)\times\overleftarrow{\nabla}_s\Big]\cdot\Big[\nabla_s\times\boldsymbol{G}(\boldsymbol{s},\boldsymbol{r}',\omega)\Big]. \end{aligned} \tag{2.19}$$

Starting from the unperturbed solution $\boldsymbol{G} = \overline{\boldsymbol{G}}$, the Dyson equation yields the linear Born expansion

$$\begin{aligned}\boldsymbol{G}(\boldsymbol{r},\boldsymbol{r}',\omega) &= \overline{\boldsymbol{G}}(\boldsymbol{r},\boldsymbol{r}',\omega) \\ &\quad - \int \mathrm{d}^3 s\,\zeta(\boldsymbol{s},\omega)\Big[\overline{\boldsymbol{G}}(\boldsymbol{r},\boldsymbol{s},\omega)\times\overleftarrow{\nabla}_s\Big]\cdot\Big[\nabla_s\times\overline{\boldsymbol{G}}(\boldsymbol{s},\boldsymbol{r}',\omega)\Big]. \end{aligned} \tag{2.20}$$

Iterative use of the Dyson equation leads to a full Born series (2.7) with terms

$$\begin{aligned}\Delta_K\boldsymbol{G}(\boldsymbol{r},\boldsymbol{r}',\omega) &= (-1)^K\int \mathrm{d}^3 s_1\,\zeta(\boldsymbol{s}_1,\omega)\cdots\int \mathrm{d}^3 s_K\,\zeta(\boldsymbol{s}_K,\omega) \\ &\quad\times\Big[\overline{\boldsymbol{G}}(\boldsymbol{r},\boldsymbol{s}_1,\omega)\times\overleftarrow{\nabla}_{s_1}\Big]\cdot\Big[\nabla_{s_1}\times\overline{\boldsymbol{G}}(\boldsymbol{s}_1,\boldsymbol{s}_2,\omega)\times\overleftarrow{\nabla}_{s_2}\Big] \\ &\quad\cdots\Big[\nabla_{s_K}\times\overline{\boldsymbol{G}}(\boldsymbol{s}_K,\boldsymbol{r}',\omega)\Big]. \end{aligned} \tag{2.21}$$

An alternative expansion can again be obtained by isolating the singular part of the Green's tensor. Applying the duality transformation (A.14) to the separation (2.9), we find

$$\begin{aligned}\nabla\times\boldsymbol{G}(\boldsymbol{r},\boldsymbol{r}',\omega)\times\overleftarrow{\nabla}' &= -\frac{\omega^2}{c^2}\,\mu(\boldsymbol{r},\omega)\boldsymbol{G}^{\circledast}(\boldsymbol{r},\boldsymbol{r}',\omega)\mu(\boldsymbol{r}',\omega) - \mu(\boldsymbol{r},\omega)\boldsymbol{\delta}(\boldsymbol{r}-\boldsymbol{r}') \\ &= -\tfrac{2}{3}\,\mu(\boldsymbol{r},\omega)\boldsymbol{\delta}(\boldsymbol{r}-\boldsymbol{r}') + \nabla\times\boldsymbol{H}(\boldsymbol{r},\boldsymbol{r}',\omega)\times\overleftarrow{\nabla}'. \end{aligned} \tag{2.22}$$

By contrast, the tensors $\nabla\times\boldsymbol{G} = \nabla\times\boldsymbol{H}$ and $\boldsymbol{G}\times\overleftarrow{\nabla}' = \boldsymbol{H}\times\overleftarrow{\nabla}'$ do not exhibit any delta-function part for the assumed independent electric and magnetic medium properties. As an intermediate step, we take the curl of the Dyson equation (2.19). Substituting the above separation into the result, we arrive at

$$\begin{aligned}&\big[1-\tfrac{2}{3}\zeta(\boldsymbol{r},\omega)\overline{\mu}(\boldsymbol{r},\omega)\big]\nabla\times\boldsymbol{G}(\boldsymbol{r},\boldsymbol{r}',\omega) \\ &\quad = \nabla\times\overline{\boldsymbol{G}}(\boldsymbol{r},\boldsymbol{r}',\omega) - \int \mathrm{d}^3 s\,\zeta(\boldsymbol{s},\omega)\Big[\nabla\times\overline{\boldsymbol{H}}(\boldsymbol{r},\boldsymbol{s},\omega)\times\overleftarrow{\nabla}_s\Big]\cdot\Big[\nabla_s\times\boldsymbol{G}(\boldsymbol{s},\boldsymbol{r}',\omega)\Big]. \end{aligned} \tag{2.23}$$

Introducing an auxiliary tensor

$$\boldsymbol{F}(\boldsymbol{r},\boldsymbol{r}',\omega)=\left[1-\tfrac{2}{3}\zeta(\boldsymbol{r},\omega)\overline{\mu}(\boldsymbol{r},\omega)\right]\nabla\times\boldsymbol{G}(\boldsymbol{r},\boldsymbol{r}',\omega)\,, \tag{2.24}$$

this equation takes the form

$$\begin{aligned}\boldsymbol{F}(\boldsymbol{r},\boldsymbol{r}',\omega)=\nabla\times\overline{\boldsymbol{G}}(\boldsymbol{r},\boldsymbol{r}',\omega)-\int \mathrm{d}^3s\,\frac{\zeta(\boldsymbol{s},\omega)}{1-\frac{2}{3}\zeta(\boldsymbol{s},\omega)\overline{\mu}(\boldsymbol{s},\omega)}\\ \times\left[\nabla\times\overline{\boldsymbol{H}}(\boldsymbol{r},\boldsymbol{s},\omega)\times\overleftarrow{\nabla}_s\right]\cdot\left[\nabla_s\times\boldsymbol{G}(\boldsymbol{s},\boldsymbol{r}',\omega)\right].\end{aligned} \tag{2.25}$$

It can easily be solved by successive iterations, leading to

$$\boldsymbol{F}(\boldsymbol{r},\boldsymbol{r}',\omega)=\nabla\times\overline{\boldsymbol{G}}(\boldsymbol{r},\boldsymbol{r}',\omega)+\sum_{K=1}^{\infty}\Delta_K\boldsymbol{F}(\boldsymbol{r},\boldsymbol{r}',\omega) \tag{2.26}$$

with

$$\begin{aligned}\Delta_K\boldsymbol{F}(\boldsymbol{r},\boldsymbol{r}',\omega)=(-1)^K\int \mathrm{d}^3s_1\,\frac{\zeta(\boldsymbol{s}_1,\omega)}{1-\frac{2}{3}\,\zeta(\boldsymbol{s}_1,\omega)\overline{\mu}(\boldsymbol{s}_1,\omega)}\\ \cdots\int \mathrm{d}^3s_K\,\frac{\zeta(\boldsymbol{s}_K,\omega)}{1-\frac{2}{3}\,\zeta(\boldsymbol{s}_K,\omega)\overline{\mu}(\boldsymbol{s}_K,\omega)}\left[\overline{\boldsymbol{H}}(\boldsymbol{r},\boldsymbol{s}_1,\omega)\times\overleftarrow{\nabla}_{s_1}\right]\\ \cdot\left[\nabla_{s_1}\times\overline{\boldsymbol{H}}(\boldsymbol{s}_1,\boldsymbol{s}_2,\omega)\times\overleftarrow{\nabla}_{s_2}\right]\cdots\left[\nabla_{s_K}\times\overline{\boldsymbol{H}}(\boldsymbol{s}_K,\boldsymbol{r}',\omega)\right].\end{aligned} \tag{2.27}$$

Substituting this result together with (2.24) into the original Dyson equation (2.19), we obtain the alternative Born expansion for $\boldsymbol{G}$. For $\boldsymbol{r}$ and $\boldsymbol{r}'$ in free space, it has the form (2.7) with

$$\begin{aligned}\Delta_K\boldsymbol{G}(\boldsymbol{r},\boldsymbol{r}',\omega)=(-1)^K\int \mathrm{d}^3s_1\,\frac{\zeta(\boldsymbol{s}_1,\omega)}{1-\frac{2}{3}\,\zeta(\boldsymbol{s}_1,\omega)\overline{\mu}(\boldsymbol{s}_1,\omega)}\\ \cdots\int \mathrm{d}^3s_K\,\frac{\zeta(\boldsymbol{s}_K,\omega)}{1-\frac{2}{3}\,\zeta(\boldsymbol{s}_K,\omega)\overline{\mu}(\boldsymbol{s}_K,\omega)}\left[\overline{\boldsymbol{H}}(\boldsymbol{r},\boldsymbol{s}_1,\omega)\times\overleftarrow{\nabla}_{s_1}\right]\\ \cdot\left[\nabla_{s_1}\times\overline{\boldsymbol{H}}(\boldsymbol{s}_1,\boldsymbol{s}_2,\omega)\times\overleftarrow{\nabla}_{s_2}\right]\cdots\left[\nabla_{s_K}\times\overline{\boldsymbol{H}}(\boldsymbol{s}_K,\boldsymbol{r}',\omega)\right].\end{aligned} \tag{2.28}$$

2.1.3 Electromagnetic Bodies

Finally, let us consider the most general case of a magnetoelectric correction

$$\varepsilon(\boldsymbol{r},\omega)=\overline{\varepsilon}(\boldsymbol{r},\omega)+\chi(\boldsymbol{r},\omega)\,,\qquad \frac{1}{\mu(\boldsymbol{r},\omega)}=\frac{1}{\overline{\mu}(\boldsymbol{r},\omega)}-\zeta(\boldsymbol{r},\omega)\,. \tag{2.29}$$

The unperturbed Green's tensor is then given by

$$\left[\boldsymbol{\nabla}\times\frac{1}{\overline{\mu}(\boldsymbol{r},\omega)}\boldsymbol{\nabla}\times-\frac{\omega^2}{c^2}\overline{\varepsilon}(\boldsymbol{r},\omega)\right]\overline{\boldsymbol{G}}(\boldsymbol{r},\boldsymbol{r}',\omega)=\boldsymbol{\delta}(\boldsymbol{r}-\boldsymbol{r}') \tag{2.30}$$

and the differential equation (1.14) reads

$$\begin{aligned}&\left[\boldsymbol{\nabla}\times\frac{1}{\overline{\mu}(\boldsymbol{r},\omega)}\boldsymbol{\nabla}\times-\frac{\omega^2}{c^2}\overline{\varepsilon}(\boldsymbol{r},\omega)\right]\boldsymbol{G}(\boldsymbol{r},\boldsymbol{r}',\omega)\\&=\boldsymbol{\delta}(\boldsymbol{r}-\boldsymbol{r}')+\frac{\omega^2}{c^2}\chi(\boldsymbol{r},\omega)\boldsymbol{G}(\boldsymbol{r},\boldsymbol{r}',\omega)+\boldsymbol{\nabla}\times\zeta(\boldsymbol{r},\omega)\boldsymbol{\nabla}\times\boldsymbol{G}(\boldsymbol{r},\boldsymbol{r}',\omega)\,.\end{aligned} \tag{2.31}$$

The corresponding Dyson equation is simply a combination of those for purely electric (2.4) or purely magnetic bodies (2.19):

$$\begin{aligned}\boldsymbol{G}(\boldsymbol{r},\boldsymbol{r}',\omega)&=\overline{\boldsymbol{G}}(\boldsymbol{r},\boldsymbol{r}',\omega)+\frac{\omega^2}{c^2}\int\mathrm{d}^3s\,\chi(\boldsymbol{s},\omega)\overline{\boldsymbol{G}}(\boldsymbol{r},\boldsymbol{s},\omega)\cdot\boldsymbol{G}(\boldsymbol{s},\boldsymbol{r}',\omega)\\&\quad-\int\mathrm{d}^3s\,\zeta(\boldsymbol{s},\omega)\left[\overline{\boldsymbol{G}}(\boldsymbol{r},\boldsymbol{s},\omega)\times\overleftarrow{\boldsymbol{\nabla}}_s\right]\cdot\left[\boldsymbol{\nabla}_s\times\boldsymbol{G}(\boldsymbol{s},\boldsymbol{r}',\omega)\right].\end{aligned} \tag{2.32}$$

Within linear order in χ and ζ, the Green's tensor can hence be approximated as

$$\begin{aligned}\boldsymbol{G}(\boldsymbol{r},\boldsymbol{r}',\omega)&=\overline{\boldsymbol{G}}(\boldsymbol{r},\boldsymbol{r}',\omega)+\frac{\omega^2}{c^2}\int\mathrm{d}^3s\,\chi(\boldsymbol{s},\omega)\overline{\boldsymbol{G}}(\boldsymbol{r},\boldsymbol{s},\omega)\cdot\overline{\boldsymbol{G}}(\boldsymbol{s},\boldsymbol{r}',\omega)\\&\quad-\int\mathrm{d}^3s\,\zeta(\boldsymbol{s},\omega)\left[\overline{\boldsymbol{G}}(\boldsymbol{r},\boldsymbol{s},\omega)\times\overleftarrow{\boldsymbol{\nabla}}_s\right]\cdot\left[\boldsymbol{\nabla}_s\times\overline{\boldsymbol{G}}(\boldsymbol{s},\boldsymbol{r}',\omega)\right].\end{aligned} \tag{2.33}$$

Obtaining the full Born series is greatly facilitated by introducing the electric–magnetic tensors $\boldsymbol{G}_{\lambda\lambda'}$ ($\lambda,\lambda'=e,m$) according to (1.172)–(1.175). In terms of these quantities, the Dyson equation takes the simple form

$$\boldsymbol{G}_{\lambda\lambda'}(\boldsymbol{r},\boldsymbol{r}',\omega)=\overline{\boldsymbol{G}}_{\lambda\lambda'}(\boldsymbol{r},\boldsymbol{r}',\omega)-\sum_{\lambda''=e,m}\int \mathrm{d}^3s\,\chi_{\lambda''}(\boldsymbol{s},\omega)\overline{\boldsymbol{G}}_{\lambda\lambda''}(\boldsymbol{r},\boldsymbol{s},\omega)\cdot\boldsymbol{G}_{\lambda''\lambda'}(\boldsymbol{s},\boldsymbol{r}',\omega) \tag{2.34}$$

where we have defined $\chi_e=\chi$, $\chi_m=\zeta$. By iterating the Dyson equation, we obtain the Born expansion

$$\boldsymbol{G}_{\lambda\lambda'}(\boldsymbol{r},\boldsymbol{r}',\omega)=\overline{\boldsymbol{G}}_{\lambda\lambda'}(\boldsymbol{r},\boldsymbol{r}',\omega)+\sum_{K=1}^{\infty}\Delta_K\boldsymbol{G}_{\lambda\lambda'}(\boldsymbol{r},\boldsymbol{r}',\omega) \tag{2.35}$$

with

$$\begin{aligned}&\Delta_K\boldsymbol{G}_{\lambda\lambda'}(\boldsymbol{r},\boldsymbol{r}',\omega)\\&=(-1)^K\sum_{\lambda_1=e,m}\int \mathrm{d}^3s_1\,\chi_{\lambda_1}(\boldsymbol{s}_1,\omega)\cdots\sum_{\lambda_K=e,m}\int \mathrm{d}^3s_K\,\chi_{\lambda_K}(\boldsymbol{s}_K,\omega)\\&\quad\times\overline{\boldsymbol{G}}_{\lambda\lambda_1}(\boldsymbol{r},\boldsymbol{s}_1,\omega)\cdot\overline{\boldsymbol{G}}_{\lambda_1\lambda_2}(\boldsymbol{s}_1,\boldsymbol{s}_2,\omega)\cdots\overline{\boldsymbol{G}}_{\lambda_K\lambda'}(\boldsymbol{s}_K,\boldsymbol{r}',\omega)\end{aligned} \tag{2.36}$$

denoting contributions of order K in χ and ζ.

To obtain the alternative Born series, we substitute the decompositions (2.9) and (2.22) into the Dyson equation (2.34) to find

$$f_\lambda(\boldsymbol{r},\omega)\boldsymbol{G}_{\lambda\lambda'}(\boldsymbol{r},\boldsymbol{r}',\omega)=\overline{\boldsymbol{G}}_{\lambda\lambda'}(\boldsymbol{r},\boldsymbol{r}',\omega)-\sum_{\lambda''=e,m}\int \mathrm{d}^3s\,\chi_{\lambda''}(\boldsymbol{s},\omega)\overline{\boldsymbol{H}}_{\lambda\lambda''}(\boldsymbol{r},\boldsymbol{s},\omega)\cdot\boldsymbol{G}_{\lambda''\lambda'}(\boldsymbol{s},\boldsymbol{r}',\omega) \tag{2.37}$$

with

$$f_e(\boldsymbol{r},\omega)=\frac{\overline{\varepsilon}(\boldsymbol{r},\omega)+\frac{1}{3}\chi(\boldsymbol{r},\omega)}{\overline{\varepsilon}(\boldsymbol{r},\omega)}\,, \tag{2.38}$$

$$f_m(\boldsymbol{r},\omega)=1-\tfrac{2}{3}\zeta(\boldsymbol{r},\omega)\overline{\mu}(\boldsymbol{r},\omega)\,. \tag{2.39}$$

Introducing the auxiliary tensors

$$\boldsymbol{F}_{\lambda\lambda'}(\boldsymbol{r},\boldsymbol{r}',\omega)=f_\lambda(\boldsymbol{r},\omega)\boldsymbol{G}_{\lambda\lambda'}(\boldsymbol{r},\boldsymbol{r}',\omega)\,, \tag{2.40}$$

these equations take the form

$$\boldsymbol{F}_{\lambda\lambda'}(\boldsymbol{r},\boldsymbol{r}',\omega) = \overline{\boldsymbol{G}}_{\lambda\lambda'}(\boldsymbol{r},\boldsymbol{r}',\omega) - \sum_{\lambda''=e,m} \int \mathrm{d}^3 s\, g_{\lambda''}(\boldsymbol{s},\omega) \overline{\boldsymbol{H}}_{\lambda\lambda''}(\boldsymbol{r},\boldsymbol{s},\omega) \cdot \boldsymbol{F}_{\lambda''\lambda'}(\boldsymbol{s},\boldsymbol{r}',\omega) \tag{2.41}$$

with

$$g_e(\boldsymbol{r},\omega) = \frac{\overline{\varepsilon}(\boldsymbol{r},\omega)\chi(\boldsymbol{r},\omega)}{\overline{\varepsilon}(\boldsymbol{r},\omega) + \frac{1}{3}\chi(\boldsymbol{r},\omega)}, \tag{2.42}$$

$$g_m(\boldsymbol{r},\omega) = \frac{\zeta(\boldsymbol{r},\omega)}{1 + \frac{2}{3}\zeta(\boldsymbol{r},\omega)\overline{\mu}(\boldsymbol{r},\omega)}. \tag{2.43}$$

We solve these equations for the auxiliary tensors iteratively. When source and field points are situated in free space, this solution coincides with the required solution for the Green's tensor, $\boldsymbol{F}_{\lambda\lambda'}(\boldsymbol{r},\boldsymbol{r}',\omega) = \boldsymbol{G}_{\lambda\lambda'}(\boldsymbol{r},\boldsymbol{r}',\omega)$. The alternative Born expansion is then given by (2.35) with

$$\Delta_K \boldsymbol{G}_{\lambda\lambda'}(\boldsymbol{r},\boldsymbol{r}',\omega) = (-1)^K \sum_{\lambda_1=e,m} \int \mathrm{d}^3 s_1\, g_{\lambda_1}(\boldsymbol{s}_1,\omega) \cdots \sum_{\lambda_K=e,m} \int \mathrm{d}^3 s_K\, g_{\lambda_K}(\boldsymbol{s}_K,\omega) \times \overline{\boldsymbol{H}}_{\lambda\lambda_1}(\boldsymbol{r},\boldsymbol{s}_1,\omega) \cdot \overline{\boldsymbol{H}}_{\lambda_1\lambda_2}(\boldsymbol{s}_1,\boldsymbol{s}_2,\omega) \cdots \overline{\boldsymbol{H}}_{\lambda_K\lambda'}(\boldsymbol{s}_K,\boldsymbol{r}',\omega). \tag{2.44}$$

2.2 Casimir–Polder Potential via Volume Integrals

The Born expansions for the Green's tensor in their various forms can be used to approximate dispersion forces involving weakly magnetoelectric, rough or inhomogeneous bodies to arbitrary order. We will restrict our attention to the CP potential of a single atom, bearing in mind that approximations of the Casimir force between bodies or body-assisted vdW potentials of two atoms can be developed in a completely analogous way.

2.2.1 Arbitrary Background

We start with an electric ground-state atom in an environment of purely electric bodies. Substituting the linear Born expansion (2.5), we find that to linear order in χ, the CP potential (1.126) or (1.132) can be approximated as [2, 3]

$$U(\boldsymbol{r}_A) = \overline{U}(\boldsymbol{r}_A) + \Delta U(\boldsymbol{r}_A). \tag{2.45}$$

Here,

$$\overline{U}(\boldsymbol{r}_A) = \frac{\hbar\mu_0}{2\pi}\int_0^\infty \mathrm{d}\xi\,\xi^2\alpha(\mathrm{i}\xi)\mathrm{tr}\,\overline{\boldsymbol{G}}^{(1)}(\boldsymbol{r}_A,\boldsymbol{r}_A,\mathrm{i}\xi) \tag{2.46}$$

is the potential associated with the background bodies and

$$\Delta U(\boldsymbol{r}_A) = -\frac{\hbar\mu_0}{2\pi c^2}\int_0^\infty \mathrm{d}\xi\,\xi^4\alpha(\mathrm{i}\xi)\int \mathrm{d}^3s\,\chi(\boldsymbol{s},\mathrm{i}\xi)\mathrm{tr}\Big[\overline{\boldsymbol{G}}(\boldsymbol{r}_A,\boldsymbol{s},\mathrm{i}\xi)\cdot\overline{\boldsymbol{G}}(\boldsymbol{s},\boldsymbol{r}_A,\mathrm{i}\xi)\Big] \tag{2.47}$$

is the first-order correction due to $\chi(\boldsymbol{r},\omega)$. Using (2.8), the full Born expansion of the CP potential for purely electric bodies reads

$$U(\boldsymbol{r}_A) = \overline{U}(\boldsymbol{r}_A) + \sum_{K=1}^{\infty}\Delta_K U(\boldsymbol{r}_A) \tag{2.48}$$

with

$$\begin{aligned}\Delta_K U(\boldsymbol{r}_A) = {} & \frac{(-1)^K\hbar\mu_0}{2\pi c^{2K}}\int_0^\infty \mathrm{d}\xi\,\xi^{2K+2}\alpha(\mathrm{i}\xi)\\ & \times\int \mathrm{d}^3s_1\,\chi(\boldsymbol{s}_1,\mathrm{i}\xi)\cdots\int \mathrm{d}^3s_K\,\chi(\boldsymbol{s}_K,\mathrm{i}\xi)\\ & \times\mathrm{tr}\Big[\overline{\boldsymbol{G}}(\boldsymbol{r}_A,\boldsymbol{s}_1,\mathrm{i}\xi)\cdot\overline{\boldsymbol{G}}(\boldsymbol{s}_1,\boldsymbol{s}_2,\mathrm{i}\xi)\cdots\overline{\boldsymbol{G}}(\boldsymbol{s}_K,\boldsymbol{r}_A,\mathrm{i}\xi)\Big]\,.\end{aligned} \tag{2.49}$$

Isolating the singular part of the Green's tensor via (2.9) and noting that the atom is always assumed to be situated in a small free-space region, we can employ the alternative expansion (2.15). It leads to an alternative Born series for the CP potential with terms

$$\begin{aligned}\Delta_K U(\boldsymbol{r}_A) = {} & \frac{(-1)^K\hbar\mu_0}{2\pi c^{2K}}\int_0^\infty \mathrm{d}\xi\,\xi^{2K+2}\alpha(\mathrm{i}\xi)\\ & \times\int \mathrm{d}^3s_1\,\frac{\chi(\boldsymbol{s}_1,\mathrm{i}\xi)\overline{\varepsilon}(\boldsymbol{s}_1,\mathrm{i}\xi)}{\overline{\varepsilon}(\boldsymbol{s}_1,\mathrm{i}\xi)+\frac{1}{3}\chi(\boldsymbol{s}_1,\mathrm{i}\xi)}\cdots\int \mathrm{d}^3s_K\,\frac{\chi(\boldsymbol{s}_K,\omega)\overline{\varepsilon}(\boldsymbol{s}_K,\mathrm{i}\xi)}{\overline{\varepsilon}(\boldsymbol{s}_K,\mathrm{i}\xi)+\frac{1}{3}\chi(\boldsymbol{s}_K,\mathrm{i}\xi)}\\ & \times\mathrm{tr}\Big[\overline{\boldsymbol{H}}(\boldsymbol{r}_A,\boldsymbol{s}_1,\mathrm{i}\xi)\cdot\overline{\boldsymbol{H}}(\boldsymbol{s}_1,\boldsymbol{s}_2,\mathrm{i}\xi)\cdots\overline{\boldsymbol{H}}(\boldsymbol{s}_K,\boldsymbol{r}_A,\mathrm{i}\xi)\Big]\end{aligned} \tag{2.50}$$

which is favourable for metals.

The linear correction for magnetic bodies can be found by substituting (2.20) for the Green's tensor into the CP potential (1.126):

$$\Delta U(\boldsymbol{r}_A) = -\frac{\hbar\mu_0}{2\pi}\int_0^\infty \mathrm{d}\xi\, \xi^2\alpha(\mathrm{i}\xi)\int \mathrm{d}^3s\,\zeta(\boldsymbol{s},\mathrm{i}\xi)$$
$$\times\,\mathrm{tr}\Big[\overline{\boldsymbol{G}}(\boldsymbol{r}_A,\boldsymbol{s},\mathrm{i}\xi)\times\overleftarrow{\nabla}_s\cdot\nabla_s\times\overline{\boldsymbol{G}}(\boldsymbol{s},\boldsymbol{r}_A,\mathrm{i}\xi)\Big]. \tag{2.51}$$

Note that within linear order, electric and magnetic corrections to the CP potential decouple, so the correction due to weakly magnetoelectric bodies is simply the sum of (2.47) and (2.51).

Finally, the CP potential (1.146) with (1.177) of an electromagnetic atom in the presence of magnetoelectric bodies can be approximated by making use of the Born expansion (2.36). We find

$$U_\lambda(\boldsymbol{r}_A) = \overline{U}_\lambda(\boldsymbol{r}_A) + \sum_{K=1}^{\infty}\Delta_K U_\lambda(\boldsymbol{r}_A) \tag{2.52}$$

with

$$\overline{U}_\lambda(\boldsymbol{r}_A) = \frac{\hbar}{2\pi\varepsilon_0}\int_0^\infty \mathrm{d}\xi\,\alpha_\lambda(\mathrm{i}\xi)\mathrm{tr}\,\overline{\boldsymbol{G}}{}^{(1)}_{\lambda\lambda}(\boldsymbol{r}_A,\boldsymbol{r}_A,\mathrm{i}\xi) \tag{2.53}$$

and

$$\Delta_K U_\lambda(\boldsymbol{r}_A) = \frac{(-1)^K\hbar}{2\pi\varepsilon_0}\int_0^\infty \mathrm{d}\xi\,\alpha_\lambda(\mathrm{i}\xi)$$
$$\times\sum_{\lambda_1=e,m}\int \mathrm{d}^3s_1\,\chi_{\lambda_1}(\boldsymbol{s}_1,\mathrm{i}\xi)\cdots\sum_{\lambda_K=e,m}\int \mathrm{d}^3s_K\,\chi_{\lambda_K}(\boldsymbol{s}_K,\mathrm{i}\xi)$$
$$\times\,\mathrm{tr}\Big[\overline{\boldsymbol{G}}_{\lambda\lambda_1}(\boldsymbol{r}_A,\boldsymbol{s}_1,\mathrm{i}\xi)\cdot\overline{\boldsymbol{G}}_{\lambda_1\lambda_2}(\boldsymbol{s}_1,\boldsymbol{s}_2,\mathrm{i}\xi)\cdots\overline{\boldsymbol{G}}_{\lambda_K\lambda}(\boldsymbol{s}_K,\boldsymbol{r}_A,\mathrm{i}\xi)\Big]. \tag{2.54}$$

With the atom being situated in a free-space region, we can make use of (2.44) to find the alternative series

$$\Delta_K U_\lambda(\boldsymbol{r}_A) = \frac{(-1)^K\hbar}{2\pi\varepsilon_0}\int_0^\infty \mathrm{d}\xi\,\alpha_\lambda(\mathrm{i}\xi)$$
$$\times\sum_{\lambda_1=e,m}\int \mathrm{d}^3s_1\,g_{\lambda_1}(\boldsymbol{s}_1,\mathrm{i}\xi)\cdots\sum_{\lambda_K=e,m}\int \mathrm{d}^3s_K\,g_{\lambda_K}(\boldsymbol{s}_K,\mathrm{i}\xi)$$
$$\times\,\mathrm{tr}\Big[\overline{\boldsymbol{H}}_{\lambda\lambda_1}(\boldsymbol{r}_A,\boldsymbol{s}_1,\mathrm{i}\xi)\cdot\overline{\boldsymbol{H}}_{\lambda_1\lambda_2}(\boldsymbol{s}_1,\boldsymbol{s}_2,\mathrm{i}\xi)\cdots\overline{\boldsymbol{H}}_{\lambda_K\lambda}(\boldsymbol{s}_K,\boldsymbol{r}_A,\mathrm{i}\xi)\Big], \tag{2.55}$$

recall (2.42) and (2.43).

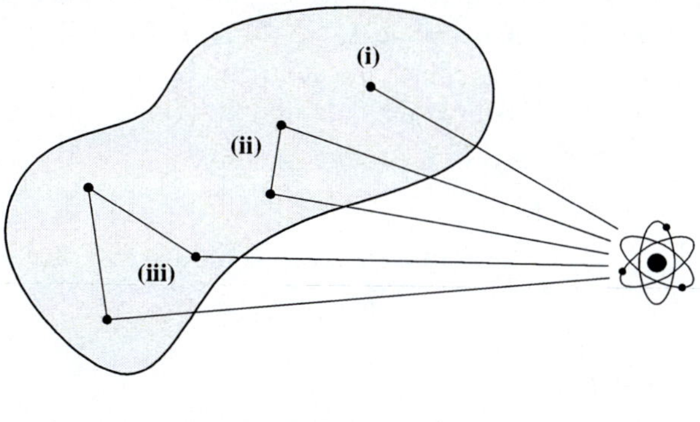

Fig. 2.2 Born expansion: **(i)** First, **(ii)** second and **(iii)** third-order contributions to CP potential of a ground-state atom near a weakly magnetoelectric body

2.2.2 Weakly Magnetodielectric Bodies in Free Space

With the aid of the various Born expansions presented above, the CP potential can be approximated as a series of multiple volume integrals over products of the respective background Green's tensors. Let us now consider the simplest and most important special case in a little more detail: When only weakly magnetodielectric bodies are present, we can choose the background to be free space, $\overline{\varepsilon}(\boldsymbol{r},\omega) \equiv 1$, $\overline{\mu}(\boldsymbol{r},\omega) \equiv 1$, so that all present bodies are characterised by $\chi(\boldsymbol{r},\omega) = \varepsilon(\boldsymbol{r},\omega) - 1$ and $\zeta(\boldsymbol{r},\omega) = 1 - 1/\mu(\boldsymbol{r},\omega)$. The background Green's tensor is then identical to the free-space one and the background potential $\overline{U}$ vanishes. The first few terms of the Born expansion are schematically represented in Fig. 2.2.

Using the explicit form (A.21) of the free-space Green's tensor and noting that the atom is well-separated from the bodies so that the delta function does not contribute, the leading-order potential (2.47) for weakly dielectric bodies reads [3]

$$U(\boldsymbol{r}_A) = -\frac{\hbar}{16\pi^3\varepsilon_0}\int_0^\infty \mathrm{d}\xi\,\alpha(\mathrm{i}\xi)\int \mathrm{d}^3s\,\frac{\chi(\boldsymbol{s},\mathrm{i}\xi)}{|\boldsymbol{r}_A-\boldsymbol{s}|^6}\,g(\xi|\boldsymbol{r}_A-\boldsymbol{s}|/c)\,, \qquad (2.56)$$

where $g(x) = \mathrm{e}^{-2x}(3 + 6x + 5x^2 + 2x^3 + x^4)$. Within this approximation, the CP force between an electric ground-state atom and purely dielectric bodies of arbitrary shapes is thus given by a single volume integral over attractive central forces, since

$$\boldsymbol{\nabla}\left[\frac{g(\xi r/c)}{r^6}\right] = -\frac{2\boldsymbol{r}}{r^8}\left[\mathrm{e}^{-2x}(9 + 18x + 16x^2 + 8x^3 + 3x^4 + x^5)\right]_{x=\xi r/c}\,. \qquad (2.57)$$

These forces closely resemble the vdW force between two electric atoms as given by (5.97) in Vol. I. In Chap. 3 of this volume, we will discuss in more detail in how CP forces and vdW forces are related in general.

The CP potential can be further simplified by considering the retarded and non-retarded limits of large and small atom–body separations. In the retarded limit $r_- \gg c/\omega_-$ (r_-: minimum atom–body distance, ω_-: minimum of all relevant atomic and medium resonance frequencies), the exponential contained in g effectively limits the ξ-integral to a range $0 \leq \xi \lesssim c/(2r_-) \ll \omega_-$, cf. Fig. 3.7(ii) of Vol. I for details. We may hence replace the atom and body response functions by their static values $\alpha(\mathrm{i}\xi) \simeq \alpha(0) \equiv \alpha$ and $\chi(\boldsymbol{r}, \mathrm{i}\xi) \simeq \chi(\boldsymbol{r}, 0) \equiv \chi(\boldsymbol{r})$ and perform the ξ-integral by means of

$$\int_0^\infty \mathrm{d}x\, g(x) = \int_0^\infty \mathrm{d}x\, (3 + 6x + 5x^2 + 2x^3 + x^4)\mathrm{e}^{-2x} = \frac{23}{4} \tag{2.58}$$

to find

$$U(\boldsymbol{r}_A) = -\frac{23\hbar c\alpha}{64\pi^3\varepsilon_0} \int \mathrm{d}^3s\, \frac{\chi(\boldsymbol{s})}{|\boldsymbol{r}_A - \boldsymbol{s}|^7}\,. \tag{2.59}$$

In the opposite nonretarded limit $r_+ \ll c/\omega_+$ (r_+: maximum atom–body distance, ω_+: maximum of all relevant atomic and medium resonance frequencies), the atom and body response functions restrict the ξ-integral to a range where $\xi|\boldsymbol{r}_A - \boldsymbol{s}|/c \leq \xi r_+/c \leq \omega_+ r_+/c \ll 1$, cf. Fig. 3.7(iii) of Vol. I. We may hence make the approximation $g(\xi|\boldsymbol{r}_A - \boldsymbol{s}|/c) \simeq g(0) = 3$, leading to

$$U(\boldsymbol{r}_A) = -\frac{3\hbar}{16\pi^3\varepsilon_0} \int_0^\infty \mathrm{d}\xi\, \alpha(\mathrm{i}\xi) \int \mathrm{d}^3s\, \frac{\chi(\boldsymbol{s}, \mathrm{i}\xi)}{|\boldsymbol{r}_A - \boldsymbol{s}|^6}\,. \tag{2.60}$$

In order to be able to judge the reliability of the approximation, it is useful to also consider the the second-order correction $\Delta_2 U(\boldsymbol{r}_A)$, which according to (2.49) consists of a double volume integral over a product of three Green's tensors. Using the decomposition of the free-space Green's tensor (2.9) and noting that the atom is well separated from all present bodies, only the middle Green's tensor contains a delta-function contribution in addition to its regular part. The second-order contribution can thus be separated into a single-point term, which results from the delta function and a two-point correlation term containing three regular Green's tensors:

$$\Delta_2 U(\boldsymbol{r}_A) = \Delta_2^1 U(\boldsymbol{r}_A) + \Delta_2^2 U(\boldsymbol{r}_A)\,. \tag{2.61}$$

With the free-space Green's tensor (A.21), the single-point term reads [3]

$$\Delta_2^1 U(\boldsymbol{r}_A) = \frac{\hbar}{48\pi^3\varepsilon_0} \int_0^\infty \mathrm{d}\xi\, \alpha(\mathrm{i}\xi) \int \mathrm{d}^3s\, \frac{\chi^2(\boldsymbol{s}, \mathrm{i}\xi)}{|\boldsymbol{r}_A - \boldsymbol{s}|^6}\, g(\xi|\boldsymbol{r}_A - \boldsymbol{s}|/c)\,. \tag{2.62}$$

It differs from the linear contribution only via the replacement $\chi \mapsto -\frac{1}{3}\chi^2$. The single-point part of the second-order correction thus leads to a reduction of the leading-order linear result. Its retarded and nonretarded limits obviously read

$$\Delta_2^1 U(\boldsymbol{r}_A) = \frac{23\hbar c\alpha}{192\pi^3\varepsilon_0} \int \mathrm{d}^3 s \, \frac{\chi^2(\boldsymbol{s})}{|\boldsymbol{r}_A - \boldsymbol{s}|^7} \tag{2.63}$$

and

$$\Delta_2^1 U(\boldsymbol{r}_A) = \frac{\hbar}{16\pi^3\varepsilon_0} \int_0^\infty \mathrm{d}\xi \, \alpha(\mathrm{i}\xi) \int \mathrm{d}^3 s \, \frac{\chi^2(\boldsymbol{s}, \mathrm{i}\xi)}{|\boldsymbol{r}_A - \boldsymbol{s}|^6}, \tag{2.64}$$

respectively.

The two-point contribution is much more complex. Performing the trace over the product of three regular free-space Green's tensors (A.21), we find [3]

$$\Delta_2^2 U(\boldsymbol{r}_A) = \frac{\hbar}{128\pi^4\varepsilon_0} \int_0^\infty \mathrm{d}\xi \, \alpha(\mathrm{i}\xi) \times \int \mathrm{d}^3 s_1 \int \mathrm{d}^3 s_2 \, \frac{\chi(s_1, \mathrm{i}\xi)\chi(s_2, \mathrm{i}\xi)}{r_1^3 r_2^3 r_3^3} \, g(\boldsymbol{r}_1, \boldsymbol{r}_2, \boldsymbol{r}_3, \xi). \tag{2.65}$$

Here, we have defined the function

$$\begin{aligned} g(\boldsymbol{r}_1, \boldsymbol{r}_2, \boldsymbol{r}_3, \xi) = \mathrm{e}^{-\xi(r_1+r_2+r_3)/c}\big[&3a(\xi r_1/c)a(\xi r_2/c)a(\xi r_3/c) \\ &- b(\xi r_1/c)a(\xi r_2/c)a(\xi r_3/c) - a(\xi r_1/c)b(\xi r_2/c)a(\xi r_3/c) \\ &- a(\xi r_1/c)a(\xi r_2/c)b(\xi r_3/c) + b(\xi r_1/c)b(\xi r_2/c)a(\xi r_3/c)(\boldsymbol{e}_1 \cdot \boldsymbol{e}_2)^2 \\ &+ a(\xi r_1/c)b(\xi r_2/c)b(\xi r_3/c)(\boldsymbol{e}_2 \cdot \boldsymbol{e}_3)^2 \\ &+ b(\xi r_1/c)a(\xi r_2/c)b(\xi r_3/c)(\boldsymbol{e}_3 \cdot \boldsymbol{e}_1)^2 \\ &- b(\xi r_1/c)b(\xi r_2/c)b(\xi r_3/c)(\boldsymbol{e}_1 \cdot \boldsymbol{e}_2)(\boldsymbol{e}_2 \cdot \boldsymbol{e}_3)(\boldsymbol{e}_3 \cdot \boldsymbol{e}_1)\big] \end{aligned} \tag{2.66}$$

with $a(x) = 1 + x + x^2$ and $b(x) = 3 + 3x + x^2$ and introduced the abbreviating notation

$$r_1 = r_A - s_1, \qquad r_2 = s_1 - s_2, \qquad r_3 = s_2 - r_A, \tag{2.67}$$

with $\boldsymbol{e}_1$, $\boldsymbol{e}_2$ and $\boldsymbol{e}_3$ being the associated unit vectors.

In the retarded limit, we may replace α and χ by their static values, so that

$$\Delta_2^2 U(\mathbf{r}_A) = \frac{\hbar\alpha}{128\pi^4\varepsilon_0}\int \mathrm{d}^3 s_1 \int \mathrm{d}^3 s_2 \,\frac{\chi(\mathbf{s}_1)\chi(\mathbf{s}_2)}{r_1^3 r_2^3 r_3^3}\int_0^\infty \mathrm{d}\xi\, g(\mathbf{r}_1, \mathbf{r}_2, \mathbf{r}_3, \xi)\,. \tag{2.68}$$

The integral over ξ can then be performed by expanding the products of polynomials in (2.66) and using

$$\int_0^\infty \mathrm{d}\xi \left(\frac{\xi r_1}{c}\right)^i \left(\frac{\xi r_2}{c}\right)^j \left(\frac{\xi r_3}{c}\right)^k \mathrm{e}^{-\xi(r_1+r_2+r_3)/c} = \frac{(i+j+k)!\,r_1^i r_2^j r_3^k c}{(r_1+r_2+r_3)^{i+j+k+1}}\,. \tag{2.69}$$

The result of this rather tedious calculation can be written in a relatively compact form with the aid of the triangle formula

$$A_\triangle \equiv 1 - (\mathbf{e}_1\cdot\mathbf{e}_2)^2 - (\mathbf{e}_2\cdot\mathbf{e}_3)^2 - (\mathbf{e}_3\cdot\mathbf{e}_1)^2 + 2(\mathbf{e}_1\cdot\mathbf{e}_2)(\mathbf{e}_2\cdot\mathbf{e}_3)(\mathbf{e}_3\cdot\mathbf{e}_1) = 0 \tag{2.70}$$

which is an immediate consequence of (2.67). Adding the expression

$$\begin{aligned}
0 = 6A_\triangle &+ \frac{6A_\triangle}{(r_1+r_2+r_3)^6}\big\{\big[r_1^5(r_2+r_3) + r_2^5(r_3+r_1) + r_3^5(r_1+r_2)\big] \\
&+ 7\big[r_1^4(r_2^2+r_3^2) + r_2^4(r_3^2+r_1^2) + r_3^4(r_1^2+r_2^2)\big] + 12\big(r_1^3r_2^3 + r_2^3r_3^3 + r_3^3r_1^3\big) \\
&+ 12 r_1 r_2 r_3 \big(r_1^3 + r_2^3 + r_3^3\big) + 138 r_1^2 r_2^2 r_3^2 \\
&+ 52 r_1 r_2 r_3\big[r_1 r_2(r_1+r_2) + r_2 r_3(r_2+r_3) + r_3 r_1(r_3+r_1)\big]\big\}
\end{aligned} \tag{2.71}$$

to our intermediate result for (2.68), we obtain

$$\begin{aligned}
\Delta_2^2 U(\mathbf{r}_A) = \frac{\hbar c\alpha}{32\pi^4\varepsilon_0}&\int \mathrm{d}^3 s_1 \int \mathrm{d}^3 s_2\, \frac{\chi(\mathbf{s}_1)\chi(\mathbf{s}_2)}{r_1^3 r_2^3 r_3^3 (r_1+r_2+r_3)} \\
&\times\big[f_1(r_1,r_2,r_3) + f_2(r_3,r_1,r_2)(\mathbf{e}_1\cdot\mathbf{e}_2)^2 + f_2(r_1,r_2,r_3)(\mathbf{e}_2\cdot\mathbf{e}_3)^2 \\
&+ f_2(r_2,r_3,r_1)(\mathbf{e}_3\cdot\mathbf{e}_1)^2 + f_3(r_1,r_2,r_3)(\mathbf{e}_1\cdot\mathbf{e}_2)(\mathbf{e}_2\cdot\mathbf{e}_3)(\mathbf{e}_3\cdot\mathbf{e}_1)\big]
\end{aligned} \tag{2.72}$$

with

$$f_1(r_1,r_2,r_3) = 9 - 39\,\frac{\sigma_2}{\sigma_1^2} + 22\,\frac{\sigma_3}{\sigma_1^3} + 54\,\frac{\sigma_2^2}{\sigma_1^4} - 65\,\frac{\sigma_2\sigma_3}{\sigma_1^5} + 20\,\frac{\sigma_3^2}{\sigma_1^6}\,, \tag{2.73}$$

$$f_2(r_1,r_2,r_3)=3\left[\frac{r_1^2}{\sigma_1^2}+\frac{3r_1^2(r_2+r_3)}{\sigma_1^3}+\frac{4r_2r_3(3r_1^2-r_2r_3)}{\sigma_1^4}-\frac{20r_1r_2^2r_3^2}{\sigma_1^5}\right], \tag{2.74}$$

$$f_3(r_1,r_2,r_3)=-1-39\frac{\sigma_2}{\sigma_1^2}+17\frac{\sigma_3}{\sigma_1^3}+72\frac{\sigma_2^2}{\sigma_1^4}-75\frac{\sigma_2\sigma_3}{\sigma_1^5}+20\frac{\sigma_3^2}{\sigma_1^6} \tag{2.75}$$

and $\sigma_i=r_1^i+r_2^i+r_3^i$.

In the opposite, nonretarded limit, we approximate

$$\begin{aligned} g(\boldsymbol{r}_1,\boldsymbol{r}_2,\boldsymbol{r}_3,\xi) &\simeq g(\boldsymbol{r}_1,\boldsymbol{r}_2,\boldsymbol{r}_3,0)\\ &=3\{-2+3[(\boldsymbol{e}_1\cdot\boldsymbol{e}_2)^2+(\boldsymbol{e}_2\cdot\boldsymbol{e}_3)^2+(\boldsymbol{e}_3\cdot\boldsymbol{e}_1)^2]-9(\boldsymbol{e}_1\cdot\boldsymbol{e}_2)(\boldsymbol{e}_2\cdot\boldsymbol{e}_3)(\boldsymbol{e}_3\cdot\boldsymbol{e}_1)\}\\ &=3[1-3(\boldsymbol{e}_1\cdot\boldsymbol{e}_2)(\boldsymbol{e}_2\cdot\boldsymbol{e}_3)(\boldsymbol{e}_3\cdot\boldsymbol{e}_1)], \end{aligned} \tag{2.76}$$

where the triangle formula (2.70) has again been used. This leads to

$$\Delta_2^2U(\boldsymbol{r}_A)=\frac{3\hbar}{128\pi^4\varepsilon_0}\int_0^\infty \mathrm{d}\xi\,\alpha(\mathrm{i}\xi)\int \mathrm{d}^3s_1\int \mathrm{d}^3s_2\,\chi(\boldsymbol{s}_1,\mathrm{i}\xi)\chi(\boldsymbol{s}_2,\mathrm{i}\xi)\times\frac{1-3(\boldsymbol{e}_1\cdot\boldsymbol{e}_2)(\boldsymbol{e}_2\cdot\boldsymbol{e}_3)(\boldsymbol{e}_3\cdot\boldsymbol{e}_1)}{r_1^3r_2^3r_3^3}. \tag{2.77}$$

The integrand of the two-point contribution Δ_2^2U can be positive or negative, depending on the relative positions of the atom and the two integration points inside the body. The magnitude of the second-order correction resulting from the double integral therefore sensitively depends on the shape of the body.

Let us next consider the leading, linear CP potential (2.51) due to weakly magnetic bodies. Calculating the left and right curls of the free-space Green's tensor (A.21),

$$\nabla\times\boldsymbol{G}(\boldsymbol{r},\boldsymbol{r}',\omega)=-\frac{\mathrm{e}^{\mathrm{i}\omega\rho/c}}{4\pi\rho^2}\left(1-\frac{\mathrm{i}\omega\rho}{c}\right)\boldsymbol{e}_\rho\times\boldsymbol{I}, \tag{2.78}$$

$$\boldsymbol{G}(\boldsymbol{r},\boldsymbol{r}',\omega)\times\overleftarrow{\nabla}'=\frac{\mathrm{e}^{\mathrm{i}\omega\rho/c}}{4\pi\rho^2}\left(1-\frac{\mathrm{i}\omega\rho}{c}\right)\boldsymbol{I}\times\boldsymbol{e}_\rho, \tag{2.79}$$

and evaluating the trace via $\mathrm{tr}[\boldsymbol{e}\times\boldsymbol{I}\times\boldsymbol{e}]=-2$, we find [3]

$$\begin{aligned} U(\boldsymbol{r}_A)&=\Delta U(\boldsymbol{r}_A)\\ &=\frac{\hbar\mu_0}{16\pi^3}\int_0^\infty \mathrm{d}\xi\,\xi^2\alpha(\mathrm{i}\xi)\int \mathrm{d}^3s\,\frac{\zeta(\boldsymbol{s},\mathrm{i}\xi)}{|\boldsymbol{r}_A-\boldsymbol{s}|^4}h(\xi|\boldsymbol{r}_A-\boldsymbol{s}|/c) \end{aligned} \tag{2.80}$$

with $h(x) = \mathrm{e}^{-2x}(1 + 2x + x^2)$. Forming the gradient

$$\nabla\left[\frac{h(\xi r/c)}{r^4}\right] = -\frac{2\boldsymbol{r}}{r^6}\left[e^{-2x}(2 + 4x + 3x^2 + x^3)\right]_{x=\xi r/c}, \tag{2.81}$$

we note that the CP force on an electric ground-state atom near weakly magnetic bodies is a volume integral over repulsive central forces. The integrand in (2.80) is very similar to the respective vdW force between an electric and a paramagnetic atom, cf. (5.124) in Vol I. In the retarded limit, we replace α and ζ by their static values and carry out the ξ-integral by means of

$$\int_0^\infty \mathrm{d}x\, x^2 h(x) = \int_0^\infty \mathrm{d}x\, x^2(1 + 2x + x^2)\mathrm{e}^{-2x} = \frac{7}{4} \tag{2.82}$$

to find

$$U(\boldsymbol{r}_A) = \frac{7\hbar c\alpha}{64\pi^3\varepsilon_0}\int \mathrm{d}^3 s\, \frac{\zeta(\boldsymbol{s})}{|\boldsymbol{r}_A - \boldsymbol{s}|^7}\,. \tag{2.83}$$

In the nonretarded limit, the approximation $h(x) \simeq h(0) = 1$ leads to

$$U(\boldsymbol{r}_A) = \frac{\hbar\mu_0}{16\pi^3}\int_0^\infty \mathrm{d}\xi\, \xi^2\alpha(\mathrm{i}\xi)\int \mathrm{d}^3 s\, \frac{\zeta(\boldsymbol{s}, \mathrm{i}\xi)}{|\boldsymbol{r}_A - \boldsymbol{s}|^4}\,. \tag{2.84}$$

2.2.3 Atom Next to a Ring

Let us apply the general results of the previous section to a specific body. We consider the CP potential of an atom placed on the symmetry axis of a thin homogeneous ring of radius R, circular cross section πa^2 ($a \ll R$) and volume $V = 2\pi^2 R a^2$, the atom being separated from the centre of the ring by a distance z_A (Fig. 2.3). For a thin ring, we have $|\boldsymbol{r}_A - \boldsymbol{s}| \simeq \sqrt{z_A^2 + R^2} = \rho_A$, so the volume integral in (2.56) becomes trivial, resulting in the attractive first-order CP potential

$$\Delta_1 U(\rho_A) = -\frac{\hbar V}{16\pi^3\varepsilon_0\rho_A^6}\int_0^\infty \mathrm{d}\xi\, \alpha(\mathrm{i}\xi)\chi(\mathrm{i}\xi)g(\xi\rho_A/c) \tag{2.85}$$

of a weakly dielectric ring of permittivity $\chi(\mathrm{i}\xi)$. Its retarded and nonretarded limits (2.59) and (2.60) are given by

Fig. 2.3 Atom next to a ring

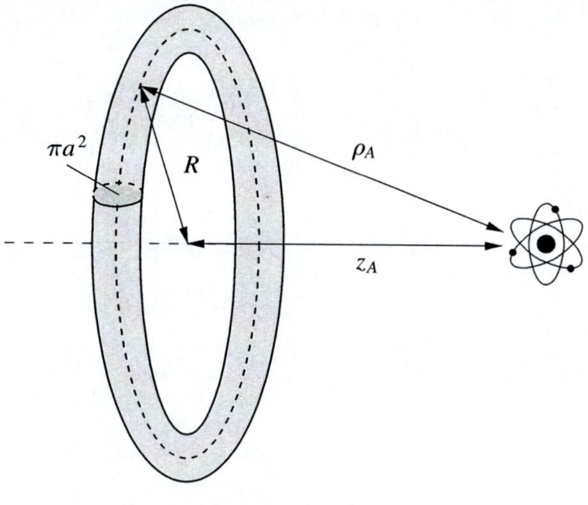

$$\Delta_1 U(\rho_A) = -\frac{23\hbar c V \alpha\chi}{64\pi^3\varepsilon_0\rho_A^7} \tag{2.86}$$

and

$$\Delta_1 U_A(\rho_A) = -\frac{3\hbar V}{16\pi^3\varepsilon_0\rho_A^6}\int_0^\infty \mathrm{d}\xi\,\alpha(\mathrm{i}\xi)\chi(\mathrm{i}\xi)\,, \tag{2.87}$$

respectively.

To assess the reliability of these first-order results, we also consider the second-order correction (2.61). The value of the single-point term $\Delta_2^1 U$ can be obtained by applying the replacement $\chi \mapsto -\frac{1}{3}\chi^2$ to the linear results, cf. the remark below (2.62). We hence have

$$\Delta_2^1 U(\rho_A) = \frac{\hbar V}{48\pi^3\varepsilon_0\rho_A^6}\int_0^\infty \mathrm{d}\xi\,\alpha(\mathrm{i}\xi)\chi^2(\mathrm{i}\xi)g(\xi\rho_A/c) \tag{2.88}$$

with retarded and nonretarded limits

$$\Delta_2^1 U(\rho_A) = \frac{23\hbar c V\alpha\chi}{192\pi^3\varepsilon_0\rho_A^7}\,, \tag{2.89}$$

$$\Delta_2^1 U_A(\rho_A) = \frac{\hbar V}{16\pi^3\varepsilon_0\rho_A^6}\int_0^\infty \mathrm{d}\xi\,\alpha(\mathrm{i}\xi)\chi(\mathrm{i}\xi)\,. \tag{2.90}$$

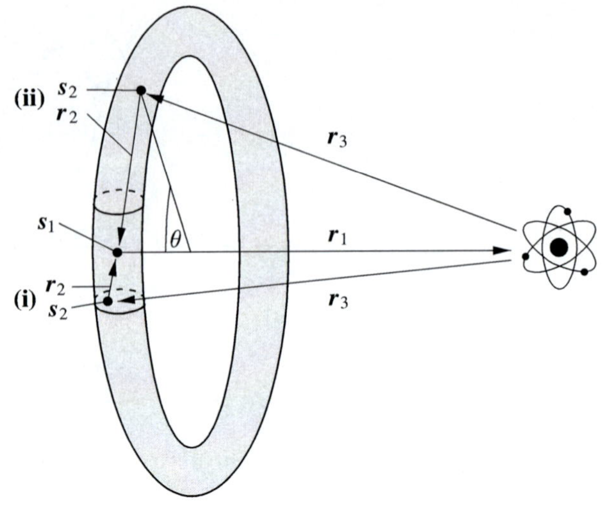

Fig. 2.4 Calculation of the two-point contributions **(i)** and **(ii)** to the CP potential of a ring

The two-point correlation term $\Delta_2^2 U$ is a lot more difficult to evaluate. In particular, it contains an apparent singularity at $\boldsymbol{s}_1 = \boldsymbol{s}_2$ that has to be treated with care. Starting with the retarded limit (2.72), we replace the variable $\boldsymbol{s}_1$ by its average across the cross section of the ring ($|\boldsymbol{r}_A - \boldsymbol{s}_1| \simeq \rho_A$) and carry out the $\boldsymbol{s}_1$-integral. In order to perform the $\boldsymbol{s}_2$-integral, we split the integration volume into two regions (i) and (ii) as illustrated in Fig. 2.4: Region (i) is an approximately cylindrical volume of cross section πa^2 and length $2l$ centred around $\boldsymbol{s}_1$; it contains the apparent singularity. In this region, the separation vector $\boldsymbol{r}_2 = \boldsymbol{s}_1 - \boldsymbol{s}_2$ may be parametrised by local cylindrical coordinates (ρ, ϕ, z). Region (ii) is the remaining open ring where $\boldsymbol{r}_2$ is adequately described by a separation angle θ. In the limit of a thin ring ($a \ll R$), we can choose the length of the cylindrical region such that $a \ll l \ll R$. With this separation of the $\boldsymbol{s}_2$-integral, the two-point term takes the form

$$\begin{aligned}
\Delta_2^2 U(\boldsymbol{r}_A) &\equiv \Delta_2^{2(\mathrm{i})} U(\boldsymbol{r}_A) + \Delta_2^{2(\mathrm{ii})} U(\boldsymbol{r}_A) \\
&= \frac{\hbar c V \alpha \chi^2}{32\pi^4 \varepsilon_0} \left\{ \int_{-l}^{l} \mathrm{d}z \int_0^a \mathrm{d}\rho\, \rho \int_0^{2\pi} \mathrm{d}\phi + \pi a^2 \int_{l/R}^{2\pi - l/R} R \mathrm{d}\theta \right\} \\
&\quad \times \frac{1}{r_1^3 r_2^3 r_3^3 (r_1 + r_2 + r_3)} \left[f_1(r_1, r_2, r_3) + f_2(r_3, r_1, r_2)(\boldsymbol{e}_1 \cdot \boldsymbol{e}_2)^2 \right. \\
&\quad + f_2(r_1, r_2, r_3)(\boldsymbol{e}_2 \cdot \boldsymbol{e}_3)^2 + f_2(r_2, r_3, r_1)(\boldsymbol{e}_3 \cdot \boldsymbol{e}_1)^2 \\
&\quad \left. + f_3(r_1, r_2, r_3)(\boldsymbol{e}_1 \cdot \boldsymbol{e}_2)(\boldsymbol{e}_2 \cdot \boldsymbol{e}_3)(\boldsymbol{e}_3 \cdot \boldsymbol{e}_1) \right],
\end{aligned} \tag{2.91}$$

For the integral over the cylindrical region (i), we may use the approximations $r_3 \simeq r_1 = \rho_A$, $r_2 = \sqrt{z^2 + \rho^2} \ll \rho_A$, $-\boldsymbol{e}_2 \cdot \boldsymbol{e}_3 \simeq \boldsymbol{e}_1 \cdot \boldsymbol{e}_2 = \rho \cos(\phi)/\sqrt{z^2 + \rho^2}$, $\boldsymbol{e}_3 \cdot \boldsymbol{e}_1 \simeq -1$ for $a \ll R$ (Fig. 2.4); and the functions (2.73)–(2.75) simplify to

$f_1(r_1, r_2, r_3) \simeq \frac{13}{8}$, $f_2(r_3, r_1, r_2) \simeq f_2(r_1, r_2, r_3) \simeq \frac{15}{8}$, $f_2(r_2, r_3, r_1) \simeq -\frac{3}{4}$ and $f_3(r_1, r_2, r_3) \simeq -\frac{51}{8}$. With these estimates, we have

$$\Delta_2^{2(\mathrm{i})} U(\boldsymbol{r}_A) = \frac{7\hbar c V \alpha \chi^2}{512\pi^4 \varepsilon_0 \rho_A^7} \int_{-l}^{l} \mathrm{d}z \int_0^a \mathrm{d}\rho\, \rho \int_0^{2\pi} \mathrm{d}\phi\, \frac{\rho^2 + z^2 - 3\rho^2 \cos^2\phi}{\sqrt{z^2+\rho^2}^{\,5}}$$
$$= \frac{7\hbar c V \alpha \chi^2}{512\pi^3 \varepsilon_0 \rho_A^7} \int_{-l}^{l} \mathrm{d}z \int_0^a \mathrm{d}\rho\, \rho\, \frac{2z^2 - \rho^2}{\sqrt{z^2+\rho^2}^{\,5}} = \frac{7\hbar c V \alpha \chi^2}{512\pi^3 \varepsilon_0 \rho_A^7} \int_{-l}^{l} \mathrm{d}z\, \frac{a^2}{\sqrt{z^2+a^2}^{\,3}}$$
$$= \frac{7\hbar c V \alpha \chi^2}{256\pi^3 \varepsilon_0 \rho_A^7} \frac{(l/a)}{\sqrt{1+(l/a)^2}} \simeq \frac{7\hbar c V \alpha \chi^2}{256\pi^3 \varepsilon_0 \rho_A^7} \tag{2.92}$$

for $a \ll l$. The integral thus remains finite although the integration region (i) contains the point $s_2 \simeq s_1$ where the denominator of the integrand vanishes.

In the open-ring region (ii), the estimates $r_3 \simeq r_1 = \rho_A$, $r_2 \simeq 2R|\sin(\theta/2)|$, $\boldsymbol{e}_1 \cdot \boldsymbol{e}_2 \simeq \boldsymbol{e}_2 \cdot \boldsymbol{e}_3 \simeq -(R/\rho_A)|\sin(\theta/2)|$, $\boldsymbol{e}_3 \cdot \boldsymbol{e}_1 \simeq 2(R^2/\rho_A^2)|\sin(\theta/2)| - 1$ hold for $a \ll R$. Due to the denominator $r_2^3 \propto \sin^3(\theta/2)$, the main contribution to the θ-integral in (2.91) comes from regions where $\sin(\theta/2) \ll 1$. We may hence apply a Taylor expansion in powers of $\sin(\theta/2)$ and retain only the terms proportional to $f_1(r_1, r_2, r_3) \simeq \frac{13}{8}$ and $f_2(r_2, r_3, r_1) \simeq -\frac{3}{4}$. With these approximations, (2.91) leads to

$$\Delta_2^{2(\mathrm{ii})} U(\boldsymbol{r}_A) = \frac{7\hbar c V \alpha \chi^2 a^2}{4096\pi^3 \varepsilon_0 \rho_A^7 R^2} \int_{l/R}^{2\pi - l/R} \frac{\mathrm{d}\theta}{|\sin^3(\theta/2)|}$$
$$\simeq \frac{7\hbar c V \alpha \chi^2 a^2}{4096\pi^3 \varepsilon_0 \rho_A^7 R^2} \times 2 \int_{l/R}^{\infty} \frac{\mathrm{d}\theta}{(\theta/2)^3} = \frac{7\hbar c V \alpha \chi^2}{512\pi^3 \varepsilon_0 \rho_A^7} \frac{a^2}{l^2} \simeq 0 \tag{2.93}$$

when $l \ll R$ and $a \ll l$. The contribution from the open-ring region hence becomes negligible for a thin ring. In this limit, the two-point term is dominated by the contribution from the cylindrical region and we have

$$\Delta_2^2 U(\rho_A) = \frac{7\hbar c V \alpha \chi^2}{256\pi^3 \varepsilon_0 \rho_A^7}\,. \tag{2.94}$$

The nonretarded limit (2.77) of the two-point term can be calculated in a completely analogous way by again using the above splitting into two regions:

$$\Delta_2^2 U(\boldsymbol{r}_A) \equiv \Delta_2^{2(\mathrm{i})} U(\boldsymbol{r}_A) + \Delta_2^{2(\mathrm{ii})} U(\boldsymbol{r}_A)$$

$$= \frac{3\hbar V}{128\pi^4\varepsilon_0}\int_0^\infty d\xi\,\alpha(i\xi)\chi^2(i\xi)$$
$$\times\left\{\int_{-l}^{l} dz\int_0^a d\rho\,\rho\int_0^{2\pi} d\phi+\pi a^2\int_{l/R}^{2\pi-l/R} R d\theta\right\}$$
$$\times\frac{1-3(\boldsymbol{e}_1\cdot\boldsymbol{e}_2)(\boldsymbol{e}_2\cdot\boldsymbol{e}_3)(\boldsymbol{e}_3\cdot\boldsymbol{e}_1)}{r_1^3r_2^3r_3^3}\,. \tag{2.95}$$

Using the approximations above (2.92), the integral over the cylindrical region is found to be

$$\Delta_2^{2(\text{i})}U(\boldsymbol{r}_A)=\frac{3\hbar V}{128\pi^4\varepsilon_0\rho_A^6}\int_0^\infty d\xi\,\alpha(i\xi)\chi^2(i\xi)\int_{-l}^{l} dz\int_0^a d\rho\,\rho\int_0^{2\pi} d\phi$$
$$\times\frac{\rho^2+z^2-3\rho^2\cos^2\phi}{\sqrt{z^2+\rho^2}^{\,5}}=\frac{3\hbar V}{64\pi^3\varepsilon_0\rho_A^6}\int_0^\infty d\xi\,\alpha(i\xi)\chi^2(i\xi)\,\frac{(l/a)}{\sqrt{1+(l/a)^2}}$$
$$\simeq\frac{3\hbar V}{64\pi^3\varepsilon_0\rho_A^6}\int_0^\infty d\xi\,\alpha(i\xi)\chi^2(i\xi) \tag{2.96}$$

for $a \ll l$. With estimates above (2.93), the open-ring integral becomes

$$\Delta_2^{2(\text{ii})}U(\boldsymbol{r}_A)=\frac{3\hbar V a^2}{1024\pi^3\varepsilon_0\rho_A^6R^2}\int_0^\infty d\xi\,\alpha(i\xi)\chi^2(i\xi)\int_{l/R}^{2\pi-l/R}\frac{d\theta}{|\sin^3(\theta/2)|}$$
$$=\frac{3\hbar V}{128\pi^3\varepsilon_0\rho_A^6}\frac{a^2}{l^2}\simeq 0 \tag{2.97}$$

when $a \ll l \ll R$, so that the total two-point term reads

$$\Delta_2^2U(\rho_A)=\frac{3\hbar V}{64\pi^3\varepsilon_0\rho_A^6}\int_0^\infty d\xi\,\alpha(i\xi)\chi^2(i\xi) \tag{2.98}$$

for a thin ring in the nonretarded limit.

Combining our retarded results (2.86), (2.89) and (2.94), the quadratic Born expansion of the retarded CP potential of an atom near a thin ring is given by [3]

$$\begin{aligned} U(\rho_A) &= \Delta_1 U(\rho_A) + \Delta_2^1 U(\rho_A) + \Delta_2^2 U(\rho_A) \\ &= -\frac{23\hbar c V\alpha}{64\pi^3\varepsilon_0\rho_A^7}\left(\chi - \tfrac{1}{3}\chi^2 - \tfrac{7}{92}\chi^2\right) \\ &= -\frac{23\hbar c V\alpha}{64\pi^3\varepsilon_0\rho_A^7}\left(\chi - 0.333\chi^2 - 0.076\chi^2\right). \end{aligned} \tag{2.99}$$

The respective Born expansion in the nonretarded limit follows from (2.87), (2.90) and (2.98):

$$\begin{aligned} U(\rho_A) &= \Delta_1 U(\rho_A) + \Delta_2^1 U(\rho_A) + \Delta_2^2 U(\rho_A) \\ &= -\frac{3\hbar V}{16\pi^3\varepsilon_0\rho_A^6}\int_0^\infty \mathrm{d}\xi\, \alpha(\mathrm{i}\xi)\left[\chi(\mathrm{i}\xi) - \tfrac{1}{3}\chi^2(\mathrm{i}\xi) - \tfrac{1}{4}\chi^2(\mathrm{i}\xi)\right] \\ &= -\frac{3\hbar V}{16\pi^3\varepsilon_0\rho_A^6}\int_0^\infty \mathrm{d}\xi\, \alpha(\mathrm{i}\xi)\left[\chi(\mathrm{i}\xi) - 0.333\chi^2(\mathrm{i}\xi) - 0.250\chi^2(\mathrm{i}\xi)\right]. \end{aligned} \tag{2.100}$$

The CP potential is seen to be attractive and proportional to $1/\rho_A^7$ and $1/\rho_A^6$ in the retarded and nonretarded limits, respectively. In both cases, the two quadratic contributions each reduce the potential in comparison to its linear approximation, but they do not change its sign. We further note that the two-point term has its strongest influence in the nonretarded limit where it leads to a very slow convergence of the Born series.

Let us next consider a weakly magnetic ring, restricting our attention to the leading-order contribution (2.80). Using $|\boldsymbol{r}_A - \boldsymbol{s}| \simeq \sqrt{z_A^2 + R^2} = \rho_A$ and carrying out the trivial volume integral, we find

$$U(\rho_A) = \frac{\hbar\mu_0 V}{16\pi^3\rho_A^4}\int_0^\infty \mathrm{d}\xi\, \xi^2\alpha(\mathrm{i}\xi)\zeta(\mathrm{i}\xi)h(\xi\rho_A/c) \tag{2.101}$$

with retarded and nonretarded limits

$$U(\rho_A) = \frac{7\hbar c\alpha\zeta V}{64\pi^3\varepsilon_0\rho_A^7} \tag{2.102}$$

and

$$U(\rho_A) = \frac{\hbar\mu_0 V}{16\pi^3\rho_A^4}\int_0^\infty \mathrm{d}\xi\,\xi^2\alpha(\mathrm{i}\xi)\zeta(\mathrm{i}\xi)\,, \tag{2.103}$$

recall (2.83) and (2.84). In contrast to the potential of a weakly dielectric ring, that of a weakly magnetic ring is repulsive. It is governed by $1/\rho_A^7$ and $1/\rho_A^4$ power laws in the retarded and nonretarded limits, respectively.

So far, we have concentrated on the CP potential of a purely electric atom. A magnetic atom could be treated by starting from the respective general Born series (2.52)–(2.54). A much simpler route is based on duality as discussed in Sect. 1.3. We have seen that CP potentials in free space are invariant under a simultaneous replacement $\alpha \mapsto \beta/c^2, \varepsilon \leftrightarrow \mu$. With $\varepsilon = 1+\chi$ and $1/\mu = 1-\zeta$, the latter replacement amounts to

$$\chi \mapsto \frac{1}{1-\zeta} - 1 = \zeta + \zeta^2 + \cdots\,, \qquad \zeta \mapsto 1 - \frac{1}{1+\chi} = \chi - \chi^2 + \cdots \tag{2.104}$$

Applying this duality transformation to (2.99) and (2.100), we find that within quadratic order in ζ, the retarded and nonretarded CP interaction of a magnetic atom with a weakly magnetic ring is given by

$$U_m(\rho_A) = -\frac{23\hbar c\mu_0 V\beta}{64\pi^3\varepsilon_0\rho_A^7}\left(\zeta + \tfrac{163}{276}\zeta^2\right) \tag{2.105}$$

and

$$U_m(\rho_A) = -\frac{3\hbar\mu_0 V}{16\pi^3\rho_A^6}\int_0^\infty \mathrm{d}\xi\,\beta(\mathrm{i}\xi)\left[\zeta(\mathrm{i}\xi) + \tfrac{5}{12}\zeta^2(\mathrm{i}\xi)\right], \tag{2.106}$$

respectively. Similarly, (2.102) and (2.103) imply that the linear CP potential of a magnetic atom interacting with a weakly dielectric ring reads

$$U_m(\rho_A) = \frac{7\hbar c\mu_0\beta\chi V}{64\pi^3\rho_A^7} \tag{2.107}$$

and

$$U_m(\rho_A) = \frac{\hbar\mu_0 V}{16\pi^3 c^2\rho_A^4}\int_0^\infty \mathrm{d}\xi\,\xi^2\beta(\mathrm{i}\xi)\chi(\mathrm{i}\xi) \tag{2.108}$$

in the retarded and nonretarded limits.

2.2.4 Atom Next to a Metal Plate or Sphere

The Born expansion of the CP potential is a rapidly converging series for weakly magnetodielectric bodies, provided that $\chi, \zeta \ll 1$. In this case, the total potential can be well approximated by calculating just the first few terms of the series, as we have done for the ring. For metals, on the contrary, we typically have $\chi \gg 1$. Even the perturbative parameter $\chi/(1+\frac{1}{3}\chi)$ of the alternative Born series (2.50) can be very close to its limiting value 3 which is realised for perfect conductors.

In order to assess the reliability of the Born expansion for metals, let us consider two geometries where the exact results are known, namely the nonretarded CP potentials of an atom next to a perfectly metal plate and sphere. Recall from (4.137) and (4.138) of Vol. I that the nonretarded potential of an atom at distance z_A from an semi-infinite metal half space of permittivity $\varepsilon(\omega)$ is given by

$$U(z_A) = -\frac{\hbar}{16\pi^2\varepsilon_0 z_A^3}\int_0^\infty \mathrm{d}\xi\,\alpha(\mathrm{i}\xi)\,\frac{\varepsilon(\mathrm{i}\xi)-1}{\varepsilon(\mathrm{i}\xi)+1}\,. \tag{2.109}$$

We invert ($\chi = \varepsilon - 1$)

$$\frac{\chi}{1+\frac{1}{3}\chi} = \frac{3(\varepsilon-1)}{\varepsilon+2} \tag{2.110}$$

to find

$$\varepsilon = \frac{3+2\chi/\left(1+\frac{1}{3}\chi\right)}{3-\chi/\left(1+\frac{1}{3}\chi\right)}\,. \tag{2.111}$$

With this relation, we can express the exact nonretarded potential in terms of the perturbative parameter $\chi/(1+\frac{1}{3}\chi)$:

$$U(z_A) = -\frac{\hbar}{16\pi^2\varepsilon_0 z_A^3}\int_0^\infty \mathrm{d}\xi\,\alpha(\mathrm{i}\xi)\,\frac{3\chi(\mathrm{i}\xi)/\left[1+\frac{1}{3}\chi(\mathrm{i}\xi)\right]}{6+\chi(\mathrm{i}\xi)/\left[1+\frac{1}{3}\chi(\mathrm{i}\xi)\right]}\,. \tag{2.112}$$

The terms (2.50) of the Born series are the unique terms of a Taylor expansion in powers of $\chi/(1+\frac{1}{3}\chi)$. With the exact solution for the total potential being known, we can thus deduce these terms without explicitly performing the spatial integrations that occur in (2.50). Instead, we simply expand (2.112) in powers of $\chi/(1+\frac{1}{3}\chi)$:

$$\begin{aligned}U(z_A) &= \Delta_1 U(z_A) + \Delta_2 U(z_A) + \Delta_3 U(z_A) + \Delta_4 U(z_A) + \cdots \\ &= -\frac{\hbar}{16\pi^2\varepsilon_0 z_A^3}\int_0^\infty \mathrm{d}\xi\,\alpha(\mathrm{i}\xi)\left\{\frac{3}{6}\frac{\chi(\mathrm{i}\xi)}{1+\frac{1}{3}\chi(\mathrm{i}\xi)} - \frac{3}{6^2}\left[\frac{\chi(\mathrm{i}\xi)}{1+\frac{1}{3}\chi(\mathrm{i}\xi)}\right]^2\right. \\ &\quad \left. + \frac{3}{6^3}\left[\frac{\chi(\mathrm{i}\xi)}{1+\frac{1}{3}\chi(\mathrm{i}\xi)}\right]^3 - \frac{3}{6^4}\left[\frac{\chi(\mathrm{i}\xi)}{1+\frac{1}{3}\chi(\mathrm{i}\xi)}\right]^4 + \cdots\right\}. \end{aligned} \tag{2.113}$$

The convergence of this series becomes slowest in the limit of a perfect conductor $\chi/(1+\frac{1}{3}\chi) \to 3$ where

$$\begin{aligned}U(z_A) &= \Delta_1 U(z_A) + \Delta_2 U(z_A) + \Delta_3 U(z_A) + \Delta_4 U(z_A) + \cdots \\ &= (\tfrac{3}{2} - \tfrac{3}{4} + \tfrac{3}{8} - \tfrac{3}{16} + \tfrac{3}{32} - \tfrac{3}{64} + \tfrac{3}{128} - \tfrac{3}{512} + \cdots)U(z_A) \\ &= (1.5 - 0.75 + 0.38 - 0.19 + 0.09 \\ &\quad - 0.05 + 0.02 - 0.01 + \cdots)U(z_A)\,. \end{aligned} \tag{2.114}$$

We see that for a perfectly conducting plate in the nonretarded limit, the Born series converges very slowly. As many as eight terms have to be included in order to reduce the error to about 1 %. Recall that for a weakly dielectric ring we had found that the series converges faster in the retarded limit, so we may expect a better convergence for larger distances for the metal plate as well. Nevertheless, this study shows that the Born expansion leads to a very poor approximation for metals. Retaining only the first one or two terms of the series, we can expect qualitatively correct results at best.

The convergence of the Born series strongly depends on the shape of the bodies under consideration. To show this, let us next consider the nonretarded potential of an atom at distance r_A from the centre of a small metal sphere of radius R, as given by (4.237) together with (4.231) in Vol. I:

$$U(r_A) = -\frac{3\hbar R^3}{4\pi^2\varepsilon_0 r_A^6}\int_0^\infty \mathrm{d}\xi\,\alpha(\mathrm{i}\xi)\,\frac{\varepsilon(\mathrm{i}\xi)-1}{\varepsilon(\mathrm{i}\xi)+2}\,. \tag{2.115}$$

Using (2.110), we can write it as

$$U(r_A) = -\frac{\hbar R^3}{4\pi^2\varepsilon_0 r_A^6}\int_0^\infty \mathrm{d}\xi\,\alpha(\mathrm{i}\xi)\,\frac{\chi(\mathrm{i}\xi)}{1+\frac{1}{3}\chi(\mathrm{i}\xi)}\,. \tag{2.116}$$

When using $\chi/(1+\frac{1}{3}\chi)$ as perturbative parameter, the sphere potential hence agrees exactly with its first-order expansion:

$$U(z_A) = \Delta_1 U(z_A)\,. \tag{2.117}$$

This perfect convergence is a result of the small-sphere geometry, where many-atom effects cancel due to the symmetry of the problem [5].

Our two examples mark the two extremes of an unbounded geometry on the one hand and a compact one on the other. For other body shapes, we may therefore expect the convergence speed to lie in between the slow convergence of the plate and the extremely rapid one found for the sphere. The nonretarded CP potential of an atom in front of a perfectly conducting plate with its slow convergence (2.114) may be viewed as a worst-case scenario with respect to distance as well as body material and shape.

2.3 Casimir–Polder Potential via Body Decomposition

As we have seen when studying the ring in the previous section, the evaluation of higher-order terms in the Born expansion can be very difficult for concrete examples. As an alternative, the Born series can be used to establish an expansion of the CP potential based on body decomposition.

2.3.1 Summation Formulae

Beginning with purely dielectric bodies, we assume that the arrangement described by $\chi(\boldsymbol{r},\omega)$ can be decomposed into a set of smaller, homogeneous bodies n with constant permittivities $\chi_n(\omega)$ and volumes V_n, so that

$$\chi(\boldsymbol{r},\omega) = \sum_n \chi_n(\omega) 1_{V_n}(\boldsymbol{r}) \tag{2.118}$$

with

$$1_V(\boldsymbol{r}) = \begin{cases} 1 \text{ if } \boldsymbol{r} \in V, \\ 0 \text{ else} \end{cases} \tag{2.119}$$

being the characteristic function of a volume V. This body decomposition applies to a number of cases as illustrated in Fig. 2.5: For instance, it can used to separate cavities into its mirror components (i) or to decompose planar (ii) or spherical (iii) stratified bodies with inhomogeneous permittivities into homogeneous layers.

Using the body decomposition, the terms (2.49) of the Born expansion take the form

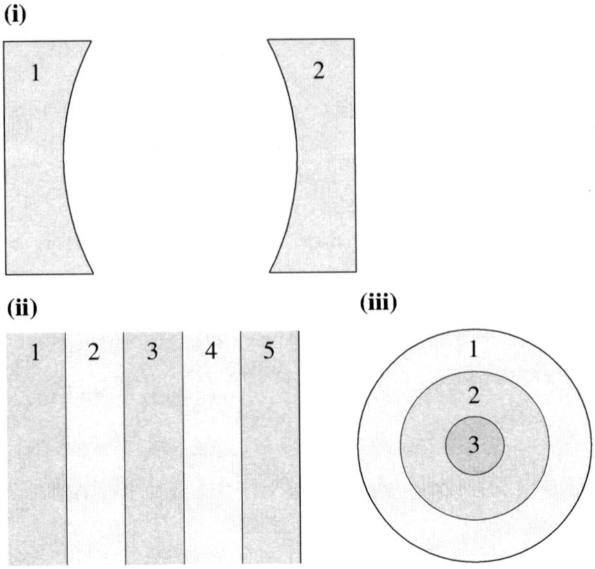

Fig. 2.5 Examples for body decompositions: **(i)** cavity, **(ii)** stratified half space, **(iii)** stratified sphere

$$\Delta_K U(\boldsymbol{r}_A) = \frac{(-1)^K \hbar\mu_0}{2\pi c^{2K}} \int_0^\infty \mathrm{d}\xi\, \xi^{2K+2} \alpha(\mathrm{i}\xi)$$
$$\times \sum_{n_1} \chi_{n_1}(\mathrm{i}\xi) \int_{V_{n_1}} \mathrm{d}^3 s_1 \cdots \sum_{n_K} \chi_{n_K}(\mathrm{i}\xi) \int_{V_{n_K}} \mathrm{d}^3 s_K$$
$$\times\, \mathrm{tr}\Big[\overline{\boldsymbol{G}}(\boldsymbol{r}_A, \boldsymbol{s}_1, \mathrm{i}\xi)\cdot\overline{\boldsymbol{G}}(\boldsymbol{s}_1, \boldsymbol{s}_2, \mathrm{i}\xi)\cdots\overline{\boldsymbol{G}}(\boldsymbol{s}_K, \boldsymbol{r}_A, \mathrm{i}\xi)\Big]\,. \tag{2.120}$$

We rearrange the multiple sums over the bodies: First, we identify how many bodies contribute by writing

$$\Delta_K U(\boldsymbol{r}_A) = \sum_{L=1}^{K} \Delta_K^L U(\boldsymbol{r}_A)\,, \tag{2.121}$$

with $\Delta_K^L U$ being the sum of all L-body contributions to the CP potential. Next, we distinguish which body contributes to which order by specifying

$$\Delta_K^L U(\boldsymbol{r}_A) = \sum_{\substack{n_1<\cdots<n_L\\ k_1+\cdots+k_L=K}} \Delta U_{n_1\cdots n_L}^{k_1\ldots k_L}(\boldsymbol{r}_A)\,, \tag{2.122}$$

where the terms $\Delta U_{n_1\ldots n_L}^{k_1\ldots k_L}$ contain contributions from the susceptibilities χ_{n_j} of each of the bodies n_j to a specific order k_j. Explicitly, they are given by

$$\Delta U_{n_1\ldots n_L}^{k_1\ldots k_L}(\boldsymbol{r}_A) = \frac{(-1)^K\hbar\mu_0}{2\pi c^{2K}}\int_0^\infty \mathrm{d}\xi\,\xi^{2K+2}\alpha(\mathrm{i}\xi)$$
$$\times \sum_{m_1,\ldots,m_L\in\{n_1,\ldots,n_L\}} \chi_{m_1}(\mathrm{i}\xi)\int_{V_{m_1}}\mathrm{d}^3s_1\cdots\chi_{m_K}(\mathrm{i}\xi)\int_{V_{m_K}}\mathrm{d}^3s_K$$
$$\times\,\mathrm{tr}\left[\overline{\boldsymbol{G}}(\boldsymbol{r}_A,\boldsymbol{s}_1,\mathrm{i}\xi)\cdot\overline{\boldsymbol{G}}(\boldsymbol{s}_1,\boldsymbol{s}_2,\mathrm{i}\xi)\cdots\overline{\boldsymbol{G}}(\boldsymbol{s}_K,\boldsymbol{r}_A,\mathrm{i}\xi)\right] \qquad (2.123)$$

where the sum runs only over those terms fulfilling the additional constraint that each n_j occurs exactly k_j times.

As an example, note that the total linear contribution in χ reads

$$\Delta_1 U(\boldsymbol{r}_A) = \Delta_1^1 U(\boldsymbol{r}_A) = \sum_n \Delta U_n^1(\boldsymbol{r}_A)\,. \qquad (2.124)$$

Within this leading order, the CP potential is additive: The total potential for a set of bodies n is simply the sum over the individual potentials ΔU_n^1 associated with these bodies.

Additivity breaks down already in the second order

$$\Delta_2 U(\boldsymbol{r}_A) = \Delta_2^1 U(\boldsymbol{r}_A) + \Delta_2^2 U(\boldsymbol{r}_A) = \sum_n \Delta U_n^2(\boldsymbol{r}_A) + \sum_{m<n} \Delta U_{mn}^{11}(\boldsymbol{r}_A) \qquad (2.125)$$

due to the presence of the two-body potentials ΔU_{mn}^{11}. In general, L-body potentials start to appear when considering the term $\Delta_L U$ of the Born expansion, provided that enough bodies are present in a given arrangement.

Our decomposition (2.121) together with (2.122) is useful in cases where the exact potentials associated with the individual bodies of the decomposition are known. In this case, the terms $\Delta U_{n_1\ldots n_L}^{k_1\ldots k_L}$ can be uniquely identified by performing a Taylor expansion in the bodies' susceptibilities. The explicit knowledge of (2.123) let alone the tedious evaluation of the integrals appearing therein is then not required. Recall our study of the metal half space in the previous section where we were also able to deduce the terms of the Born expansion without explicitly calculating them, by expanding the known exact solution instead.

The developed body decomposition can also be used for metal bodies. To that end, we apply exactly the same steps as before to the respective Born expansion (2.50). The result is again of the form (2.121) with (2.122), $\Delta U_{n_1\ldots n_L}^{k_1\ldots k_L}$ now denotes contributions from the alternative expansion parameters $\chi_{n_j}/(1+\frac{1}{3}\chi_{n_j})$ of bodies n_j with powers k_j. Finally, the body decomposition can be generalised to electromagnetic atoms in magnetoelectric environments. Starting from (2.118) together with

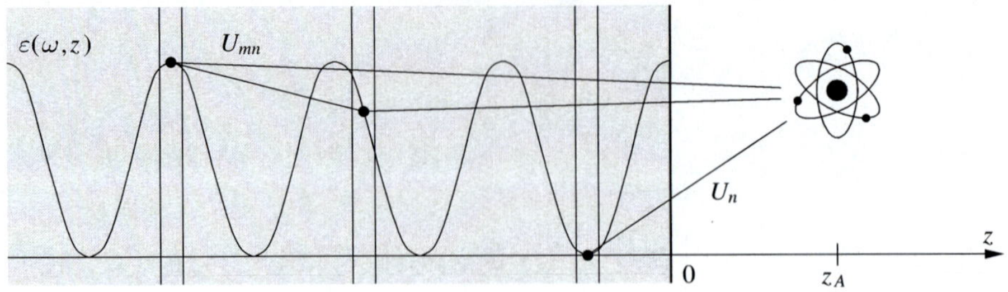

Fig. 2.6 Atom in front of an inhomogeneous half space

$$\zeta(\boldsymbol{r},\omega)=\sum_n \zeta_n(\omega)1_{V_n}(\boldsymbol{r}) \tag{2.126}$$

the respective Born expansion (2.54), we again arrive at (2.121) and (2.122) where $\Delta U^{k_1..k_L}_{n_1...n_L}$ represents contributions containing χ_{n_j}, ζ_{n_j} with total powers k_j.

2.3.2 *Atom in Front of an Inhomogeneous Half Space*

As an example for the use of body decomposition, let us consider an atom at a distance z_A from an inhomogeneous half space as sketched in Fig. 2.6 whose susceptibility only depends on z,

$$\varepsilon(\boldsymbol{r},\omega)=1+\chi(\omega)p(-z)\,. \tag{2.127}$$

Note that such a body whose material properties only change in one direction is commonly known as a stratified body. The profile function is normalised such that $0\le p(-z)\le 1$ and we have $p(-z)=0$ for $z<0$.

We decompose the half space into a number of plates of asymptotically small thickness d such that the susceptibility is approximately constant for each plate. The CP potential associated with such plate is known: As seen from (4.101) in Vol. I, the potential for a plate of permittivity $\varepsilon(\omega)$ and thickness d at separation s from an atom reads

$$U(s)=\frac{\hbar\mu_0}{8\pi^2}\int\limits_0^\infty \mathrm{d}\xi\,\xi^2\alpha(\mathrm{i}\xi)\int\limits_{\xi/c}^\infty \mathrm{d}\kappa^\perp\,\mathrm{e}^{-2\kappa^\perp s}\left[r_s+\left(1-2\,\frac{\kappa^{\perp 2}c^2}{\xi^2}\right)r_p\right] \tag{2.128}$$

where

$$r_s = \frac{\left[\kappa^{\perp 2} - \kappa_1^{\perp 2}\right] \tanh\left(\kappa_1^{\perp} d\right)}{2\kappa^{\perp}\kappa_1^{\perp} + \left[\kappa^{\perp 2} + \kappa_1^{\perp 2}\right] \tanh\left(\kappa_1^{\perp} d\right)}, \tag{2.129}$$

$$r_p = \frac{\left[\varepsilon^2(\mathrm{i}\xi)\kappa^{\perp 2} - \kappa_1^{\perp 2}\right] \tanh\left(\kappa_1^{\perp} d\right)}{2\varepsilon(\mathrm{i}\xi)\kappa^{\perp}\kappa_1^{\perp} + \left[\varepsilon^2(\mathrm{i}\xi)\kappa^{\perp 2} + \kappa_1^{\perp 2}\right] \tanh\left(\kappa_1^{\perp} d\right)} \tag{2.130}$$

with $\kappa_1^{\perp} = \sqrt{[\varepsilon(\mathrm{i}\xi) - 1]\xi^2/c^2 + \kappa^{\perp 2}}$ being the reflection coefficients of the plate for s- and p-polarised waves. In the limit of an asymptotically thin plate, $\sqrt{\varepsilon}d \ll s$, we may replace the reflection coefficients by their leading-order Taylor expansion in $\kappa_1^{\perp} d$,

$$r_s \simeq \frac{[\kappa^{\perp 2} - \kappa_1^{\perp 2}]d}{2\kappa^{\perp}} = -\frac{[\varepsilon(\mathrm{i}\xi) - 1]\kappa^{\perp} d}{2} \frac{\xi^2}{\kappa^{\perp 2} c^2}, \tag{2.131}$$

$$r_p \simeq \frac{[\varepsilon^2(\mathrm{i}\xi)\kappa^{\perp 2} - \kappa_1^{\perp 2}]d}{2\varepsilon(\mathrm{i}\xi)\kappa^{\perp}} = \frac{[\varepsilon^2(\mathrm{i}\xi) - 1]\kappa^{\perp} d}{2\varepsilon(\mathrm{i}\xi)} - \frac{[\varepsilon(\mathrm{i}\xi) - 1]\kappa^{\perp} d}{2\varepsilon(\mathrm{i}\xi)} \frac{\xi^2}{\kappa^{\perp 2} c^2}, \tag{2.132}$$

so that

$$U(s) = -\frac{\hbar d}{16\pi^2\varepsilon_0} \int_0^{\infty} \mathrm{d}\xi\, \alpha(\mathrm{i}\xi)[\varepsilon(\mathrm{i}\xi) - 1] \int_{\xi/c}^{\infty} \mathrm{d}\kappa^{\perp}\, \kappa^{\perp 3} \mathrm{e}^{-2\kappa^{\perp} s}$$
$$\times \left[\frac{2\varepsilon(\mathrm{i}\xi) + 2}{\varepsilon(\mathrm{i}\xi)} - \frac{\varepsilon(\mathrm{i}\xi) + 3}{\varepsilon(\mathrm{i}\xi)} \frac{\xi^2}{\kappa^{\perp 2} c^2} + \frac{\varepsilon(\mathrm{i}\xi) + 1}{\varepsilon(\mathrm{i}\xi)} \frac{\xi^4}{\kappa^{\perp 4} c^4}\right]. \tag{2.133}$$

Labelling the plates by n such that each plate is at position $z = -nd$ and hence at a distance $s = z_A + nd$ from the atom, the potential of plate n reads

$$U_n(z_A) = -\frac{\hbar d}{16\pi^2\varepsilon_0} \int_0^{\infty} \mathrm{d}\xi\, \alpha(\mathrm{i}\xi)[\varepsilon_n(\mathrm{i}\xi) - 1] \int_{\xi/c}^{\infty} \mathrm{d}\kappa\, \kappa^{\perp 3} \mathrm{e}^{-2\kappa(z_A + nd)}$$
$$\times \left[\frac{2\varepsilon_n(\mathrm{i}\xi) + 2}{\varepsilon_n(\mathrm{i}\xi)} - \frac{\varepsilon_n(\mathrm{i}\xi) + 3}{\varepsilon_n(\mathrm{i}\xi)} \frac{\xi^2}{\kappa^{\perp 2} c^2} + \frac{\varepsilon_n(\mathrm{i}\xi) + 1}{\varepsilon_n(\mathrm{i}\xi)} \frac{\xi^4}{\kappa^{\perp 4} c^4}\right] \tag{2.134}$$

where $\varepsilon_n(\omega) = 1 + \chi_n(\omega)$ and $\chi_n(\omega) = \chi(\omega)p(nd)$. Applying a leading-order Taylor expansion in χ_n, we find

$$\Delta U_n^1(z_A) = -\frac{\hbar d}{8\pi^2\varepsilon_0}\int\limits_0^\infty \mathrm{d}\xi\,\alpha(\mathrm{i}\xi)\chi_n(\mathrm{i}\xi)\int\limits_{\xi/c}^\infty \mathrm{d}\kappa^\perp\,\kappa^{\perp 3}\mathrm{e}^{-2\kappa^\perp(z_A+nd)} \times\left(2-2\,\frac{\xi^2}{\kappa^{\perp 2}c^2}+\frac{\xi^4}{\kappa^{\perp 4}c^4}\right). \tag{2.135}$$

According to (2.124), the leading-order potential $\Delta_1 U(z_A)$ of the inhomogeneous half space can be obtained by summing over these thin-plate potentials. In the limit of asymptotically thin plates, the sum becomes an integral ($d\sum_n = \int_{-\infty}^0 \mathrm{d}z$ with $z=-nd$), so we have

$$\Delta_1 U(z_A) = -\frac{\hbar}{8\pi^2\varepsilon_0}\int\limits_{-\infty}^0 \mathrm{d}z\int\limits_0^\infty \mathrm{d}\xi\,\alpha(\mathrm{i}\xi)\chi(\mathrm{i}\xi)p(-z)\int\limits_{\xi/c}^\infty \mathrm{d}\kappa^\perp\,\kappa^{\perp 3} \times\mathrm{e}^{-2\kappa^\perp(z_A-z)}\left(2-2\,\frac{\xi^2}{\kappa^{\perp 2}c^2}+\frac{\xi^4}{\kappa^{\perp 4}c^4}\right). \tag{2.136}$$

After making the substitution $z\mapsto -z$, we have

$$\Delta_1 U(z_A) = -\frac{\hbar}{8\pi^2\varepsilon_0}\int\limits_0^\infty \mathrm{d}\xi\,\alpha(\mathrm{i}\xi)\chi(\mathrm{i}\xi)\int\limits_{\xi/c}^\infty \mathrm{d}\kappa^\perp\,\kappa^{\perp 3}\mathrm{e}^{-2\kappa^\perp z_A} \times\left(2-2\,\frac{\xi^2}{\kappa^{\perp 2}c^2}+\frac{\xi^4}{\kappa^{\perp 4}c^4}\right)\int\limits_0^\infty \mathrm{d}z\,\mathrm{e}^{-2\kappa^\perp z}p(z). \tag{2.137}$$

The quadratic correction $\Delta_2 U$ to this potential consists of a single-plate term $\Delta_2^1 U$ and a two-plate contribution $\Delta_2^2 U$, recall (2.124). The single-plate term can be easily found by performing a second-order Taylor expansion in χ_n of the single-plate potential U_n given above,

$$\Delta U_n^2(z_A) = \frac{\hbar d}{16\pi^2\varepsilon_0}\int\limits_0^\infty \mathrm{d}\xi\,\alpha(\mathrm{i}\xi)\chi_n(\mathrm{i}\xi)\int\limits_{\xi/c}^\infty \mathrm{d}\kappa^\perp\,\kappa^{\perp 3}\mathrm{e}^{-2\kappa^\perp(z_A+nd)} \times\left(2-3\,\frac{\xi^2}{\kappa^{\perp 2}c^2}+\frac{\xi^4}{\kappa^{\perp 4}c^4}\right), \tag{2.138}$$

followed by an integration over all plates:

$$\Delta_2^1 U(z_A) = \frac{\hbar}{16\pi^2\varepsilon_0}\int_0^\infty \mathrm{d}\xi\,\alpha(\mathrm{i}\xi)\chi^2(\mathrm{i}\xi)\int_{\xi/c}^\infty \mathrm{d}\kappa^\perp\,\kappa^{\perp 3}\mathrm{e}^{-2\kappa^\perp z_A}$$
$$\times\left(2-3\frac{\xi^2}{\kappa^{\perp 2}c^2}+\frac{\xi^4}{\kappa^{\perp 4}c^4}\right)\int_0^\infty \mathrm{d}z\,\mathrm{e}^{-2\kappa^\perp z}p(z)\,. \qquad (2.139)$$

For the two-plate contribution, we require the CP potential of an atom at a distance s from two plates of permittivities $\varepsilon(\omega)$, $\varepsilon'(\omega)$, thicknesses d, d' and separation l. This potential is again of the form (2.128) where the reflection coefficients have to be replaced by those of the two-plate system, $r_\sigma \mapsto \overline{r}_\sigma$. These coefficients can be obtained by repeated use of the recursion relations (A.39) and (A.40) in App. A.3.2. To leading order in $\kappa_1^\perp d$, $\kappa_1^\perp d'$, one finds

$$\begin{aligned}\overline{r}_s &\simeq r_s+\left(1-\frac{\kappa^{\perp 2}+\kappa_1^{\perp 2}}{\kappa^{\perp 2}}\,\kappa^\perp d\right)\mathrm{e}^{-2\kappa l}r'_s\\ &= r_s+\left\{1-2\kappa^\perp d-[\varepsilon(\mathrm{i}\xi)-1]\kappa^\perp d\,\frac{\xi^2}{\kappa^{\perp 2}c^2}\right\}\mathrm{e}^{-2\kappa^\perp l}r'_s\,,\end{aligned} \qquad (2.140)$$

$$\begin{aligned}\overline{r}_p &\simeq r_p+\left[1-\frac{\varepsilon^2(\mathrm{i}\xi)\kappa^{\perp 2}+\kappa_1^{\perp 2}}{\varepsilon(\mathrm{i}\xi)\kappa^{\perp 2}}\,\kappa^\perp d\right]\mathrm{e}^{-2\kappa^\perp l}r'_p\\ &= r_p+\left\{1-2\kappa^\perp d-\frac{[\varepsilon(\mathrm{i}\xi)-1]^2\kappa^\perp d}{\varepsilon(\mathrm{i}\xi)}\right.\\ &\qquad\left.-\frac{[\varepsilon(\mathrm{i}\xi)-1]\kappa^\perp d}{\varepsilon(\mathrm{i}\xi)}\,\frac{\xi^2}{\kappa^{\perp 2}c^2}\right\}\mathrm{e}^{-2\kappa^\perp l}r'_p\,,\end{aligned} \qquad (2.141)$$

with r_σ, r'_σ being the single-plate reflection coefficients as given by (2.131) and (2.132). The first term in (2.140), (2.141) describes reflection at the front plate, while the second term is associated with transmission through the front plate, propagation to the rear plate, reflection at the rear plate, propagation back to and transmission through the front plate. Substituting the reflection coefficients into (2.128), we obtain single-plate terms which depend on ε or ε' only and a two-plate term that depends on both ε and ε'. The single-plate potential has already been treated. Noting that the plates are at positions $z=-md$ and $z'=-nd$ so that $s=z_A+md$ and $l=nd-md$, the two-plate potential reads

$$U_{mn}(z_A) = \frac{\hbar dd'}{16\pi^2\varepsilon_0} \int_0^\infty d\xi\, \alpha(i\xi)[\varepsilon_m(i\xi)-1][\varepsilon_n(i\xi)-1] \int_{\xi/c}^\infty d\kappa^\perp\, \kappa^{\perp 4} e^{-2\kappa^\perp(z_A+nd)}$$
$$\times \left\{ 2\,\frac{[\varepsilon_m(i\xi)-1][\varepsilon_m(i\xi)+1]}{\varepsilon_m(i\xi)\varepsilon_n(i\xi)} - \frac{\varepsilon_m(i\xi)\varepsilon_n(i\xi)+3\varepsilon_m(i\xi)-3\varepsilon_n(i\xi)-1}{\varepsilon_m(i\xi)\varepsilon_n(i\xi)}\,\frac{\xi^2}{\kappa^{\perp 2}c^2} \right.$$
$$\left. - \frac{\varepsilon_m(i\xi)+\varepsilon_n(i\xi)+2}{\varepsilon_m(i\xi)\varepsilon_n(i\xi)}\,\frac{\xi^4}{\kappa^{\perp 4}c^4} + \frac{\varepsilon_m(i\xi)\varepsilon_n(i\xi)+1}{\varepsilon_m(i\xi)\varepsilon_n(i\xi)}\,\frac{\xi^6}{\kappa^{\perp 6}c^6} \right\}. \tag{2.142}$$

The required leading-order, linear term in χ_m, χ_n reads

$$\Delta U^{11}_{mn}(z_A) = \frac{\hbar\mu_0 dd'}{8\pi^2} \int_0^\infty d\xi\, \xi^2\alpha(i\xi)\chi_m(i\xi)\chi_n(i\xi) \int_{\xi/c}^\infty d\kappa^\perp\, \kappa^{\perp 2}$$
$$\times e^{-2\kappa^\perp(z_A+nd)} \left(2 - 2\,\frac{\xi^2}{\kappa^{\perp 2}c^2} + \frac{\xi^4}{\kappa^{\perp 4}c^4}\right). \tag{2.143}$$

Summing over all plates in accordance with (2.125), the total two-plate contribution is given by ($dd'\sum_{m<n} = \int_{-\infty}^0 dz \int_{-\infty}^z dz'$ with $z = -md$, $z' = -nd$)

$$\Delta_2^2 U(z_A) = \frac{\hbar\mu_0}{8\pi^2} \int_0^\infty d\xi\, \xi^2\alpha(i\xi)\chi^2(i\xi) \int_{\xi/c}^\infty d\kappa^\perp\, \kappa^{\perp 2} e^{-2\kappa^\perp z_A}$$
$$\times \left(2 - 2\,\frac{\xi^2}{\kappa^{\perp 2}c^2} + \frac{\xi^4}{\kappa^{\perp 4}c^4}\right) \int_0^\infty dz\, p(z) \int_0^\infty dz'\, e^{-2\kappa^\perp z'} p(z') \tag{2.144}$$

where we have made the substitutions $z \mapsto -z$, $z' \mapsto -z'$.

As an example, let us consider a dielectric medium whose permittivity oscillates as a function of z, with the profile function being given by

$$p(z) = \cos^2(kz)\Theta(-z) \tag{2.145}$$

The parameter k determines the period λ of the permittivity oscillations according to $\lambda = \pi/k$.

With this choice of profile function, the z- and z'-integrals in (2.137), (2.139) and (2.144) can be evaluated explicitly,

$$\int_0^\infty \mathrm{d}z\, \mathrm{e}^{-2\kappa^\perp z} \cos^2(kz) = \frac{2\kappa^{\perp 2}+k^2}{4\kappa^\perp(\kappa^{\perp 2}+k^2)}, \tag{2.146}$$

$$\int_0^\infty \mathrm{d}z\, \cos^2(kz) \int_z^\infty \mathrm{d}z'\, \mathrm{e}^{-2\kappa^\perp z'} \cos^2(kz') = \frac{2\kappa^{\perp 6}+8\kappa^{\perp 4}k^2+5\kappa^{\perp 2}k^4+2k^6}{8\kappa^{\perp 2}(\kappa^{\perp 2}+k^2)^2(\kappa^{\perp 2}+4k^2)}, \tag{2.147}$$

and we find

$$\Delta_1 U(z_A) = -\frac{\hbar}{32\pi^2\varepsilon_0} \int_0^\infty \mathrm{d}\xi\, \alpha(\mathrm{i}\xi)\chi(\mathrm{i}\xi) \int_{\xi/c}^\infty \mathrm{d}\kappa^\perp\, \kappa^{\perp 2} \mathrm{e}^{-2\kappa^\perp z_A} \frac{2\kappa^{\perp 2}+k^2}{\kappa^{\perp 2}+k^2} \times \left(2 - 2\frac{\xi^2}{\kappa^{\perp 2}c^2} + \frac{\xi^4}{\kappa^{\perp 4}c^4}\right), \tag{2.148}$$

$$\Delta_2^1 U(z_A) = \frac{\hbar}{64\pi^2\varepsilon_0} \int_0^\infty \mathrm{d}\xi\, \alpha(\mathrm{i}\xi)\chi^2(\mathrm{i}\xi) \int_{\xi/c}^\infty \mathrm{d}\kappa^\perp\, \kappa^{\perp 2} \mathrm{e}^{-2\kappa^\perp z_A} \frac{2\kappa^{\perp 2}+k^2}{\kappa^{\perp 2}+k^2} \times \left(2 - 3\frac{\xi^2}{\kappa^{\perp 2}c^2} + \frac{\xi^4}{\kappa^{\perp 4}c^4}\right), \tag{2.149}$$

$$\Delta_2^2 U(z_A) = \frac{\hbar\mu_0}{64\pi^2} \int_0^\infty \mathrm{d}\xi\, \xi^2\alpha(\mathrm{i}\xi)\chi^2(\mathrm{i}\xi) \int_{\xi/c}^\infty \mathrm{d}\kappa^\perp\, \mathrm{e}^{-2\kappa^\perp z_A} \times \frac{2\kappa^{\perp 6}+8\kappa^{\perp 4}k^2+5\kappa^{\perp 2}k^4+2k^6}{(\kappa^{\perp 2}+k^2)^2(\kappa^{\perp 2}+4k^2)} \left(2 - 2\frac{\xi^2}{\kappa^{\perp 2}c^2} + \frac{\xi^4}{\kappa^{\perp 4}c^4}\right). \tag{2.150}$$

The potential simplifies considerably in the retarded and nonretarded limits. In the retarded limit $z_A \gg c/\omega_-$, the exponential $\exp^{-2\kappa^\perp z_A}$ restricts the ξ-integral to a range where $0 \le \xi \lesssim c/(2z_A) \ll \omega_-$, so that we may make the approximations $\alpha(\mathrm{i}\xi) \simeq \alpha$ and $\chi(\mathrm{i}\xi) \simeq \chi$. Introducing the new integration variables $v = \kappa^\perp c/\xi$ and $s = \kappa^\perp z_A = \xi z_A v/c$, we subsequently transform the integrals according to $\int_0^\infty \mathrm{d}\xi \int_{\xi/c}^\infty \mathrm{d}\kappa^\perp = \int_0^\infty \mathrm{d}\xi\, \xi \int_1^\infty \mathrm{d}v/c = (c/z_A^2) \int_0^\infty \mathrm{d}s\, s \int_1^\infty \mathrm{d}v/v^2$. The integrals can then be carried out to give

$$\Delta_1 U(z_A) = -\frac{23\hbar c\alpha\chi}{480\pi^2\varepsilon_0 z_A^4}\int_0^\infty ds\, s^3 e^{-2s}\,\frac{2s^2+(kz_A)^2}{s^2+(kz_A)^2}\,, \tag{2.151}$$

$$\Delta_2^1 U(z_A) = \frac{3\hbar c\alpha\chi^2}{160\pi^2\varepsilon_0 z_A^4}\int_0^\infty ds\, s^3 e^{-2s}\,\frac{2s^2+(kz_A)^2}{s^2+(kz_A)^2}\,, \tag{2.152}$$

$$\Delta_2^2 U(z_A) = \frac{43\hbar c\alpha\chi^2}{6720\pi^2\varepsilon_0 z_A^4}\int_0^\infty ds\, s^3 e^{-2s}$$
$$\times\frac{2s^6+8s^4(kz_A)^2+5s^2(kz_A)^4+2(kz_A)^6}{[s^2+(kz_A)^2]^2[s^2+4(kz_A)^2]}\,. \tag{2.153}$$

In the nonretarded limit $z_A \ll c/\omega_+$, the atom and medium response functions restrict the ξ-integral to values such that $\xi/\kappa^\perp c \lesssim \xi z_A/c \le \omega_+ z_A/c \ll 1$. We may hence set the lower limit of the $\kappa^\perp$-integral to zero and discard higher-order terms in $\xi/\kappa^\perp c$ in its integrand. After again using $s = \kappa^\perp z_A$, we find

$$\Delta_1 U(z_A) = -\frac{\hbar}{16\pi^2\varepsilon_0 z_A^3}\int_0^\infty d\xi\,\alpha(i\xi)\chi(i\xi)$$
$$\times\int_0^\infty ds\, s^2 e^{-sx}\,\frac{2s^2+(kz_A)^2}{s^2+(kz_A)^2}\,, \tag{2.154}$$

$$\Delta_2^1 U(z_A) = \frac{\hbar}{32\pi^2\varepsilon_0 z_A^3}\int_0^\infty d\xi\,\alpha(i\xi)\chi^2(i\xi)$$
$$\times\int_0^\infty ds\, s^2 e^{-2s}\,\frac{2s^2+(kz_A)^2}{s^2+(kz_A)^2}\,, \tag{2.155}$$

$$\Delta_2^2 U(z_A) = \frac{\hbar\mu_0}{32\pi^2 z_A}\int_0^\infty d\xi\,\xi^2\alpha(i\xi)\chi^2(i\xi)\int_0^\infty ds\, e^{-2s}$$
$$\times\frac{2s^6+8s^4(kz_A)^2+5s^2(kz_A)^4+2(kz_A)^6}{[s^2+(kz_A)^2]^2[s^2+4(kz_A)^2]}\,. \tag{2.156}$$

It is instructive to consider limits of small- and large-scale oscillations of the half-space susceptibility. When the period of the oscillations is much larger than the atom–surface distance, $\lambda \gg z_A$, i.e., $kz_A \ll 1$, then the s-integrals in (2.151)–(2.153) can be performed easily. Adding the results, we obtain the total retarded CP potential

$$U(z_A) = \frac{C_4}{z_A^4}\,. \tag{2.157}$$

It agrees with that of a homogeneous semi-infinite half space of susceptibility χ, where

$$C_4 = -\frac{\hbar c\alpha}{\pi^2\varepsilon_0}\left(\tfrac{23}{640}\,\chi - \tfrac{9}{640}\,\chi^2 - \tfrac{43}{8960}\,\chi^2\right) \tag{2.158}$$

is simply the second-order approximation in χ to the exact half-space coefficient

$$C_4 = -\frac{3\hbar c\alpha}{64\pi^2\varepsilon_0}\int_1^\infty \mathrm{d}v\left[\left(\frac{2}{v^2}-\frac{1}{v^4}\right)\frac{\varepsilon v-\sqrt{\varepsilon\mu-1+v^2}}{\varepsilon v+\sqrt{\varepsilon\mu-1+v^2}} - \frac{1}{v^4}\,\frac{\mu v-\sqrt{\varepsilon\mu-1+v^2}}{\mu v+\sqrt{\varepsilon\mu-1+v^2}}\right], \tag{2.159}$$

cf. (4.133), (4.134) and (4.143) in Vol. I. In the opposite limit of the oscillation period being much smaller than the atom-surface distance $\lambda \ll z_A$, i.e., $kz_A \gg 1$, the integrals in (2.151)–(2.153) can again be performed and we find a total potential (2.157) with coefficient

$$C_4 = -\frac{\hbar c\alpha}{\pi^2\varepsilon_0}\left(\tfrac{23}{1280}\,\chi - \tfrac{9}{1280}\,\chi^2 - \tfrac{43}{35840}\,\chi^2\right). \tag{2.160}$$

We see that the linear contribution in χ is simply given by one half its value for the homogeneous half space. This is in accordance with the simple intuition that the potential for a half space with rapid permittivity oscillations should be determined by the average permittivity. However, the two-plate contribution is reduced to one quarter of its homogeneous-case value for the rapidly oscillating half space, leading to a failure of this simple intuition. Due to many-body correlations, the potential of a half space with rapidly oscillating permittivity is hence slightly more than one half the value for a corresponding homogeneous half space. The value of the potential for intermediate values of z_A/λ can be given as ($k = \pi/\lambda$)

$$U(z_A) = \frac{C_4 f(z_A/\lambda)}{z_A^4} \tag{2.161}$$

where C_4 is the homogeneous-case coefficient (2.158) and the normalised potential $f(z_A/\lambda) = U(z_A)/U(z_A)|_{\lambda\to\infty}$ with

$$f(x) = \frac{6720}{322-169\chi}\left\{\frac{23-9\chi}{360}\int_0^\infty \mathrm{d}s\, s^2 e^{-2s}\,\frac{2s^2+\pi^2x^2}{s^2+\pi^2x^2} - \frac{43\chi}{5040}\int_0^\infty \mathrm{d}s\, e^{-2s}\,\frac{2s^6+8s^4\pi^2x^2+5s^2\pi^4x^4+2\pi^6x^6}{[s^2+\pi^2x^2]^2[s^2+4\pi^2x^2]}\right\} \tag{2.162}$$

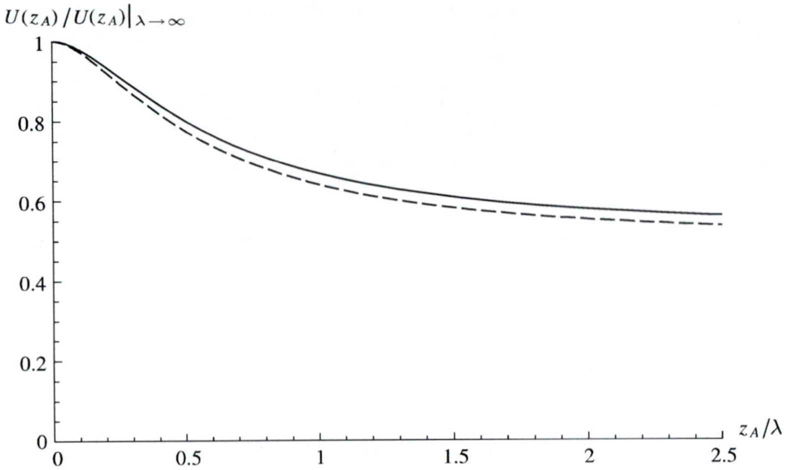

Fig. 2.7 Normalised retarded (*solid line*) and nonretarded (*dashed line*) CP potentials of an atom in front of a half space with spatially oscillating permittivity (where $\chi = 0.5$ for the retarded potential)

depends on the dimensionless parameter z_A/λ. As we will see in Sect. 3.2 below, $f(x)$ is an example of a scaling function. It is depicted in Fig. 2.7 and describes the effect of the permittivity-oscillations. As seen, the CP potential is gradually reduced from its homogeneous-case value as the atom-surface distance increases. As a consequence, it decreases more strongly with distance than $1/z_A^4$ in the transition region $z_A \simeq \lambda$.

Asymptotes for the nonretarded potential can be found in a similar way. Note that the two-plate contribution (2.156) becomes negligible in comparison to (2.154) and (2.155) in the nonretarded limit, because it increases less strongly with decreasing z_A. For large-scale oscillations $\lambda \gg z_A$, we may perform the s-integrals to find

$$U(z_A) = \frac{C_3}{z_A^3} \tag{2.163}$$

where

$$C_3 = -\frac{\hbar}{16\pi^2\varepsilon_0}\int_0^\infty \mathrm{d}\xi\,\alpha(\mathrm{i}\xi)\left[\chi(\mathrm{i}\xi) - \tfrac{1}{2}\,\chi^2(\mathrm{i}\xi)\right] \tag{2.164}$$

is simply the second-order approximation to the coefficient

$$C_3 = -\frac{\hbar}{16\pi^2\varepsilon_0}\int_0^\infty \mathrm{d}\xi\,\alpha(\mathrm{i}\xi)\,\frac{\varepsilon(\mathrm{i}\xi)-1}{\varepsilon(\mathrm{i}\xi)+1} \tag{2.165}$$

for a homogeneous half space. In the opposite limit of small-scale oscillations $\lambda \ll z_A$, the potential is governed by

$$C_3 = -\frac{\hbar}{32\pi^2\varepsilon_0}\int_0^\infty \mathrm{d}\xi\,\alpha(\mathrm{i}\xi)\left[\chi(\mathrm{i}\xi) - \tfrac{1}{2}\chi^2(\mathrm{i}\xi)\right] \tag{2.166}$$

and hence equal to one half the homogeneous-case result. This is a consequence of the fact that the two-plate term does not contribute in the nonretarded limit. The behaviour of the potential for between the two extremes reads

$$U(z_A) = \frac{C_3 f(z_A/\lambda)}{z_A^3}\,. \tag{2.167}$$

The nonretarded coefficient C_3 is given according to (2.164) and we find a normalised potential $f(z_A/\lambda) = U(z_A)/U(z_A)|_{\lambda\to\infty}$ as given by the function

$$f(x) = \frac{4}{3}\int_0^\infty \mathrm{d}s\, s^2 \mathrm{e}^{-2s}\,\frac{2s^2+\pi^2x^2}{s^2+\pi^2x^2}\,. \tag{2.168}$$

It is displayed in Fig. 2.7. Again, we see that the CP potential is reduced from its homogeneous-case value as the distance increases in comparison to the oscillation period. In contrast to the retarded case, the normalised potential is reduced to exactly one half in the limit of small-scale oscillations; this is due to the absence of a two-plate contribution in the nonretarded limit.

References

1. W.C. Chew, *Waves and Fields in Inhomogeneous Media* (IEEE, New York, 1995)
2. S.Y. Buhmann, D.G. Welsch, Prog. Quantum Electron. **31**(2), 51 (2007)
3. S.Y. Buhmann, D.G. Welsch, Appl. Phys. B **82**(2), 189 (2006)
4. R. Golestantian, Phys. Rev. A **80**(1), 012519 (2009)
5. H.Y. Kim, J. Sofo, D. Velegol, M.W. Cole, A.A. Lucas, Langmuir **23**(4), 1735 (2007)

Chapter 3
Common Properties of Dispersion Forces

Having established the general theory of dispersion forces, applied them to simple examples (see Vol. I) and developed strategies for treating more complex geometries, we will now discuss common properties of all ground-state dispersion forces. We begin by summarising the asymptotic retarded and nonretarded power laws found for dispersion forces between objects of simple shapes. These power laws are revealed to be special cases of universal scaling laws. Finally, we show that the three types of dispersion forces are closely related where forces involving bodies have their microscopic origin in many-atom forces on the atoms contained therein.

3.1 Asymptotic Power Laws

Casimir forces, CP forces and vdW forces in specific scenarios can be calculated by using the appropriate Green's tensors. Green's tensors are known for highly symmetric geometries, so that the respective dispersion forces can be found immediately, as demonstrated in Vol. I. For more involved scenarios, approximative methods based on the Born series can be used. In this way, forces involving objects of various shapes have been studied. In all of these examples, it has been found that the distance-dependence of the forces reduces to simple power laws in the retarded and nonretarded limits, i.e., when the object separation is much larger or smaller than the relevant atomic and medium wavelengths.

The asymptotic power laws are summarised in Table 3.1 [1–3], where we list the retarded and nonretarded forces between two atoms [cf. (5.101), (5.104), (5.109), (5.110) and (5.128)–(5.130) in Sect. 5.4 of Vol. I]; an atom and a small sphere [(4.236), (4.237), (4.239) and (4.240) in Sect. 4.7.2 of Vol. I]; an atom and a thin ring [(2.99), (2.100), (2.102), (2.103) and (2.105)–(2.108) in Sect. 2.2.3 of this volume]; an atom an a thin plate [(4.166), (4.168), (4.182) and (4.184) in Sect. 4.6.3 of Vol. I]; an atom and a half space [(4.133), (4.137), (4.156) and (4.158) in Sect. 4.6.2 of Vol. I]; and two half spaces [(3.86), (3.94), (3.99), (3.100), (3.105) and (3.112)

S. Y. Buhmann, *Dispersion Forces II*, Springer Tracts in Modern Physics 248,
DOI: 10.1007/978-3-642-32466-6_3, © Springer-Verlag Berlin Heidelberg 2012

Table 3.1 Asymptotic power laws for the forces between (**i**) two atoms, (**ii**) an atom and a small sphere, (**iii**) an atom and a thin ring, (**iv**) an atom an a thin plate, (**v**) an atom and a half space and (**vi**) the force per unit area between two half spaces

Distance →		Retarded		Nonretarded	
Object combination →		$e \leftrightarrow e$	$e \leftrightarrow m$	$e \leftrightarrow e$	$e \leftrightarrow m$
Dual object combination →		$m \leftrightarrow m$	$m \leftrightarrow e$	$m \leftrightarrow m$	$m \leftrightarrow e$
(**i**)	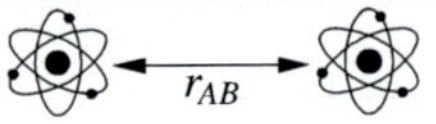	$-\frac{1}{r_{AB}^8}$	$+\frac{1}{r_{AB}^8}$	$-\frac{1}{r_{AB}^7}$	$+\frac{1}{r_{AB}^5}$
(**ii**)	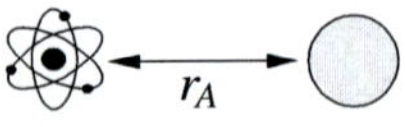	$-\frac{1}{r_A^8}$	$+\frac{1}{r_A^8}$	$-\frac{1}{r_A^7}$	$+\frac{1}{r_A^5}$
(**iii**)	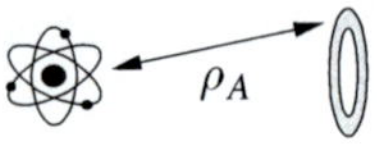	$-\frac{1}{\rho_A^8}$	$+\frac{1}{\rho_A^8}$	$-\frac{1}{\rho_A^7}$	$+\frac{1}{\rho_A^5}$
(**iv**)	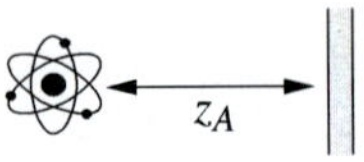	$-\frac{1}{z_A^6}$	$+\frac{1}{z_A^6}$	$-\frac{1}{z_A^5}$	$+\frac{1}{z_A^3}$
(**v**)	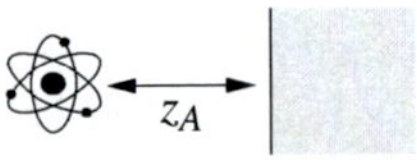	$-\frac{1}{z_A^5}$	$+\frac{1}{z_A^5}$	$-\frac{1}{z_A^4}$	$+\frac{1}{z_A^2}$
(**vi**)	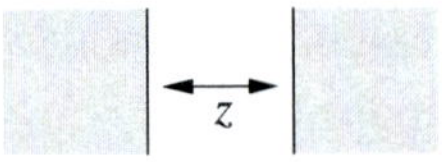	$-\frac{1}{z^4}$	$+\frac{1}{z^4}$	$-\frac{1}{z^3}$	$+\frac{1}{z}$

In the table heading, e stands for an electric object and m for a magnetic one. The signs $-$ and $+$ denote attractive and repulsive forces, respectively

in Sect. 3.3.2 of Vol. I]. In order to be able to compare with the Casimir force, the CP and vdW interactions have also been represented via the forces rather than the potentials.

In the table, we have distinguished between purely *e*lectric objects and purely (para)*m*agnetic ones, such that for each pair of interacting objects, four possible combinations $e \leftrightarrow e$, $e \leftrightarrow m$, $m \leftrightarrow e$, and $m \leftrightarrow m$ need to be considered. As discussed in Sect. 1.3, dispersion forces on objects in free space are invariant under a duality transformation $\alpha \leftrightarrow \beta/c^2, \varepsilon \leftrightarrow \mu$ [4, 5]. As a consequence, the combination $m \leftrightarrow m$ gives rise to the same signs and asymptotic power laws as the combination $e \leftrightarrow e$; a similar statement holds for the combinations $e \leftrightarrow m$ and $m \leftrightarrow e$. In fact, we have extensively made use of duality invariance throughout Vol. I. In this way, we have derived dispersion forces for the combinations $m \leftrightarrow e$ and $m \leftrightarrow m$ from the results for $e \leftrightarrow e$ and $e \leftrightarrow m$, without the need to calculate them explicitly.

For all of the examples studied in the table, dispersion forces between two purely electric or magnetic objects are attractive while those between mixed combinations of electric and magnetic objects are repulsive. For some of the examples, we have made this difference plausible by physical arguments. For two half-spaces (vi), we recall the discussion from Sect. 3.3.1 of Vol. I [6]: According to (3.68) in Vol. I, the Casimir force is attractive if the signs of the reflection coefficients r_σ, r'_σ of the two half spaces agree; it is repulsive if the signs are opposite. For two perfectly conducting plates ($e \leftrightarrow e$), the reflection coefficients follow from the requirement that the tangential component of the electric field must vanish on the plate surfaces. A geometric construction (Fig. 3.6 in Vol. I) reveals that $r_s = r'_s = -1$ for s-polarised waves and $r_p = r'_p = +1$ for p-polarisation. For both polarisations, the reflection coefficients of the two plates have the same sign, leading to an attractive force. For the mixed case ($e \leftrightarrow m$) of a perfectly conducting plate ($r_s = -1, r_p = +1$) interacting with an infinitely permeable one, we note that the tangential components of the magnetic field vanish on the surface of the latter (hence $r'_s = +1, r'_p = -1$). The reflection coefficients of the two plates have opposite signs and hence the force is repulsive.

The signs of the CP force on an atom in front of a half space (v) becomes plausible from the results of Sect. 4.6.1 of Vol. I: As shown, the CP potential of an electric (e) or paramagnetic (m) atom in front of a perfectly conducting plate (e) can be derived from the interaction of the atomic electric or magnetic dipole moments with their images in the plane [7]. For an electric atom, the dipole behaves like a vector under reflection and the dipole–image interaction ($e \leftrightarrow e$) turns out to be attractive when averaging over all dipole orientations. For a magnetic atom, the dipole is a pseudo-vector, so the respective interaction ($e \leftrightarrow m$) has the opposite, repulsive sign.

As explained in Sect. 5.4 of Vol. I, the signs of the nonretarded vdW force between two atoms (i) can also be understood from the interaction of two dipoles [8]: For the pure $e \leftrightarrow e$ case, the electric dipole of an atom gives rise to an electric field which induces an electric dipole moment of a second atom, giving rise to a force. Averaging over all orientations of original and induced dipoles, attractive configurations dominate, leading to an attractive total force. The mixed $e \leftrightarrow m$ interaction follows from the magnetic field created by the electric dipole of the first atom, which then gives rise to a magnetic dipole of the second atom. Magnetic field and induced magnetic moment are proportional to the time derivative of the electric dipole moment, and this additional phase leads to a repulsive total force after rotational averaging. Recall from the examples studied in Sect. 5.5 of Vol. I that the presence of magnetoelectric bodies may modify the strength of the vdW interaction, but it does not change its sign.

Let us next discuss the power laws as listed in Table 3.1. We observe that in the retarded limit, the attractive dispersion forces between two purely electric or magnetic objects ($e \leftrightarrow e$, $m \leftrightarrow m$) follow the same power law as the repulsive ones between mixed combinations ($e \leftrightarrow m$, $m \leftrightarrow e$). In the nonretarded limit, the forces between mixed object combinations are weaker than those in the purely electric or magnetic cases by two powers in the object separation. For the vdW force between two atoms (i), this can again be understood from the dipole–dipole interaction: While the electric and magnetic far fields created by an oscillating electric dipole (e) display the same distance dependence, the electric near field (which interacts with a second electric atom e) is stronger than the magnetic near field (which interacts with a magnetic atom m) by one power in the object separation. The vdW force being quadratic in the dipole–dipole interaction, this implies the observed difference of two powers between the different nonretarded forces. The differences between the retarded and nonretarded power laws governing the vdW interaction of two electric atoms may be understood by requiring the retarded interaction to be proportional to the speed of light c: Multiplying the nonretarded $1/r_{AB}^7$ force by the dimensionless factor $c/(r_{AB}\omega_k)$, we obtain the more rapidly decreasing $1/r_{AB}^8$ dependence of the retarded limit.

As a note of caution, recall from the examples studied in Sects. 3.3 and 4.6 of Vol. I that the perfect reflector limits lead to the correct signs, but may fail to reproduce the corresponding power laws for dispersion forces of objects with realistic magnetoelectric properties: The Casimir force between two perfectly reflecting plates is predicted to be proportional $1/z^4$ in all cases, in contrast to the different power laws for magnetoelectric half spaces as listed in row (vi) of Table 3.1. Similarly, the nonretarded CP force between a perfectly reflecting plate and an atom is found to be always proportional to $1/z_A^4$, in contrast to the two different nonretarded asymptotes given in row (v) of the table. These differences are due to the unrealistic assumption of frequency-independent reflectivities inherent in the perfect-reflector limit.

When comparing the different rows of Table 3.1, we note that each of them displays exactly the same sequence of signs and power laws. In the retarded limit, pure and mixed object combinations lead to attractive and repulsive forces with the same power law (first and second columns). The asymptotic behaviour changes by one inverse power when going to the attractive nonretarded force between two purely electric or magnetic objects (third column). It is reduced by two further inverse powers when considering the repulsive nonretarded interaction for the mixed case (fourth column).

These similarities between the different rows can be easily understood from the common microscopic origin of all three types of dispersion forces. As shown in detail in Sect. 3.3.1 below, dispersion forces involving bodies that consist of a dilute gas of atoms are entirely due to the vdW forces on the gas atoms. In other words, CP forces between an atom and a gas body are pairwise sums over the vdW forces between the atom and the gas atoms [9, 10]. A summation of the vdW forces (i) between a single atom and the atoms inside the compact volumes of a small sphere (ii) or a thin ring (iii) does not change the respective power law, which is why the asymptotes given in rows (i)–(iii) agree. Summation over a non-compact volume lowers the leading inverse power according to the number of non-compact dimensions. So, the leading inverse powers are lowered by two and three for the interaction of an atom with a thin plate of infinite lateral extension (iv) and a half space (v), respectively. Similarly, the Casimir force between two bodies in the dilute-gas limit is simply the sum over the CP forces on the atoms of one body due to the second one. The power laws for the force between two half spaces (vi) can thus be obtained from the force between an atom and an half space (v) by summing over three non-compact dimensions. Summation in the direction perpendicular to the plate surface lowers the leading inverse powers by one, while the trivial sums along the plate surface yield an infinite force, i.e., a finite force per unit area.

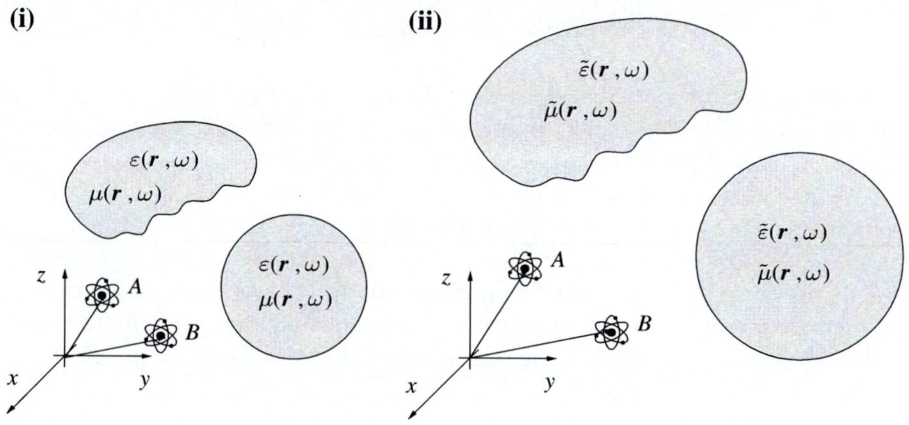

Fig. 3.1 (**i**) Original and (**ii**) scaled configurations of bodies and atoms ($a = 1.4$)

As discussed in Sect. 3.3.2, the simple additive approach breaks down for bodies containing a denser arrangement of atoms. In this case, many-atom vdW forces need to be taken into account, making the relation to the macroscopic CP and Casimir forces more complex [9, 10]. However, the table shows that many-atom interactions do not change the leading power laws compared to those found from the pairwise summation approach. They only modify the proportionality factors (which are not listed in the table).

Before we study the microscopic origin of dispersion forces in more detail, let us address the following question: Are the observed simple asymptotic power laws just a consequence of the very simple object shapes considered, or can we expect them to hold even for more complex geometries? The answer to this question can be given by studying the behaviour of dispersion interactions under a general scaling transformation.

3.2 Universal Scaling Laws

A general scaling transformation can be introduced as follows: We start from an arbitrary arrangement of bodies characterised by their permittivity $\varepsilon(\boldsymbol{r}, \omega)$ and permeability $\mu(\boldsymbol{r}, \omega)$ in the possible presence of one or two atoms at positions $\boldsymbol{r}_A$ and $\boldsymbol{r}_B$; an example is shown in Fig. 3.1(i). The corresponding scaled arrangement with a scaling factor $a > 0$ is described by the new permittivity and permeability

$$\tilde{\varepsilon}(\tilde{\boldsymbol{r}}, \omega) = \varepsilon(\boldsymbol{r}, \omega) , \qquad \tilde{\mu}(\tilde{\boldsymbol{r}}, \omega) = \mu(\boldsymbol{r}, \omega) \tag{3.1}$$

with $\tilde{\boldsymbol{r}} = a\boldsymbol{r}$, i.e.,

$$\tilde{\varepsilon}(\boldsymbol{r},\omega)=\varepsilon(\boldsymbol{r}/a,\omega)\,,\quad \tilde{\mu}(\boldsymbol{r},\omega)=\mu(\boldsymbol{r}/a,\omega)\,. \tag{3.2}$$

The atomic positions are scaled accordingly: $\tilde{\boldsymbol{r}}_A = a\boldsymbol{r}_A$, $\tilde{\boldsymbol{r}}_B = a\boldsymbol{r}_B$, see Fig. 3.1(ii). In other words, in the scaled arrangement, all sizes and distances are globally increased ($a > 1$) or reduced ($a < 1$) by the scaling factor a.

We now want to investigate how this global scaling transformation affects dispersion interactions, e.g., we want to see whether the respective potentials and forces in the scaled arrangement can be related to the original one via some power of a. As discussed in the previous section, dispersion forces are governed by different asymptotic power laws in the retarded versus nonretarded limits. To find generally valid scaling laws, we thus need to distinguish these two asymptotic regimes.

3.2.1 Retarded Dispersion Forces

As seen for the various examples studied, retarded dispersion interactions depend on the static properties of atoms and bodies (cf. Chap. 2 or the detailed discussion in Sect. 3.3.2 of Vol. I). With these simplifications $\alpha(\mathrm{i}\xi)\simeq\alpha(0)\equiv\alpha$, $\varepsilon(\boldsymbol{r},\mathrm{i}\xi)\simeq\varepsilon(\boldsymbol{r},0)\equiv\varepsilon(\boldsymbol{r})$ and $\mu(\boldsymbol{r},\mathrm{i}\xi)\simeq\mu(\boldsymbol{r},0)\equiv\mu(\boldsymbol{r})$, the CP and vdW potentials (1.128) and (1.153) of purely electric atoms are given by

$$U_e(\boldsymbol{r}_A)=\frac{\hbar\mu_0\alpha}{2\pi}\int_0^\infty \mathrm{d}\xi\,\xi^2\,\mathrm{tr}\,\boldsymbol{G}^{(1)}(\boldsymbol{r}_A,\boldsymbol{r}_A,\mathrm{i}\xi) \tag{3.3}$$

$$U_{ee}(\boldsymbol{r}_A,\boldsymbol{r}_B)=-\frac{\hbar\mu_0^2\alpha_A\alpha_B}{2\pi}\int_0^\infty \mathrm{d}\xi\,\xi^4\mathrm{tr}\big[\boldsymbol{G}(\boldsymbol{r}_A,\boldsymbol{r}_B,\mathrm{i}\xi)\cdot\boldsymbol{G}(\boldsymbol{r}_B,\boldsymbol{r}_A,\mathrm{i}\xi)\big] \tag{3.4}$$

with the Green's tensor (1.14) for the retarded limit being determined by the simplified differential equation

$$\left[\nabla\times\frac{1}{\mu(\boldsymbol{r})}\nabla\times-\frac{\omega^2}{c^2}\,\varepsilon(\boldsymbol{r})\right]\boldsymbol{G}(\boldsymbol{r},\boldsymbol{r}',\omega)=\boldsymbol{\delta}(\boldsymbol{r}-\boldsymbol{r}')\,. \tag{3.5}$$

Similarly, the Casimir force (1.117) can be given in terms of this retarded-limit Green's tensor as

$$\boldsymbol{F}=\int_{\partial V}\mathrm{d}\boldsymbol{A}\cdot\boldsymbol{T} \tag{3.6}$$

with a Casimir stress (i.e., directed force per unit area)

$$\boldsymbol{T}(\boldsymbol{r}) = -\frac{\hbar}{\pi}\int_0^\infty \mathrm{d}\xi \left\{ \frac{\xi^2}{c^2}\,\boldsymbol{G}^{(1)}(\boldsymbol{r},\boldsymbol{r},\mathrm{i}\xi) + \nabla\times\boldsymbol{G}^{(1)}(\boldsymbol{r},\boldsymbol{r}',\mathrm{i}\xi)\times\overleftarrow{\nabla}'|_{\boldsymbol{r}'=\boldsymbol{r}} \right.$$
$$\left. - \tfrac{1}{2}\,\mathrm{tr}\left[\frac{\xi^2}{c^2}\,\boldsymbol{G}^{(1)}(\boldsymbol{r},\boldsymbol{r},\mathrm{i}\xi) + \nabla\times\boldsymbol{G}^{(1)}(\boldsymbol{r},\boldsymbol{r}',\mathrm{i}\xi)\times\overleftarrow{\nabla}'|_{\boldsymbol{r}'=\boldsymbol{r}} \right]\boldsymbol{I} \right\}. \tag{3.7}$$

To investigate the scaling behaviour of these retarded interactions, we hence need to determine how the scaling transformation affects the Green's tensor. The retarded-limit Green's tensor for the scaled arrangement of bodies is the solution to the differential equation

$$\left[\tilde{\nabla}\times\frac{1}{\tilde{\mu}(\tilde{\boldsymbol{r}})}\tilde{\nabla}\times - \frac{\omega^2}{c^2}\,\tilde{\varepsilon}(\tilde{\boldsymbol{r}})\right]\tilde{\boldsymbol{G}}(\tilde{\boldsymbol{r}},\tilde{\boldsymbol{r}}',\omega) = \boldsymbol{\delta}(\tilde{\boldsymbol{r}}-\tilde{\boldsymbol{r}}')\,. \tag{3.8}$$

Using $\tilde{\boldsymbol{r}}=a\boldsymbol{r}, \tilde{\boldsymbol{r}}'=a\boldsymbol{r}', \tilde{\nabla}=\nabla/a$ and $\delta(a\boldsymbol{r})=\delta(\boldsymbol{r})/a^3$, replacing $\omega\mapsto\tilde{\omega}=\omega/a$ and invoking the scaling relation (3.1), we can rewrite this equation as

$$\left[\nabla\times\frac{1}{\mu(\boldsymbol{r})}\nabla\times - \frac{\omega^2}{c^2}\,\varepsilon(\boldsymbol{r})\right]a\tilde{\boldsymbol{G}}(a\boldsymbol{r},a\boldsymbol{r}',\omega/a) = \boldsymbol{\delta}(\boldsymbol{r}-\boldsymbol{r}')\,. \tag{3.9}$$

Comparison with (3.5) reveals the scaling

$$\tilde{\boldsymbol{G}}(\tilde{\boldsymbol{r}},\tilde{\boldsymbol{r}}',\tilde{\omega}) = \frac{1}{a}\,\boldsymbol{G}(\boldsymbol{r},\boldsymbol{r}',\omega) \tag{3.10}$$

of the Green's tensor appropriate for the retarded limit. Note that a frequency argument scaling $\tilde{\omega}=\omega/a$ also occurs which is opposite to the scaling of the position arguments, $\tilde{\boldsymbol{r}}=a\boldsymbol{r}$, $\tilde{\boldsymbol{r}}'=a\boldsymbol{r}'$. This can be intuitively understood from the fact that the retarded-limit Green's tensor involves the propagation of waves. Here, $|\boldsymbol{r}-\boldsymbol{r}'|\omega=|\tilde{\boldsymbol{r}}-\tilde{\boldsymbol{r}}'|\tilde{\omega}=2\pi c$ describes the propagation of a wave front and its scaled version in vacuum.

The scaling properties of the dispersion interactions follow immediately by using this relation. Substitution into the CP potential (3.3) yields ($\tilde{\xi}=\xi/a$)

$$\begin{aligned}
\tilde{U}_e(\tilde{\boldsymbol{r}}_A) &= \frac{\hbar\mu_0\alpha}{2\pi}\int_0^\infty \mathrm{d}\tilde{\xi}\,\tilde{\xi}^2\,\mathrm{tr}\,\tilde{\boldsymbol{G}}^{(1)}(\tilde{\boldsymbol{r}}_A,\tilde{\boldsymbol{r}}_A,\mathrm{i}\tilde{\xi}) \\
&= \frac{\hbar\mu_0\alpha}{2\pi}\int_0^\infty \mathrm{d}\tilde{\xi}\,\tilde{\xi}^2\,\frac{1}{a}\,\mathrm{tr}\,\boldsymbol{G}^{(1)}(\boldsymbol{r}_A,\boldsymbol{r}_A,\mathrm{i}\xi) \\
&= \frac{1}{a^4}\,\frac{\hbar\mu_0\alpha}{2\pi}\int_0^\infty \mathrm{d}\xi\,\xi^2\mathrm{tr}\,\boldsymbol{G}^{(1)}(\boldsymbol{r}_A,\boldsymbol{r}_A,\mathrm{i}\xi)\,.
\end{aligned} \tag{3.11}$$

Comparing with the unscaled potential (3.3), we find the scaling law [11]

$$\tilde{U}_e(\tilde{\boldsymbol{r}}_A) = \frac{1}{a^4}\,U_e(\boldsymbol{r}_A) \tag{3.12}$$

for the CP potential. The CP force (1.119) thus scales as ($\tilde{\boldsymbol{\nabla}}_A = \boldsymbol{\nabla}_A/a$)

$$\tilde{\boldsymbol{F}}(\tilde{\boldsymbol{r}}_A) = -\tilde{\boldsymbol{\nabla}}_A\tilde{U}(\tilde{\boldsymbol{r}}_A) = -\frac{1}{a^4}\,\tilde{\boldsymbol{\nabla}}_A U(\boldsymbol{r}_A) = \frac{1}{a^5}\,\boldsymbol{F}(\boldsymbol{r}_A)\,. \tag{3.13}$$

Similarly, using the Green's tensor scaling (3.10) in (3.4), we find

$$\begin{aligned}
\tilde{U}_{ee}(\tilde{\boldsymbol{r}}_A, \tilde{\boldsymbol{r}}_B) &= -\frac{\hbar\mu_0^2\alpha_A\alpha_B}{2\pi}\int_0^\infty \mathrm{d}\tilde{\xi}\,\tilde{\xi}^4\frac{1}{a^2}\mathrm{tr}\big[\boldsymbol{G}(\boldsymbol{r}_A, \boldsymbol{r}_B, \mathrm{i}\xi)\cdot\boldsymbol{G}(\boldsymbol{r}_B, \boldsymbol{r}_A, \mathrm{i}\xi)\big]\\
&= -\frac{1}{a^7}\,\frac{\hbar\mu_0^2\alpha_A\alpha_B}{2\pi}\int_0^\infty \mathrm{d}\xi\,\xi^4\mathrm{tr}\big[\boldsymbol{G}(\boldsymbol{r}_A, \boldsymbol{r}_B, \mathrm{i}\xi)\cdot\boldsymbol{G}(\boldsymbol{r}_B, \boldsymbol{r}_A, \mathrm{i}\xi)\big]
\end{aligned} \tag{3.14}$$

so that the vdW potential obeys the scaling law [11]

$$\tilde{U}_{ee}(\tilde{\boldsymbol{r}}_A, \tilde{\boldsymbol{r}}_B) = \frac{1}{a^7}\,U_{ee}(\boldsymbol{r}_A, \boldsymbol{r}_B)\,. \tag{3.15}$$

This implies a scaling

$$\tilde{\boldsymbol{F}}(\tilde{\boldsymbol{r}}_A, \tilde{\boldsymbol{r}}_B) = -\frac{1}{a^7}\,\tilde{\boldsymbol{\nabla}}_A U(\boldsymbol{r}_A, \boldsymbol{r}_B) = \frac{1}{a^8}\,\boldsymbol{F}(\boldsymbol{r}_A, \boldsymbol{r}_B)\,. \tag{3.16}$$

of the vdW force. Finally, for the Casimir stress we find ($\tilde{\boldsymbol{\nabla}} = \boldsymbol{\nabla}/a$)

$$\begin{aligned}
\tilde{\boldsymbol{T}}(\tilde{\boldsymbol{r}}) = &-\frac{\hbar}{\pi}\int_0^\infty \mathrm{d}\tilde{\xi}\bigg\{\frac{\tilde{\xi}^2}{c^2}\,\frac{1}{a}\,\boldsymbol{G}^{(1)}(\boldsymbol{r}, \boldsymbol{r}, \mathrm{i}\xi) + \frac{1}{a}\,\tilde{\boldsymbol{\nabla}}\times\boldsymbol{G}^{(1)}(\boldsymbol{r}, \boldsymbol{r}', \mathrm{i}\xi)\times\overleftarrow{\tilde{\boldsymbol{\nabla}}}'|_{\boldsymbol{r}'=\boldsymbol{r}}\\
&-\tfrac{1}{2}\,\mathrm{tr}\bigg[\frac{\tilde{\xi}^2}{c^2}\,\frac{1}{a}\,\boldsymbol{G}^{(1)}(\boldsymbol{r}, \boldsymbol{r}, \mathrm{i}\xi) + \frac{1}{a}\,\tilde{\boldsymbol{\nabla}}\times\boldsymbol{G}^{(1)}(\boldsymbol{r}, \boldsymbol{r}', \mathrm{i}\xi)\times\overleftarrow{\tilde{\boldsymbol{\nabla}}}'|_{\boldsymbol{r}'=\boldsymbol{r}}\bigg]\boldsymbol{I}\bigg\}\\
= &-\frac{1}{a^4}\,\frac{\hbar}{\pi}\int_0^\infty \mathrm{d}\xi\bigg\{\frac{\xi^2}{c^2}\,\boldsymbol{G}^{(1)}(\tilde{\boldsymbol{r}}, \boldsymbol{r}, \mathrm{i}\xi) + \boldsymbol{\nabla}\times\boldsymbol{G}^{(1)}(\boldsymbol{r}, \boldsymbol{r}', \mathrm{i}\xi)\times\overleftarrow{\boldsymbol{\nabla}}'|_{\boldsymbol{r}'=\boldsymbol{r}}\\
&-\tfrac{1}{2}\,\mathrm{tr}\bigg[\frac{\xi^2}{c^2}\,\boldsymbol{G}^{(1)}(\boldsymbol{r}, \boldsymbol{r}, \mathrm{i}\xi) + \boldsymbol{\nabla}\times\boldsymbol{G}^{(1)}(\boldsymbol{r}, \boldsymbol{r}', \mathrm{i}\xi)\times\overleftarrow{\boldsymbol{\nabla}}'|_{\boldsymbol{r}'=\boldsymbol{r}}\bigg]\boldsymbol{I}\bigg\}\,,
\end{aligned} \tag{3.17}$$

so that [11]

$$\tilde{\boldsymbol{T}}(\tilde{\boldsymbol{r}}) = \frac{1}{a^4}\,\boldsymbol{T}(\boldsymbol{r})\,. \tag{3.18}$$

The Casimir force (3.6) hence scales as ($\mathrm{d}\tilde{\boldsymbol{A}} = a^2\mathrm{d}\boldsymbol{A}$)

$$\tilde{\boldsymbol{F}} = \int_{\partial\tilde{V}} \mathrm{d}\tilde{\boldsymbol{A}}\cdot\tilde{\boldsymbol{T}}(\tilde{\boldsymbol{r}}) = \frac{1}{a^2}\int_{\partial V} \mathrm{d}\boldsymbol{A}\cdot\boldsymbol{T}(\boldsymbol{r}) = \frac{1}{a^2}\,\boldsymbol{F}\,. \tag{3.19}$$

These results can be generalised to electromagnetic atoms. In the retarded limit, the respective atomic potentials (1.177) and (1.178) take the forms

$$U_\lambda(\boldsymbol{r}_A) = \frac{\hbar\alpha_\lambda}{2\pi\varepsilon_0}\int_0^\infty \mathrm{d}\xi\,\mathrm{tr}\,\boldsymbol{G}^{(1)}_{\lambda\lambda}(\boldsymbol{r}_A,\boldsymbol{r}_A,\mathrm{i}\xi)\,, \tag{3.20}$$

$$U_{\lambda\lambda'}(\boldsymbol{r}_A,\boldsymbol{r}_B) = -\frac{\hbar\alpha^A_\lambda\alpha^B_{\lambda'}}{2\pi\varepsilon_0^2}\int_0^\infty \mathrm{d}\xi\,\mathrm{tr}\big[\boldsymbol{G}_{\lambda\lambda'}(\boldsymbol{r}_A,\boldsymbol{r}_B,\mathrm{i}\xi)\cdot\boldsymbol{G}_{\lambda'\lambda}(\boldsymbol{r}_B,\boldsymbol{r}_A,\mathrm{i}\xi)\big] \tag{3.21}$$

(λ, $\lambda' = e, m$). With the tensors $\boldsymbol{G}_{\lambda\lambda'}$ being defined by (1.172)–(1.175), their scaling behaviour follows from (3.10) together with $\tilde{\boldsymbol{\nabla}} = \boldsymbol{\nabla}/a$ and $\tilde{\omega} = \omega/a$:

$$\tilde{\boldsymbol{G}}_{\lambda\lambda'}(\tilde{\boldsymbol{r}},\tilde{\boldsymbol{r}}',\tilde{\omega}) = \frac{1}{a^3}\,\boldsymbol{G}_{\lambda\lambda'}(\boldsymbol{r},\boldsymbol{r}',\omega)\,. \tag{3.22}$$

Following similar steps as above, substitution into (3.20) and (3.21) above immediately implies the required scaling laws for CP and vdW potentials of electric as well as magnetic atoms,

$$\tilde{U}_\lambda(\tilde{\boldsymbol{r}}_A) = \frac{1}{a^4}\,U_\lambda(\boldsymbol{r}_A)\,, \tag{3.23}$$

$$\tilde{U}_{\lambda\lambda'}(\tilde{\boldsymbol{r}}_A,\tilde{\boldsymbol{r}}_B) = \frac{1}{a^7}\,U_{\lambda\lambda'}(\boldsymbol{r}_A,\boldsymbol{r}_B)\,. \tag{3.24}$$

the scaling laws (3.13) and (3.16) above for the associated dispersion forces hence remain valid for electromagnetic atoms as well.

3.2.2 Nonretarded Dispersion Forces

As we have just seen, retarded-limit dispersion interactions are given by scaling laws which do not depend on whether the bodies or atoms involved are electric or magnetic. This in agreement with the examples listed in the first two columns of

Table 3.1. As seen from the last two columns of the table, nonretarded interactions follow different power laws, depending on whether the atoms or bodies are electric or magnetic. A unique scaling law valid for all cases can hence not be expected and we have to distinguish between electric and magnetic objects.

For nonretarded distances, the CP and vdW potentials (1.128) and (1.153) for purely electric atoms as well as the Casimir stress (3.7) depend on the simplified nonretarded Green's tensor. To determine the scaling behaviour of the latter, we represent it as the unique solution to the Dyson equation (2.4). For purely electric bodies, it can be given as

$$\begin{aligned}&\boldsymbol{G}(\boldsymbol{r},\boldsymbol{r}',\omega)\\&\quad=\boldsymbol{G}^{(0)}(\boldsymbol{r},\boldsymbol{r}',\omega)+\frac{\omega^2}{c^2}\int\mathrm{d}^3s\,[\varepsilon(\boldsymbol{s},\omega)-1]\boldsymbol{G}^{(0)}(\boldsymbol{r},\boldsymbol{s},\omega)\cdot\boldsymbol{G}(\boldsymbol{s},\boldsymbol{r}',\omega)\end{aligned} \tag{3.25}$$

where

$$\boldsymbol{G}^{(0)}(\boldsymbol{r},\boldsymbol{r}',\omega)=-\frac{c^2}{3\omega^2}\,\delta(\boldsymbol{\rho})-\frac{c^2}{4\pi\omega^2\rho^3}[\boldsymbol{I}-3\boldsymbol{e}_\rho\boldsymbol{e}_\rho] \tag{3.26}$$

is the nonretarded free-space Green's tensor (A.25) from App. A.2. Recalling that $\delta(a\boldsymbol{r})=\delta(\boldsymbol{r})/a^3$, we immediately find that the latter scales as

$$\tilde{\boldsymbol{G}}^{(0)}(\tilde{\boldsymbol{r}},\tilde{\boldsymbol{r}}',\omega)=-\frac{c^2}{3\omega^2}\,\delta(a\boldsymbol{\rho})-\frac{c^2}{4\pi\omega^2(a\rho)^3}[\boldsymbol{I}-3\boldsymbol{e}_\rho\boldsymbol{e}_\rho]=\frac{1}{a^3}\,\boldsymbol{G}^{(0)}(\boldsymbol{r},\boldsymbol{r}',\omega)\,. \tag{3.27}$$

Using this property and invoking the scaling transformation (3.1), the Dyson equation for the total scaled Green's tensor can be written as ($\tilde{\boldsymbol{s}}=a\boldsymbol{s}$)

$$\begin{aligned}&\tilde{\boldsymbol{G}}(\tilde{\boldsymbol{r}},\tilde{\boldsymbol{r}}',\omega)\\&\quad=\tilde{\boldsymbol{G}}^{(0)}(\tilde{\boldsymbol{r}},\tilde{\boldsymbol{r}}',\omega)+\frac{\omega^2}{c^2}\int\mathrm{d}^3\tilde{s}\,[\tilde{\varepsilon}(\tilde{\boldsymbol{s}},\omega)-1]\tilde{\boldsymbol{G}}^{(0)}(\tilde{\boldsymbol{r}},\tilde{\boldsymbol{s}},\omega)\cdot\boldsymbol{G}(\tilde{\boldsymbol{s}},\tilde{\boldsymbol{r}}',\omega)\\&\quad=\frac{1}{a^3}\,\boldsymbol{G}^{(0)}(\boldsymbol{r},\boldsymbol{r}',\omega)+\frac{\omega^2}{c^2}\int\mathrm{d}^3s\,[\varepsilon(\boldsymbol{s},\omega)-1]\boldsymbol{G}^{(0)}(\boldsymbol{r},\boldsymbol{s},\omega)\cdot\boldsymbol{G}(\tilde{\boldsymbol{s}},\tilde{\boldsymbol{r}}',\omega)\,.\end{aligned} \tag{3.28}$$

Multiplying this equation with a^3 and comparing with the original Dyson equation (3.25), we deduce the scaling law

$$\tilde{\boldsymbol{G}}(\tilde{\boldsymbol{r}},\tilde{\boldsymbol{r}}',\omega)=\frac{1}{a^3}\,\boldsymbol{G}(\boldsymbol{r},\boldsymbol{r}',\omega) \tag{3.29}$$

for the nonretarded Green's tensor in the presence of electric bodies. Note that in contrast to the retarded case, the frequency dependence of the Green's tensor remains unscaled.

With this scaling of the Green's tensor, (1.128) leads to

$$\tilde{U}(\tilde{\boldsymbol{r}}_A) = \frac{\hbar\mu_0}{2\pi}\int_0^\infty \mathrm{d}\xi\,\xi^2\alpha(\mathrm{i}\xi)\,\mathrm{tr}\,\tilde{\boldsymbol{G}}^{(1)}(\tilde{\boldsymbol{r}}_A,\tilde{\boldsymbol{r}}_A,\mathrm{i}\xi)$$
$$= \frac{\hbar\mu_0}{2\pi}\int_0^\infty \mathrm{d}\xi\,\xi^2\alpha(\mathrm{i}\xi)\,\frac{1}{a^3}\,\mathrm{tr}\,\boldsymbol{G}^{(1)}(\boldsymbol{r}_A,\boldsymbol{r}_A,\mathrm{i}\xi)\,. \tag{3.30}$$

The nonretarded CP potential of an electric atom in the presence of electric bodies hence obeys the scaling law [11]

$$\tilde{U}(\tilde{\boldsymbol{r}}_A) = \frac{1}{a^3}\,U(\boldsymbol{r}_A) \tag{3.31}$$

and the respective CP force scales as

$$\tilde{\boldsymbol{F}}(\tilde{\boldsymbol{r}}_A) = \frac{1}{a^4}\,\boldsymbol{F}(\boldsymbol{r}_A)\,. \tag{3.32}$$

Similarly, combining the Green's-tensor scaling with (1.153), the vdW potential and force are found to behave as [11]

$$\tilde{U}(\tilde{\boldsymbol{r}}_A,\tilde{\boldsymbol{r}}_B) = \frac{1}{a^6}\,U(\boldsymbol{r}_A,\boldsymbol{r}_B) \tag{3.33}$$

and

$$\tilde{\boldsymbol{F}}(\tilde{\boldsymbol{r}}_A,\tilde{\boldsymbol{r}}_B) = \frac{1}{a^7}\,\boldsymbol{F}(\boldsymbol{r}_A,\boldsymbol{r}_B) \tag{3.34}$$

under a scaling transformation.

To treat the Casimir stress, we also require the scaling of $\nabla\times\boldsymbol{G}^{(1)}\times\overleftarrow{\nabla}'$. Taking the left and right curls of (3.25) and decomposing the Green's tensor into its bulk and scattering parts according to (1.111), we can represent $\nabla\times\boldsymbol{G}^{(1)}\times\overleftarrow{\nabla}'$ as the solution to the Dyson equation

$$\nabla\times\boldsymbol{G}^{(1)}(\boldsymbol{r},\boldsymbol{r}',\omega)\times\overleftarrow{\nabla}' = \frac{\omega^2}{c^2}\int \mathrm{d}^3s\,[\varepsilon(\boldsymbol{s},\omega)-1]\nabla\times\boldsymbol{G}^{(0)}(\boldsymbol{r},\boldsymbol{s},\omega)$$
$$\cdot\Big[\boldsymbol{G}^{(0)}(\boldsymbol{s},\boldsymbol{r}',\omega)\times\overleftarrow{\nabla}' + \boldsymbol{G}^{(1)}(\boldsymbol{s},\boldsymbol{r}',\omega)\times\overleftarrow{\nabla}'\Big] \tag{3.35}$$

where $\boldsymbol{G}^{(1)}\times\overleftarrow{\nabla}'$ in turn is the solution to

$$\boldsymbol{G}^{(1)}(\boldsymbol{r},\boldsymbol{r}',\omega)\times\overleftarrow{\nabla}' = \frac{\omega^2}{c^2}\int \mathrm{d}^3s\,[\varepsilon(\boldsymbol{s},\omega)-1]\boldsymbol{G}^{(0)}(\boldsymbol{r},\boldsymbol{s},\omega)$$
$$\cdot\Big[\boldsymbol{G}^{(0)}(\boldsymbol{s},\boldsymbol{r}',\omega)\times\overleftarrow{\nabla}' + \boldsymbol{G}^{(1)}(\boldsymbol{s},\boldsymbol{r}',\omega)\times\overleftarrow{\nabla}'\Big]. \tag{3.36}$$

In the nonretarded limit, the left and right curls (2.78) and (2.79) of the free-space Green's tensor reduce to

$$\nabla\times\boldsymbol{G}^{(0)}(\boldsymbol{r},\boldsymbol{r}',\omega)=-\frac{\boldsymbol{e}_\rho\times\boldsymbol{I}}{4\pi\rho^2}\,,\qquad \boldsymbol{G}^{(0)}(\boldsymbol{r},\boldsymbol{r}',\omega)\times\overleftarrow{\nabla}'=\frac{\boldsymbol{I}\times\boldsymbol{e}_\rho}{4\pi\rho^2}\,;\tag{3.37}$$

they obviously scale as

$$\tilde{\nabla}\times\tilde{\boldsymbol{G}}^{(0)}(\tilde{\boldsymbol{r}},\tilde{\boldsymbol{r}}',\omega)=\frac{1}{a^2}\nabla\times\boldsymbol{G}^{(0)}(\boldsymbol{r},\boldsymbol{r}',\omega)\,,\tag{3.38}$$

$$\tilde{\boldsymbol{G}}^{(0)}(\tilde{\boldsymbol{r}},\tilde{\boldsymbol{r}}',\omega)\times\overleftarrow{\tilde{\nabla}}'=\frac{1}{a^2}\boldsymbol{G}^{(0)}(\boldsymbol{r},\boldsymbol{r}',\omega)\times\overleftarrow{\nabla}'\,.\tag{3.39}$$

Using these results together with (3.27) and the transformation (3.1), the scaled version of (3.36) reads

$$\begin{aligned}\tilde{\boldsymbol{G}}^{(1)}(\tilde{\boldsymbol{r}},\tilde{\boldsymbol{r}}',\omega)\times\overleftarrow{\tilde{\nabla}}'&=\frac{\omega^2}{c^2}\int\mathrm{d}^3\tilde{s}\,[\tilde{\varepsilon}(\tilde{\boldsymbol{s}},\omega)-1]\tilde{\boldsymbol{G}}^{(0)}(\tilde{\boldsymbol{r}},\tilde{\boldsymbol{s}},\omega)\\&\quad\cdot\Big[\tilde{\boldsymbol{G}}^{(0)}(\tilde{\boldsymbol{s}},\tilde{\boldsymbol{r}}',\omega)\times\overleftarrow{\tilde{\nabla}}'+\tilde{\boldsymbol{G}}^{(1)}(\tilde{\boldsymbol{s}},\tilde{\boldsymbol{r}}',\omega)\times\overleftarrow{\tilde{\nabla}}'\Big]\\&=\frac{\omega^2}{c^2}\int\mathrm{d}^3s\,[\varepsilon(\boldsymbol{s},\omega)-1]\boldsymbol{G}^{(0)}(\boldsymbol{r},\boldsymbol{s},\omega)\\&\quad\cdot\left[\frac{1}{a^2}\boldsymbol{G}^{(0)}(\boldsymbol{s},\boldsymbol{r}',\omega)\times\overleftarrow{\nabla}'+\tilde{\boldsymbol{G}}^{(1)}(\tilde{\boldsymbol{s}},\tilde{\boldsymbol{r}}',\omega)\times\overleftarrow{\tilde{\nabla}}'\right],\end{aligned}\tag{3.40}$$

implying

$$\tilde{\boldsymbol{G}}^{(1)}(\tilde{\boldsymbol{r}},\tilde{\boldsymbol{r}}',\omega)\times\overleftarrow{\tilde{\nabla}}'=\frac{1}{a^2}\boldsymbol{G}^{(1)}(\boldsymbol{r},\boldsymbol{r}',\omega)\times\overleftarrow{\nabla}'\,.\tag{3.41}$$

Substituting this together with (3.38) and (3.39) back into the Dyson equation (3.35),

$$\begin{aligned}&\tilde{\nabla}\times\tilde{\boldsymbol{G}}^{(1)}(\tilde{\boldsymbol{r}},\tilde{\boldsymbol{r}}',\omega)\times\overleftarrow{\tilde{\nabla}}'\\&=\frac{\omega^2}{c^2}\int\mathrm{d}^3\tilde{s}\,[\tilde{\varepsilon}(\tilde{\boldsymbol{s}},\omega)-1]\tilde{\nabla}\times\tilde{\boldsymbol{G}}^{(0)}(\tilde{\boldsymbol{r}},\tilde{\boldsymbol{s}},\omega)\cdot\tilde{\boldsymbol{G}}(\tilde{\boldsymbol{s}},\tilde{\boldsymbol{r}}',\omega)\times\overleftarrow{\tilde{\nabla}}'\\&=\frac{1}{a}\frac{\omega^2}{c^2}\int\mathrm{d}^3s\,[\varepsilon(\boldsymbol{s},\omega)-1]\nabla\times\boldsymbol{G}^{(0)}(\boldsymbol{r},\boldsymbol{s},\omega)\cdot\boldsymbol{G}(\boldsymbol{s},\boldsymbol{r}',\omega)\times\overleftarrow{\nabla}'\,,\end{aligned}\tag{3.42}$$

we find the required scaling

$$\tilde{\nabla}\times\tilde{\boldsymbol{G}}^{(1)}(\tilde{\boldsymbol{r}},\tilde{\boldsymbol{r}}',\omega)\times\overleftarrow{\tilde{\nabla}}'=\frac{1}{a}\nabla\times\boldsymbol{G}^{(1)}(\boldsymbol{r},\boldsymbol{r}',\omega)\times\overleftarrow{\nabla}'\,.\tag{3.43}$$

The Casimir stress (3.7) contains both $\boldsymbol{G}^{(1)}$ and $\nabla \times \boldsymbol{G}^{(1)} \times \overleftarrow{\nabla}'$. The former with its $1/a^3$ scaling (3.29) is seen to dominate over the latter with its $1/a$ behaviour and we find [11]

$$\tilde{\boldsymbol{T}}(\tilde{\boldsymbol{r}}) = \frac{1}{a^3}\,\boldsymbol{T}(\boldsymbol{r}) \tag{3.44}$$

for the Casimir stress. Consequently, the Casimir force (3.6) hence scales as

$$\tilde{\boldsymbol{F}} = \frac{1}{a}\,\boldsymbol{F}\,. \tag{3.45}$$

The case of purely magnetic bodies can be treated by means of a duality transformation $\varepsilon \to \mu$. Combining the transformation (A.31) with the known scaling (3.43), we find

$$\tilde{\boldsymbol{G}}^{(1)}(\tilde{\boldsymbol{r}}, \tilde{\boldsymbol{r}}', \omega) = \frac{1}{a}\,\boldsymbol{G}^{(1)}(\boldsymbol{r}, \boldsymbol{r}', \omega) \tag{3.46}$$

Similarly, the transformation (A.32) together with (3.29) implies

$$\tilde{\nabla} \times \tilde{\boldsymbol{G}}(\tilde{\boldsymbol{r}}, \tilde{\boldsymbol{r}}', \omega) \times \overleftarrow{\tilde{\nabla}}' = \frac{1}{a^3}\,\nabla \times \boldsymbol{G}(\boldsymbol{r}, \boldsymbol{r}', \omega) \times \overleftarrow{\nabla}'\,. \tag{3.47}$$

With these laws, the CP potential (1.128) of electric atoms scales as [11]

$$\tilde{U}(\tilde{\boldsymbol{r}}_A) = \frac{1}{a}\,U(\boldsymbol{r}_A) \tag{3.48}$$

for purely magnetic bodies in the nonretarded limit; and the respective force behaves as

$$\tilde{\boldsymbol{F}}(\tilde{\boldsymbol{r}}_A) = \frac{1}{a^2}\,\boldsymbol{F}(\boldsymbol{r}_A)\,. \tag{3.49}$$

The vdW potential (1.153) contains contributions from the bulk and scattering Green's tensors with different scalings (3.27) and (3.46). Recalling (1.154), we separate the potential into a free-space part $U^{(0)}$ and a body-induced part $U^{(1)}$. The free-space interaction contains only $\boldsymbol{G}^{(0)}$ and hence scales as (3.33) and (3.34). The body-induced potential and force are dominated by mixed terms $\boldsymbol{G}^{(0)}\boldsymbol{G}^{(1)}$ which scale as [11]

$$\tilde{U}^{(1)}(\tilde{\boldsymbol{r}}_A, \tilde{\boldsymbol{r}}_B) = \frac{1}{a^4}\,U^{(1)}(\boldsymbol{r}_A, \boldsymbol{r}_B) \tag{3.50}$$

and

$$\tilde{\boldsymbol{F}}^{(1)}(\tilde{\boldsymbol{r}}_A, \tilde{\boldsymbol{r}}_B) = \frac{1}{a^5}\,\boldsymbol{F}^{(1)}(\boldsymbol{r}_A, \boldsymbol{r}_B)\,, \tag{3.51}$$

respectively. Finally, the Casimir stress (3.7) and force are dominated by the $1/a^3$ scaling of $\boldsymbol{\nabla} \times \boldsymbol{G}^{(0)} \times \overleftarrow{\boldsymbol{\nabla}}'$ in the nonretarded limit. Their scaling behaviour is hence given by (3.44) and (3.45) also for magnetic bodies [11].

Again, our investigations can be extended to the case of magnetoelectric atoms whose potentials (1.177) and (1.178) are given in terms of the tensors $\boldsymbol{G}_{\lambda\lambda'}$ as given by (1.172)–(1.175). Recalling (3.27) and its duality transform (A.14) as well as (3.38) and (3.39), the free-space parts of these tensors scale as

$$\tilde{\boldsymbol{G}}^{(0)}_{\lambda\lambda'}(\tilde{\boldsymbol{r}}, \tilde{\boldsymbol{r}}', \tilde{\omega}) = \begin{cases} \dfrac{1}{a^3}\,\boldsymbol{G}^{(0)}_{\lambda\lambda'}(\boldsymbol{r}, \boldsymbol{r}', \omega) & \text{for } \lambda = \lambda'\,, \\[2ex] \dfrac{1}{a^2}\,\boldsymbol{G}^{(0)}_{\lambda\lambda'}(\boldsymbol{r}, \boldsymbol{r}', \omega) & \text{for } \lambda \neq \lambda' \end{cases} \tag{3.52}$$

in the nonretarded limit. Recalling the Dyson equation (2.34), the scattering parts are defined by

$$\boldsymbol{G}^{(1)}_{\lambda\lambda'}(\boldsymbol{r}, \boldsymbol{r}', \omega) = -\sum_{\lambda''=e,m} \int \mathrm{d}^3 s\, \chi_{\lambda''}(\boldsymbol{s}, \omega)\,\boldsymbol{G}^{(0)}_{\lambda\lambda''}(\boldsymbol{r}, \boldsymbol{s}, \omega) \cdot\left[\boldsymbol{G}^{(0)}_{\lambda''\lambda'}(\boldsymbol{s}, \boldsymbol{r}', \omega) + \boldsymbol{G}^{(1)}_{\lambda''\lambda'}(\boldsymbol{s}, \boldsymbol{r}', \omega)\right] \tag{3.53}$$

with $\chi_e = \varepsilon - 1$, $\chi_m = 1 - 1/\mu$. We again distinguish between purely electric ($\lambda'' = e$) and purely magnetic bodies ($\lambda'' = m$) and follow the same steps as described below (3.35). For purely electric bodies, (3.52) and (3.53) lead to

$$\tilde{\boldsymbol{G}}^{(1)}_{\lambda\lambda}(\tilde{\boldsymbol{r}}, \tilde{\boldsymbol{r}}', \tilde{\omega}) = \begin{cases} \dfrac{1}{a^3}\,\boldsymbol{G}^{(1)}_{\lambda\lambda}(\boldsymbol{r}, \boldsymbol{r}', \omega) & \text{for } \lambda = e\,, \\[2ex] \dfrac{1}{a}\,\boldsymbol{G}^{(1)}_{\lambda\lambda}(\boldsymbol{r}, \boldsymbol{r}', \omega) & \text{for } \lambda = m\,, \end{cases} \tag{3.54}$$

while for purely magnetic ones we obtain

$$\tilde{\boldsymbol{G}}^{(1)}_{\lambda\lambda}(\tilde{\boldsymbol{r}}, \tilde{\boldsymbol{r}}', \tilde{\omega}) = \begin{cases} \dfrac{1}{a}\,\boldsymbol{G}^{(1)}_{\lambda\lambda}(\boldsymbol{r}, \boldsymbol{r}', \omega) & \text{for } \lambda = e\,, \\[2ex] \dfrac{1}{a^3}\,\boldsymbol{G}^{(1)}_{\lambda\lambda}(\boldsymbol{r}, \boldsymbol{r}', \omega) & \text{for } \lambda = m\,. \end{cases} \tag{3.55}$$

The mixed tensors are found to scale as

$$\tilde{\boldsymbol{G}}^{(1)}_{\lambda\lambda'}(\tilde{\boldsymbol{r}}, \tilde{\boldsymbol{r}}', \tilde{\omega}) = \frac{1}{a^2}\,\boldsymbol{G}^{(1)}_{\lambda\lambda'}(\boldsymbol{r}, \boldsymbol{r}', \omega) \quad (\lambda \neq \lambda') \tag{3.56}$$

regardless of whether the bodies are electric or magnetic.

The scaling laws for nonretarded dispersion potentials (1.177) and (1.178) of electromagnetic atoms follow immediately. We have

$$\tilde{U}_e(\tilde{\boldsymbol{r}}_A) = \frac{1}{a^3}\,U_e(\boldsymbol{r}_A)\,, \qquad \tilde{U}_m(\tilde{\boldsymbol{r}}_A) = \frac{1}{a}\,U_m(\boldsymbol{r}_A)\,, \tag{3.57}$$

$$\tilde{U}_{ee}(\tilde{\boldsymbol{r}}_A, \tilde{\boldsymbol{r}}_B) = \frac{1}{a^6}\,U_{ee}(\boldsymbol{r}_A, \boldsymbol{r}_B)\,, \tag{3.58}$$

$$\tilde{U}^{(0)}_{mm}(\tilde{\boldsymbol{r}}_A, \tilde{\boldsymbol{r}}_B) = \frac{1}{a^6}\,U^{(0)}_{mm}(\boldsymbol{r}_A, \boldsymbol{r}_B)\,, \tag{3.59}$$

$$\tilde{U}^{(1)}_{mm}(\tilde{\boldsymbol{r}}_A, \tilde{\boldsymbol{r}}_B) = \frac{1}{a^4}\,U^{(1)}_{mm}(\boldsymbol{r}_A, \boldsymbol{r}_B)\,, \tag{3.60}$$

$$\tilde{U}_{\lambda\lambda'}(\tilde{\boldsymbol{r}}_A, \tilde{\boldsymbol{r}}_B) = \frac{1}{a^4}\,U_{\lambda\lambda'}(\boldsymbol{r}_A, \boldsymbol{r}_B) \quad (\lambda \neq \lambda') \tag{3.61}$$

for electric bodies and

$$\tilde{U}_e(\tilde{\boldsymbol{r}}_A) = \frac{1}{a}\,U_e(\boldsymbol{r}_A)\,, \qquad \tilde{U}_m(\tilde{\boldsymbol{r}}_A) = \frac{1}{a^3}\,U_m(\boldsymbol{r}_A)\,, \tag{3.62}$$

$$\tilde{U}^{(0)}_{ee}(\tilde{\boldsymbol{r}}_A, \tilde{\boldsymbol{r}}_B) = \frac{1}{a^6}\,U^{(0)}_{ee}(\boldsymbol{r}_A, \boldsymbol{r}_B)\,, \tag{3.63}$$

$$\tilde{U}^{(1)}_{ee}(\tilde{\boldsymbol{r}}_A, \tilde{\boldsymbol{r}}_B) = \frac{1}{a^4}\,U^{(1)}_{ee}(\boldsymbol{r}_A, \boldsymbol{r}_B)\,, \tag{3.64}$$

$$\tilde{U}_{mm}(\tilde{\boldsymbol{r}}_A, \tilde{\boldsymbol{r}}_B) = \frac{1}{a^6}\,U_{mm}(\boldsymbol{r}_A, \boldsymbol{r}_B)\,, \tag{3.65}$$

$$\tilde{U}_{\lambda\lambda'}(\tilde{\boldsymbol{r}}_A, \tilde{\boldsymbol{r}}_B) = \frac{1}{a^4}\,U_{\lambda\lambda'}(\boldsymbol{r}_A, \boldsymbol{r}_B) \quad (\lambda \neq \lambda') \tag{3.66}$$

for magnetic bodies.

3.2.3 Applications

In the previous two sections, we have derived general scaling laws for dispersion forces and potentials. They are summarised in Table 3.2. It is worth recalling that these laws owe their existence to entirely different reasons in the retarded versus nonretarded regimes.

In the retarded regime, the Green's tensor describes electromagnetic waves, so a scaling of distances is always accompanied by a frequency scaling. This frequency scaling will prohibit simple scaling laws for quantities depending on the Green's tensor at specific frequencies, such as CP potentials of excited atoms (Chap. 4) or rates of spontaneous decay (Chap. 5). Dispersion interactions, on the contrary, are

Table 3.2 Scaling laws for the CP potentials (**i, ii**), free-space (**iii, iv**) and body-induced vdW potentials (**iv–vi**) of electric and magnetic atoms, Casimir pressure (**vii**) and Casimir force (**viii**)

Distance →	Retarded	Nonretarded	
Bodies →	Magnetoelectric	Electric	Magnetic
(i) $U_e(\boldsymbol{r}_A)$	$\frac{1}{a^4}$	$\frac{1}{a^3}$	$\frac{1}{a}$
(ii) $U_m(\boldsymbol{r}_A)$	$\frac{1}{a^4}$	$\frac{1}{a}$	$\frac{1}{a^3}$
(iii) $U_{ee}^{(0)}(\boldsymbol{r}_A, \boldsymbol{r}_B)$, $U_{mm}^{(0)}(\boldsymbol{r}_A, \boldsymbol{r}_B)$	$\frac{1}{a^7}$	$\frac{1}{a^6}$	$\frac{1}{a^6}$
(iv) $U_{em}(\boldsymbol{r}_A, \boldsymbol{r}_B)$, $U_{me}(\boldsymbol{r}_A, \boldsymbol{r}_B)$	$\frac{1}{a^7}$	$\frac{1}{a^4}$	$\frac{1}{a^4}$
(v) $U_{ee}^{(1)}(\boldsymbol{r}_A, \boldsymbol{r}_B)$	$\frac{1}{a^7}$	$\frac{1}{a^6}$	$\frac{1}{a^4}$
(vi) $U_{mm}^{(1)}(\boldsymbol{r}_A, \boldsymbol{r}_B)$	$\frac{1}{a^7}$	$\frac{1}{a^4}$	$\frac{1}{a^6}$
(vii) $\boldsymbol{T}(\boldsymbol{r})$	$\frac{1}{a^4}$	$\frac{1}{a^3}$	$\frac{1}{a^3}$
(viii) F	$\frac{1}{a^2}$	$\frac{1}{a}$	$\frac{1}{a}$

an integral effect and depend on the Green's tensor at all frequencies. In the retarded regime, only the static, zero-frequency properties of the atomic polarisabilities and magnetisabilities, body permittivities and permeabilities are relevant. The frequency scaling in the Green's tensor can then be accounted for by a simple redefinition of the integration frequency. A consequence of the intertwined position and frequency scalings is the fact that dispersion interactions are subject to the same scaling for electric or magnetic atoms or bodies.

The nonretarded Green's tensor, on the contrary, describes the near-field behaviour of the electromagnetic field. In this limit, positions and frequencies decouple and a scaling law can be formulated for the position-dependence of the Green's tensor alone. Scaling laws for dispersion interactions follow without recourse to their integral nature. However, since electric and magnetic fields exhibit different power laws in the near-field limit, we have to distinguish electric versus magnetic atoms and bodies, which lead to different scaling laws.

Scaling laws indicate the absence of a characteristic length scale of the system under investigation. For dispersion interactions, there are in fact two such length scales: the typical interatomic distances or lattice constants of the atoms contained in the bodies and the wavelengths of atomic and body response functions. As a consequence, the derived scaling laws are not universally valid, but only within certain ranges. The retarded scaling laws are only valid as long as all distances are well above the atomic and medium wave lengths, while the nonretarded laws are restricted to distances well between the lattice constants and the atomic and medium wave lengths.

Let us give some examples on how to apply the general scaling laws to concrete geometries, making contact to the examples listed in Table 3.1 at the beginning of this chapter. The simplest geometries are those where the dispersion interaction

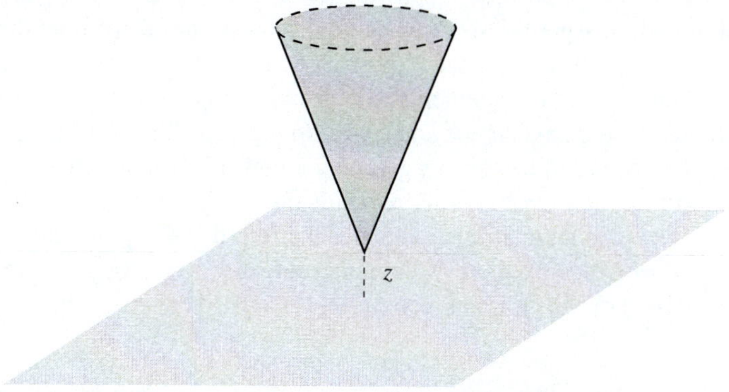

Fig. 3.2 Casimir force between a conical tip at distance z from a plane surface

in question depends on a single length parameter. In such cases, the scaling laws directly determine the full dependence of the interaction on that parameter. In this way, the power laws for the force between two atoms listed in row (i) of Table 3.1 are equivalent to the scaling laws of free-space vdW potentials (iii, iv) given in Table 3.2. The CP force (v) between an atom and a half space given in Table 3.1 is another example of such a simple geometry; the listed power laws follow immediately from the general scaling laws (i, ii) for CP potentials (Table 3.2). Finally, the Casimir forces (vi) between two electric or two magnetic half spaces (Table 3.1) are a consequence of the general scaling laws (vii) for the Casimir pressure (Table 3.2). Note that the power laws for Casimir forces between electric and magnetic half spaces do not follow in such a simple way, because we have not formulated a scaling law in the simultaneous presence of electric and magnetic bodies.

An example of a single-parameter geometry that is particularly relevant to atomic force microscopy [12] is the Casimir force between a sharp conical tip and a flat surface. As seen from Fig. 3.2, such an arrangement is completely characterised by the distance z between the surface and the tip. This is true when neglecting the finite extent of the tip, i.e., as long as the tip length is much larger than the tip-surface separation. Applying the scaling law (vii) for the Casimir force as given in Table 3.2, we can conclude that the total force between the tip and the surface is proportional to $1/z^2$ in the retarded limit and to $1/z$ in the physically more relevant nonretarded distance regime. The respective proportionality constants do not follow from the scaling laws. They depend on the surface and tip materials as well as the opening angle of the tip and have been calculated numerically for a perfectly conducting tip [13]. Note that the derived distance laws do not apply to spherical tips or conical tips with a rounded edge. In such cases, the respective curvature radius defines a second characteristic length parameter of the system, so that the scaling laws are insufficient to determine the full distance-dependence of the force. Further examples of simple single-parameter problems are, for instance, the CP potentials of single atoms at the centres of planar, spherical or cylindrical cavities.

For more complex geometries, the scaling laws do not determine the respective forces completely, but they provide an important constraint. An important class of geometries consists of arrangements which are completely characterised by two length parameters, e.g., a distance and a size. In this case, the scaling laws imply that dispersion potentials and forces can given by a power law in the distance multiplied by a scaling function $f(x)$ that only depends on the dimensionless ratio x between distance and size. As an example, consider the retarded CP potential of an atom at distance z_A from dielectric plate of static permittivity $\varepsilon \equiv \varepsilon(0)$ and thickness d. As seen from (4.164) of Vol. I, it is given by [1, 2, 14]

$$U(z_A) = \frac{\hbar c\alpha}{8\pi^2\varepsilon_0 z_A^4} \int_1^\infty \mathrm{d}v \int_0^\infty \mathrm{d}y\, y^3 \mathrm{e}^{-2vy} \times \left\{ (1-2v^2) \frac{(\varepsilon^2 v^2 - v_1^2)\tanh(v_1 y d/z_A)}{2\varepsilon v v_1 + (\varepsilon^2 v^2 + v_1^2)\tanh(v_1 y d/z_A)} + \frac{(v^2 - v_1^2)\tanh(v_1 y d/z_A)}{2 v v_1 + (v^2 + v_1^2)\tanh(v_1 y d/z_A)} \right\}, \tag{3.67}$$

with $v_1 = \sqrt{\varepsilon - 1 + v^2}$, cf. the steps leading to (4.132) in Vol. I. In the limit $d \gg z_A$ of a thick plate, we have $\tanh(v_1 y d/z_A) \simeq 1$. The y-integral can then be performed to give

$$U(z_A) = -\frac{C_4}{z_A^4}, \tag{3.68}$$

$$C_4 = \frac{3\hbar c\alpha}{64\pi^2\varepsilon_0} \int_1^\infty \mathrm{d}v \left[\left(\frac{2}{v^2} - \frac{1}{v^4} \right) \frac{\varepsilon v - v_1}{\varepsilon v + v_1} - \frac{1}{v^4} \frac{v - v_1}{v + v_1} \right], \tag{3.69}$$

in agreement with the result for a semi-infinite half space as given by (4.134) and (4.134) in Vol. I. For an asymptotically thin plate $d \ll z_A$, the approximation $\tanh(v_1 y d/z_A) \simeq v_1 y d/z_A$ leads to

$$U(z_A) = -\frac{C_5}{z_A^5} \tag{3.70}$$

with

$$C_5 = \frac{\hbar c\alpha d}{160\pi^2\varepsilon_0} \frac{14\varepsilon^2 - 5\varepsilon - 9}{\varepsilon}, \tag{3.71}$$

as already found in Vol. I, see (4.166) together with (4.167) therein.

In order to identify the scaling function, we introduce the dimensionless parameter as $x = d/z_A$. Combining (3.67)–(3.69), the CP potential for arbitrary values of z_A and d can given in the form [11]

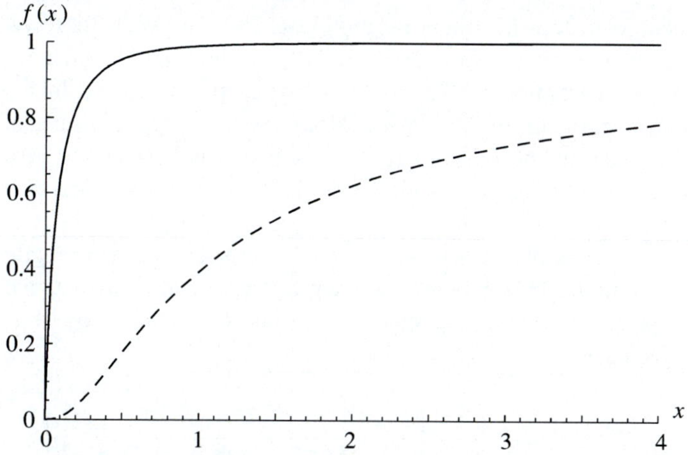

Fig. 3.3 Scaling functions for the retarded CP potential of a Si plate (*solid line*) and for the nonretarded CP potential of a perfectly conducting sphere (*dashed line*)

$$U(z_A) = -\frac{C_4 f(d/z_A)}{z_A^4}, \tag{3.72}$$

in accordance wi th the scaling law (3.12). We haveintroduced the scaling function as $f(d/z_A) = U(z_A)/U(z_A)|_{d\to\infty}$. Explicitly, it reads

$$\begin{aligned} f(x) =& \frac{8}{3}\int_1^\infty \mathrm{d}v \int_0^\infty \mathrm{d}y\, y^3 \mathrm{e}^{-2vy}\bigg[(2v^2-1)\frac{(\varepsilon^2v^2-v_1^2)\tanh(v_1yx)}{2\varepsilon vv_1+(\varepsilon^2v^2+v_1^2)\tanh(v_1yx)} \\ &- \frac{(v^2-v_1^2)\tanh(v_1yx)}{2vv_1+(v^2+v_1^2)\tanh(v_1yx)}\bigg] \\ &\times \left\{\int_1^\infty \mathrm{d}v \left[\left(\frac{2}{v^2}-\frac{1}{v^4}\right)\frac{\varepsilon v - v_1}{\varepsilon v + v_1} - \frac{1}{v^4}\frac{v-v_1}{v+v_1}\right]\right\}^{-1}. \end{aligned} \tag{3.73}$$

The scaling function contains all the relevant geometric information on the potential; it governs the transition between the thin and thick-plate limits given above. Being determined by the plate permittivity alone, it is universally valid for any atom interacting with plates of a given material.

As an example, we display the scaling function for Si plates ($\varepsilon = 11.68$) in Fig. 3.3. For large arguments $x = d/z_A \gg 1$ the scaling function tends to unity, indicating that the power law (3.72) for the CP potential coincides with that of a semi-infinite half space in the thick-plate limit. In the opposite extreme $x = d/z_A \ll 1$, the scaling function becomes linear in x, so (3.72) implies a $x/z_A^4 \propto 1/z_A^5$ potential, in agreement with the thin-plate asymptote (3.70). Between these two extremes, the scaling function changes rather abruptly from the linear behaviour to a plateau. Saturation sets in around $x \approx 0.5$, showing that even moderately thick plates can be modelled as a semi-infinity half space. Note that the specific profile of the scaling function depends on the plate material. For plates with a larger permittivity, the saturation sets in more rapidly, with the profile becoming rectangular in the limit of metals with $\varepsilon = \infty$.

As a second example of a geometry characterisable by two length parameters, consider the nonretarded CP potential of a ground-state atom at distance r_A from the centre of a perfectly conducting sphere of radius R. As shown by (4.215) in Vol. I, it can be given as [15]

$$U(r_A) = -\frac{\langle \hat{\boldsymbol{d}}^2 \rangle}{24\pi\varepsilon_0}\left[\frac{4R^3}{(r_A^2 - R^2)^3} + \frac{R}{(r_A^2 - R^2)^2} - \frac{R}{r_A^4}\right]. \tag{3.74}$$

Introducing the dimensionless parameter $x = R/z_A$ with $z_A = r_A - R$, we can rewrite this potential in the form [11]

$$U(z_A) = -\frac{C_3 f(R/z_A)}{z_A^3} \tag{3.75}$$

with

$$C_3 = \frac{\langle \hat{\boldsymbol{d}}^2 \rangle}{48\pi\varepsilon_0}. \tag{3.76}$$

The scaling function $f(R/z_A) = U(z_A)/U(z_A)|_{R\to\infty}$ for this geometry is given by

$$f(x) = \frac{8x^3}{(2x^2+1)^3} + \frac{2x}{(2x^2+1)^2} - \frac{x}{(x+1)^4}. \tag{3.77}$$

The scaling function for the nonretarded CP potential of an arbitrary atom next to a perfectly conducting sphere is also shown in Fig. 3.3. For large arguments $x \gg 1$, it approaches unity, indicating that the potential (3.76) reaches its half-space asymptote as governed by a $1/z_A^3$ power law. The scale function of the sphere potential is cubic for small x, corresponding to a $x^3/z_A^3 \propto 1/z_A^6$ dependence. The transition between the two asymptotes is much more gradual than for the case of the plate. This indicates that proximity force approximations [16, 17], which model a curved surface by a collection of flat sections, should be used with care.

A third example of a two-parameter geometry has already been studied in Sect. 2.3.2, where we had calculated the CP potential of an atom in front of a weakly dielectric half space whose permittivity exhibits periodic spatial oscillations. In the retarded and nonretarded limits, the potential depends on two geometric parameters only, namely the atom–surface distance z_A and the period of the permittivity-oscillations λ. As required by the scaling laws, the potential can then be given in the forms (2.161) and (2.167). The associated scaling functions are given by (2.162) and (2.168) and are displayed in Fig. 2.7. Note that in contrast to the previous two examples, we have here introduced the dimensionless quantity $x = z_A/\lambda$ as the ratio of distance to size (of the oscillations). As a consequence, the scale functions tend to unity for small rather than large x.

Scale functions are thus a universal tool to describe CP potentials in the retarded and nonretarded limits for geometries which depend on two length parameters. Returning to Table 3.1 with its asymptotic power laws, we note that they bridge the gap between different rows of the table: They continuously describe the potential of a plate of finite thickness, interpolating between the two extremes of an asymptotically thin plate as given in row (iv) of the table and the semi-infinite half space shown in row (v). Similarly, the describe the transition between an asymptotically small sphere (ii) on the one hand and a half space (v) on the other.

Even more complicated geometries can be described by scaling functions, which for k characteristic length parameters will depend on $k-1$ dimensionless quantities. As an example, consider the nonretarded vdW potential of two atoms A and B in front of a perfectly conducting plate. The potential can be given in terms of three distance parameters, e.g., the distances z_A and z_B of each atom from the plate surface and the distance r between the atoms. As given by (5.184) of Vol. I, it reads [18, 19]

$$U(\mathbf{r}_A,\mathbf{r}_B) = \left[1 \mp \frac{64}{23}\frac{3r^6r_+^2 + r^4(r^2+5rr_+ + r_+^2)x^2}{r_+^3(r+r_+)^5} + \frac{r^7}{r_+^7}\right] U^{(0)}(\mathbf{r}_A,\mathbf{r}_B) \tag{3.78}$$

($x = \sqrt{r^2-(z_A-z_B)^2}$, interatomic distance in the direction parallel to the plate; $r_+ = \sqrt{x^2+(z_A+z_B)^2}$, distance between one atom and the image of the other one behind the plate) where

$$U^{(0)}(\mathbf{r}_A,\mathbf{r}_B) = -\frac{23\hbar\alpha_A\alpha_B}{64\pi^3\varepsilon_0^2 r^7} \tag{3.79}$$

is the vdW potential of the two atoms in free space. One could find a scaling function by introducing the dimensionless parameters z_A/r and z_B/r.

Instead, let us illustrate another important consequence of the scaling laws. They imply that lines of constant potential are transformed into new equipotential lines under the scaling transformation. In other words, equipotential lines will be stretched or shrunk while retaining their shape. The actual value of the potential on such a line will of course change according to the scaling law (3.15). This complication does not occur when we consider the normalised potential $U/U^{(0)}$, i.e., the plate-induced enhancement of the potential with respect to its free-space value. The usefulness of the equipotential-line preserving property of scaling transformations is demonstrated in Fig. 3.4. In Fig. 3.4(i), we display equipotential lines of atom A corresponding to different plate-induced enhancements of the interatomic potential for a fixed position of atom B. The plate is seen to enhance the interatomic interaction in two lobe-shaped

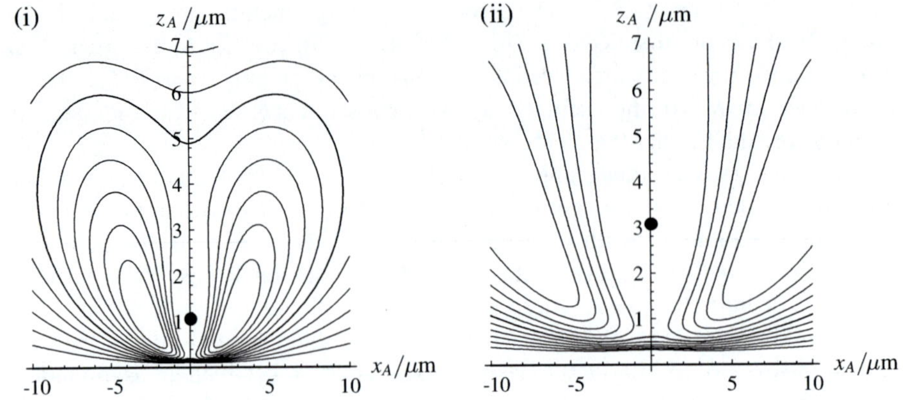

Fig. 3.4 Retarded vdW potential next to a perfectly conducting plate. Atom B is held at different fixed positions (**i, ii**) (*large dot*). The *contours* denote positions of atom A with constant enhancements of the potential U with respect to its free-space value $U^{(0)}$. The *thick contour* corresponds to $U/U^{(0)} = 1$, values are increasing towards the exterior of this contour in steps of 0.02

regions to the left and right of atom B. Equipotential lines for a different position of atom B can be easily obtained by making use of the scaling law for the retarded vdW potential: As shown in Fig. 3.4(ii), one may simply stretch or shrink all distances to obtain the desired atom–surface separation for atom B. Our observations regarding the plate-induced enhancement hence remain valid for all distances compatible with the retarded limit.

3.3 Microscopic Origin

As seen in the previous section, simple asymptotic power laws for Casimir, CP and vdW forces are a consequence of the individual scaling laws for these forces that hold in the retarded and nonretarded limits. What we have not yet discussed in detail is the relation between the three types of dispersion forces and the associated power laws. As we will show in the following, both Casimir and CP forces have their common origin in the microscopic vdW forces between individual atoms. We will first establish this relation for bodies consisting of a dilute gas of atoms and then show how many-atom interactions start to contribute for denser media.

3.3.1 Dilute-Gas Limit

Let us begin with the Casimir force acting on a dielectric body of volume V in the presence of an arbitrary arrangement of other bodies. Assuming the body to consist of a dilute gas of atoms, we will relate the Casimir force on the body as whole to the CP forces on the individual gas atoms. With the gaseous body being situated in a free-space region, the Casimir force can be given in the form

$$\boldsymbol{F} = -\frac{\hbar}{2\pi}\int_V \mathrm{d}^3r \int_0^\infty \mathrm{d}\xi \left(\frac{\xi^2}{c^2}\chi(\boldsymbol{r},\mathrm{i}\xi)\nabla \mathrm{tr}\,\boldsymbol{G}^{(1)}(\boldsymbol{r},\boldsymbol{r},\mathrm{i}\xi)\right.$$
$$\left. - \nabla\left\{\zeta(\boldsymbol{r},\mathrm{i}\xi)\mathrm{tr}\left[\nabla\times\boldsymbol{G}^{(1)}(\boldsymbol{r},\boldsymbol{r}',\mathrm{i}\xi)\times\overleftarrow{\nabla}'\right]_{\boldsymbol{r}'=\boldsymbol{r}}\right\}\right), \quad (3.80)$$

recall (1.113). We separate the electric susceptibility $\chi(\boldsymbol{r},\omega)$ of the gaseous body from the permittivity $\overline{\varepsilon}(\boldsymbol{r},\omega)$ of all other bodies, so that the total permittivity of the system can be written in the form

$$\varepsilon(\boldsymbol{r},\omega) = \overline{\varepsilon}(\boldsymbol{r},\omega) + \chi(\boldsymbol{r},\omega)\,. \quad (3.81)$$

As shown in Sect. 2.1.1, the Green's tensor $\boldsymbol{G}$ in the presence of the gaseous body can be related to the Green's tensor $\overline{\boldsymbol{G}}$ in its absence via a Born expansion in powers of χ. With the susceptibility of the gaseous body being small for a sufficiently dilute gas ($\chi \ll 1$), we are interested in the leading, linear expansion of the Casimir force. With the above expression already containing an explicit factor of χ, we employ the zero-order expansion $\boldsymbol{G} = \overline{\boldsymbol{G}}$ of the Green's tensor to obtain

$$\boldsymbol{F} = -\frac{\hbar}{2\pi}\int_V \mathrm{d}^3r \int_0^\infty \mathrm{d}\xi\, \frac{\xi^2}{c^2}\chi(\boldsymbol{r},\mathrm{i}\xi)\nabla \mathrm{tr}\,\overline{\boldsymbol{G}}^{(1)}(\boldsymbol{r},\boldsymbol{r},\mathrm{i}\xi)\,. \quad (3.82)$$

Note that the second term in (3.80) above does not contribute for a non-magnetic body with $\zeta(\boldsymbol{r},\mathrm{i}\xi) = 0$.

The gap between the macroscopic susceptibility $\chi(\boldsymbol{r},\omega)$ of the gaseous body and the microscopic polarisabilities $\alpha(\omega)$ of the atoms contained therein may be bridged by means of the Clausius–Mosotti law. In the dilute-gas limit, it reads [20]

$$\chi(\boldsymbol{r},\omega) = \frac{\eta(\boldsymbol{r})\alpha(\omega)}{\varepsilon_0} \quad (3.83)$$

where $\eta(\boldsymbol{r})$ denotes the number density of the atoms. Using this relation, we find [2, 3, 21]

$$\boldsymbol{F} = \int_V \mathrm{d}^3r_A\, \eta(\boldsymbol{r}_A)\boldsymbol{F}(\boldsymbol{r}_A) \quad (3.84)$$

with $\boldsymbol{F}(\boldsymbol{r}_A) = -\nabla_A U(\boldsymbol{r}_A)$ and $U(\boldsymbol{r}_A) = U_e(\boldsymbol{r}_A)$ with

$$U_e(\boldsymbol{r}_A) = \frac{\hbar\mu_0}{2\pi}\int_0^\infty \mathrm{d}\xi\, \xi^2\alpha(\mathrm{i}\xi)\mathrm{tr}\,\overline{\boldsymbol{G}}^{(1)}(\boldsymbol{r}_A,\boldsymbol{r}_A,\mathrm{i}\xi)\,. \quad (3.85)$$

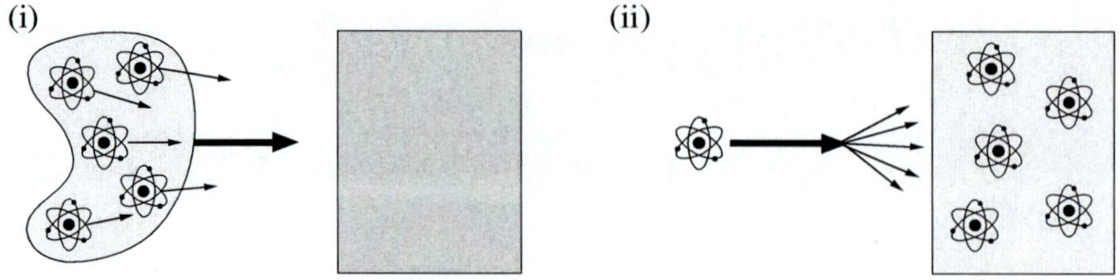

Fig. 3.5 Microscopic origins of the (**i**) Casimir and (**ii**) CP forces in the dilute-gas limit

The total Casimir force on the gaseous body is hence simply the sum of the CP forces acting on the gas atoms. The respective CP potential agrees exactly with the result (1.128) of a microscopic calculation as given in Sect. 1.2.2. The considerations of this section may in fact be regarded as an alternative, macroscopic derivation of the CP potential. Lifshitz made use of such an approach to obtain CP and vdW potentials from his famous expression for the Casimir force between two plates in the dilute-gas limit [22, 23]. The microscopic origin of the Casimir force is schematically illustrated in Fig. 3.5(i).

Having reduced the macroscopic Casimir force on a dilute-gas body to the sum of the microscopic CP forces on the gas atoms, let us proceed to show that the CP force in turn is a consequence of the even more microscopic vdW forces. To this end, we start from the CP potential $U(\boldsymbol{r}_A) = U_e(\boldsymbol{r}_A)$ with

$$U_e(\boldsymbol{r}_A) = \frac{\hbar\mu_0}{2\pi}\int_0^\infty \mathrm{d}\xi\,\xi^2\,\alpha_A(\mathrm{i}\xi)\mathrm{tr}\,\boldsymbol{G}^{(1)}(\boldsymbol{r}_A,\boldsymbol{r}_A,\mathrm{i}\xi) \tag{3.86}$$

of a single atom A where $\boldsymbol{G}$ represents all present bodies. Again, we assume that one of these bodies of volume V is weakly dielectric so that its susceptibility $\chi(\boldsymbol{r},\omega)$ may be separated from the permittivity $\overline{\varepsilon}(\boldsymbol{r},\omega)$ of all other bodies according to (3.81). Employing the linear Born expansion (2.5) of the Green's tensor $\boldsymbol{G}$ in terms of χ, the leading-order CP potential of the atom due to its interaction with the gaseous body takes the form

$$U_e(\boldsymbol{r}_A) = -\frac{\hbar\mu_0}{2\pi}\int_V \mathrm{d}^3r\int_0^\infty \mathrm{d}\xi\,\frac{\xi^4}{c^2}\,\alpha_A(\mathrm{i}\xi)\chi(\boldsymbol{r},\mathrm{i}\xi)\mathrm{tr}\Big[\overline{\boldsymbol{G}}(\boldsymbol{r}_A,\boldsymbol{r},\mathrm{i}\xi)\cdot\overline{\boldsymbol{G}}(\boldsymbol{r},\boldsymbol{r}_A,\mathrm{i}\xi)\Big]. \tag{3.87}$$

Note that we have discarded the zero-order potential (2.46) which is due to the interaction of atom A with the background bodies alone. We now assume the weakly body to consist of a dilute gas of atoms B such that the Clausius–Mosotti law (3.83) holds. As a consequence, the CP potential reads [2, 3, 9, 10]

$$U(\boldsymbol{r}_A) = \int_V \mathrm{d}^3 r_B\, \eta(\boldsymbol{r}_B) U(\boldsymbol{r}_A, \boldsymbol{r}_B) \tag{3.88}$$

where $U(\boldsymbol{r}_A, \boldsymbol{r}_B) = U_{ee}(\boldsymbol{r}_A, \boldsymbol{r}_B)$ with

$$U_{ee}(\boldsymbol{r}_A, \boldsymbol{r}_B) = -\frac{\hbar\mu_0^2}{2\pi} \int_0^\infty \mathrm{d}\xi\, \xi^4 \alpha_A(\mathrm{i}\xi)\, \alpha_B(\mathrm{i}\xi) \mathrm{tr}\Big[\overline{\boldsymbol{G}}(\boldsymbol{r}_A, \boldsymbol{r}_B, \mathrm{i}\xi) \cdot \overline{\boldsymbol{G}}(\boldsymbol{r}_B, \boldsymbol{r}_A, \mathrm{i}\xi)\Big] \tag{3.89}$$

coincides with the microscopic vdW potential (1.153) as given in Sect. 1.2.3. The CP interaction of an atom with a gaseous body is hence the sum of its vdW interactions with the gas atoms. The microscopic origin of the CP force is represented in Fig. 3.5(ii). Again, the explicit form for the vdW potential has been found as a by-product our considerations [24].

Combining the two results (3.84) and (3.88) and recalling the relations (1.119) and (1.148) between dispersion potentials and the associated forces, we find [2, 3, 21]

$$\boldsymbol{F} = \int_{V_1} \mathrm{d}^3 r_A\, \eta(\boldsymbol{r}_A) \int_{V_2} \mathrm{d}^3 r_B\, \eta(\boldsymbol{r}_B) \boldsymbol{F}(\boldsymbol{r}_A, \boldsymbol{r}_B)\,. \tag{3.90}$$

In other words, the Casimir force between two dilute gaseous bodies of volumes V_1 and V_2 is the sum of the pairwise vdW forces between the gas atoms contained in the bodies.

Our considerations can be easily generalised to the magnetodielectric case. Starting from the Casimir force (3.80) on a gaseous, magnetodielectric body, we apply a decomposition

$$\frac{1}{\mu(\boldsymbol{r}, \omega)} = \frac{1}{\overline{\mu}(\boldsymbol{r}, \omega)} - \zeta(\boldsymbol{r}, \omega)\,. \tag{3.91}$$

of the permeability into the magnetic susceptibility $\zeta(\boldsymbol{r}, \omega)$ of the gaseous body and the permeability $\overline{\mu}(\boldsymbol{r}, \omega)$ of all other bodies. Substituting the Clausius–Mosotti law (3.83) together with its magnetic counterpart [24]

$$\zeta(\boldsymbol{r}, \omega) = \mu_0 \eta(\boldsymbol{r}) \beta(\omega)\,, \tag{3.92}$$

into the Casimir force (3.80), we find that the relation (3.84) stating the microscopic origin of the Casimir force remains valid for a dilute magnetodielectric body, where $U(\boldsymbol{r}_A) = U_e(\boldsymbol{r}_A) + U_m(\boldsymbol{r}_A)$ with (3.85) and

$$U_m(\boldsymbol{r}_A) = \frac{\hbar\mu_0}{2\pi} \int_0^\infty \mathrm{d}\xi\, \beta(\mathrm{i}\xi) \mathrm{tr}\big[\boldsymbol{\nabla} \times \overline{\boldsymbol{G}}^{(1)}(\boldsymbol{r}_A, \boldsymbol{r}_A, \mathrm{i}\xi) \times \overleftarrow{\boldsymbol{\nabla}}'\big] \tag{3.93}$$

is now the full CP potential of the electromagnetic gas atoms. The magnetic CP potential (3.93) obtained as a by-product of our derivation again agrees with the known microscopic result (1.145) [25]. Note that the Casimir force on a dilute gas of magnetoelectric atoms can only be written in the form (3.84) if the gas is homogeneous, $\eta(\boldsymbol{r}) \equiv \eta$. This is due to the fact that the gradient on the second line of (3.80) acts on $\zeta(\boldsymbol{r}, \mathrm{i}\xi)$.

To formulate the microscopic origin of the CP potential in the magnetodielectric case, we start from the potential of a single electromagnetic atom in the form $U(\boldsymbol{r}_A) = U_e(\boldsymbol{r}_A) + U_m(\boldsymbol{r}_A)$ with

$$U_\lambda(\boldsymbol{r}_A) = \frac{\hbar}{2\pi\varepsilon_0} \int_0^\infty \mathrm{d}\xi\, \alpha_\lambda^A(\mathrm{i}\xi) \mathrm{tr}\, \boldsymbol{G}_{\lambda\lambda}^{(1)}(\boldsymbol{r}_A, \boldsymbol{r}_A, \mathrm{i}\xi) \quad (\lambda = e, m) \tag{3.94}$$

with $\alpha_e = \alpha$, $\alpha_m = \beta/c^2$. Separating the atom's electromagnetic environment into contributions from a weakly magnetodielectric gaseous body plus those from additional magnetoelectric bodies according to (3.81) and (3.91), we may use the Born expansion (2.35) with (2.36) to find the linear potential due to the gaseous body

$$U_\lambda(\boldsymbol{r}_A) = -\frac{\hbar}{2\pi\varepsilon_0} \sum_{\lambda'=e,m} \int_V \mathrm{d}^3 r \int_0^\infty \mathrm{d}\xi\, \alpha_\lambda^A(\mathrm{i}\xi) \chi_{\lambda'}(\boldsymbol{r}, \omega)$$
$$\times \mathrm{tr}\Big[\overline{\boldsymbol{G}}_{\lambda\lambda'}(\boldsymbol{r}_A, \boldsymbol{r}, \mathrm{i}\xi) \cdot \overline{\boldsymbol{G}}_{\lambda'\lambda}(\boldsymbol{r}, \boldsymbol{r}_A, \mathrm{i}\xi)\Big] \quad (\lambda = e, m) \tag{3.95}$$

with $\chi_e = \chi$, $\chi_m = \zeta$. Using the two Clausius–Mosotti relations (3.83) and (3.92) in the comprehensive form

$$\chi_\lambda(\boldsymbol{r}, \omega) = \frac{\eta(\boldsymbol{r}) \alpha_\lambda(\omega)}{\varepsilon_0} \quad (\lambda = e, m)\,; \tag{3.96}$$

we find that the relation (3.88) between CP and vdW potentials remains valid where $U(\boldsymbol{r}_A, \boldsymbol{r}_B) = U_{ee}(\boldsymbol{r}_A, \boldsymbol{r}_B) + U_{em}(\boldsymbol{r}_A, \boldsymbol{r}_B) + U_{me}(\boldsymbol{r}_A, \boldsymbol{r}_B) + U_{mm}(\boldsymbol{r}_A, \boldsymbol{r}_B)$ is the full electromagnetic vdW potential. Its components [24]

$$U_{\lambda\lambda'}(\boldsymbol{r}_A, \boldsymbol{r}_B) = -\frac{\hbar}{2\pi\varepsilon_0^2} \int_0^\infty \mathrm{d}\xi\, \alpha_\lambda^A(\mathrm{i}\xi) \alpha_{\lambda'}^B(\mathrm{i}\xi)$$
$$\times \mathrm{tr}\Big[\overline{\boldsymbol{G}}_{\lambda\lambda'}(\boldsymbol{r}_A, \boldsymbol{r}_B, \mathrm{i}\xi) \cdot \overline{\boldsymbol{G}}_{\lambda'\lambda}(\boldsymbol{r}_B, \boldsymbol{r}_A, \mathrm{i}\xi)\Big] \quad (\lambda, \lambda' = e, m) \tag{3.97}$$

agree with the microscopic result (1.178).

In the dilute-gas limit, dispersion forces are thus additive. The Casimir force on a body is simply a volume integral over the CP forces of the atoms contained therein; which in turn is a simple volume integral over vdW interactions with atoms contained in a second body. This additivity immediately explains the relations between the asymptotic power laws for dispersion forces between objects of simple shapes given in the different rows of Table 3.1. Performing the respective volume integrals, one finds that the results for Casimir and CP forces can all be traced back to the fundamental vdW force between two atoms.

3.3.2 *Many-Atom Contributions*

For bodies which are not sufficiently dilute, linear expansions in terms of the body susceptibilities become an inappropriate description of Casimir and CP forces. Higher-order terms in χ and ζ need to be taken into account. They correspond to many-atom contributions and lead to a breakdown of additivity. We will illustrate their significance by considering the microscopic origin of the CP potential in more detail.

As before, we start from the CP potential (3.86) of a single electric atom in the presence of an arbitrary arrangement of bodies. We concentrate on a weakly dielectric body of volume V whose susceptibility $\chi(\boldsymbol{r},\omega)$ may be separated from the permittivity $\overline{\varepsilon}(\boldsymbol{r},\omega)$ of all other bodies in accordance with (3.81). As a consequence of this separation, the Green's tensor of the total system of bodies can be expanded in terms χ using a Born series. In contrast to our treatment for the dilute-gas limit, we now make use of the full Born expansion. For for reasons that will become clear in an instant, we make use of the alternative Born series (2.50) in terms of the parameter $\chi\overline{\varepsilon}/(\overline{\varepsilon}+\frac{1}{3}\chi)$ rather than χ alone. Requiring the weakly dielectric body to be well separated from all other bodies, so that $\overline{\varepsilon}(\boldsymbol{r},\omega)\equiv 1$ for $\boldsymbol{r}\in V$, and the CP potential due to the weakly dielectric body takes the form

$$
\begin{aligned}
U(\boldsymbol{r}_A) = \sum_{K=1}^{\infty} \frac{(-1)^K \hbar\mu_0}{2\pi c^{2K}} \int_0^{\infty} \mathrm{d}\xi\, \xi^{2K+2}\alpha(\mathrm{i}\xi) \\
\times \int_V \mathrm{d}^3 r_1 \frac{\chi(\boldsymbol{r}_1,\mathrm{i}\xi)}{1+\frac{1}{3}\chi(\boldsymbol{r}_1,\mathrm{i}\xi)} \cdots \int_V \mathrm{d}^3 r_K \frac{\chi(\boldsymbol{r}_K,\mathrm{i}\xi)}{1+\frac{1}{3}\chi(\boldsymbol{r}_K,\mathrm{i}\xi)} \\
\times \operatorname{tr}\Big[\overline{\boldsymbol{H}}(\boldsymbol{r}_A,\boldsymbol{r}_1,\mathrm{i}\xi)\cdot\overline{\boldsymbol{H}}(\boldsymbol{r}_1,\boldsymbol{r}_2,\mathrm{i}\xi)\cdots\overline{\boldsymbol{H}}(\boldsymbol{r}_K,\boldsymbol{r}_A,\mathrm{i}\xi)\Big]\,. \qquad (3.98)
\end{aligned}
$$

Once more, we have discarded the zero-order potential (2.46) due to the background bodies alone.

For simplicity, we assume that the weakly dielectric body consists of atoms which are of the same species of the single atom. Its susceptibility can be related to the atoms' polarisability via the full Clausius–Mosotti law [20]

$$\frac{\chi(\boldsymbol{r},\omega)}{1+\frac{1}{3}\chi(\boldsymbol{r},\omega)} = 3\,\frac{\varepsilon(\boldsymbol{r},\omega)-1}{\varepsilon(\boldsymbol{r},\omega)+2} = \frac{\eta(\boldsymbol{r})\alpha(\omega)}{\varepsilon_0} \tag{3.99}$$

which generalises (3.83) beyond the dilute-gas limit. The reason for employing $\chi/(1+\frac{1}{3}\chi)$ as our expansion parameter is now clear, as it directly proportional to the atomic polarisability. The Clausius–Mosotti law leads to

$$U(\boldsymbol{r}_A) = \sum_{K=1}^{\infty} \frac{(-1)^K \hbar\mu_0^{K+1}}{2\pi} \int_V \mathrm{d}^3 r_1\, \eta(\boldsymbol{r}_1) \cdots \int_V \mathrm{d}^3 r_K\, \eta(\boldsymbol{r}_K) \int_0^{\infty} \mathrm{d}\xi\, \xi^{2K+2} \\ \times \alpha^{K+1}(\mathrm{i}\xi)\mathrm{tr}\Big[\overline{\boldsymbol{H}}(\boldsymbol{r}_A,\boldsymbol{r}_1,\mathrm{i}\xi)\cdot\overline{\boldsymbol{H}}(\boldsymbol{r}_1,\boldsymbol{r}_2,\mathrm{i}\xi)\cdots\overline{\boldsymbol{H}}(\boldsymbol{r}_K,\boldsymbol{r}_A,\mathrm{i}\xi)\Big]. \tag{3.100}$$

The integrands of the multiple integrals over the volume of the weakly dielectric body cannot immediately be identified as vdW potentials, which must be symmetric with respect to an exchange of atomic positions for $N = K+1$ identical atoms. A function $f(\boldsymbol{r}_1,\ldots,\boldsymbol{r}_N)$ with N position arguments can be symmetrised by summing over all $N!$ permutations $\Pi \in P(N)$, i.e., over all possible orders of these arguments to occur. These permutations may be graphically represented by different paths connecting the positions of the interacting atoms, as shown in Fig. 3.6(i), (ii) for the cases of $N = 2$ and $N = 3$ atoms.

In our case, the relevant function

$$f(\boldsymbol{r}_1,\ldots,\boldsymbol{r}_N) = \mathrm{tr}\Big[\overline{\boldsymbol{H}}(\boldsymbol{r}_1,\boldsymbol{r}_2,\omega)\cdot\overline{\boldsymbol{H}}(\boldsymbol{r}_2,\boldsymbol{r}_3,\omega)\cdots\overline{\boldsymbol{H}}(\boldsymbol{r}_N,\boldsymbol{r}_1,\omega)\Big] \tag{3.101}$$

is a trace of a product of Green's tensors. Due to the cyclic property of the trace, this function is already symmetric under a cyclic permutation of the position arguments, i.e.,

$$f(\boldsymbol{r}_1,\ldots,\boldsymbol{r}_N) = f(\boldsymbol{r}_{\Pi(1)},\boldsymbol{r}_{\Pi(2)},\ldots,\boldsymbol{r}_{\Pi(N)}) \tag{3.102}$$

if Π a cyclic permutation. In addition, the Onsager reciprocity (A.4) of the Green's tensor implies that $f(\boldsymbol{r}_1,\ldots,\boldsymbol{r}_N)$ remains unchanged if the order of the arguments is reversed, or more generally, (3.102) holds for any permutation Π that is the reverse of a cyclic permutation. In the representations of Fig. 3.6(i), (ii), this means that two paths give the same result if they only differ by their starting point or orientation. We group the $N!$ possible permutations of position arguments into classes giving the same result; they are graphically represented in Fig. 3.6 by paths without starting point and orientation. For $N \geq 3$, each class has $2N$ members, so there are $(N-1)!/2$ classes in total. As seen in the figure, the $3! = 2\times 3$ permutations for $N = 3$ atoms all belong to a single class, whereas three classes need to be distinguished for $N = 4$. The case $N = 2$ is special because cyclic permutation and reverse coincide; the single existing class thus has only 2 rather than $2\times 2 = 4$ members.

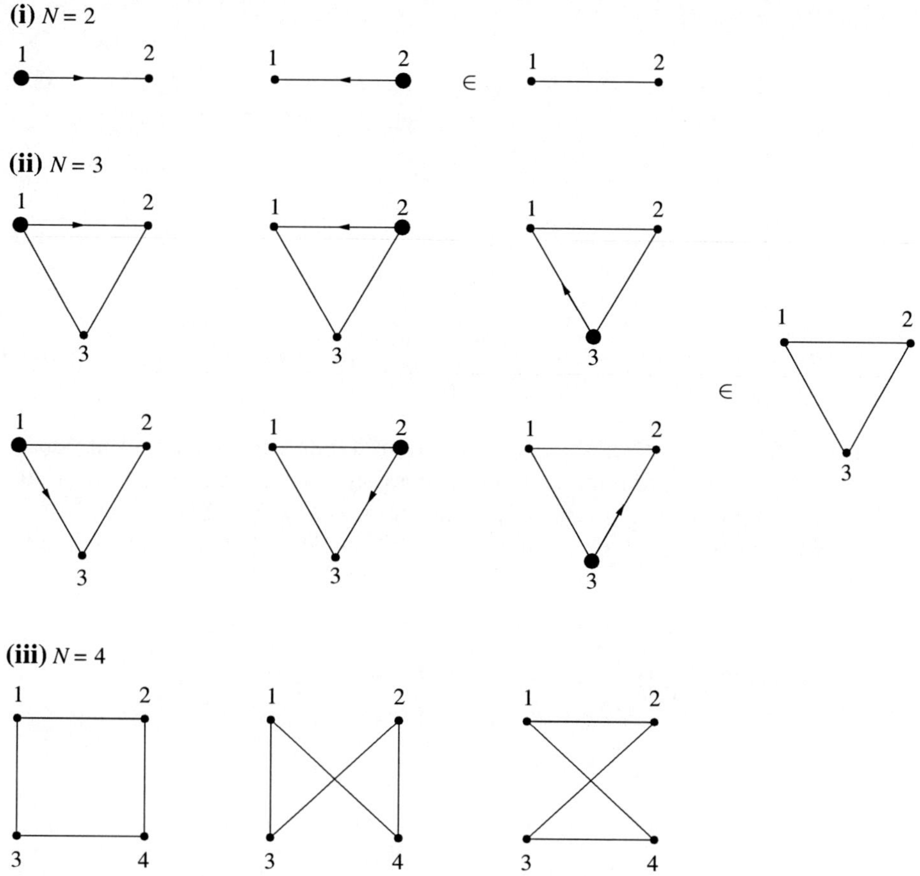

Fig. 3.6 Contributions to the vdW interactions of **(i)** $N = 2$, **(ii)** $N = 3$ and **(iii)** $N = 4$ atoms. For $N = 2, 3$, we represent the individual permutations on the *left* (the starting point and orientation being indicated by the *thick dot* and the *arrow*) and the permutation class on the *right*. For $N = 4$, only the classes are shown

With $f(\boldsymbol{r}_1, \ldots, \boldsymbol{r}_N)$ thus already being partially symmetric, a full symmetrisation can be achieved by summing over those permutations giving distinct results. We form a set $\overline{P}(N) \subsetneq P(N)$ containing exactly one representative of each class and introduce the symmetrisation operator as

$$\mathcal{S}\, f(\boldsymbol{r}_1, \ldots, \boldsymbol{r}_N) = \sum_{\Pi \in \overline{P}(N)} f(\boldsymbol{r}_{\Pi(1)}, \boldsymbol{r}_{\Pi(2)}, \ldots, \boldsymbol{r}_{\Pi(N)}) \,. \tag{3.103}$$

Note that this definition involves a sum rather than an average over the classes. The sum has more than one term for $N \geq 4$. Alternatively, it is of course possible to perform symmetrisation by summing over all $N!$ possible permutations and removing the redundancy via division by the number of members of each class:

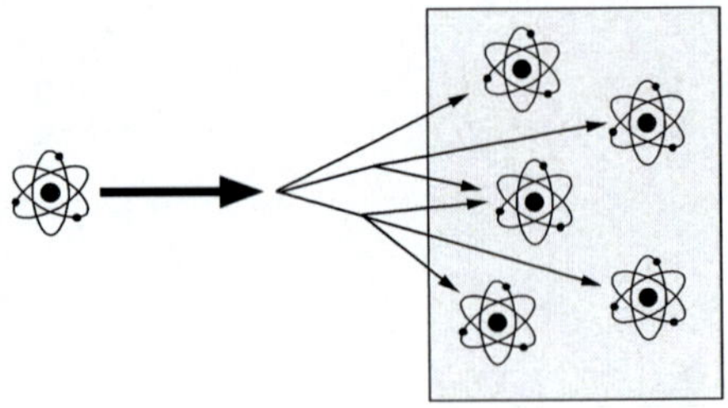

Fig. 3.7 Microscopic origin of the CP force for a non-dilute body

$$\mathcal{S}\, f(\boldsymbol{r}_1,\ldots,\boldsymbol{r}_N) = \frac{1}{(2-\delta_{2N})N} \sum_{\Pi\in P(N)} f(\boldsymbol{r}_{\Pi(1)},\boldsymbol{r}_{\Pi(2)},\ldots,\boldsymbol{r}_{\Pi(N)})\,. \quad (3.104)$$

The integral in (3.100) can be cast in the required symmetrised form by introducing the factor $1/K!$ and summing over all $K!$ possible ways of renaming the variables $\boldsymbol{r}_{A_1}\ldots\boldsymbol{r}_{A_K}$. With $N = K+1$, this procedure generates exactly $K!/[(N-1)!/2] = 2$ representatives from each class [only for $K!/(N-1)! = 1$ for $N = 2$], so we find [2, 3, 9, 10]

$$U(\boldsymbol{r}_A) = \sum_{K=1}^{\infty} \frac{1}{K!} \int_V \mathrm{d}^3 r_1\, \eta(\boldsymbol{r}_1) \cdots \int_V \mathrm{d}^3 r_K\, \eta(\boldsymbol{r}_K) U(\boldsymbol{r}_A,\boldsymbol{r}_1,\ldots,\boldsymbol{r}_K) \quad (3.105)$$

with

$$U(\boldsymbol{r}_1,\ldots,\boldsymbol{r}_N) = \frac{(-1)^{N-1}\hbar\mu_0^N}{(1+\delta_{2N})\pi} \int_0^\infty \mathrm{d}\xi\, \xi^{2N} \alpha^N(\mathrm{i}\xi) \times \mathcal{S}\,\mathrm{tr}\Big[\overline{\boldsymbol{H}}(\boldsymbol{r}_1,\boldsymbol{r}_2,\mathrm{i}\xi)\cdot\overline{\boldsymbol{H}}(\boldsymbol{r}_2,\boldsymbol{r}_3,\mathrm{i}\xi)\cdots\overline{\boldsymbol{H}}(\boldsymbol{r}_N,\boldsymbol{r}_1,\mathrm{i}\xi)\Big]\,. \quad (3.106)$$

Our result generalises the microscopic origin (3.88) of the CP potential beyond the dilute-gas limit. It shows that the CP interaction of a single atom with a macroscopic dielectric body is the result of its microscopic N-atom vdW interactions with the atoms contained in the body, as illustrated in Fig. 3.7. A relation of this type was first derived for the special case of a homogeneous dielectric half space filled with harmonic-oscillator atoms [26] and later extended to homogeneous dielectric bodies of arbitrary shapes [27] and beyond the harmonic-oscillator model [28].

We had already seen in Chap. 2 that the CP potential between an atom and a weakly dielectric body is no longer additive when taking into account higher-order contributions in the electric susceptibility, recall Fig. 2.2. This breakdown of additivity is now seen to be due to microscopic N-atom vdW interactions. As a by-product of our derivation, we conclude that the vdW potential of N identical atoms of polarisability $\alpha(\omega)$ at mutually distinct positions $\boldsymbol{r}_1,\ldots,\boldsymbol{r}_N$ ($\overline{\boldsymbol{H}} = \overline{\boldsymbol{G}}$) in the presence of arbitrary magnetoelectric bodies or permittivity $\varepsilon(\boldsymbol{r},\omega)$ and permeability $\mu(\boldsymbol{r},\omega)$ reads ($\overline{\boldsymbol{G}} \to \boldsymbol{G}$)

$$U(\boldsymbol{r}_1,\ldots,\boldsymbol{r}_N)=\frac{(-1)^{N-1}\hbar\mu_0^N}{(1+\delta_{2N})\pi}\int_0^\infty \mathrm{d}\xi\,\xi^{2N}\alpha^N(\mathrm{i}\xi) \times \mathcal{S}\,\mathrm{tr}\Big[\boldsymbol{G}(\boldsymbol{r}_1,\boldsymbol{r}_2,\mathrm{i}\xi)\cdot\boldsymbol{G}(\boldsymbol{r}_2,\boldsymbol{r}_3,\mathrm{i}\xi)\cdots\boldsymbol{G}(\boldsymbol{r}_N,\boldsymbol{r}_1,\mathrm{i}\xi)\Big]. \tag{3.107}$$

For $N=2$, we find the expected agreement with the microscopic result (1.153) from Sect. 1.2.3. For higher N, microscopic calculations only exist for atoms in free space [29, 30], where the findings are again consistent with (3.107).

Our results can be generalised with several respects. First, we have only considered the interaction of a single atom with a dielectric body consisting of the same species of atoms. As a result, we have only obtained N-atom potentials for identical atoms. In order to derive vdW potentials for N different atoms, we assume that the body characterised by the susceptibility χ contains different atoms with polarisabilities $\alpha_i(\omega)$ and number densities $\eta_i(\boldsymbol{r})$. The Clausius–Mosotti law then generalises to

$$\frac{\chi(\boldsymbol{r},\omega)}{1+\frac{1}{3}\chi(\boldsymbol{r},\omega)}=3\,\frac{\varepsilon(\boldsymbol{r},\omega)-1}{\varepsilon(\boldsymbol{r},\omega)+2}=\sum_i\frac{\eta_i(\boldsymbol{r})\alpha_i(\omega)}{\varepsilon_0}\,. \tag{3.108}$$

We again start from the Born series (3.98) for the CP potential, but now using the more general Clausius–Mosotti law. After symmetrisation, we find that the CP interaction of an atom with a dielectric body containing different atoms can be given as [2, 9]

$$U(\boldsymbol{r}_A)=\sum_{K=1}^{\infty}\frac{1}{K!}\sum_{i_1}\int_V \mathrm{d}^3r_{i_1}\,\eta_{i_1}(\boldsymbol{r}_{i_1})\cdots\sum_{i_K}\int_V \mathrm{d}^3r_{i_K}\,\eta_{i_K}(\boldsymbol{r}_{i_K}) \times U(\boldsymbol{r}_A,\boldsymbol{r}_{i_1},\ldots,\boldsymbol{r}_{i_K}) \tag{3.109}$$

with

$$U(\boldsymbol{r}_1,\ldots,\boldsymbol{r}_N)=\frac{(-1)^{N-1}\hbar\mu_0^N}{(1+\delta_{2N})\pi}\int_0^\infty \mathrm{d}\xi\,\xi^{2N}\alpha_1(\mathrm{i}\xi)\cdots\alpha_N(\mathrm{i}\xi) \times \mathcal{S}\,\mathrm{tr}\Big[\overline{\boldsymbol{H}}(\boldsymbol{r}_1,\boldsymbol{r}_2,\mathrm{i}\xi)\cdot\overline{\boldsymbol{H}}(\boldsymbol{r}_2,\boldsymbol{r}_3,\mathrm{i}\xi)\cdots\overline{\boldsymbol{H}}(\boldsymbol{r}_N,\boldsymbol{r}_1,\mathrm{i}\xi)\Big]. \tag{3.110}$$

We conclude that the vdW potential of N different atoms at mutually distinct positions in the possible presence of magnetoelectric bodies is given by

$$U(\boldsymbol{r}_1,\ldots,\boldsymbol{r}_N)=\frac{(-1)^{N-1}\hbar\mu_0^N}{(1+\delta_{2N})\pi}\int_0^\infty d\xi\,\xi^{2N}\alpha_1(i\xi)\cdots\alpha_N(i\xi)$$
$$\times\,\mathcal{S}\,\mathrm{tr}\Big[\boldsymbol{G}(\boldsymbol{r}_1,\boldsymbol{r}_2,i\xi)\cdot\boldsymbol{G}(\boldsymbol{r}_2,\boldsymbol{r}_3,i\xi)\cdots\boldsymbol{G}(\boldsymbol{r}_N,\boldsymbol{r}_1,i\xi)\Big]\,. \tag{3.111}$$

Next, let us generalise our investigations of the microscopic origin of the CP potential to the electromagnetic case. To that end, consider the interaction of an electromagnetic atom with a weakly magnetodielectric body of volume V and electric and magnetic susceptibilities $\chi(\boldsymbol{r},\omega)$ and $\zeta(\boldsymbol{r},\omega)$ in the possible presence of other bodies with permittivity $\overline{\varepsilon}(\boldsymbol{r},\omega)$ and permeability $\overline{\mu}(\boldsymbol{r},\omega)$. Using the decomposition (2.29) for the total permittivity and permeability, the CP potential $U(\boldsymbol{r}_A)=U_e(\boldsymbol{r}_A)+U_m(\boldsymbol{r}_A)$ can be given by the Born series (2.55). Assuming the weakly magnetodielectric body to be well separated from all other bodies, it reads

$$U_\lambda(\boldsymbol{r}_A)=\sum_K\frac{(-1)^K\hbar}{2\pi\varepsilon_0}\int_0^\infty d\xi\,\alpha_\lambda(i\xi)\sum_{\lambda_1=e,m}\int d^3s_1\,g_{\lambda_1}(\boldsymbol{s}_1,i\xi)$$
$$\cdots\sum_{\lambda_K=e,m}\int d^3s_K\,g_{\lambda_K}(\boldsymbol{s}_K,i\xi)$$
$$\times\,\mathrm{tr}\Big[\overline{\boldsymbol{H}}_{\lambda\lambda_1}(\boldsymbol{r}_A,\boldsymbol{s}_1,i\xi)\cdot\overline{\boldsymbol{H}}_{\lambda_1\lambda_2}(\boldsymbol{s}_1,\boldsymbol{s}_2,i\xi)\cdots\overline{\boldsymbol{H}}_{\lambda_K\lambda}(\boldsymbol{s}_K,\boldsymbol{r}_A,i\xi)\Big]\,, \tag{3.112}$$

with

$$g_e(\boldsymbol{r},\omega)=\frac{\chi(\boldsymbol{r},\omega)}{1+\frac{1}{3}\chi(\boldsymbol{r},\omega)}\,,\qquad g_m(\boldsymbol{r},\omega)=\frac{\zeta(\boldsymbol{r},\omega)}{1-\frac{2}{3}\zeta(\boldsymbol{r},\omega)}\,, \tag{3.113}$$

recall (2.42) and (2.43).

Assuming the single atom and those contained in the weakly magnetodielectric body to be of the same species, we can summarise the Clausius–Mosotti law (4.140) and its magnetic counterpart

$$\frac{\zeta(\boldsymbol{r},\omega)}{1+\frac{2}{3}\zeta(\boldsymbol{r},\omega)}=3\,\frac{\mu(\boldsymbol{r},\omega)-1}{\mu(\boldsymbol{r},\omega)+2}=\mu_0\eta(\boldsymbol{r})\beta(\omega) \tag{3.114}$$

in the compact form

$$g_\lambda(\boldsymbol{r},\omega)=\frac{\eta(\boldsymbol{r})\alpha_\lambda(\omega)}{\varepsilon_0}\quad(\lambda=e,m) \tag{3.115}$$

with $\alpha_e = \alpha$, $\alpha_m = \beta/c^2$. After substitution and symmetrisation, the microscopic origin of the CP potential can still be given in the form (3.105) where

$$U(\boldsymbol{r}_1, \ldots, \boldsymbol{r}_N) = \sum_{\lambda_1 \ldots \lambda_N = e,m} U_{\lambda_1 \ldots \lambda_N}(\boldsymbol{r}_1, \ldots, \boldsymbol{r}_N) \tag{3.116}$$

with

$$\begin{aligned} &U_{\lambda_1 \ldots \lambda_N}(\boldsymbol{r}_1, \ldots, \boldsymbol{r}_N) \\ &= \frac{(-1)^{N-1}\hbar}{(1+\delta_{2N})\pi\varepsilon_0^N} \int_0^\infty \mathrm{d}\xi \, \alpha_{\lambda_1}(\mathrm{i}\xi) \cdots \alpha_{\lambda_N}(\mathrm{i}\xi) \\ &\times \mathcal{S} \operatorname{tr}\Big[\overline{\boldsymbol{H}}_{\lambda_1\lambda_2}(\boldsymbol{r}_1, \boldsymbol{r}_2, \mathrm{i}\xi) \cdot \overline{\boldsymbol{H}}_{\lambda_2\lambda_3}(\boldsymbol{r}_2, \boldsymbol{r}_3, \mathrm{i}\xi) \cdots \overline{\boldsymbol{H}}_{\lambda_N\lambda_1}(\boldsymbol{r}_N, \boldsymbol{r}_1, \mathrm{i}\xi)\Big] \end{aligned} \tag{3.117}$$

are now the total vdW potentials for electromagnetic atoms. We conclude that the vdW potential of N identical atoms at mutually distinct positions in the possible presence of magnetoelectric bodies characterised by $\varepsilon(\boldsymbol{r}, \omega)$ and $\mu(\boldsymbol{r}, \omega)$ is given by (3.116) with

$$\begin{aligned} U_{\lambda_1 \ldots \lambda_N}(\boldsymbol{r}_1, \ldots, \boldsymbol{r}_N) = &\frac{(-1)^{N-1}\hbar}{(1+\delta_{2N})\pi\varepsilon_0^N} \int_0^\infty \mathrm{d}\xi \, \alpha_{\lambda_1}(\mathrm{i}\xi) \cdots \alpha_{\lambda_N}(\mathrm{i}\xi) \\ &\times \mathcal{S} \operatorname{tr}\Big[\mathbf{G}_{\lambda_1\lambda_2}(\boldsymbol{r}_1, \boldsymbol{r}_2, \mathrm{i}\xi) \cdot \mathbf{G}_{\lambda_2\lambda_3}(\boldsymbol{r}_2, \boldsymbol{r}_3, \mathrm{i}\xi) \\ &\cdots \mathbf{G}_{\lambda_N\lambda_1}(\boldsymbol{r}_N, \boldsymbol{r}_1, \mathrm{i}\xi)\Big]. \end{aligned} \tag{3.118}$$

In order to obtain the vdW interaction of N distinct electromagnetic atoms, one uses the Clausius–Mosotti law

$$g_\lambda(\boldsymbol{r}, \omega) = \sum_i \frac{\eta_i(\boldsymbol{r})\alpha_{\lambda,i}(\omega)}{\varepsilon_0} \tag{3.119}$$

($\lambda = e, m$) for a body consisting of different atomic species. The microscopic expansion of the CP potential then takes the form (3.109). The sought-after vdW potential of N atoms with electric/magnetic polarisabilities $\alpha_{1,\lambda}(\omega)$, …, $\alpha_{N,\lambda}(\omega)$ is given by (3.116) with

$$U_{\lambda_1\ldots\lambda_N}(\boldsymbol{r}_1,\ldots,\boldsymbol{r}_N)=\frac{(-1)^{N-1}\hbar}{(1+\delta_{2N})\pi\varepsilon_0^N}\int_0^\infty d\xi\,\alpha_{\lambda_1,1}(i\xi)\cdots\alpha_{\lambda_N,N}(i\xi)$$
$$\times\,\mathcal{S}\,\mathrm{tr}\Big[\boldsymbol{G}_{\lambda_1\lambda_2}(\boldsymbol{r}_1,\boldsymbol{r}_2,i\xi)\cdot\boldsymbol{G}_{\lambda_2\lambda_3}(\boldsymbol{r}_2,\boldsymbol{r}_3,i\xi)$$
$$\cdots\boldsymbol{G}_{\lambda_N\lambda_1}(\boldsymbol{r}_N,\boldsymbol{r}_1,i\xi)\Big]\,. \tag{3.120}$$

As examples, let us consider N-atom vdW potentials in free space. The vdW potential of $N=2$ atoms has been calculated and discussed in Sect. 5.4 of Vol. I, recall the asymptotic distance laws given in row (i) of Table 3.1. It is attractive for two electric or two magnetic atoms and repulsive for mixed combinations of electric and magnetic atoms. It depends on the distance r between the atoms as the only geometric parameter, being proportional to $1/r^7$ in the retarded limit. In the nonretarded limit, the two-atom vdW potential follows a $1/r^6$ law for attractive combinations while exhibiting a $1/r^4$ asymptote for repulsive ones.

The vdW potential of three atoms A, B and C is a lot more complex even for purely electric atoms in free space. It can be calculated explicitly by substituting the free-space Green's tensor (A.21) into (3.111) for $N=3$,

$$U(\boldsymbol{r}_A,\boldsymbol{r}_B,\boldsymbol{r}_C)=\frac{\hbar\mu_0^3}{\pi}\int_0^\infty d\xi\,\xi^6\alpha_A(i\xi)\alpha_B(i\xi)\alpha_C(i\xi)$$
$$\times\,\mathrm{tr}\Big[\boldsymbol{G}(\boldsymbol{r}_A,\boldsymbol{r}_B,i\xi)\cdot\boldsymbol{G}(\boldsymbol{r}_B,\boldsymbol{r}_C,i\xi)\cdot\boldsymbol{G}(\boldsymbol{r}_C,\boldsymbol{r}_A,i\xi)\Big] \tag{3.121}$$

recall that the sum in the symmetrisation operator (3.103) consists of only one term in this case. Instead, we can simply deduce it by comparing the second-order Born expansion (2.65) of the CP potential with the $K=2$ term in the microscopic expansion (3.109) by means of the linearised version

$$\chi(\boldsymbol{r},\omega)=\sum_i\frac{\eta_i(\boldsymbol{r})\alpha_i(\omega)}{\varepsilon_0}\,. \tag{3.122}$$

of the Clausius–Mosotti law (3.108). We find

$$U(\boldsymbol{r}_A,\boldsymbol{r}_B,\boldsymbol{r}_C)=\frac{\hbar}{64\pi^4\varepsilon_0^3}\int_0^\infty d\xi\,\alpha_A(i\xi)\alpha_B(i\xi)\alpha_C(i\xi)$$
$$\times\,\frac{g(\boldsymbol{r}_{AB},\boldsymbol{r}_{BC},\boldsymbol{r}_{CA},\xi)}{r_{AB}^3r_{BC}^3r_{CA}^3}\,, \tag{3.123}$$

with $\boldsymbol{r}_{ij} = \boldsymbol{r}_i - \boldsymbol{r}_j$ and $r_{ij} = |\boldsymbol{r}_{ij}|$ recall definition (2.66) of the function g. This is in agreement with results [29, 30] obtained from free-space QED. In a similar way, we can deduce from (2.72) and (2.77) that the three-atom potential simplifies to

$$
\begin{aligned}
U(\boldsymbol{r}_A, \boldsymbol{r}_B, \boldsymbol{r}_C) = {} & \frac{\hbar c \alpha_A \alpha_B \alpha_C}{16\pi^4 \varepsilon_0^3 r_{AB}^3 r_{BC}^3 r_{CA}^3 (r_{AB}+r_{BC}+r_{CA})} \\
& \times \big[f_1(r_{AB}, r_{BC}, r_{CA}) + f_2(r_{CA}, r_{AB}, r_{BC})(\boldsymbol{e}_{AB}\cdot\boldsymbol{e}_{BC})^2 \\
& + f_2(r_{AB}, r_{BC}, r_{CA})(\boldsymbol{e}_{BC}\cdot\boldsymbol{e}_{CA})^2 \\
& + f_2(r_{BC}, r_{CA}, r_{AB})(\boldsymbol{e}_{CA}\cdot\boldsymbol{e}_{AB})^2 \\
& + f_3(r_{AB}, r_{BC}, r_{CA})(\boldsymbol{e}_{AB}\cdot\boldsymbol{e}_{BC})(\boldsymbol{e}_{BC}\cdot\boldsymbol{e}_{CA})(\boldsymbol{e}_{CA}\cdot\boldsymbol{e}_{AB}) \big]
\end{aligned}
\tag{3.124}
$$

with $\boldsymbol{e}_{ij} = \boldsymbol{r}_{ij}/r_{ij}$ [recall definitions (2.73)–(2.75)] and

$$
\begin{aligned}
U(\boldsymbol{r}_A, \boldsymbol{r}_B, \boldsymbol{r}_C) = {} & \frac{3\hbar}{64\pi^4 \varepsilon_0^3} \int_0^\infty \mathrm{d}\xi\, \alpha_A(\mathrm{i}\xi)\alpha_B(\mathrm{i}\xi)\alpha_C(\mathrm{i}\xi) \\
& \times \frac{1 - 3(\boldsymbol{e}_{AB}\cdot\boldsymbol{e}_{BC})(\boldsymbol{e}_{BC}\cdot\boldsymbol{e}_{CA})(\boldsymbol{e}_{CA}\cdot\boldsymbol{e}_{AB})}{r_{AB}^3 r_{BC}^3 r_{CA}^3}
\end{aligned}
\tag{3.125}
$$

in the retarded and nonretarded limits, respectively. The nonretarded potential of three atoms is commonly known as the Axilrod–Teller potential [31–33]. Comparison with Sect. 5.5.2 of Vol. I reveals that the nonretarded potential (5.300) of two atoms interacting with a small electric sphere is also of the Axilrod–Teller form where the polarisability of the sphere occurs in place of that of atom C.

The three-atom potential depends on the three distances between the atoms. In contrast to the two-atom potential with its $1/r^7$ and $1/r^6$ retarded and nonretarded asymptotes, it depends on distance to the inverse 10th and 9th powers in these limits. Even for purely electric atoms, the three-atom potential may be attractive or repulsive, depending on the shape of the triangle formed by the three atoms. This is illustrated in Fig. 3.8 where we show the sign of the Axilrod–Teller potential (3.125) depending on the position of atom A for fixed positions of atoms B and C. As seen, the potential is attractive when the atoms lie on a straight line and repulsive when they form an equilateral triangle. In between these extremes, it is attractive whenever the triangle formed by the atoms has two sufficiently sharp angles and repulsive otherwise. The angle-dependence of the three-atom potential is responsible for the shape-dependence of the second-order contribution $\Delta_2^2 U(\boldsymbol{r}_A)$ to the Born expansion of the CP interaction of an atom with a weakly dielectric body, as noted in Sect. 2.2.2.

Potentials of four or more atoms exhibit an even more complex geometry-dependence. As an additional complication, symmetrisation must be performed in

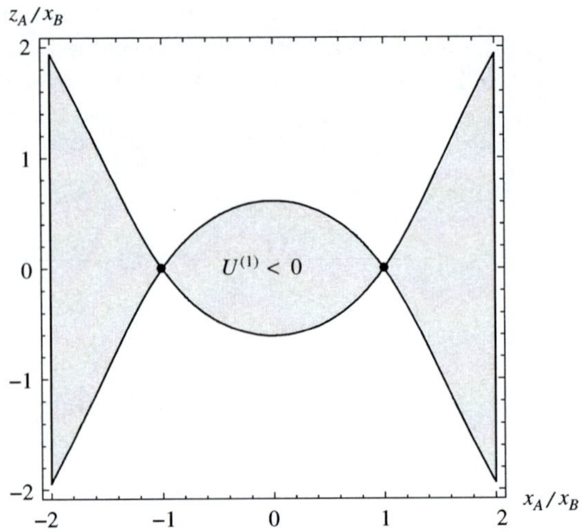

Fig. 3.8 Sign of the Axilrod–Teller potential. The *shaded area* indicates the positions of atom A corresponding to attractive potentials for fixed positions of atoms B and C at $\boldsymbol{r}_B = (x_B, 0, 0)$ and $\boldsymbol{r}_C = (-x_B, 0, 0)$

accordance with (3.103). The four-atom potential is calculated explicitly in [29, 30] for the special case of the atoms being arranged as a regular tetrahedron.

References

1. S.Y. Buhmann, D.T. Ho, T. Kampf, D.G. Welsch, Eur. Phys. J. D **35**(1), 15 (2005)
2. S.Y. Buhmann, D.G. Welsch, Prog. Quantum Electron. **31**(2), 51 (2007)
3. S. Scheel, S.Y. Buhmann, Acta Phys. Slovaca **58**(5), 675 (2008)
4. S.Y. Buhmann, S. Scheel, H. Safari, D.G. Welsch, Int. J. Mod. Phys. A **24**(8–9), 1796 (2009)
5. S.Y. Buhmann, S. Scheel, Phys. Rev. Lett. **102**(14), 140404 (2009)
6. M. Schaden, L. Spruch, Phys. Rev. A **58**(2), 935 (1998)
7. J.E. Lennard-Jones, Trans. Faraday Soc. **28**, 333 (1932)
8. C. Farina, F.C. Santos, A.C. Tort, Am. J. Phys. **70**(4), 421 (2002)
9. S.Y. Buhmann, H. Safari, D.G. Welsch, D.T. Ho, Open Syst. Inf. Dyn. **13**(4), 427 (2006)
10. S.Y. Buhmann, D.G. Welsch, Appl. Phys. B **82**(2), 189 (2006)
11. S.Y. Buhmann, S. Scheel, J. Babington, Phys. Rev. Lett. **104**(7), 070404 (2010)
12. F.J. Giessibl, Rev. Mod. Phys. **75**(3), 949 (2003)
13. M.F. Maghrebi, S.J. Rahi, T. Emig, N. Graham, R.L. Jaffe, M. Kardar, Proc. Natl. Acad. Sci. USA **108**(17), 6867 (2011)
14. S.Y. Buhmann, D.G. Welsch, T. Kampf, Phys. Rev. A **72**(3), 032112 (2005)
15. M.M. Taddei, T.N.C. Mendes, C. Farina, Eur. J. Phys. **31**(1), 89 (2010)
16. B.V. Derjaguin, I.I. Abrikosova, E.M. Lifshitz, Q. Rev, Chem. Soc. **10**(3), 295 (1956)
17. J. Błoccki, J. Randrup, W.J. Swiatecki, C.F. Tsang, Ann. Phys. **105**(2), 427 (1977)
18. E.A. Power, T. Thirunamachandran, Phys. Rev. A **25**(5), 2473 (1982)
19. S. Spagnolo, R. Passante, L. Rizzuto, Phys. Rev. A **73**(6), 062117 (2006)
20. J.D. Jackson, *Classical Electrodynamics*, 3rd edn. (Wiley, New York, 1998)
21. C. Raabe, D.G. Welsch, Phys. Rev. A **73**(6), 063822 (2006)
22. E.M. Lifshitz, Sov. Phys. JETP **2**(1), 73 (1956)
23. I.E. Dzyaloshinskii, E.M. Lifshitz, L.P. Pitaevskii, Adv. Phys. **10**(38), 165 (1961)
24. S.Y. Buhmann, H. Safari, D.T. Ho, D.G. Welsch, Opt. Spectrosc. (USSR) **103**(3), 374 (2006)

25. A. Sambale, D.G. Welsch, S.Y. Buhmann, T.D. Ho, Opt. Spectrosc. (USSR) **108**(3), 391 (2010)
26. M.J. Renne, B.R.A. Nijboer, Chem. Phys. Lett. **1**(8), 317 (1967)
27. J. Mahanty, B.W. Ninham, J. Chem. Phys. **59**(11), 6157 (1973)
28. P.W. Milonni, P.B. Lerner, Phys. Rev. A **46**(3), 1185 (1992)
29. E.A. Power, T. Thirunamachandran, Proc. R. Soc. Lond. Ser. A **401**(1821), 267 (1985)
30. E.A. Power, T. Thirunamachandran, Phys. Rev. A **50**(5), 3929 (1994)
31. B.M. Axilrod, E. Teller, J. Chem. Phys. **11**(6), 299 (1943)
32. B.M. Axilrod, J. Chem. Phys. **17**(12), 1349 (1949)
33. B.M. Axilrod, J. Chem. Phys. **19**(6), 719 (1951)

Chapter 4
Casimir–Polder Forces on Excited Atoms: Static Theory

So far, we have been exclusively concerned with dispersion forces involving ground-state objects. Being due to the entire fluctuation spectrum of the electromagnetic field and the objects' polarisation and magnetisation, such forces depend on electromagnetic response functions over a wide range of frequencies. In particular, the CP potential of a ground-state atom is due to atomic transitions to an excited energy eigenstate, accompanied by the emission of a photon. Violating energy conservation, such a process is purely virtual; the atom must immediately reabsorb the photon while returning to its ground state. Virtual photons from a continuous range of energies and frequencies contribute.

The situation is fundamentally different for an excited atom: It can undergo a transition to a lower-lying state while releasing its energy in the form of a real photon. This energy-conserving process involves photons of certain discrete and well-defined frequencies, the atomic transition frequencies. Real photons give rise to resonant contributions to the CP potential. They often dominate over the non-resonant contributions from virtual photons, leading to a new and radically altered behaviour. Interference phenomena accompanied by strongly enhanced and spatially oscillating potentials may be observed.

Furthermore, real transitions to lower lying states imply that the internal state of the atom will not be constant over time. Due to such spontaneous decay, the atom will continuously evolve from its initial excited state until it finally reaches its ground state. We may expect this internal dynamics to manifest itself in a time-dependent CP force.

In this chapter, we will study the CP force by generalising the ideas of Casimir and Polder [1]. Their approach is able to capture the phenomenon of resonant potentials. Being based on a time-independent calculation, it neglects the internal atomic dynamics and the resulting dynamics of the force, to be addressed in the next Chap. 5. Instead, Casimir and Polder's concept provides a snapshot of the potential at a given instant of time with the atom being in a well-defined energy eigenstate.

We will begin by calculating the CP potential using perturbation theory, working within the alternative minimal and multipolar coupling schemes. Applying our results, we will calculate resonant CP potentials of an excited atom in front of a perfectly conducting plate, a magnetoelectric half space and a meta-material perfect lens.

S. Y. Buhmann, *Dispersion Forces II*, Springer Tracts in Modern Physics 248,
DOI: 10.1007/978-3-642-32466-6_4, © Springer-Verlag Berlin Heidelberg 2012

4.1 Perturbation Theory

Recall from Sect. 1.2.2 that the CP force (1.119) can be derived from the CP potential. Following Casimir and Polder, the potential may be identified as position-dependent energy shift arising from the atom-field coupling. We calculate this shift using perturbation theory, starting with the minimal coupling scheme.

4.1.1 Minimal Coupling

We assume the atom to be prepared in an arbitrary internal energy-eigenstate $|n\rangle$ and the body-assisted field in its ground state $|\{0\}\rangle$. Starting from the uncoupled state $|\psi\rangle = |n\rangle|\{0\}\rangle$ of the atom–field system within the minimal coupling scheme, we calculate the leading-order energy shift (1.120). For a non-magnetic atom, the atom–field interaction Hamiltonian in long-wavelength approximation (1.88)

$$\hat{H}_{\mathrm{AF}} = -\hat{\boldsymbol{d}}\cdot\hat{\boldsymbol{E}}^{\parallel}(\boldsymbol{r}_A) - \sum_{\alpha\in A}\frac{q_\alpha}{m_\alpha}\hat{\boldsymbol{p}}_\alpha\cdot\hat{\mathbf{A}}(\boldsymbol{r}_A) + \sum_{\alpha\in A}\frac{q_\alpha^2}{2m_\alpha}\hat{\mathbf{A}}^2(\boldsymbol{r}_A) \tag{4.1}$$

contains two terms linear and one term quadratic in the field operators. The latter contributes to the first-order energy shift as given by (1.121) with $|0\rangle \mapsto |\psi\rangle$,

$$\Delta_1 E = \langle\{0\}|\langle n|\sum_{\alpha\in A}\frac{q_\alpha^2}{2m_\alpha}\hat{\mathbf{A}}^2(\boldsymbol{r}_A)|n\rangle|\{0\}\rangle\,. \tag{4.2}$$

This contribution may be evaluated explicitly by using the expansion (1.50) of the vector potential, the commutation relations (1.17) and (1.18) and the integral relation (1.25),

$$\begin{aligned}\Delta_1 E &= \sum_{\alpha\in A}\frac{q_\alpha^2}{2m_\alpha}\int_0^\infty\frac{\mathrm{d}\omega}{\omega^2}\sum_{\lambda=e,m}\int\mathrm{d}^3s\,\mathrm{tr}\big[{}^\perp\boldsymbol{G}_\lambda(\boldsymbol{r}_A,\boldsymbol{s},\omega)\cdot\boldsymbol{G}_\lambda^{*\mathrm{T}\perp}(\boldsymbol{r}_A,\boldsymbol{s},\omega)\big]\\ &= \frac{\hbar\mu_0}{\pi}\sum_{\alpha\in A}\frac{q_\alpha^2}{2m_\alpha}\int_0^\infty\mathrm{d}\omega\,\mathrm{tr}\big[\mathrm{Im}\,{}^\perp\boldsymbol{G}^\perp(\boldsymbol{r}_A,\boldsymbol{r}_A,\omega)\big]\,.\end{aligned} \tag{4.3}$$

Note that the first-order contributions is independent of the internal state $|n\rangle$ in which the atom is prepared.

The first-order contribution is quadratic in the particle charges q_α which are a measure of the electromagnetic coupling strength. To be consistent, the leading-order energy shift must contain all contributions quadratic in q_α. We hence need to include the second-order energy shift (1.122) ($|0\rangle \mapsto |\psi\rangle$) due to the first two terms of the interaction Hamiltonian. The relevant intermediate states $|I\rangle = |k\rangle|\mathbf{1}_\lambda(\boldsymbol{r},\omega)\rangle$ consist of the atom in state k and the field in its single-photon Fock state (1.35), so that the energy denominators read $E_0 - E_I = E_n - (E_k + \hbar\omega) = -\hbar(\omega - \omega_{nk})$ with $\omega_{mn} = (E_m - E_n)/\hbar$ and the energy shift takes the form

$$\Delta_2 E = -\sum_k \sum_{\lambda=e,m} \int \mathrm{d}^3 r \, \mathcal{P}\int_0^\infty \frac{\mathrm{d}\omega}{\hbar(\omega-\omega_{nk})}$$
$$\times \langle\{0\}|\langle n|\left[-\hat{\boldsymbol{d}}\cdot\hat{\boldsymbol{E}}^{\parallel}(\boldsymbol{r}_A) - \sum_{\alpha\in A}\frac{q_\alpha}{m_\alpha}\hat{\overline{\boldsymbol{p}}}_\alpha\cdot\hat{\boldsymbol{A}}(\boldsymbol{r}_A)\right]|k\rangle|\mathbf{1}_\lambda(\boldsymbol{r},\omega)\rangle$$
$$\cdot\langle\mathbf{1}_\lambda(\boldsymbol{r},\omega)|\langle k|\left[-\hat{\boldsymbol{d}}\cdot\hat{\boldsymbol{E}}^{\parallel}(\boldsymbol{r}_A) - \sum_{\alpha\in A}\frac{q_\alpha}{m_\alpha}\hat{\overline{\boldsymbol{p}}}_\alpha\cdot\hat{\boldsymbol{A}}(\boldsymbol{r}_A)\right]|n\rangle|\{0\}\rangle \tag{4.4}$$

($\mathcal{P}$: principal value). The required matrix elements can be calculated using the field expansions (1.22) and (1.50) and the commutation relations (1.17) and (1.18). One finds

$$\langle\mathbf{1}_\lambda(\boldsymbol{r},\omega)|\langle k|-\hat{\boldsymbol{d}}\cdot\hat{\boldsymbol{E}}^{\parallel}(\boldsymbol{r}_A)|n\rangle|\{0\}\rangle = -\boldsymbol{d}_{kn}\cdot{}^{\parallel}\boldsymbol{G}^*_\lambda(\boldsymbol{r}_A,\boldsymbol{r},\omega)\,, \tag{4.5}$$
$$\langle\mathbf{1}_\lambda(\boldsymbol{r},\omega)|\langle k|-\sum_{\alpha\in A}\frac{q_\alpha}{m_\alpha}\hat{\overline{\boldsymbol{p}}}_\alpha\cdot\hat{\boldsymbol{A}}(\boldsymbol{r}_A)|n\rangle|\{0\}\rangle = \frac{\omega_{kn}}{\omega}\boldsymbol{d}_{kn}\cdot{}^{\perp}\boldsymbol{G}^*_\lambda(\boldsymbol{r}_A,\boldsymbol{r}\,,\omega) \tag{4.6}$$

with $\boldsymbol{d}_{mn} = \langle m|\hat{\boldsymbol{d}}|n\rangle$, where definition (1.51) has been recalled and the atomic identity (1.73) has been employed for the second matrix element. Using these results and invoking the integral relation (1.25), the second-order energy shift reads

$$\begin{aligned}
\Delta_2 E
&= -\frac{1}{\hbar}\sum_k \mathcal{P}\int_0^\infty \frac{\mathrm{d}\omega}{\omega-\omega_{nk}}\sum_{\lambda=e,m}\int \mathrm{d}^3 s\,\boldsymbol{d}_{nk}\cdot\Big\{{}^{\parallel}\boldsymbol{G}_\lambda(\boldsymbol{r}_A,\boldsymbol{s},\omega)\cdot\boldsymbol{G}_\lambda^{*\mathrm{T}\parallel}(\boldsymbol{r}_A,\boldsymbol{s},\omega)\\
&\quad -\frac{\omega_{kn}}{\omega}\Big[{}^{\parallel}\boldsymbol{G}_\lambda(\boldsymbol{r}_A,\boldsymbol{s},\omega)\cdot\boldsymbol{G}_\lambda^{*\mathrm{T}\perp}(\boldsymbol{r}_A,\boldsymbol{s},\omega) + {}^{\perp}\boldsymbol{G}_\lambda(\boldsymbol{r}_A,\boldsymbol{s},\omega)\cdot\boldsymbol{G}_\lambda^{*\mathrm{T}\parallel}(\boldsymbol{r}_A,\boldsymbol{s},\omega)\Big]\\
&\quad +\frac{\omega_{kn}^2}{\omega^2}{}^{\perp}\boldsymbol{G}_\lambda(\boldsymbol{r}_A,\boldsymbol{s},\omega)\cdot\boldsymbol{G}_\lambda^{*\mathrm{T}\perp}(\boldsymbol{r}_A,\boldsymbol{s},\omega)\Big\}\cdot\boldsymbol{d}_{kn}\\
&= -\frac{\mu_0}{\pi}\sum_k \mathcal{P}\int_0^\infty \frac{\mathrm{d}\omega}{\omega-\omega_{nk}}\boldsymbol{d}_{nk}\cdot\mathrm{Im}\Big\{\omega^2\,{}^{\parallel}\boldsymbol{G}^{\parallel}(\boldsymbol{r}_A,\boldsymbol{r}_A,\omega) + \omega_{kn}^2\,{}^{\perp}\boldsymbol{G}^{\perp}(\boldsymbol{r}_A,\boldsymbol{r}_A,\omega)\\
&\quad -\omega_k\omega\Big[{}^{\parallel}\boldsymbol{G}^{\perp}(\boldsymbol{r}_A,\boldsymbol{r}_A,\omega) + {}^{\perp}\boldsymbol{G}^{\parallel}(\boldsymbol{r}_A,\boldsymbol{r}_A,\omega)\Big]\Big\}\cdot\boldsymbol{d}_{kn}\,.
\end{aligned} \tag{4.7}$$

In order to combine $\Delta_1 E$ and $\Delta_2 E$ in accordance with (1.120), we make use of the Thomas–Reiche–Kuhn sum rule (1.74) to cast the former into the alternative form

$$\Delta_1 E = \frac{\mu_0}{\pi} \sum_k \omega_{kn} \int_0^\infty \mathrm{d}\omega\, \boldsymbol{d}_{nk} \cdot \mathrm{Im}\, {}^{\perp}\boldsymbol{G}^{\perp}(\boldsymbol{r}_A, \boldsymbol{r}_A, \omega) \cdot \boldsymbol{d}_{kn} \,. \tag{4.8}$$

Use of the identity $\boldsymbol{G} = {}^{\perp}\boldsymbol{G}^{\perp} + {}^{\perp}\boldsymbol{G}^{\parallel} + {}^{\parallel}\boldsymbol{G}^{\perp} + {}^{\parallel}\boldsymbol{G}^{\parallel}$ then leads to

$$\begin{aligned} \Delta E = {} & \frac{\mu_0}{\pi} \sum_k \mathcal{P} \int_0^\infty \frac{\mathrm{d}\omega}{\omega - \omega_{nk}} \\ & \times \boldsymbol{d}_{nk} \cdot \left\{ \omega_{kn}\omega \left[\mathrm{Im}\boldsymbol{G}(\boldsymbol{r}_A, \boldsymbol{r}_A, \omega) - \mathrm{Im}^{\parallel}\boldsymbol{G}^{\parallel}(\boldsymbol{r}_A, \boldsymbol{r}_A, \omega) \right] \right. \\ & \left. - \omega^2 \mathrm{Im}^{\parallel}\boldsymbol{G}^{\parallel}(\boldsymbol{r}_A, \boldsymbol{r}_A, \omega) \right\} \cdot \boldsymbol{d}_{kn} \,. \end{aligned} \tag{4.9}$$

In order to extract the position-dependent CP potential (1.118) from the leading-order energy shift, we decompose the Green's tensor into its bulk and scattering parts $\boldsymbol{G}^{(0)}$ and $\boldsymbol{G}^{(1)}$ according to (1.111). The energy shift associated with the bulk part is a position-independent infinite self-energy, which is part of the free-space Lamb shift. It is discussed in some detail in Sect. 4.1 of Vol. I. We discard it by making the replacement $\boldsymbol{G} \mapsto \boldsymbol{G}^{(1)}$, hence obtaining the CP potential as the position-dependent part of the energy shift. Noting that the scattering Green's tensor is purely transverse (A.28), writing $\mathrm{Im}\,\boldsymbol{G} = (\boldsymbol{G} - \boldsymbol{G}^*)/(2\mathrm{i})$ and using the Schwarz reflection principle (A.3), we find

$$\begin{aligned} U(\boldsymbol{r}_A) = {} & \frac{\mu_0}{2\mathrm{i}\pi} \sum_k \left[\mathcal{P} \int_0^\infty \frac{\mathrm{d}\omega}{\omega - \omega_{nk}} + \mathcal{P} \int_0^{-\infty} \frac{\mathrm{d}\omega}{\omega + \omega_{nk}} \right] \omega_{kn}\omega \\ & \times \boldsymbol{d}_{nk} \cdot \boldsymbol{G}^{(1)}(\boldsymbol{r}_A, \boldsymbol{r}_A, \omega) \cdot \boldsymbol{d}_{kn} \,. \end{aligned} \tag{4.10}$$

So far, the calculation has been completely analogous to that for the ground-state CP potential as detailed in Sect. 4.1 of Vol. I and summarised in Sect. 1.2.2 of this volume. A change of the initial state from the ground state 0 to an arbitrary state n leads to the replacements $\boldsymbol{d}_{0k} \mapsto \boldsymbol{d}_{nk}$ and $\omega_k \equiv \omega_{k0} \mapsto \omega_{kn}$. However, this simple modification results in an important qualitative change of the potential. For an atom in an excited state n, transitions to states of lower energy $E_k < E_n$ exist such that the transition energy ω_{nk} is positive. The integrands of the positive- and negative-frequency integrals above then exhibit poles at $\omega = \omega_{nk}$ and $\omega = -\omega_{nk}$, respectively; which is why we have introduced the principal value integral. The consequences of these poles can be made more explicit by applying contour-integral techniques. To that end, we note that the integrand is analytic in the upper half of the complex frequency plane including the real axis and according to (A.30) is also finite at the origin. Applying the integration contour shown in Fig. 4.1 to transform the integrals along the positive/negative real frequency axis into integrals along the positive imaginary axis plus integrals along infinite quarter-circles plus, in

Fig. 4.1 Integration contours used for transforming real-frequency integrals into ones along the positive imaginary axis plus contributions from the poles

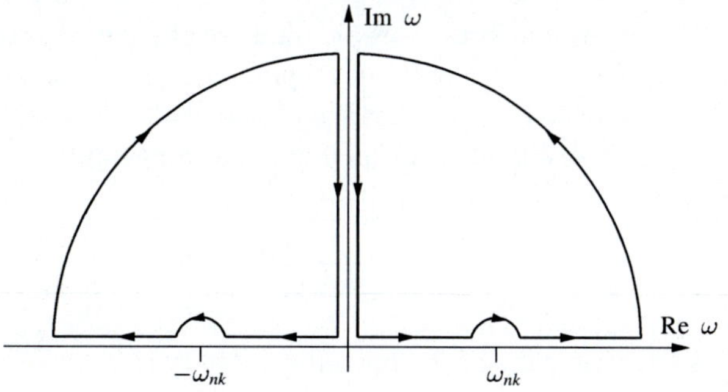

the case $\omega_{nk} > 0$, integrals along infinitesimally small semi-circles around the poles at $\omega = \omega_{nk}$ and $\omega = -\omega_{nk}$. The integrals along the infinite quarter-circles vanish as a consequence of the high-frequency asymptote (A.29), so we obtain [2–4]

$$U_n(\boldsymbol{r}_A) = U_n^{\mathrm{nres}}(\boldsymbol{r}_A) + U_n^{\mathrm{res}}(\boldsymbol{r}_A) \tag{4.11}$$

where

$$U_n^{\mathrm{nres}}(\boldsymbol{r}_A) = \frac{\mu_0}{\pi} \sum_k \int_0^\infty \mathrm{d}\xi \, \frac{\omega_{kn}\xi^2}{\omega_{kn}^2 + \xi^2} \boldsymbol{d}_{nk} \cdot \boldsymbol{G}^{(1)}(\boldsymbol{r}_A, \boldsymbol{r}_A, \mathrm{i}\xi) \cdot \boldsymbol{d}_{kn} \tag{4.12}$$

is the non-resonant part of the CP potential and

$$U_n^{\mathrm{res}}(\boldsymbol{r}_A) = -\mu_0 \sum_{k<n} \omega_{nk}^2 \boldsymbol{d}_{nk} \cdot \mathrm{Re}\, \boldsymbol{G}^{(1)}(\boldsymbol{r}_A, \boldsymbol{r}_A, \omega_{nk}) \cdot \boldsymbol{d}_{kn} \tag{4.13}$$

is the resonant part arising from the residua at the poles. The former can be cast into the more compact form

$$U_n^{\mathrm{nres}}(\boldsymbol{r}_A) = \frac{\hbar\mu_0}{2\pi} \int_0^\infty \mathrm{d}\xi \, \xi^2 \mathrm{tr}\left[\boldsymbol{\alpha}_n(\mathrm{i}\xi) \cdot \boldsymbol{G}^{(1)}(\boldsymbol{r}_A, \boldsymbol{r}_A, \mathrm{i}\xi)\right] \tag{4.14}$$

by introducing the polarisability tensor

$$\boldsymbol{\alpha}_n(\omega) = \lim_{\epsilon \to 0+} \frac{1}{\hbar} \sum_k \left(\frac{\boldsymbol{d}_{nk}\boldsymbol{d}_{kn}}{\omega_{kn} - \omega - \mathrm{i}\epsilon} + \frac{\boldsymbol{d}_{kn}\boldsymbol{d}_{nk}}{\omega_{kn} + \omega + \mathrm{i}\epsilon} \right) \tag{4.15}$$

of the atom in state $|n\rangle$.

For an atom in an isotropic state, we group the sum over k into sums over manifolds of states $k' \in \{\bar{k}\}$ with degenerate energies. For instance, such a manifold could consist of states with identical principal and total angular momenta which only differ by the z-components of their angular momentum. On each manifold, we have

$$\sum_{k' \in \{\bar{k}\}} \boldsymbol{d}_{nk'} \boldsymbol{d}_{k'n} = \tfrac{1}{3} \sum_{k' \in \{\bar{k}\}} |\boldsymbol{d}_{nk'}|^2 \boldsymbol{I}, \tag{4.16}$$

so the two parts of the potential simplify to

$$U_n^{\text{nres}}(\boldsymbol{r}_A) = \frac{\hbar\mu_0}{2\pi} \int_0^\infty \mathrm{d}\xi\, \xi^2 \alpha_n(\mathrm{i}\xi)\, \mathrm{tr}\, \boldsymbol{G}^{(1)}(\boldsymbol{r}_A, \boldsymbol{r}_A, \mathrm{i}\xi)\,, \tag{4.17}$$

$$U_n^{\text{res}}(\boldsymbol{r}_A) = -\frac{\mu_0}{3} \sum_{k<n} \omega_{nk}^2 |\boldsymbol{d}_{nk}|^2\, \mathrm{tr}\big[\mathrm{Re}\, \boldsymbol{G}^{(1)}(\boldsymbol{r}_A, \boldsymbol{r}_A, \omega_{nk})\big] \tag{4.18}$$

with

$$\alpha_n(\omega) = \lim_{\epsilon \to 0+} \frac{2}{3\hbar} \sum_k \frac{\omega_{kn} |\boldsymbol{d}_{nk}|^2}{\omega_{kn}^2 - \omega^2 - \mathrm{i}\omega\epsilon}\,. \tag{4.19}$$

We have thus obtained the CP potential (4.11) together with (4.13) and (4.14) of an electric atom in an arbitrary internal energy eigenstate $|n\rangle$, generalising the ground-state potential (1.126) given in Sect. 1.2.2. An alternative derivation of this result may be given on the basis of linear-response theory by employing the fluctuation–dissipation theorem (1.28) for the electromagnetic field [5].

In contrast to the ground-state case, the CP potential of an excited atom contains two distinct contributions. The non-resonant part is similar in form to the ground-state potential, with the only difference being a replacement of the ground-state polarisability with its excited-state counterpart. It depends on the atomic polarisability as well as the Green's tensor for the electromagnetic field in an integral form, where all positive imaginary frequencies contribute. As shown in Sect. 4.1 of Vol. I, the ground-state potential is a pure quantum effect. Its physical origin lies in the correlated fluctuations of the atomic dipole moment and the electromagnetic field, where the latter may be simply referred to as virtual photons. This interpretation remains valid for the non-resonant part of the potential for an atom in an arbitrary state.

The resonant potential, on the contrary, depends on the Green's tensor at discrete frequencies corresponding to possible real transitions of the atom to a lower-energy eigenstate. It is only present for excited atoms and may be attributed to real photons. The resonant potential often dominates over the non-resonant potential. In contrast to the non-resonant potential, it reacts very sensitive to the magnetoelectric properties of the atomic environment at specific frequencies. The resonant potential can partly

Table 4.1 Dominant contributions to the non-resonant and resonant parts of the CP potential

Distance →	Retarded	Intermediate	Nonretarded
Non-resonant	$\hat{\boldsymbol{A}}^2$, $\hat{\boldsymbol{p}}\cdot\hat{\boldsymbol{A}}$	$\hat{\boldsymbol{A}}^2$	$\hat{\boldsymbol{d}}\cdot\hat{\boldsymbol{E}}^{\parallel}$
Resonant	$\hat{\boldsymbol{p}}\cdot\hat{\boldsymbol{A}}$	$\hat{\boldsymbol{p}}\cdot\hat{\boldsymbol{A}}$	$\hat{\boldsymbol{d}}\cdot\hat{\boldsymbol{E}}^{\parallel}$

be understood in classical terms. To see this, consider a classical oscillating dipole

$$\boldsymbol{d}(t) = \frac{\boldsymbol{d}\mathrm{e}^{-\mathrm{i}\omega t}}{\sqrt{2}} + \mathrm{C.c.} \tag{4.20}$$

placed at $\boldsymbol{r}_A$ within an arbitrary arrangement of magnetoelectric bodies (where the normalisation has been chosen such that the long-time average of the dipole moment is $\overline{\boldsymbol{d}^2(t)} = |\boldsymbol{d}|^2$). The dipole emits an electric field which is reflected at the surface of present bodies. Writing the current density associated with the dipole in the form

$$\boldsymbol{j}(\boldsymbol{r},t) = \dot{\boldsymbol{d}}(t)\delta(\boldsymbol{r}-\boldsymbol{r}_A) = \frac{-\mathrm{i}\omega\boldsymbol{d}\mathrm{e}^{-\mathrm{i}\omega t}}{\sqrt{2}}\delta(\boldsymbol{r}-\boldsymbol{r}_A) + \mathrm{C.c.}\,, \tag{4.21}$$

the reflected field (1.15) reads

$$\boldsymbol{E}^{(1)}(\boldsymbol{r},t) = \frac{\mu_0\omega^2\mathrm{e}^{-\mathrm{i}\omega t}}{\sqrt{2}}\boldsymbol{\mathsf{G}}^{(1)}(\boldsymbol{r},\boldsymbol{r}_A,\omega)\cdot\boldsymbol{d} + \mathrm{C.c.} \tag{4.22}$$

The interaction energy of the classical dipole in its own reflected field is hence on the long-time average given by

$$\overline{W(\boldsymbol{r}_A,t)} = -\tfrac{1}{2}\overline{\boldsymbol{d}(t)\cdot\boldsymbol{E}^{(1)}(\boldsymbol{r}_A,t)} = -\tfrac{1}{2}\mu_0\omega^2\boldsymbol{d}^*\cdot\mathrm{Re}\,\boldsymbol{\mathsf{G}}^{(1)}(\boldsymbol{r}_A,\boldsymbol{r}_A,\omega)\cdot\boldsymbol{d}\,. \tag{4.23}$$

Comparison with (4.13) reveals that our classical model renders one half of the corresponding quantum contributions to the resonant CP potential. The discrepancy can be understood from the fact that the other half of the resonant CP interaction is due to fluctuations of the electromagnetic field, which are absent from the classical description [6].

It is worth discussing which terms in the interaction Hamiltonian (4.1) give the dominant contributions to the non-resonant and resonant CP potentials. For a ground-state atom, the relative importance of the terms was analysed in Sect. 4.1 of Vol. I, with the results being summarised in the first row of Table 4.1. In the retarded limit, the contributions due to the $\hat{\boldsymbol{A}}^2$- and the $\hat{\boldsymbol{p}}\cdot\hat{\boldsymbol{A}}$-interactions dominate and strongly cancel each other. For intermediate, moderately retarded distances, the $\hat{\boldsymbol{A}}^2$-interaction begins to dominate over the $\hat{\boldsymbol{p}}\cdot\hat{\boldsymbol{A}}$-term. In the nonretarded limit, the potential is dominated by the electrostatic Coulomb interaction $\hat{\boldsymbol{d}}\cdot\hat{\boldsymbol{E}}^{\parallel}$. These findings remain valid for the non-resonant part of the CP potential of an excited atom.

For the resonant part, the situation is different: The $\hat{\boldsymbol{A}}^2$-interaction does not contribute at all, because it does not exhibit any poles. To see this explicitly, we start from the energy shift (4.7) without the $\hat{\boldsymbol{A}}^2$-contribution. Discarding the self-energy by making the replacement $\boldsymbol{G} \mapsto \boldsymbol{G}^{(1)}$, exploiting the fact that the scattering Green's tensor is purely transverse and following similar steps as above, we then obtain a CP potential

$$U(\boldsymbol{r}_A) = \frac{\mu_0}{2\mathrm{i}\pi} \sum_k \left[\mathcal{P} \int_0^{\infty} \frac{\mathrm{d}\omega}{\omega - \omega_{nk}} - \mathcal{P} \int_0^{-\infty} \frac{\mathrm{d}\omega}{\omega + \omega_{nk}} \right] \omega_{kn}^2 \\ \times \boldsymbol{d}_{nk} \cdot \boldsymbol{G}^{(1)}(\boldsymbol{r}_A, \boldsymbol{r}_A, \omega) \cdot \boldsymbol{d}_{kn} \,. \tag{4.24}$$

After using the integration contour displayed in Fig. 4.1, we find a total CP potential (4.11) whose non-resonant part

$$U_n^{\mathrm{nres}}(\boldsymbol{r}_A) = \frac{\mu_0}{\pi} \sum_k \int_0^{\infty} \mathrm{d}\xi \, \frac{\omega_{kn}^3}{\omega_{kn}^2 + \xi^2} \boldsymbol{d}_{nk} \cdot \boldsymbol{G}^{(1)}(\boldsymbol{r}_A, \boldsymbol{r}_A, \mathrm{i}\xi) \cdot \boldsymbol{d}_{kn} \tag{4.25}$$

differs from (4.12) due to the neglect of the $\hat{\boldsymbol{A}}^2$-contribution, but whose resonant part (4.13) remains unchanged. Having hence established that the $\hat{\boldsymbol{A}}^2$-term does not contribute to the resonant potential, we proceed to identify which of the other two dominates. To that end we return to the first lines of (4.7), where the second-order energy shift is represented by a spatial integral over Green's tensors $\boldsymbol{G}(\boldsymbol{r}_A, \boldsymbol{s}, \omega)$ connecting the atomic position with points inside the present bodies. The first term ${}^{\parallel}\boldsymbol{G}_\lambda \cdot \boldsymbol{G}_\lambda^{*\mathrm{T}\parallel}$ is due to the $\hat{\boldsymbol{d}} \cdot \hat{\boldsymbol{E}}^{\parallel}$-interaction, the next two terms represent mixed contributions involving both the $\hat{\boldsymbol{d}} \cdot \hat{\boldsymbol{E}}^{\parallel}$- and $\hat{\boldsymbol{p}} \cdot \hat{\boldsymbol{A}}$-couplings, while the last term ${}^{\perp}\boldsymbol{G}_\lambda \cdot \boldsymbol{G}_\lambda^{*\mathrm{T}\perp}$ is entirely due to the $\hat{\boldsymbol{p}} \cdot \hat{\boldsymbol{A}}$-interaction. In the retarded limit, the Green's tensor $\boldsymbol{G}(\boldsymbol{r}_A, \boldsymbol{s}, \omega)$ becomes purely transverse (cf. Appendix A.2) and hence only the term ${}^{\perp}\boldsymbol{G}_\lambda \cdot \boldsymbol{G}_\lambda^{*\mathrm{T}\perp}$ due to the $\hat{\boldsymbol{p}} \cdot \hat{\boldsymbol{A}}$-interaction is relevant. For nonretarded distances, the Green's tensor is purely longitudinal, so the $\hat{\boldsymbol{d}} \cdot \hat{\boldsymbol{E}}^{\parallel}$-interaction dominates. The two cases are displayed in the second row of Table 4.1.

To summarise, the $\hat{\boldsymbol{d}} \cdot \hat{\boldsymbol{E}}^{\parallel}$-interaction dominates the entire potential in the nonretarded limit; the $\hat{\boldsymbol{p}} \cdot \hat{\boldsymbol{A}}$-term contributes predominantly in the retarded limit; and the $\hat{\boldsymbol{A}}^2$-interaction only contributes to the non-resonant potential, being relevant for retarded and particularly for intermediate distances. These results agree with earlier observations made for the example of an atom in front of a perfectly reflecting plate [6–9].

The total CP potential of an atom in an energy eigenstate takes a particularly simple form for perfectly conducting bodies in the nonretarded limit. To see this, we make use of the Born expansion (2.7) with (2.15). On a free-space background

$\overline{\varepsilon}(\boldsymbol{r},\omega) \equiv 1$, we have

$$\boldsymbol{G}^{(1)}(\boldsymbol{r},\boldsymbol{r}',\omega) = \sum_{K=1}^{\infty} \frac{\omega^{2K}}{c^{2K}} \int_V \mathrm{d}^3 s_1 \frac{\chi(\boldsymbol{s}_1,\omega)}{1+\frac{1}{3}\chi(\boldsymbol{s}_1,\omega)} \cdots \int_V \mathrm{d}^3 s_K \frac{\chi(\boldsymbol{s}_K,\omega)}{1+\frac{1}{3}\chi(\boldsymbol{s}_K,\omega)}$$
$$\times \boldsymbol{H}_{\mathrm{free}}(\boldsymbol{r},\boldsymbol{s}_1,\omega)\cdot\boldsymbol{H}_{\mathrm{free}}(\boldsymbol{s}_1,\boldsymbol{s}_2,\omega)\cdots\boldsymbol{H}_{\mathrm{free}}(\boldsymbol{s}_K,\boldsymbol{r}',\omega) \tag{4.26}$$

where V is the volume occupied by the present bodies, $\chi(\boldsymbol{r},\omega) = \varepsilon(\boldsymbol{r},\omega) - 1$ is their electric susceptibility and

$$\boldsymbol{H}_{\mathrm{free}}(\boldsymbol{r},\boldsymbol{r}',\omega) = -\frac{c^2 e^{\mathrm{i}\omega\rho/c}}{4\pi\omega^2\rho^3}\left\{\left[1 - \mathrm{i}\frac{\omega\rho}{c} - \left(\frac{\omega\rho}{c}\right)^2\right]\boldsymbol{I}\right.$$
$$\left. - \left[3 - 3\mathrm{i}\frac{\omega\rho}{c} - \left(\frac{\omega\rho}{c}\right)^2\right]\boldsymbol{e}_\rho\boldsymbol{e}_\rho\right\} \tag{4.27}$$

is the non-singular part of the free-space Green's tensor, see (A.21) in App. A. In the nonretarded limit $|(\boldsymbol{r}-\boldsymbol{s}_1)\omega|/c, |(\boldsymbol{s}_1-\boldsymbol{s}_2)\omega|/c \ldots |(\boldsymbol{s}_K-\boldsymbol{r}')\omega|/c \ll 1$ and for perfectly conducting bodies $\chi(\boldsymbol{r},\omega) \to \infty$, we then have

$$\frac{\omega^2}{c^2}\boldsymbol{G}^{(1)}(\boldsymbol{r},\boldsymbol{r}',\omega) \simeq \boldsymbol{G}^{(1)}(\boldsymbol{r},\boldsymbol{r}') \tag{4.28}$$

where the electrostatic Green's tensor

$$\boldsymbol{G}^{(1)}(\boldsymbol{r},\boldsymbol{r}') = \lim_{\omega\to 0}\left[\frac{\omega^2}{c^2}\boldsymbol{G}^{(1)}(\boldsymbol{r},\boldsymbol{r}',\omega)\right]$$
$$= \frac{1}{4\pi}\sum_{K=1}^{\infty}\left(\frac{3}{4\pi}\right)^K \int_V \mathrm{d}^3 s_1 \cdots \int_V \mathrm{d}^3 s_K$$
$$\times \frac{\boldsymbol{e}_{r1}\boldsymbol{e}_{r1} - \boldsymbol{I}}{|\boldsymbol{r}-\boldsymbol{s}_1|^3}\cdot\frac{\boldsymbol{e}_{12}\boldsymbol{e}_{12} - \boldsymbol{I}}{|\boldsymbol{s}_1-\boldsymbol{s}_2|^3}\cdots\frac{\boldsymbol{e}_{Kr'}\boldsymbol{e}_{Kr'} - \boldsymbol{I}}{|\boldsymbol{s}_K-\boldsymbol{r}'|^3} \tag{4.29}$$

with direction unit vectors $\boldsymbol{e}_{r1} = (\boldsymbol{r}-\boldsymbol{s}_1)/|\boldsymbol{r}-\boldsymbol{s}_1|$, $\boldsymbol{e}_{IJ} = (\boldsymbol{s}_I-\boldsymbol{s}_J)/|\boldsymbol{s}_I-\boldsymbol{s}_J|$ and $\boldsymbol{e}_{Kr'} = (\boldsymbol{s}_K-\boldsymbol{r}')/|\boldsymbol{s}_K-\boldsymbol{r}'|$ is frequency-independent.

We apply this result to the CP potential whose nonretarded limit can be defined as $r_+ \ll c/\omega_+$ (r_+: maximum of all atom-body distances, ω_+: maximum of all relevant atomic resonance frequencies). The non-resonant potential (4.14) consists of an integral over the Green's tensor at imaginary frequencies. The factor $\boldsymbol{\alpha}(\mathrm{i}\xi)$ effectively limits the ξ-integral to a range where $|\boldsymbol{r}_A-\boldsymbol{s}_1|\xi/c, |\boldsymbol{s}_K-\boldsymbol{r}_A|\xi/c \ll 1$. Furthermore, the dominant contribution to the $\boldsymbol{s}$-integrals comes from regions where $|\boldsymbol{s}_I-\boldsymbol{s}_J|\xi/c \ll 1$. Using the electrostatic Green's tensor, we find

$$U_n^{\mathrm{nres}}(\boldsymbol{r}_A) = -\frac{\hbar}{2\pi\varepsilon_0}\int_0^\infty \mathrm{d}\xi\,\mathrm{tr}\big[\boldsymbol{\alpha}_n(\mathrm{i}\xi)\cdot\boldsymbol{G}^{(1)}(\boldsymbol{r}_A,\boldsymbol{r}_A)\big]\,. \tag{4.30}$$

With the definition (4.15) of the atomic polarisability as given above, the ξ-integral can then be performed according to

$$\int_0^\infty \mathrm{d}\xi\, \alpha_n(\mathrm{i}\xi) = \frac{\pi}{\hbar} \sum_k \mathrm{sgn}(\omega_{kn}) \boldsymbol{d}_{nk} \boldsymbol{d}_{kn} \tag{4.31}$$

[sgn(x): sign function], resulting in

$$U_n^{\mathrm{nres}}(\boldsymbol{r}_A) = -\frac{1}{2\varepsilon_0} \sum_k \mathrm{sgn}(\omega_{kn}) \boldsymbol{d}_{nk} \cdot \mathbf{G}^{(1)}(\boldsymbol{r}_A, \boldsymbol{r}_A) \cdot \boldsymbol{d}_{kn} \,. \tag{4.32}$$

For the resonant potential (4.13), we have $|\boldsymbol{r} - \boldsymbol{s}_1|\omega_{nk}/c$, $|\boldsymbol{s}_K - \boldsymbol{r}'|\omega_{nk}/c \ll 1$ in the nonretarded limit. In addition, we assume that no atomic transition ω_{nk} is close to a geometric body resonance (such as the cavity resonances discussed in Chap. 6). The $\boldsymbol{s}$-integrals are then dominated by region where $|\boldsymbol{s}_I - \boldsymbol{s}_J|\omega_{nk}/c \ll 1$. The purely real electrostatic Green's tensor (4.29) hence applies once more and we find

$$U_n^{\mathrm{res}}(\boldsymbol{r}_A) = -\frac{1}{\varepsilon_0} \sum_{k<n} \boldsymbol{d}_{nk} \cdot \mathbf{G}^{(1)}(\boldsymbol{r}_A, \boldsymbol{r}_A) \cdot \boldsymbol{d}_{kn} \,. \tag{4.33}$$

Combining the non-resonant and resonant potentials and invoking the completeness relation $\sum_k \boldsymbol{d}_{nk}\boldsymbol{d}_{kn} = \langle \hat{\boldsymbol{d}}\hat{\boldsymbol{d}} \rangle_n$, the total CP potential of an atom at nonretarded distance from perfectly conducting bodies takes the simple form

$$U_n(\boldsymbol{r}_A) = -\frac{\langle \hat{\boldsymbol{d}} \cdot \mathbf{G}^{(1)}(\boldsymbol{r}_A, \boldsymbol{r}_A) \cdot \hat{\boldsymbol{d}} \rangle_n}{2\varepsilon_0} \,. \tag{4.34}$$

It reduces to

$$U_n(\boldsymbol{r}_A) = -\frac{\langle \hat{\boldsymbol{d}}^2 \rangle_n \mathrm{tr}\, \mathbf{G}^{(1)}(\boldsymbol{r}_A, \boldsymbol{r}_A)}{2\varepsilon_0} \tag{4.35}$$

for an isotropic atom. A potential of this form was first derived from the Coulomb interaction of the atom with the bodies [10]. Note that $\langle \hat{\boldsymbol{d}}^2 \rangle_n$ is always positive, irrespective of n. In the isotropic case, we may hence conclude that the nonretarded potential of an excited atom near perfectly conducting bodies has the same sign as the corresponding potential of a ground-state atom.

An important difference between the CP potentials of ground-state vs excited atoms concerns isotropy. The ground states of atoms and molecules are generically isotropic in the absence of applied electric and magnetic fields. On the contrary, an excited atom is often in an anisotropic state. For instance, such a breaking of the orientational symmetry may be due to a state preparation via polarised electromagnetic waves or internal dynamics in an anisotropic environment (see Sect. 7.3.3).

4.1.2 Multipolar Coupling

It is instructive to also calculate the CP potential of an excited atom using the alternative, multipolar coupling scheme. To that end, we assume the atom to be in an energy eigenstate $|n'\rangle$ and the body-assisted field in its ground state $|\{0'\}\rangle$ so that the uncoupled state of the system in the multipolar coupling scheme reads $|\psi\rangle = |n'\rangle|\{0'\}\rangle$. Neglecting the velocity-dependent Röntgen interaction, the respective interaction Hamiltonian for a non-magnetic atom in electric-dipole approximation (1.98) consists of a single term only,

$$\hat{H}'_{\mathrm{AF}} = -\hat{\boldsymbol{d}}' \cdot \hat{\boldsymbol{E}}'(\boldsymbol{r}_A) \,. \tag{4.36}$$

This term being linear in the field operators, the leading energy shift is the second-order shift (1.22) with $|0\rangle \mapsto |\psi\rangle$. With intermediate states $|I\rangle = |k'\rangle|\mathbf{1}'_\lambda(\boldsymbol{r},\omega)\rangle$ and transition frequencies $\omega'_{nk} = (E'_n - E'_k)/\hbar$, it takes the form

$$\begin{aligned}\Delta E' = \Delta_2 E' = -\frac{1}{\hbar}\sum_k \sum_{\lambda=e,m} \int \mathrm{d}^3 r \int_0^\infty \frac{\mathrm{d}\omega}{\omega - \omega'_{nk}} \\ \times \langle n'|\langle\{0'\}| - \hat{\boldsymbol{d}}' \cdot \hat{\boldsymbol{E}}'(\boldsymbol{r}_A)|\mathbf{1}'_\lambda(\boldsymbol{r},\omega)\rangle|k'\rangle \\ \cdot \langle k'|\langle\{\mathbf{1}'_\lambda(\boldsymbol{r},\omega)\}| - \hat{\boldsymbol{d}}' \cdot \hat{\boldsymbol{E}}'(\boldsymbol{r}_A)|\{0'\}\rangle|n'\rangle \,.\end{aligned} \tag{4.37}$$

The transition matrix elements can be found by noting that the field expansion (1.22) and the commutation relations (1.17) and (1.18) remain valid for the primed variables of the multipolar scheme, so that

$$\langle k'|\langle\mathbf{1}'_\lambda(\boldsymbol{r},\omega)| -\hat{\boldsymbol{d}}' \cdot \hat{\boldsymbol{E}}'(\boldsymbol{r}_A)|\{0'\}\rangle|n'\rangle = -\boldsymbol{d}'_{kn} \cdot \boldsymbol{G}^*_\lambda(\boldsymbol{r}_A,\boldsymbol{r},\omega) \tag{4.38}$$

with $\boldsymbol{d}'_{mn} = \langle m'|\hat{\boldsymbol{d}}'|n'\rangle$. With these explicit matrix elements and the integral relation (1.25), the energy shift becomes

$$\Delta E' = -\frac{\mu_0}{\pi}\sum_k \mathcal{P}\int_0^\infty \frac{\mathrm{d}\omega}{\omega - \omega'_{nk}}\, \omega^2 \boldsymbol{d}'_{nk} \cdot \mathrm{Im}\,\boldsymbol{G}(\boldsymbol{r}_A,\boldsymbol{r}_A,\omega) \cdot \boldsymbol{d}'_{kn} \,. \tag{4.39}$$

Again, this result differs from the ground-state one only via the replacements $\boldsymbol{d}'_{0k} \mapsto \boldsymbol{d}'_{nk}$ and $\omega'_{k0} \mapsto \omega'_{kn}$. The latter of these modifications leads to poles in the frequency integrand for transitions to lower-lying states.

We now follow exactly the same steps as for the minimal-coupling calculation. Extracting the CP potential by discarding the bulk Green's tensor and using the integration contour as depicted in Fig. 4.1, we arrive at [2–4]

$$U'_n(\boldsymbol{r}_A) = U'^{\text{nres}}_n(\boldsymbol{r}_A) + U'^{\text{res}}_n(\boldsymbol{r}_A) \tag{4.40}$$

with

$$U'^{\text{nres}}_n(\boldsymbol{r}_A) = \frac{\hbar\mu_0}{2\pi}\int_0^\infty \mathrm{d}\xi\,\xi^2 \mathrm{tr}\left[\boldsymbol{\alpha}'_n(\mathrm{i}\xi)\cdot\boldsymbol{G}^{(1)}(\boldsymbol{r}_A,\boldsymbol{r}_A,\mathrm{i}\xi)\right], \tag{4.41}$$

$$U'^{\text{res}}_n(\boldsymbol{r}_A) = -\mu_0\sum_{k<n}\omega'^2_{nk}\boldsymbol{d}'_{nk}\cdot\mathrm{Re}\,\boldsymbol{G}^{(1)}(\boldsymbol{r}_A,\boldsymbol{r}_A,\omega'_{nk})\cdot\boldsymbol{d}'_{kn} \tag{4.42}$$

and

$$\boldsymbol{\alpha}'_n(\omega) = \lim_{\epsilon\to 0+}\frac{1}{\hbar}\sum_k\left(\frac{\boldsymbol{d}'_{nk}\boldsymbol{d}'_{kn}}{\omega'_{kn}-\omega-\mathrm{i}\epsilon}+\frac{\boldsymbol{d}'_{kn}\boldsymbol{d}'_{nk}}{\omega'_{kn}+\omega+\mathrm{i}\epsilon}\right). \tag{4.43}$$

For an isotropic atom, the two components of the CP potential reduce to

$$U'^{\text{nres}}_n(\boldsymbol{r}_A) = \frac{\hbar\mu_0}{2\pi}\int_0^\infty \mathrm{d}\xi\,\xi^2\alpha'_n(\mathrm{i}\xi)\,\mathrm{tr}\,\boldsymbol{G}^{(1)}(\boldsymbol{r}_A,\boldsymbol{r}_A,\mathrm{i}\xi), \tag{4.44}$$

$$U'^{\text{res}}_n(\boldsymbol{r}_A) = -\frac{\mu_0}{3}\sum_{k<n}\omega'^2_{nk}|\boldsymbol{d}'_{nk}|^2\,\mathrm{tr}\left[\mathrm{Re}\,\boldsymbol{G}^{(1)}(\boldsymbol{r}_A,\boldsymbol{r}_A,\omega'_{nk})\right] \tag{4.45}$$

with

$$\alpha'_n(\omega) = \lim_{\epsilon\to 0+}\frac{2}{3\hbar}\sum_k\frac{\omega'_{kn}|\boldsymbol{d}'_{nk}|^2}{\omega'^2_{kn}-\omega^2-\mathrm{i}\omega\epsilon}. \tag{4.46}$$

As in the case of the minimal coupling calculation, we find that the CP potential of an atom in an excited energy eigenstate has two components: a non-resonant part due to virtual photons which depends on the atomic polarisability and the Green's tensor in an integral form and a resonant part due to real photons that depends on the Green's tensor at downward atomic transition frequencies. In fact, the multipolar result has exactly the same form as the minimal-coupling one. The only difference lies in the different atomic transition frequencies and dipole matrix elements, which are determined by the atomic Hamiltonians (1.72) and (1.92) in the two cases. The two results represent two perturbative approximations to the same exact CP potential. Note that the multipolar coupling scheme is a lot simpler to work with, as the interaction Hamiltonian consists of a single term only, which is able to produce the full non-resonant and resonant potentials for all distance regimes. Needless to say that a result of the form (4.34) for an atom at nonretarded distance from perfectly conducting bodies also holds in the multipolar coupling scheme,

$$U_n'(\boldsymbol{r}_A) = -\frac{\langle \hat{\boldsymbol{d}}' \cdot \boldsymbol{G}^{(1)}(\boldsymbol{r}_A, \boldsymbol{r}_A) \cdot \hat{\boldsymbol{d}}' \rangle_{n'}}{2\varepsilon_0} = -\frac{\langle \hat{\boldsymbol{d}}'^2 \rangle_{n'} \mathrm{tr}\, \boldsymbol{G}^{(1)}(\boldsymbol{r}_A, \boldsymbol{r}_A)}{2\varepsilon_0}, \tag{4.47}$$

where the second equality is valid for an isotropic atom.

In the following Sect. 4.2, we will discard the primes distinguishing the minimal and multipolar coupling schemes, bearing in mind that our results apply equally for both schemes.

4.2 Excited Atom in Front of a Plate

In order to illustrate the differences between the non-resonant vs resonant CP potentials, we consider the example of an atom at distance z_A from the surface of different plates. We will begin with the simplest case of a perfectly conducting plate, followed by the more complicated examples of a semi-infinite magnetoelectric half space and a meta-material superlens in front of a perfectly conducting plate.

According to App. A.3.2, the scattering Green's tensor of the plate (A.35) is for each of these examples given by

$$\boldsymbol{G}^{(1)}(\boldsymbol{r}, \boldsymbol{r}, \omega) = \frac{\mathrm{i}}{8\pi^2} \int \frac{\mathrm{d}^2 k^{\parallel}}{k^{\perp}} \sum_{\sigma=s,p} \boldsymbol{e}_{\sigma+} \boldsymbol{e}_{\sigma-} r_\sigma \mathrm{e}^{2\mathrm{i}k^{\perp} z} \tag{4.48}$$

($\boldsymbol{k}^{\parallel} \perp \boldsymbol{e}_z$). The components $\boldsymbol{k}^{\parallel}$ and $k^{\perp}$ of the wave vector parallel and perpendicular to the plate satisfy the dispersion relation

$$k^{\perp} = \sqrt{\frac{\omega^2}{c^2} - k^{\parallel 2}}\,, \qquad \mathrm{Im}\, k^{\perp} > 0\,. \tag{4.49}$$

The polarisation unit vectors for s- and p-polarised waves are given by

$$\boldsymbol{e}_{s\pm} = \boldsymbol{e}_{k^{\parallel}} \times \boldsymbol{e}_z\,, \qquad \boldsymbol{e}_{p\pm} = \frac{c}{\omega}\left(k^{\parallel} \boldsymbol{e}_z \mp k^{\perp} \boldsymbol{e}_{k^{\parallel}}\right) \tag{4.50}$$

with r_σ denoting the respective reflection coefficients of the plate.

Introducing polar coordinates in the $\boldsymbol{k}^{\|}$-plane according to $\boldsymbol{e}_{k^{\|}} = (\cos\phi, \sin\phi, 0)$ and noting that $\boldsymbol{e}_z = (0, 0, 1)$, the unit vectors take the forms

$$\boldsymbol{e}_{s\pm} = (\sin\phi, -\cos\phi, 0)\,, \qquad \boldsymbol{e}_{p\pm} = \frac{c}{\omega} = \left(\mp k^{\perp}\cos\phi, \mp k^{\perp}\sin\phi, k^{\|}\right), \tag{4.51}$$

and hence their products read

$$\boldsymbol{e}_{s+}\boldsymbol{e}_{s-} = \begin{pmatrix} \sin^2\phi & -\sin\phi\cos\phi & 0 \\ -\sin\phi\cos\phi & \cos^2\phi & 0 \\ 0 & 0 & 0 \end{pmatrix}, \tag{4.52}$$

$$\boldsymbol{e}_{p+}\boldsymbol{e}_{p-} = \frac{c^2}{\omega^2}\begin{pmatrix} -k^{\perp 2}\cos^2\phi & -k^{\perp 2}\sin\phi\cos\phi & -k^{\|}k^{\perp}\cos\phi \\ -k^{\perp 2}\sin\phi\cos\phi & -k^{\perp 2}\sin^2\phi & -k^{\|}k^{\perp}\sin\phi \\ k^{\|}k^{\perp}\cos\phi & k^{\|}k^{\perp}\sin\phi & k^{\|2} \end{pmatrix}. \tag{4.53}$$

The angular integration over these products can be easily performed to give

$$\int_0^{2\pi} \mathrm{d}\phi\, \boldsymbol{e}_{s+}\boldsymbol{e}_{s-} = \pi \begin{pmatrix} 1 & 0 & 0 \\ 0 & 1 & 0 \\ 0 & 0 & 0 \end{pmatrix}, \tag{4.54}$$

$$\int_0^{2\pi} \mathrm{d}\phi\, \boldsymbol{e}_{p+}\boldsymbol{e}_{p-} = \frac{\pi c^2}{\omega^2}\begin{pmatrix} -k^{\perp 2} & 0 & 0 \\ 0 & -k^{\perp 2} & 0 \\ 0 & 0 & 2k^{\|2} \end{pmatrix}. \tag{4.55}$$

With these results and $\int \mathrm{d}^2k^{\|} = \int_0^{\infty} k^{\|}\mathrm{d}k^{\|} \int_0^{2\pi}\mathrm{d}\phi$, the scattering Green's tensor becomes

$$\boldsymbol{\mathsf{G}}^{(1)}(\boldsymbol{r}, \boldsymbol{r}, \omega) = \frac{\mathrm{i}}{8\pi}\int_0^{\infty} \mathrm{d}k^{\|}\, \frac{k^{\|}}{k^{\perp}}\, \mathrm{e}^{2\mathrm{i}k^{\perp}z} \times \left[\begin{pmatrix} 1 & 0 & 0 \\ 0 & 1 & 0 \\ 0 & 0 & 0 \end{pmatrix} r_s + \frac{c^2}{\omega^2}\begin{pmatrix} -k^{\perp 2} & 0 & 0 \\ 0 & -k^{\perp 2} & 0 \\ 0 & 0 & 2k^{\|2} \end{pmatrix} r_p\right]. \tag{4.56}$$

For real frequencies ω, the wave vector (4.49) in the direction parallel to the plate is real and positive for $0 \le k^{\|} < \omega/c$, corresponding to propagating waves. For $k^{\|} > \omega/c$, it becomes purely imaginary and is given by $k^{\perp} = \mathrm{i}\kappa^{\perp}$ with

$$\kappa^{\perp} = \sqrt{k^{\|2} - \frac{\omega^2}{c^2}}\,, \tag{4.57}$$

so that the associated waves are evanescent or exponentially damped. To separate the contributions from propagating vs evanescent waves, we use the integration variables $k^{\perp}$ ($\int_0^{\omega/c} \mathrm{d}k^{\parallel}\, k^{\parallel}/k^{\perp} = \int_0^{\omega/c} \mathrm{d}k^{\parallel}$) and $\kappa^{\perp}$ ($\int_{\omega/c}^{\infty} \mathrm{d}k^{\parallel}\, k^{\parallel}/k^{\perp} = -\mathrm{i}\int_0^{\infty} \mathrm{d}\kappa^{\perp}$) for the two intervals. The Green's tensor then reads

$$
\begin{aligned}
&\boldsymbol{G}^{(1)}(\boldsymbol{r},\boldsymbol{r},\omega)\\
&= \frac{\mathrm{i}}{8\pi}\int_0^{\omega/c} \mathrm{d}k^{\perp}\, \mathrm{e}^{2\mathrm{i}k^{\perp}z}\left[\begin{pmatrix}1&0&0\\0&1&0\\0&0&0\end{pmatrix} r_s + \frac{c^2}{\omega^2}\begin{pmatrix}-k^{\perp 2}&0&0\\0&-k^{\perp 2}&0\\0&0&2k^{\parallel 2}\end{pmatrix} r_p\right]\\
&\quad + \frac{1}{8\pi}\int_0^{\infty} \mathrm{d}\kappa^{\perp}\, \mathrm{e}^{-2\kappa^{\perp}z}\left[\begin{pmatrix}1&0&0\\0&1&0\\0&0&0\end{pmatrix} r_s + \frac{c^2}{\omega^2}\begin{pmatrix}\kappa^{\perp 2}&0&0\\0&\kappa^{\perp 2}&0\\0&0&2k^{\parallel 2}\end{pmatrix} r_p\right]. \qquad (4.58)
\end{aligned}
$$

For purely imaginary frequencies $\omega = \mathrm{i}\xi$, the wave vector in the direction perpendicular to the plate is always purely imaginary, $k^{\perp} = \mathrm{i}\kappa^{\perp}$ with

$$
\kappa^{\perp} = \sqrt{\frac{\xi^2}{c^2} + k^{\parallel 2}}\,. \qquad (4.59)
$$

Using the integration variable $\kappa^{\perp}$ ($\int_0^{\infty} \mathrm{d}k^{\parallel}\, k^{\parallel}/\kappa^{\perp} = \int_{\xi/c}^{\infty} \mathrm{d}\kappa^{\perp}$), the Green's tensor assumes the form

$$
\begin{aligned}
&\boldsymbol{G}^{(1)}(\boldsymbol{r},\boldsymbol{r},\mathrm{i}\xi)\\
&= \frac{1}{8\pi}\int_{\xi/c}^{\infty} \mathrm{d}\kappa^{\perp}\, \mathrm{e}^{-2\kappa^{\perp}z}\left[\begin{pmatrix}1&0&0\\0&1&0\\0&0&0\end{pmatrix} r_s - \frac{c^2}{\xi^2}\begin{pmatrix}\kappa^{\perp 2}&0&0\\0&\kappa^{\perp 2}&0\\0&0&2k^{\parallel 2}\end{pmatrix} r_p\right]. \qquad (4.60)
\end{aligned}
$$

Using these results, the CP potential (4.11) with (4.13) and (4.14) of an excited atom in front of a plate can be given as

$$
U_n(z_A) = U_n^{\mathrm{nres}}(z_A) + U_n^{\mathrm{prop}}(z_A) + U_n^{\mathrm{evan}}(z_A)\,. \qquad (4.61)
$$

Here,

$$
\begin{aligned}
U_n^{\mathrm{nres}}(z_A) = \frac{\hbar\mu_0}{8\pi^2}\int_0^{\infty} \mathrm{d}\xi\, \xi^2 \int_{\xi/c}^{\infty} \mathrm{d}\kappa^{\perp}\, \mathrm{e}^{-2\kappa^{\perp}z_A}\Big\{\alpha_n^{\parallel}(\mathrm{i}\xi) r_s\\
- \left[\frac{\kappa^{\perp 2}c^2}{\xi^2}\alpha_n^{\parallel}(\mathrm{i}\xi) + \left(\frac{\kappa^{\perp 2}c^2}{\xi^2} - 1\right)\alpha_n^{\perp}(\mathrm{i}\xi)\right] r_p\Big\} \qquad (4.62)
\end{aligned}
$$

with $\alpha^{\|} = \frac{1}{2}(\alpha_{xx} + \alpha_{yy})$ and $\alpha^{\perp} = \alpha_{zz}$ is the non-resonant potential due to virtual photons;

$$U_n^{\text{prop}}(z_A) = \frac{\mu_0}{8\pi} \sum_{k<n} \omega_{nk}^2 \int_0^{\omega_{nk}/c} \mathrm{d}k^{\perp} \left\{ |\boldsymbol{d}_{nk}^{\|}|^2 \mathrm{Im}\left(\mathrm{e}^{2\mathrm{i}k^{\perp} z_A} r_s\right) - \left[\frac{k^{\perp 2} c^2}{\omega_{nk}^2} |\boldsymbol{d}_{nk}^{\|}|^2 + 2\left(\frac{k^{\perp 2} c^2}{\omega_{nk}^2} - 1\right) |\boldsymbol{d}_{nk}^{\perp}|^2\right] \mathrm{Im}\left(\mathrm{e}^{2\mathrm{i}k^{\perp} z_A} r_p\right) \right\} \tag{4.63}$$

with $\boldsymbol{d}^{\|} = d_x \boldsymbol{e}_x + d_y \boldsymbol{e}_y$ and $\boldsymbol{d}^{\perp} = d_z \boldsymbol{e}_z$ is the resonant potential due to real propagating photons; and

$$U_n^{\text{evan}}(z_A) = -\frac{\mu_0}{8\pi} \sum_{k<n} \omega_{nk}^2 \int_0^{\infty} \mathrm{d}\kappa^{\perp}\, \mathrm{e}^{-2\kappa^{\perp} z_A} \left\{ |\boldsymbol{d}_{nk}^{\|}|^2 \mathrm{Re}(r_s) + \left[\frac{\kappa^{\perp 2} c^2}{\omega_{nk}^2} |\boldsymbol{d}_{nk}^{\|}|^2 + 2\left(\frac{\kappa^{\perp 2} c^2}{\omega_{nk}^2} + 1\right) |\boldsymbol{d}_{nk}^{\perp}|^2\right] \mathrm{Re}(r_p) \right\} \tag{4.64}$$

is the resonant potential due to real evanescent photons. For an atom in an isotropic state, the potentials simplify to ($\alpha^{\|} = \alpha^{\perp} = \alpha$, $|\boldsymbol{d}^{\|}|^2 = \frac{2}{3}|\boldsymbol{d}|^2$, $|\boldsymbol{d}^{\perp}|^2 = \frac{1}{3}|\boldsymbol{d}|^2$)

$$U_n^{\text{nres}}(z_A) = \frac{\hbar\mu_0}{8\pi^2} \int_0^{\infty} \mathrm{d}\xi\, \xi^2 \alpha_n(\mathrm{i}\xi) \int_{\xi/c}^{\infty} \mathrm{d}\kappa^{\perp}\, \mathrm{e}^{-2\kappa^{\perp} z_A} \times \left[r_s - \left(2\frac{\kappa^{\perp 2} c^2}{\xi^2} - 1\right) r_p \right], \tag{4.65}$$

$$U_n^{\text{prop}}(z_A) = \frac{\mu_0}{12\pi} \sum_{k<n} \omega_{nk}^2 |\boldsymbol{d}_{nk}|^2 \int_0^{\omega_{nk}/c} \mathrm{d}k^{\perp} \times \left[\mathrm{Im}\left(\mathrm{e}^{2\mathrm{i}k^{\perp} z_A} r_s\right) - \left(2\frac{k^{\perp 2} c^2}{\omega_{nk}^2} - 1\right) \mathrm{Im}\left(\mathrm{e}^{2\mathrm{i}k^{\perp} z_A} r_p\right) \right], \tag{4.66}$$

$$U_n^{\text{evan}}(z_A) = -\frac{\mu_0}{12\pi} \sum_{k<n} \omega_{nk}^2 |\boldsymbol{d}_{nk}|^2 \int_0^{\infty} \mathrm{d}\kappa^{\perp}\, \mathrm{e}^{-2\kappa^{\perp} z_A} \times \left[\mathrm{Re}(r_s) + \left(2\frac{\kappa^{\perp 2} c^2}{\omega_{nk}^2} + 1\right) \mathrm{Re}(r_p) \right]. \tag{4.67}$$

Note that while the separation (4.61) provides a physical interpretation in terms of contributions due to virtual vs real propagating and evanescent photons, these components cannot be observed individually.

4.2.1 Perfectly Conducting Plate

Let us begin with the simplest case of a perfectly conducting plate. Requiring the parallel component of the electric field to vanish on the plate surface, one can show that the reflection coefficients of such a plate are $r_s = -1$ and $r_p = +1$, cf. Sect. 3.3.1 in Vol. I. The non-resonant potential for a perfectly conducting plate thus reads

$$U_n^{\text{nres}}(z_A) = -\frac{\hbar\mu_0}{8\pi^2\varepsilon_0}\int_0^\infty \mathrm{d}\xi \int_{\xi/c}^\infty \mathrm{d}\kappa^\perp\, \mathrm{e}^{-2\kappa^\perp z_A}\left[\left(\kappa^{\perp 2} + \frac{\xi^2}{c^2}\right)\alpha_n^\parallel(\mathrm{i}\xi) + \left(\kappa^{\perp 2} - \frac{\xi^2}{c^2}\right)\alpha_n^\perp(\mathrm{i}\xi)\right] \tag{4.68}$$

and after performing the $\kappa^\perp$-integral, we find

$$U_n^{\text{nres}}(z_A) = -\frac{\hbar}{32\pi^2\varepsilon_0 z_A^3}\int_0^\infty \mathrm{d}\xi\, \mathrm{e}^{-2\xi z_A/c}\left[\alpha_n^\parallel(\mathrm{i}\xi)\left(1 + 2\frac{\xi z_A}{c} + 4\frac{\xi^2 z_A^2}{c^2}\right) + \alpha_n^\perp(\mathrm{i}\xi)\left(1 + 2\frac{\xi z_A}{c}\right)\right]. \tag{4.69}$$

This result may be further simplified in the retarded and nonretarded limits. In the retarded limit $z_A \gg c/\omega_-$ (ω_-: minimum of the relevant atomic transition frequencies) of large atom–plate distances, the main contribution to the ξ-integral comes from a range where the approximation $\boldsymbol{\alpha}_n(\mathrm{i}\xi) \simeq \boldsymbol{\alpha}_n(0) \equiv \boldsymbol{\alpha}_n$ is valid, and an evaluation of the ξ-integral results in

$$U_n^{\text{nres}}(z_A) = -\frac{\hbar c(2\alpha_n^\parallel + \alpha_n^\perp)}{32\pi^2\varepsilon_0 z_A^4}\,. \tag{4.70}$$

In the opposite nonretarded limit $z_A \ll c/\omega_+$ (ω_+: maximum of the relevant atomic transition frequencies), the factors $\alpha_n(\mathrm{i}\xi)$ limit the ξ-integral to an interval where we may approximately set $\mathrm{e}^{-2\xi z_A/c} \simeq 1$ and neglect the second and third terms in the large round brackets. Recalling the definition (4.15) of the atomic polarisability, the ξ-integral can then be performed by means of

$$\int_0^\infty \mathrm{d}\xi\, \alpha_n^\parallel(\mathrm{i}\xi) = \frac{\pi}{2\hbar}\sum_k \mathrm{sgn}(\omega_{kn})|\boldsymbol{d}_{nk}^\parallel|^2\,, \tag{4.71}$$

$$\int_0^\infty \mathrm{d}\xi\, \alpha_n^\perp(\mathrm{i}\xi) = \frac{\pi}{\hbar}\sum_k \mathrm{sgn}(\omega_{kn})|\boldsymbol{d}_{nk}^\perp|^2\,, \tag{4.72}$$

resulting in

$$U_n^{\mathrm{nres}}(z_A) = -\frac{1}{64\pi\varepsilon_0 z_A^3}\sum_k \mathrm{sgn}(\omega_{kn})\left(|\boldsymbol{d}_{nk}^{\parallel}|^2 + 2|\boldsymbol{d}_{nk}^{\perp}|^2\right). \tag{4.73}$$

In general, the non-resonant potential contains attractive contributions due to virtual transitions to higher-energy levels as well as repulsive ones due to downward transitions. For an isotropic ground-state atom, the CP full potential is purely non-resonant, $U_0(z_A) = U_0^{\mathrm{nres}}(z_A)$, and all transitions are downward, $\mathrm{sgn}(\omega_{k0}) = +1$ for all k. Invoking the completeness relation $\sum_k |\boldsymbol{d}_{0k}|^2 = \langle\hat{\boldsymbol{d}}^2\rangle$, we then recover our earlier results (4.103)–(4.105) of Vol. I: The ground-state potential of an atom next to a perfectly conducting plate is purely attractive and proportional to $1/z_A^4$ and $1/z_A^3$ in the retarded and nonretarded limits.

For an excited atom, we need to consider the resonant potentials due to propagating and evanescent waves as well. For a perfectly reflecting plate ($r_s = -1$ and $r_p = +1$) the propagating-wave potential (4.63) reads

$$U_n^{\mathrm{prop}}(z_A) = -\frac{1}{8\pi\varepsilon_0}\sum_{k<n}\int_0^{\omega_{nk}/c} \mathrm{d}k^{\perp}\,\sin(2k^{\perp}z_A)\left[\left(k^{\perp 2} + \frac{\omega_{nk}^2}{c^2}\right)|\boldsymbol{d}_{nk}^{\parallel}|^2 + 2\left(k^{\perp 2} - \frac{\omega_{nk}^2}{c^2}\right)|\boldsymbol{d}_{nk}^{\perp}|^2\right]. \tag{4.74}$$

The $k^{\perp}$-integral can be performed, resulting in

$$U_n^{\mathrm{prop}}(z_A) = -\frac{1}{32\pi\varepsilon_0 z_A^3} \times \sum_{k<n}\{|\boldsymbol{d}_{nk}^{\parallel}|^2[\cos(2x) + 2x\sin(2x) - 4x^2\cos(2x) - 1 + 2x^2] + 2|\boldsymbol{d}_{nk}^{\perp}|^2[\cos(2x) + 2x\sin(2x) - 1 - 2x^2]\}_{x=\omega_{nk}z_A/c}. \tag{4.75}$$

The propagating-wave potential simplifies to

$$U_n^{\mathrm{prop}}(z_A) = \frac{\mu_0}{16\pi z_A}\sum_{k<n}\omega_{nk}^2\left\{|\boldsymbol{d}_{nk}^{\parallel}|^2\left[2\cos\left(\frac{2\omega_{nk}z_A}{c}\right) - 1\right] + 2|\boldsymbol{d}_{nk}^{\perp}|^2\right\} \tag{4.76}$$

in the retarded limit, while becoming negligible in the nonretarded limit,

$$U_n^{\mathrm{prop}}(z_A) = 0\,. \tag{4.77}$$

The evanescent-wave potential (4.64) for a perfectly conducting plate

$$U_n^{\mathrm{evan}}(z_A) = -\frac{1}{8\pi\varepsilon_0}\sum_{k<n}\int_0^\infty \mathrm{d}\kappa^\perp\, \mathrm{e}^{-2\kappa^\perp z_A}\left[\left(\kappa^{\perp 2} - \frac{\omega_{nk}}{c^2}\right)|\boldsymbol{d}_{nk}^{\parallel}|^2 + 2\left(\kappa^{\perp 2} + \frac{\omega_{nk}}{c^2}\right)|\boldsymbol{d}_{nk}^{\perp}|^2\right] \tag{4.78}$$

can be found by carrying out the $\kappa^\perp$-integral,

$$U_n^{\mathrm{evan}}(z_A) = -\frac{1}{32\pi\varepsilon_0 z_A^3}\sum_{k<n}\left[|\boldsymbol{d}_{nk}^{\parallel}|^2\left(1 - 2\frac{\omega_{nk}^2 z_A^2}{c^2}\right) + 2|\boldsymbol{d}_{nk}^{\perp}|^2\left(1 + 2\frac{\omega_{nk}^2 z_A^2}{c^2}\right)\right]. \tag{4.79}$$

In the retarded and nonretarded limits, it simplifies to

$$U_n^{\mathrm{evan}}(z_A) = \frac{\mu_0}{16\pi z_A}\sum_{k<n}\omega_{nk}^2\left(|\boldsymbol{d}_{nk}^{\parallel}|^2 - 2|\boldsymbol{d}_{nk}^{\perp}|^2\right) \tag{4.80}$$

and

$$U_n^{\mathrm{evan}}(z_A) = -\frac{1}{32\pi\varepsilon_0 z_A^3}\sum_{k<n}\left(|\boldsymbol{d}_{nk}^{\parallel}|^2 + 2|\boldsymbol{d}_{nk}^{\perp}|^2\right), \tag{4.81}$$

respectively.

The total CP potential of an excited atom in front of a perfectly conducting plate is given by the sum of the non-resonant (4.69), propagating-wave (4.75) and evanescent-wave (4.79) potentials. Note that the result has originally been obtained and later been re-examined on the basis of two alternative approaches: normal-mode QED in free space [6, 7, 11, 12] and linear-response theory [5, 13]. The CP potential takes particularly simple forms in the retarded and nonretarded limits. In the retarded limit, the non-resonant potential (4.70) becomes negligible in comparison with the other two components. The evanescent-wave potential (4.79) cancels the non-oscillating component of the propagating-wave potential (4.76) to yield

$$U_n(z_A) = \frac{\mu_0}{8\pi z_A}\sum_{k<n}\omega_{nk}^2|\boldsymbol{d}_{nk}^{\parallel}|^2\cos\left(\frac{2\omega_{nk}z_A}{c}\right) - \frac{1}{32\pi\varepsilon_0 z_A^3}\sum_{k<n}\left(|\boldsymbol{d}_{nk}^{\parallel}|^2 + 2|\boldsymbol{d}_{nk}^{\perp}|^2\right). \tag{4.82}$$

The retarded CP potential is hence dominated by spatially oscillating contributions for each downward transition with periods $\lambda_{nk}/2$, $\lambda_{nk} = 2\pi c/\omega_{nk}$ denoting the wavelength of the emitted real photons. This can be easily understood from an interference effect. The main contribution to the interaction in the retarded limit is due to normal-incident waves ($\boldsymbol{k}^{\parallel} = 0$). Such waves are emitted by the atom and they travel a distance $2z_A$ to the plate and back. To understand the sign of the potential, we note that normal-incident electric waves acquire a minus sign upon reflection from the surface of a perfectly reflecting plate. The reflected field is hence anti-parallel to the emitting dipole moment whenever the path length is equal to an integer multiple to the wavelength, $2z_A = m\lambda_{nk}$ with $m \in \mathbb{N}$. According to our classical interpretation (4.23), the interaction energy is equal to minus the product of dipole moment and reflected field, so that $U_n \propto +\cos(2\omega_{nk}z_A/c)$, in agreement with the above result. Note that dipole moments perpendicular to the plate surface do not contribute to the oscillating potential, as they cannot emit transverse electric waves in the direction normal to the plate. For an isotropic atom, we find the rotationally averaged result ($|\boldsymbol{d}^{\parallel}|^2 = 2|\boldsymbol{d}^{\perp}_{nk}|^2 = \frac{2}{3}|\boldsymbol{d}|^2$)

$$U_n(z_A) = \frac{\mu_0}{12\pi z_A}\sum_{k<n}\omega_{nk}^2|\boldsymbol{d}_{nk}|^2\cos\left(\frac{2\omega_{nk}z_A}{c}\right) - \frac{1}{24\pi\varepsilon_0 z_A^3}\sum_{k<n}|\boldsymbol{d}_{nk}|^2\,. \tag{4.83}$$

For large atomic transition frequencies, the period of the spatial oscillations may become very small; this is true in particular for atoms in low-lying excited states. When the oscillations cannot be resolved experimentally, then the first term averages to zero and only the attractive $1/z_A^3$ potential due to evanescent waves remains.

In the nonretarded limit, the propagating-waves contribution (4.77) to the total potential vanishes. The non-resonant potential (4.73) with its attractive and repulsive components combines with the attractive evanescent-wave potential (4.73) to yield the purely attractive total potential

$$U_n(z_A) = -\frac{1}{64\pi\varepsilon_0 z_A^3}\sum_k\left(|\boldsymbol{d}^{\parallel}_{nk}|^2 + 2|\boldsymbol{d}^{\perp}_{nk}|^2\right)\,. \tag{4.84}$$

Invoking the completeness relation $\sum_k \boldsymbol{d}_{nk}\boldsymbol{d}_{kn} = \langle\hat{\boldsymbol{d}}\hat{\boldsymbol{d}}\rangle_n$, we can represent it in the compact form

$$U_n(z_A) = -\frac{\langle\hat{\boldsymbol{d}}^{\parallel 2} + 2\hat{\boldsymbol{d}}^{\perp 2}\rangle_n}{64\pi\varepsilon_0 z_A^3}\,, \tag{4.85}$$

which reduces to

$$U_n(z_A) = -\frac{\langle\hat{\boldsymbol{d}}^2\rangle_n}{48\pi\varepsilon_0 z_A^3} \tag{4.86}$$

for an isotropic atom. These results are a special case of the general nonretarded potential for perfectly conducting bodies as obtained in Sect. 4.1. They immediately follow from (4.34) or (4.35) by using the electrostatic Green's tensor of the plate, which according to (4.29) and (A.48) is given by

$$\boldsymbol{G}^{(1)}(\boldsymbol{r},\boldsymbol{r}) = \frac{1}{32\pi z^3}\begin{pmatrix}1&0&0\\0&1&0\\0&0&2\end{pmatrix}. \tag{4.87}$$

Just like the ground-state potential, the nonretarded potential of an excited atom in front of a perfectly conducting plate is attractive and proportional to $1/z_A^3$. Its strength is governed by the state-dependent average of atom's electric dipole moment squared. The nonretarded potential can be easily understood using the image-dipole model: An electric dipole moment $\hat{\boldsymbol{d}} = (\hat{d}_x, \hat{d}_y, \hat{d}_z)$ situated at a distance z_A from a perfectly conducting plate produces an image $\hat{\boldsymbol{d}}^\star = (-\hat{d}_x, -\hat{d}_y, \hat{d}_z)$ at position $-z_A$ behind the plate. The CP potential (4.85) is simply the average interaction energy of the dipole and its image [14]

$$U_n(z_A) = \frac{1}{2}\frac{\langle\hat{\boldsymbol{d}}\cdot\hat{\boldsymbol{d}}^\star - 3\hat{d}_z\hat{d}_z^\star\rangle_n}{4\pi\varepsilon_0(2z_A)^3} = -\frac{\langle\hat{\boldsymbol{d}}^{\parallel 2} + 2\hat{\boldsymbol{d}}^{\perp 2}\rangle_n}{64\pi\varepsilon_0 z_A^3}. \tag{4.88}$$

4.2.2 Half Space

We next consider a semi-infinite half space with finite magnetoelectric properties. Using the reflection coefficients (A.41) and (A.42) from App. A.3.2, the non-resonant CP potential (4.62) takes the explicit form

$$U_n^{\mathrm{nres}}(z_A) = \frac{\hbar\mu_0}{8\pi^2}\int_0^\infty \mathrm{d}\xi\,\xi^2 \int_{\xi/c}^\infty \mathrm{d}\kappa^\perp\,\mathrm{e}^{-2\kappa^\perp z_A}\left\{\alpha_n^\parallel(\mathrm{i}\xi)\frac{\mu(\mathrm{i}\xi)\kappa^\perp - \kappa_1^\perp}{\mu(\mathrm{i}\xi)\kappa^\perp + \kappa_1^\perp}\right.$$
$$\left. - \left[\frac{\kappa^{\perp 2}c^2}{\xi^2}\alpha_n^\parallel(\mathrm{i}\xi) + \left(\frac{\kappa^{\perp 2}c^2}{\xi^2} - 1\right)\alpha_n^\perp(\mathrm{i}\xi)\right]\frac{\varepsilon(\mathrm{i}\xi)\kappa^\perp - \kappa_1^\perp}{\varepsilon(\mathrm{i}\xi)\kappa^\perp + \kappa_1^\perp}\right\} \tag{4.89}$$

with

$$\kappa_1^\perp = \sqrt{\kappa^{\perp 2} + [\varepsilon(\mathrm{i}\xi)\mu(\mathrm{i}\xi) - 1]\frac{\xi^2}{c^2}}\,. \tag{4.90}$$

The retarded limit $z_A \gg c/\omega_-$ (ω_-: minimum of all relevant atomic and medium resonance frequencies) can be treated by introducing the variable $v = \kappa^\perp c/\xi$ to transform the integral according to $\int_0^\infty \mathrm{d}\xi \int_{\xi/c}^\infty \mathrm{d}\kappa^\perp = \int_1^\infty \mathrm{d}v \int_0^\infty \mathrm{d}\xi\,\xi/c$ where now $\kappa_1^\perp = (\xi/c)\sqrt{\varepsilon\mu - 1 + v^2}$. The exponential effectively limits the ξ-integral to a range where $\boldsymbol{\alpha}_n(\mathrm{i}\xi) \simeq \boldsymbol{\alpha}_n(0) \equiv \boldsymbol{\alpha}_n$, $\varepsilon(\mathrm{i}\xi) \simeq \varepsilon(0) \equiv \varepsilon$, $\mu(\mathrm{i}\xi) \simeq \mu(0) \equiv \mu$, With these approximations, the ξ-integral can be performed to yield the retarded potential

$$U_n^{\mathrm{nres}}(z_A) = -\frac{3\hbar c}{64\pi^2\varepsilon_0 z_A^4}\int_1^\infty \mathrm{d}v\left[\left(\frac{\alpha_n^{\parallel}+\alpha_n^{\perp}}{v^2}-\frac{\alpha_n^{\perp}}{v^4}\right)\frac{\varepsilon v-\sqrt{\varepsilon\mu-1+v^2}}{\varepsilon v+\sqrt{\varepsilon\mu-1+v^2}}\right.$$
$$\left.-\frac{\alpha_n^{\parallel}}{v^4}\frac{\mu v-\sqrt{\varepsilon\mu-1+v^2}}{\mu v+\sqrt{\varepsilon\mu-1+v^2}}\right]. \tag{4.91}$$

In the opposite, nonretarded limit $\sqrt{\varepsilon\mu}\,z_A \ll c/\omega_+$ (ω_+: maximum of all relevant atomic and medium resonance frequencies), the factors $\boldsymbol{\alpha}_n(\mathrm{i}\xi)$, $\varepsilon(\mathrm{i}\xi)\kappa^{\perp}-\kappa_1^{\perp}$ and $\mu(\mathrm{i}\xi)\kappa^{\perp}-\kappa_1^{\perp}$ limit the ξ-integral to a range where we can apply a Taylor expansion in $\xi\sqrt{\varepsilon(\mathrm{i}\xi)\mu(\mathrm{i}\xi)-1}/(c\kappa^{\perp})$. Carrying out the $\kappa^{\perp}$-integral while retaining terms quadratic in ξ, one finds

$$U_n^{\mathrm{nres}}(z_A) = -\frac{\hbar}{32\pi^2\varepsilon_0 z_A^3}\int_0^\infty \mathrm{d}\xi\left[\alpha_n^{\parallel}(\mathrm{i}\xi)+\alpha_n^{\perp}(\mathrm{i}\xi)\right]\frac{\varepsilon(\mathrm{i}\xi)-1}{\varepsilon(\mathrm{i}\xi)+1}$$
$$+\frac{\hbar\mu_0}{16\pi^2 z_A}\int_0^\infty \mathrm{d}\xi\,\xi^2\left\{\alpha_n^{\parallel}(\mathrm{i}\xi)\frac{\mu(\mathrm{i}\xi)-1}{\mu(\mathrm{i}\xi)+1}+\alpha_n^{\perp}(\mathrm{i}\xi)\frac{\varepsilon(\mathrm{i}\xi)-1}{\varepsilon(\mathrm{i}\xi)+1}\right.$$
$$\left.+\left[\alpha_n^{\parallel}(\mathrm{i}\xi)+\alpha_n^{\perp}(\mathrm{i}\xi)\right]\frac{\varepsilon(\mathrm{i}\xi)[\varepsilon(\mathrm{i}\xi)\mu(\mathrm{i}\xi)-1]}{[\varepsilon(\mathrm{i}\xi)+1]^2}\right\}. \tag{4.92}$$

The permittivity of a metal can be described by the Drude model

$$\varepsilon(\omega) = 1-\frac{\omega_{\mathrm{P}}^2}{\omega(\omega+\mathrm{i}\gamma)} \tag{4.93}$$

(ω_{P}: plasma frequency, γ: damping constant). It becomes infinite in the limit of small frequencies $|\omega|\to 0$. The perfect-conductor model is hence an excellent approximation for the small-frequency response of a metal and hence the large-distance behaviour of the non-resonant potential: For larger distances, the non-resonant CP potential increasingly dominated by low-frequency contributions. In the retarded limit, the v-integral can be carried out for a metal with $\varepsilon(0)=\infty$ and we recover the perfect-conductor result (4.70). The dissipation-less currents inside a superconductor at very small temperatures is more accurately described by a plasma model [15]

$$\varepsilon(\omega) = 1-\frac{\omega_{\mathrm{P}}^2}{\omega^2}. \tag{4.94}$$

Again, we have $\varepsilon(0)=\infty$ and the perfect-conductor result holds. Due to the strongly reduced absorption, this agreement with the ideal case is valid over a even larger range of distances [16].

For an isotropic ground-state atom, the CP potential is purely non-resonant, $U_0(z_A) = U_0^{\mathrm{nres}}(z_A)$. Using $\alpha^{\parallel} = \alpha^{\perp} = \alpha$, we recover our earlier results (4.133), (4.134), (4.137)–(4.139) as given in Vol. I. With the ground-state polarisability being strictly positive, the sign of the potential purely depends on the relative strengths of the electric vs magnetic properties of the half space. The retarded potential is proportional to $1/z_A^4$ and attractive or repulsive for dominantly electric or magnetic half spaces, respectively. The nonretarded potential is attractive with a $1/z_A^3$ asymptote for a dominantly electric half space and repulsive with an $1/z_A$ power law for a purely magnetic one. For details, see the discussion in Sect. 4.6.2 of Vol. I.

For an excited atom, the behaviour of the non-resonant CP potential is more complex, because upward and downward transitions contribute to the polarisability with different signs. In addition, we need to include the resonant contributions to the potential. With the reflection coefficients (A.41) and (A.42) of the half space, the resonant potential (4.63) due to propagating waves reads

$$U_n^{\mathrm{prop}}(z_A) = \frac{\mu_0}{8\pi} \sum_{k<n} \omega_{nk}^2 \int_0^{\omega_{nk}/c} \mathrm{d}k^{\perp} \times \left\{ |\boldsymbol{d}_{nk}^{\parallel}|^2 \mathrm{Im}\left[\mathrm{e}^{2\mathrm{i}k^{\perp}z_A} \frac{\mu(\omega_{nk})k^{\perp} - k_1^{\perp}}{\mu(\omega_{nk})k^{\perp} + k_1^{\perp}} \right] - \left[\frac{k^{\perp 2}c^2}{\omega_{nk}^2} |\boldsymbol{d}_{nk}^{\parallel}|^2 + 2\left(\frac{k^{\perp 2}c^2}{\omega_{nk}^2} - 1 \right) |\boldsymbol{d}_{nk}^{\perp}|^2 \right] \mathrm{Im}\left[\mathrm{e}^{2\mathrm{i}k^{\perp}z_A} \frac{\varepsilon(\omega_{nk})k^{\perp} - k_1^{\perp}}{\varepsilon(\omega_{nk})k^{\perp} + k_1^{\perp}} \right] \right\} \tag{4.95}$$

with

$$k_1^{\perp} = \sqrt{[\varepsilon(\omega_{nk})\mu(\omega_{nk}) - 1] \frac{\omega_{nk}^2}{c^2} + k^{\perp 2}}\,, \qquad \mathrm{Im}\, k_1^{\perp} > 0\,. \tag{4.96}$$

In the retarded limit $z_A \gg c/\omega_{nk}$, the main contribution to the integral with its oscillating integrand comes from the stationary-phase point $k^{\parallel} = 0$ where $\mathrm{d}k^{\perp}/\mathrm{d}k^{\parallel} = 0$, cf. (4.49). We may hence set $k^{\perp} \simeq \omega_{nk}/c$ and $k_1^{\perp} \simeq \sqrt{\varepsilon(\omega_{nk})\mu(\omega_{nk})}\omega_{nk}/c$ in the reflection coefficients. The $k^{\perp}$-integral can then be solved. Retaining only the leading order in $c/(\omega_{nk}z_A)$, we find

$$U_n^{\mathrm{prop}}(z_A) = \frac{\mu_0}{16\pi z_A} \sum_{k<n} \omega_{nk}^2 \left\{ 2|\boldsymbol{d}_{nk}^{\perp}|^2 \mathrm{Re}\left[\frac{\sqrt{\varepsilon(\omega_{nk})} - \sqrt{\mu(\omega_{nk})}}{\sqrt{\varepsilon(\omega_{nk})} + \sqrt{\mu(\omega_{nk})}} \right] + |\boldsymbol{d}_{nk}^{\parallel}|^2 \mathrm{Re}\left[\left(2\mathrm{e}^{2\mathrm{i}\omega_{nk}z_A/c} - 1 \right) \frac{\sqrt{\varepsilon(\omega_{nk})} - \sqrt{\mu(\omega_{nk})}}{\sqrt{\varepsilon(\omega_{nk})} + \sqrt{\mu(\omega_{nk})}} \right] \right\} \tag{4.97}$$

where the roots have to be taken such that $\mathrm{Im}\sqrt{\varepsilon}$, $\mathrm{Im}\sqrt{\mu} > 0$ for absorbing media. In the nonretarded limit $z_A \ll c/\omega_{nk}$, we approximate $\mathrm{e}^{2\mathrm{i}k^{\perp}z_A} \simeq 1$ and let $k_1^{\perp} \simeq k^{\perp}$

in the reflection coefficients. Performing the $k^{\perp}$-integral and noting that

$$\mathrm{Im}\left(\frac{\varepsilon-1}{\varepsilon+1}\right)=\frac{2\mathrm{Im}\,\varepsilon}{|\varepsilon+1|^2}\,,\qquad \mathrm{Im}\left(\frac{\mu-1}{\mu+1}\right)=\frac{2\mathrm{Im}\,\mu}{|\mu+1|^2}\,,\tag{4.98}$$

we obtain

$$U_n^{\mathrm{prop}}(z_A)=\frac{\mu_0}{4\pi c}\sum_{k<n}\omega_{nk}^3\left[|\boldsymbol{d}_{nk}^{\parallel}|^2\,\frac{\mathrm{Im}\,\mu(\omega_{nk})}{|\mu(\omega_{nk})+1|^2}\right.$$
$$\left.+\left(\tfrac{4}{3}|\boldsymbol{d}_{nk}^{\perp}|^2-\tfrac{1}{3}|\boldsymbol{d}_{nk}^{\parallel}|^2\right)\frac{\mathrm{Im}\,\varepsilon(\omega_{nk})}{|\varepsilon(\omega_{nk})+1|^2}\right].\tag{4.99}$$

In contrast to the perfect-conductor result (4.77), the propagating-wave potential of a magnetoelectric half space hence takes a non-vanishing value on the surface.

The evanescent-wave potential (4.64) takes the form

$$U_n^{\mathrm{evan}}(z_A)=-\frac{\mu_0}{8\pi}\sum_{k<n}\omega_{nk}^2\int_0^\infty \mathrm{d}\kappa^{\perp}\,\mathrm{e}^{-2\kappa^{\perp}z_A}\left\{|\boldsymbol{d}_{nk}^{\parallel}|^2\mathrm{Re}\left[\frac{\mu(\omega_{nk})\kappa^{\perp}-\kappa_1^{\perp}}{\mu(\omega_{nk})\kappa^{\perp}+\kappa_1^{\perp}}\right]\right.$$
$$\left.+\left[\frac{\kappa^{\perp 2}c^2}{\omega_{nk}^2}|\boldsymbol{d}_{nk}^{\parallel}|^2+2\left(\frac{\kappa^{\perp 2}c^2}{\omega_{nk}^2}+1\right)|\boldsymbol{d}_{nk}^{\perp}|^2\right]\mathrm{Re}\left[\frac{\varepsilon(\omega_{nk})\kappa^{\perp}-\kappa_1^{\perp}}{\varepsilon(\omega_{nk})\kappa^{\perp}+\kappa_1^{\perp}}\right]\right\}\tag{4.100}$$

for the magnetoelectric half space with

$$\kappa_1^{\perp}=\sqrt{\kappa^{\perp 2}-[\varepsilon(\omega_{nk})\mu(\omega_{nk})-1]\,\frac{\omega_{nk}^2}{c^2}}\,,\qquad \mathrm{Re}\,\kappa_1^{\perp}>0\,.\tag{4.101}$$

In the retarded limit, we put $\kappa^{\perp}\simeq\omega_{nk}/c$ and $\kappa_1^{\perp}\simeq\sqrt{\varepsilon(\omega_{nk})\mu(\omega_{nk})}\omega_{nk}/c$ in the reflection coefficients. Solving the integral and retaining only the leading orders in $c/(\omega_{nk}z_A)$, we obtain

$$U_n^{\mathrm{evan}}(z_A)=\frac{\mu_0}{16\pi z_A}\sum_{k<n}\omega_{nk}^2\left(|\boldsymbol{d}_{nk}^{\parallel}|^2-2|\boldsymbol{d}_{nk}^{\perp}|^2\right)\mathrm{Re}\left[\frac{\sqrt{\varepsilon(\omega_{nk})}-\sqrt{\mu(\omega_{nk})}}{\sqrt{\varepsilon(\omega_{nk})}+\sqrt{\mu(\omega_{nk})}}\right].\tag{4.102}$$

In the nonretarded limit, we set $\kappa_1^{\perp}\simeq\kappa^{\perp}$ in the reflection coefficients and integrate to find

$$U_n^{\mathrm{evan}}(z_A) = -\frac{1}{32\pi\varepsilon_0 z_A^3}\sum_{k<n}\left(|\boldsymbol{d}_{nk}^{\parallel}|^2 + 2|\boldsymbol{d}_{nk}^{\perp}|^2\right)\frac{|\varepsilon(\omega_{nk})|^2-1}{|\varepsilon(\omega_{nk})+1|^2}$$
$$-\frac{\mu_0}{16\pi z_A}\sum_{k<n}\omega_{nk}^2\left[|\boldsymbol{d}_{nk}^{\parallel}|^2\,,\frac{|\mu(\omega_{nk})|^2-1}{|\mu(\omega_{nk})+1|^2} + 2|\boldsymbol{d}_{nk}^{\perp}|^2\frac{|\varepsilon(\omega_{nk})|^2-1}{|\varepsilon(\omega_{nk})+1|^2}\right], \tag{4.103}$$

note that

$$\mathrm{Re}\left(\frac{\varepsilon-1}{\varepsilon+1}\right) = \frac{|\varepsilon|^2-1}{|\varepsilon+1|^2}\,,\qquad \mathrm{Re}\left(\frac{\mu-1}{\mu+1}\right) = \frac{|\mu|^2-1}{|\mu+1|^2}\,. \tag{4.104}$$

The total CP potential of an excited atom in front of a magnetoelectric half space is the sum of the non-resonant, propagating-wave and evanescent-wave components as found above. Alternative derivations of our result include normal-mode QED [11, 17, 18] linear-response theory [5] and a microscopic, dilute-gas model [19, 20]. In the retarded limit, the non-resonant contribution (4.91) is negligible. To leading order in $1/z_A$, the evanescent-wave potential (4.100) cancels the non-oscillating parts of the propagating-wave potential (4.97) to yield a oscillating potential [21]

$$U_n(z_A) = \frac{\mu_0}{8\pi z_A}\sum_{k<n}\omega_{nk}^2|\boldsymbol{d}_{nk}^{\parallel}|^2\mathrm{Re}\left[\mathrm{e}^{2\mathrm{i}\omega_{nk}z_A/c}\frac{\sqrt{\varepsilon(\omega_{nk})}-\sqrt{\mu(\omega_{nk})}}{\sqrt{\varepsilon(\omega_{nk})}+\sqrt{\mu(\omega_{nk})}}\right]. \tag{4.105}$$

For a metal (4.93) or a superconductor (4.94), we have $|\varepsilon(\omega_{nk})| \gg 1$, and the total retarded potential is well approximated by the perfect-conductor result (4.82). For a magnetodielectric half space, one finds a spatially oscillating potential whose amplitude and phase are determined by the permittivity and permeability. Note that we have discarded the non-oscillating $1/z_A^3$ contribution to the retarded potential.

In the nonretarded limit, the finite contribution (4.99) from propagating waves can be ignored in comparison to the non-resonant and evanescent-wave potentials (4.92) and (4.103) which are governed by inverse power laws. It is instructive to distinguish between electric and magnetic half spaces. For a dominantly electric half space, the nonretarded CP potential reads [21–23]

$$U_n(z_A) = -\frac{\hbar}{32\pi^2\varepsilon_0 z_A^3}\int_0^\infty \mathrm{d}\xi\left[\alpha_n^{\parallel}(\mathrm{i}\xi)+\alpha_n^{\perp}(\mathrm{i}\xi)\right]\frac{\varepsilon(\mathrm{i}\xi)-1}{\varepsilon(\mathrm{i}\xi)+1}$$
$$-\frac{1}{32\pi\varepsilon_0 z_A^3}\sum_{k<n}\left(|\boldsymbol{d}_{nk}^{\parallel}|^2+2|\boldsymbol{d}_{nk}^{\perp}|^2\right)\frac{|\varepsilon(\omega_{nk})|^2-1}{|\varepsilon(\omega_{nk})+1|^2}\,. \tag{4.106}$$

It consists of various contributions which all follow $1/z_A^3$ power laws. For the case of a superconductor, the plasma model (4.94) together with the atomic polarisability (4.15) leads to

$$U_n(z_A) = -\frac{1}{64\pi\varepsilon_0 z_A^3}\sum_k \operatorname{sgn}(\omega_{kn})\left(|\boldsymbol{d}_{nk}^{\parallel}|^2 + 2|\boldsymbol{d}_{nk}^{\perp}|^2\right)\frac{\omega_S}{\omega_S + |\omega_{kn}|}$$
$$-\frac{1}{32\pi\varepsilon_0 z_A^3}\sum_{k<n}\left(|\boldsymbol{d}_{nk}^{\parallel}|^2 + 2|\boldsymbol{d}_{nk}^{\perp}|^2\right)\frac{\omega_S^2}{\omega_S^2 + \omega_{kn}^2}\,. \tag{4.107}$$

The non-resonant potential (first term) contains attractive contributions from upward transitions as well as well as repulsive ones from downward transitions. On the contrary, the contributions to the evanescent-wave potential (second term) are purely attractive. The total potential of the superconductor is attractive and well approximated by the perfect-conductor result (4.85) provided that $\omega_S \gg |\omega_{kn}|$. Similar results can be found for an ordinary metal (4.93), where the potential is slightly reduced due to material absorption.

For a dielectric, the attractive and repulsive contributions to the non-resonant potential are smaller than the respective perfect-conductor results (4.73). However, the contributions to the evanescent-wave potential (second term) can exceed their perfect-conductor counterparts (4.81) in magnitude. In particular, the evanescent-wave contribution associated with an atomic transition is strongly enhanced near the surface-plasmon resonance $\varepsilon(\omega_{nk}) \simeq -1$ where the denominator $|\varepsilon(\omega_{nk}) + 1|^2$ becomes very small. Furthermore, it may take different signs for a dielectric half space. Being attractive for $|\varepsilon(\omega_{nk})| > 1$, repulsive potentials can be realised when $|\varepsilon(\omega_{nk})| < 1$. Note that the resonant evanescent potential dominates over the non-resonant one near the surface-plasmon resonance, hence determining the sign of the total potential. For an example, see Fig. 7.4 in Sect. 7.2.2.

For a purely magnetic half space, the total CP potential reads [21]

$$U_n(z_A) = \frac{\hbar\mu_0}{16\pi^2 z_A}\int_0^\infty \mathrm{d}\xi\,\xi^2\left\{\alpha_n^{\parallel}(\mathrm{i}\xi)\,\frac{\mu(\mathrm{i}\xi)-1}{\mu(\mathrm{i}\xi)+1}\right.$$
$$\left.+\left[\alpha_n^{\parallel}(\mathrm{i}\xi)+\alpha_n^{\perp}(\mathrm{i}\xi)\right]\frac{\mu(\mathrm{i}\xi)-1}{4}\right\}$$
$$-\frac{\mu_0}{16\pi z_A}\sum_{k<n}\omega_{nk}^2|\boldsymbol{d}_{nk}^{\parallel}|^2\,\frac{|\mu(\omega_{nk})|^2-1}{|\mu(\omega_{nk})+1|^2}\,. \tag{4.108}$$

It is similar in structure to the result for an electric half space, but exhibits a weaker, $1/z_A$ power law. In addition, the signs of the non-resonant contributions are reversed, now being repulsive for upward transitions and attractive for downward ones. In close analogy to the electric case, the evanescent-wave contributions are attractive if $|\mu(\omega_{nk})| > 1$ and repulsive if $|\mu(\omega_{nk})| < 1$.

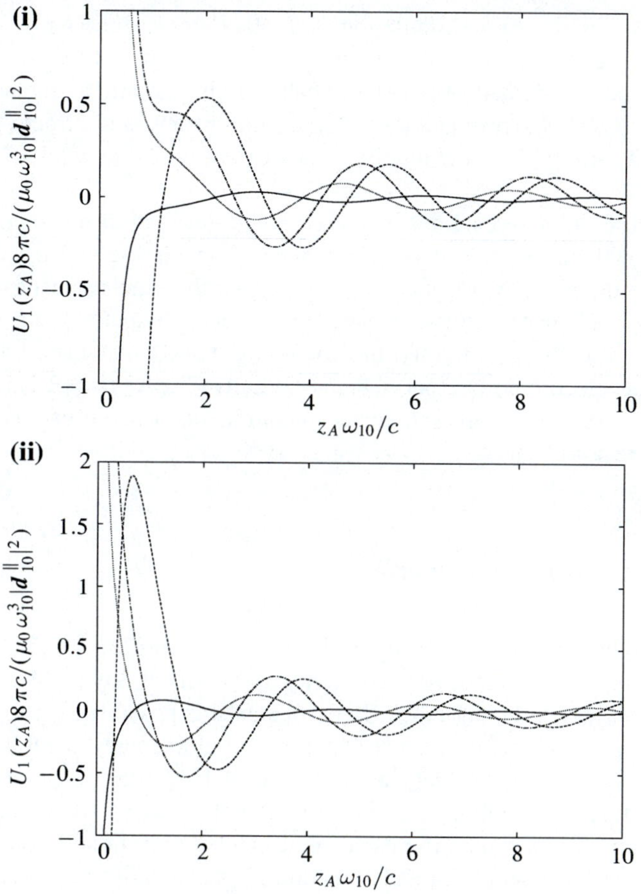

Fig. 4.2 CP potential of an excited two-level atom with parallel dipole moment in front of (**i**) electric or (**ii**) magnetic half spaces with different strengths of the electric/magnetic properties. The chosen parameters for ε or μ are 1.5 (*solid lines*), -1.5 (*dashed lines*), 0.2 (*dotted lines*) and -0.2 (*dash-dotted lines*). All half spaces are weakly absorbing, $\mathrm{Im}\,\varepsilon$, $\mathrm{Im}\,\mu = 10^{-3}$

The transition between the oscillating retarded potential and the strictly monotonous nonretarded regime is illustrated in Fig. 4.2(i) where we display the CP potential of an excited two-level atom in front of semi-infinite half spaces consisting of four different weakly absorbing dielectrics with permittivities $\varepsilon(\omega_{10}) \equiv \varepsilon$. Neglecting the non-resonant contribution, the potential (4.61) has been obtained by numerically integrating the resonant potentials (4.95) and (4.100) due to propagating and evanescent waves.

In agreement with our analytical result (4.105), the potential oscillates at large distances with a period $\lambda_{10}/2 = \pi c/\omega_{10}$. For $\mathrm{Re}\,\varepsilon = 1.5$ or 0.2, the fraction in (4.105) is almost purely real and consequently the potential is proportional to $\pm\cos(2\omega_{10}z_A/c)/z_A$. As the numerator is smaller than the denominator in both cases, the amplitudes of the oscillations are rather small. For $\mathrm{Re}\,\varepsilon = -1.5$ or -0.2, the fraction is genuinely complex, resulting in phase shifts of the oscillations. With numerator and denominator being more balanced in magnitude, the oscillations are more pronounced. The resonant potential at short distances is attractive for the cases $\mathrm{Re}\,\varepsilon = \pm 1.5$ with $|\varepsilon| > 1$ and repulsive for $\mathrm{Re}\,\varepsilon = \pm 0.2$ due to $|\varepsilon| < 1$, as expected from (4.106).

The corresponding potentials of an excited two-level atom above purely magnetic half spaces with permeabilities $\mu(\omega_{10}) \equiv \mu$ are shown in Fig. 4.2(ii). The oscillating retarded potentials for the magnetic half spaces are the exact opposites of the respective electric half-space results, as may have been anticipated from our analytical findings (4.105). On the contrary, the potentials at short distances exhibit the same signs as their electric counterparts, being attractive for $\mathrm{Re}\,\mu = \pm 1.5$ and repulsive for $\mathrm{Re}\,\mu = \pm 0.2$. They are smaller in magnitude as a result of the weaker, $1/z_A$ power law, recall (4.106).

4.2.3 Meta-Material Superlens

Meta-materials are recently proposed [24] and fabricated [25–30] micro- or nanostructures whose effective permittivity and permeability can be efficiently tailored over a wide parameter range. In particular, it is possible to realize left-handed media whose permittivity and permeability simultaneously exhibit negative real parts. As pointed out by Veselago as early as 1968 [31], Maxwell's equations dictate that the vectors $\boldsymbol{E}$, $\boldsymbol{B}$ and $\boldsymbol{k}$ of an electromagnetic wave in such a medium form a left-handed triad rather than the usual right-handed one. In addition, the refractive index $n = \sqrt{\varepsilon\mu}$ inside a left-handed medium has a negative real part, as can easily be seen: In an absorbing medium, both permittivity and permeability must have a positive imaginary part. For a left-handed medium, they furthermore have a negative real part and are situated in the second quadrant of the complex plane. The two possible choices for the square root defining the refractive index hence lie in the second and fourth quadrants. With the physical requirement that the refractive index must have a positive imaginary part inside an absorbing medium, n must be situated in the second quadrant, hence exhibiting a negative real part. This leads to the phenomenon of negative refraction. According to Snell's law [14]

$$\sin\theta_{\mathrm{in}} = \mathrm{Re}(n)\sin\theta_{\mathrm{trans}}\,, \tag{4.109}$$

a light beam incident on the surface of an ordinary right-handed medium of refractive index n is refracted towards the axis of incidence, as shown in Fig. 4.3(i). For a left-handed material with $\mathrm{Re}(n) < 0$, Snell's law requires the angle of the transmitted beam to be negative, so that the beam is refracted across the axis of incidence.

Pendry suggested to exploit the effect negative refraction for the construction of a planar superlens [32]. A plate of thickness d consisting of an idealised left-handed

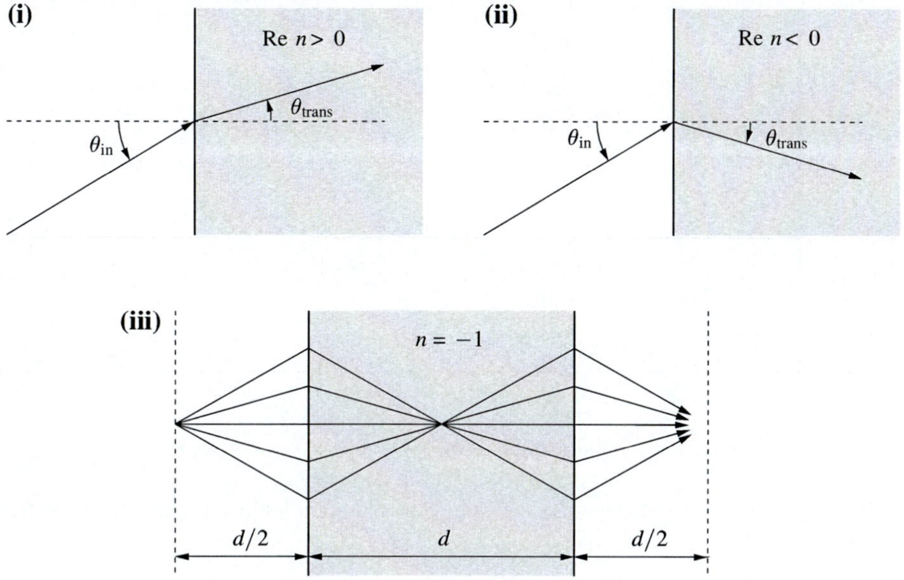

Fig. 4.3 (**i**) Positive versus (**ii**) negative refraction of light incident on right- and left-handed media. (**iii**) Superlens consisting of a left-handed medium

meta-material with refractive index $n = -1$ will exhibit two focal planes at distance $d/2$ on either side. As illustrated in Fig. 4.3(iii), light emitted from an arbitrary point on one of the focal planes will be perfectly focussed into a corresponding point on the opposite focal plane. Such an ideal superlens would not only focus propagating waves, but also transmit evanescent waves without any loss of amplitude. The field at the image point would hence be a complete and faithful reconstruction of the field at the object point. This is possible due to negative refraction. In a manner of speaking, optical path lengths inside a left-handed medium are negative, so that the lens is able to undo the exponential damping which the evanescent field experiences in vacuum and restore it at the image point.

Two atoms situated at corresponding focal points on either side of a superlens will experience perfect coupling as if they were situated in the same place [33, 34]. To mimic this effect with a single atom, we replace the second atom by a perfect mirror, considering the scenario depicted in Fig. 4.4: An excited two-level atom is placed at a distance z_A from a left-handed plate of thickness d with permittivity $\varepsilon(\omega_{10}) \equiv \varepsilon = -1$ and permeability $\mu(\omega_{10}) \equiv \mu = -1$ with a perfectly conducting mirror placed at the far end.

The effect of the superlens plus perfect mirror on the atom can be anticipated by an image construction: As shown in Fig. 4.4, an atom at position $\overline{z}_A = z_A - d$ relative to the focal plane gives rise to an image at $z^\star_A = d - \overline{z}_A$. The atom and its image are situated at opposite sides of the focal plane and are separated by a distance $2|\overline{z}_A|$. The image created by the superlens plus perfectly conducting mirror is exactly the same

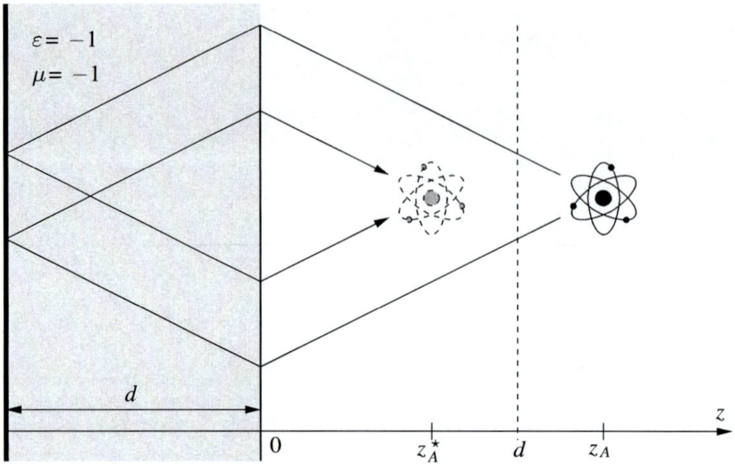

Fig. 4.4 Excited atom in front of a superlens consisting of a left-handed plate and a perfectly conducting mirror

as that which would be created by a perfectly conducting mirror placed at the focal plane. We can hence expect a strong attraction of the atom towards the focal plane.

For a more quantitative analysis, let us calculate the two dominant, resonant contributions to the CP potential of the excited atom. The propagating-wave potential is given by (4.63) where according to (A.39) and (A.40) in App. A.3.2, the reflection coefficients of a magnetoelectric plate plus perfect mirror read

$$r_s = \frac{(\mu k^\perp - k_1^\perp) - (\mu k^\perp + k_1^\perp)\,\mathrm{e}^{2\mathrm{i}k_1^\perp d}}{(\mu k^\perp + k_1^\perp) - (\mu k^\perp - k_1^\perp)\,\mathrm{e}^{2\mathrm{i}k_1^\perp d}}, \tag{4.110}$$

$$r_p = \frac{(\varepsilon k^\perp - k_1^\perp) + (\varepsilon k^\perp + k_1^\perp)\,\mathrm{e}^{2\mathrm{i}k_1^\perp d}}{(\varepsilon k^\perp + k_1^\perp) + (\varepsilon k^\perp - k_1^\perp)\,\mathrm{e}^{2\mathrm{i}k_1^\perp d}} \tag{4.111}$$

with $k_1^\perp$ defined as in (4.96). For an ideal superlens with $\varepsilon = -1$ and $\mu = -1$, we have $k_1^\perp = -k^\perp$ and the reflection coefficients simplify to

$$r_s = -\mathrm{e}^{-2\mathrm{i}k^\perp d}, \qquad r_p = \mathrm{e}^{-2\mathrm{i}k^\perp d}\,. \tag{4.112}$$

Substituting them into the propagating-wave potential, we find

$$U_1^{\text{prop}}(z_A) = -\frac{1}{8\pi\varepsilon_0}\int_0^{\omega_{10}/c} \mathrm{d}k^\perp \, \sin(2k^\perp \bar{z}_A)\left[\left(k^{\perp 2} + \frac{\omega_{10}^2}{c^2}\right)|\boldsymbol{d}_{10}^{\parallel}|^2 + 2\left(k^{\perp 2} - \frac{\omega_{10}^2}{c^2}\right)|\boldsymbol{d}_{10}^{\perp}|^2\right]. \tag{4.113}$$

This result exactly agrees with the propagating-wave potential (4.74) of an atom at distance $\bar{z}_A = z_A - d$ from a perfectly conducting mirror. By comparison with (4.75), the $k^\perp$-integral leads to

$$U_1^{\text{prop}}(z_A) = -\frac{1}{32\pi\varepsilon_0\bar{z}_A^3} \times \left\{|\boldsymbol{d}_{10}^{\parallel}|^2\left[\cos(2x) + 2x\sin(2x) - 4x^2\cos(2x) - 1 + 2x^2\right] + 2|\boldsymbol{d}_{10}^{\perp}|^2\left[\cos(2x) + 2x\sin(2x) - 1 - 2x^2\right]\right\}_{x=\omega_{10}\bar{z}_A/c}. \tag{4.114}$$

The evanescent-wave potential of a magnetoelectric plate plus perfectly conducting mirror reads is given by (4.64) with reflection coefficients

$$r_s = \frac{(\mu\kappa^\perp - \kappa_1^\perp) - (\mu\kappa^\perp + \kappa_1^\perp)\,\mathrm{e}^{-2\kappa_1^\perp d}}{(\mu\kappa^\perp + \kappa_1^\perp) - (\mu\kappa^\perp - \kappa_1^\perp)\,\mathrm{e}^{-2\kappa_1^\perp d}}, \tag{4.115}$$

$$r_p = \frac{(\varepsilon\kappa^\perp - \kappa_1^\perp) + (\varepsilon\kappa^\perp + \kappa_1^\perp)\,\mathrm{e}^{-2\kappa_1^\perp d}}{(\varepsilon\kappa^\perp + \kappa_1^\perp) + (\varepsilon\kappa^\perp - \kappa_1^\perp)\,\mathrm{e}^{-2\kappa_1^\perp d}} \tag{4.116}$$

with $\kappa_1^\perp$ being given by (4.101). In particular, for a superlens, $\varepsilon = -1$, $\mu = -1$ and $\kappa_1^\perp = \kappa^\perp$ lead to

$$r_s = -\mathrm{e}^{2\kappa^\perp d}, \qquad r_p = \mathrm{e}^{2\kappa^\perp d}. \tag{4.117}$$

Substitution into (4.64) yields

$$U_1^{\text{evan}}(z_A) = -\frac{1}{8\pi\varepsilon_0}\int_0^\infty \mathrm{d}\kappa^\perp \, \mathrm{e}^{-2\kappa^\perp \bar{z}_A}\left[\left(\kappa^{\perp 2} - \frac{\omega_{10}}{c^2}\right)|\boldsymbol{d}_{10}^{\parallel}|^2 + 2\left(\kappa^{\perp 2} + \frac{\omega_{10}}{c^2}\right)|\boldsymbol{d}_{10}^{\perp}|^2\right], \tag{4.118}$$

again in agreement with respective potential (4.78) of an atom at distance $\bar{z}_A$ from perfectly conducting plate. Note, however, that the position $\bar{z}_A$ of the atom relative to the focal plane takes negative values when the atom is between the superlens and the focal plane, $z_A < d$. In this case, the $\kappa^\perp$-integral diverges. The convergent result for $z_A > d$ can be read off by comparison with (4.79), so that we have

$$U_1^{\mathrm{evan}}(z_A) = \begin{cases} -\dfrac{1}{32\pi\varepsilon_0\bar{z}_A^3}\left[|\boldsymbol{d}_{10}^{\parallel}|^2\left(1-2\dfrac{\omega_{10}^2\bar{z}_A^2}{c^2}\right)+2|\boldsymbol{d}_{10}^{\perp}|^2\left(1+2\dfrac{\omega_{10}^2\bar{z}_A^2}{c^2}\right)\right], \\ \qquad z_A > d\,, \\ -\infty\,, \qquad z_A \leq d\,. \end{cases} \tag{4.119}$$

The total resonant potential of the excited atom is the sum of the propagating- and evanescent-wave contributions. By comparison with the corresponding perfect-conductor results (4.97), (4.99), (4.102) and (4.103), it reduces to [35]

$$U_1(z_A) = \frac{\mu_0\omega_{10}^2|\boldsymbol{d}_{nk}^{\parallel}|^2}{8\pi\bar{z}_A}\cos\left(\frac{2\omega_{10}\bar{z}_A}{c}\right) \tag{4.120}$$

in the retarded limit $\bar{z}_A \gg c/\omega_{10}$, and to

$$U_1(z_A) = -\frac{|\boldsymbol{d}_{10}^{\parallel}|^2+2|\boldsymbol{d}_{10}^{\perp}|^2}{32\pi\varepsilon_0\bar{z}_A^3} \tag{4.121}$$

in the nonretarded limit $0 < \bar{z}_A \ll c/\omega_{10}$.

The left-handed superlens hence effectively moves the perfectly conducting mirror from its original position to the focal plane. In a manner of speaking, the negative optical path inside the lens cancels the optical path in the free-space region between the lens and its focal plane. As a result, the oscillatory potential in the retarded regime is slightly enhanced due to the replacement $1/z_A \mapsto 1/\bar{z}_A$; and it acquires a phase shift. Furthermore, the lens-induced enhancement of the evanescent field leads to a strongly attracting, $1/\bar{z}_A^3$ potential when approaching the focal plane. Strikingly, this diverging power-law potential occurs at a position in free space which is not in close proximity to a physical surface.

The results of our idealised setup require two notes of caution. Firstly, while the propagating-wave potential is in agreement with our expectations from the image construction of Fig. 4.4 for all atomic positions, the evanescent-wave potential gives the expected result only for positions beyond the focal plane. In the region between the focal plane and the superlens surface, the potential is infinite due to an over-enhancement of evanescent waves by the lens. Secondly, recall the dispersive, frequency-dependent nature of permittivity and permittivity. No material can act as a superlens at all frequencies, so our configuration is unable to enhance the non-resonant CP potential, which depends on the magnetoelectric response over a wide range of frequencies. The non-resonant contribution (4.73) is therefore absent from the $1/\bar{z}_A^3$ potential (4.121).

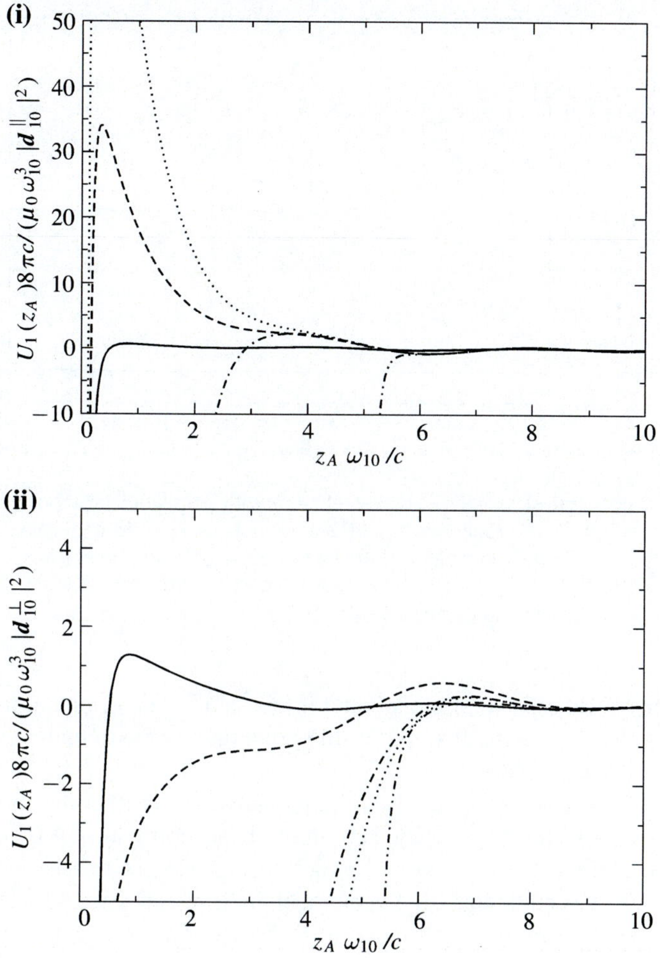

Fig. 4.5 CP potential of an excited two-level atom with **(i)** parallel and **(ii)** perpendicular dipole moment in front of a meta-material superlens ($\varepsilon = \mu = -1 + \mathrm{i}\delta$) of thickness $d = 5c/\omega_{10}$, backed by a perfectly conducting mirror. The chosen value for the absorption strength are $\delta = 10^{-1}$ (*solid lines*), 10^{-3} (*dashed lines*), 10^{-4} (*dotted lines*), 10^{-5} (*dash-dotted lines*) and 0 (*dash-double dotted lines*)

The divergence of the potential between the superlens and its focal plane is a result of the unphysical assumptions of a perfectly conducting mirror and a strictly non-absorbing superlens. In order to investigate the impact of absorption on the superlensing effect, we calculate the resonant CP potential of a two-level atom according to (4.63) and (4.64) with reflection coefficients (4.110), (4.111), (4.115) and (4.116) for lenses with various degrees of absorption, $\varepsilon = \mu = -1 + \mathrm{i}\delta$. The results are displayed in Fig. 4.5 [35]. The figure reveals that the potential reacts very sensitively to absorption. The ideal non-absorbing result with its strongly attractive potential near the focal plane is approached only for very small values of the absorption coeffi-

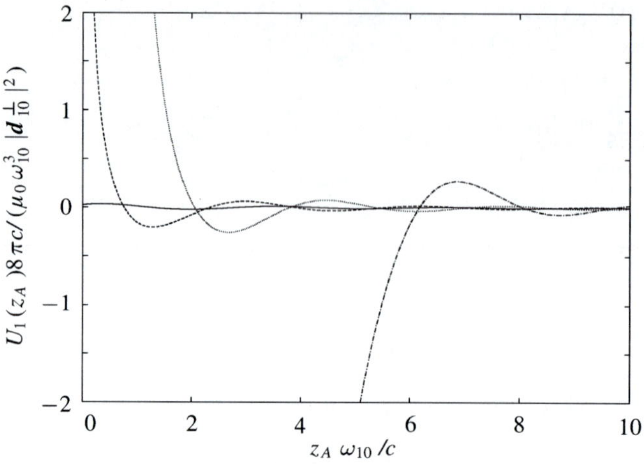

Fig. 4.6 CP potential of an excited two-level atom with perpendicular dipole moment in front of a left- versus right-handed meta-material slabs of thickness $d = 5c/\omega_{10}$, backed by a perfectly conducting mirror. The chosen parameters are $\mathrm{Re}\,\varepsilon = \mathrm{Re}\,\mu = 1$ (*solid lines*), $\mathrm{Re}\,\varepsilon = 1$, $\mathrm{Re}\,\mu = -1$ (*dashed lines*), $\mathrm{Re}\,\varepsilon = -1$, $\mathrm{Re}\,\mu = 1$ (*dotted lines*) and $\mathrm{Re}\,\varepsilon = \mathrm{Re}\,\mu = -1$ (*dash-dotted lines*). All half spaces are weakly absorbing, $\mathrm{Im}\,\varepsilon$, $\mathrm{Im}\,\mu = 10^{-4}$

cient δ. The convergence is slightly better for perpendicular orientation of the atomic dipole moment. For finite absorption, the potential always remains finite between the lens and its focal plane.

These observations from Fig. 4.5 raise the question whether the enhanced attractive potential near the focal plane for finite absorption can still be regarded as a superlensing effect due to negative refraction. To address this issue, we consider an atom with perpendicular dipole moment in front of a weakly absorbing lens in Fig. 4.6, comparing the outcomes for the four different possible choices for the signs of $\mathrm{Re}\,\varepsilon$, $\mathrm{Re}\,\mu$ [21]. The plate with $\varepsilon = \mu = 1 + 10^{-4}\mathrm{i}$ is almost transparent. As a result, we observe the weak potential of the distant mirror alone. The two plates where either the permittivity or the permeability exhibit a negative real part exhibit the typical behaviour of dominantly electric ($\varepsilon = -1 + 10^{-4}\mathrm{i}$) or magnetic ($\mu = -1 + 10^{-4}\mathrm{i}$) plates as known from Fig. 4.2. In both cases, an oscillating potential in the retarded regime turns into a repulsive short-range potential. Only the left-handed plate ($\varepsilon = \mu = -1 + 10^{-4}\mathrm{i}$) shows a strongly attractive potential in the focal region, which can hence be identified as a genuine superlensing effect.

References

1. H.B.G. Casimir, D. Polder, Phys. Rev. **73**(4), 360 (1948)
2. S.Y. Buhmann, D.T. Ho, D.G. Welsch, J. Opt. B: Quantum Semiclass. Opt. **6**(3), S127 (2004)
3. S.Y. Buhmann, L. Knöll, D.G. Welsch, D.T. Ho, Phys. Rev. A **70**(5), 052117 (2004)
4. S.Y. Buhmann, D.G. Welsch, Prog. Quantum Electron. **31**(2), 51 (2007)

5. J.M. Wylie, J.E. Sipe, Phys. Rev. A **32**(4), 2030 (1985)
6. D. Meschede, W. Jhe, E.A. Hinds, Phys. Rev. A **41**(3), 1587 (1990)
7. G. Barton, J. Phys. B: At. Mol. Opt. Phys. **7**(16), 2134 (1974)
8. S. Haroche, in *Fundamental Systems in Quantum Optics, Les Houches Summer School Session LIII*, ed. by J. Dalibard, J.M. Raimond, J. Zinn-Justin (North-Holland, Amsterdam, 1992), p. 767
9. E.A. Hinds, Adv. At. Mol. Opt. Phys. Suppl. **2**, 1 (1994)
10. C. Eberlein, R. Zietal, Phys. Rev. A **75**(3), 032516 (2007)
11. M. Babiker, G. Barton, J. Phys. A: Math. Gen. **9**(1), 129 (1976)
12. E.A. Power, T. Thirunamachandran, Phys. Rev. A **25**(5), 2473 (1982)
13. G.S. Agarwal, Phys. Rev. Lett. **32**(13), 703 (1974)
14. J.D. Jackson, *Classical Electrodynamics*, 3rd edn. (Wiley, New York, 1998)
15. J. Schrieffer, *Theory of Superconductivity* (W.A. Benjamin, New York, 1964)
16. H. Haakh, F. Intravaia, C. Henkel, S. Spagnolo, R. Passante, B. Power, F. Sols, Phys. Rev. A **80**(6), 062905 (2009)
17. H.F. Arnoldus, Surf. Sci. **444**(1–3), 221 (2000)
18. C. Eberlein, S.T. Wu, Phys. Rev. A **68**(3), 033813 (2003)
19. Y. Sherkunov, Phys. Rev. A **72**(5), 052703 (2005)
20. Y. Sherkunov, Phys. Rev. A **75**(1), 012705 (2007)
21. A. Sambale, S.Y. Buhmann, T.D. Ho, D.G. Welsch, Phys. Scr. T **135**, 015019 (2009)
22. F. Schuller, Z. Naturforsch, A **49**(9), 885 (1994)
23. M. Fichet, F. Schuller, D. Bloch, M. Ducloy, Phys. Rev. A **51**(2), 1553 (1995)
24. J.B. Pendry, A.J. Holden, D.J. Robbins, W.J. Stewart, IEEE Trans. Microw. Theory Tech. **47**(11), 2075 (1999)
25. D.R. Smith, W.J. Padilla, D.C. Vier, S.C. Nemat-Nasser, S. Schultz, Phys. Rev. Lett. **84**(18), 4184 (2000)
26. S. Linden, C. Enkrich, M. Wegener, J. Zhou, T. Koschny, C.M. Soukoulis, Science **306**(5700), 1351 (2004)
27. C. Enkrich, M. Wegener, S. Linden, S. Burger, L. Zschiedrich, F. Schmidt, J.F. Zhou, T. Koschny, C.M. Soukoulis, Phys. Rev. Lett. **95**(20), 203901 (2005)
28. G. Dolling, C. Enkrich, M. Wegener, J.F. Zhou, C.M. Soukoulis, S. Linden, Opt. Lett. **30**(23), 3198 (2005)
29. A.N. Grigorenko, A.K. Geim, H.F. Gleeson, Y. Zhang, A.A. Firsov, I.Y. Khrushchev, J. Petrovic, Nature **438**(7066), 335 (2005)
30. S. Zhang, W. Fan, N.C. Panoiu, K.J. Malloy, R.M. Osgood, S.R.J. Brueck, Phys. Rev. Lett. **95**(13), 137404 (2005)
31. V.G. Veselago, Sov. Phys. Uspekhi **10**(4), 509 (1968)
32. J.B. Pendry, Phys. Rev. Lett. **85**(18), 3966 (2000)
33. J. Kästel, M. Fleischhauer, Laser Phys. **15**(1), 135 (2005)
34. J. Kästel, M. Fleischhauer, Phys. Rev. A 71(1), 011804(R) (2005)
35. A. Sambale, D.G. Welsch, D.T. Ho, S.Y. Buhmann, Phys. Rev. A **78**(5), 053828 (2008)

Chapter 5
Casimir–Polder Forces on Excited Atoms: Dynamical Approach

As seen in the previous chapter, excited atoms are subject to resonant CP potentials due to the emission of real photons. However, these enhanced potentials intrinsically have a limited lifetime. As the excited atom emits a real photon, it decays to a lower-lying internal state. Spontaneous decay will eventually return the atom to its ground state where it is subject to the purely non-resonant ground-state potential. The static approach employed in the previous chapter fails to describe this dynamics of the CP force. Instead, it only provides us with snapshots of the force at given instants of time, with the atom being prepared in a given state.

In this chapter, we will develop a dynamical description of the CP force on an excited atom. Rather than calculating energy shifts, we will identify the CP force as the quantum average of the time-dependent total Lorentz force acting on the atom. We will begin by deriving equivalent expressions for this force using the alternative minimal and multipolar coupling schemes. Solving the coupled atom–field dynamics, we will then evaluate the CP force as the average quantum Lorentz force. In addition to being time-dependent, our result will reveal another aspect not grasped by leading-order perturbation theory: the atomic transitions determining the force are themselves shifted and broadened in the presence of magnetoelectric bodies. We will illustrate the impact of this effect by studying an atom above a dielectric half space.

5.1 Lorentz Force

As a foundation for our dynamical approach, we require the total quantum Lorentz force acting on an atom. We derive suitable equivalent forms for this force using the minimal and multipolar coupling schemes.

5.1.1 Minimal Coupling

Recall that an atom is simply a collection of point particles α with charges q_α and magnetic moments $\hat{\boldsymbol{m}}_\alpha$. Writing the Newton equation (1.84), in the form

S. Y. Buhmann, *Dispersion Forces II*, Springer Tracts in Modern Physics 248,
DOI: 10.1007/978-3-642-32466-6_5, © Springer-Verlag Berlin Heidelberg 2012

$$m_\alpha \ddot{\hat{\boldsymbol{r}}}_\alpha = \hat{\boldsymbol{f}}_\alpha\,, \tag{5.1}$$

we see that each particle experiences a force

$$\hat{\boldsymbol{f}}_\alpha = q_\alpha \hat{\boldsymbol{\mathcal{E}}}(\hat{\boldsymbol{r}}_\alpha) + q_\alpha \mathcal{S}\Big[\dot{\hat{\boldsymbol{r}}}_\alpha \times \hat{\boldsymbol{\mathcal{B}}}(\hat{\boldsymbol{r}}_\alpha)\Big] + \nabla_\alpha\big[\hat{\boldsymbol{m}}_\alpha \cdot \hat{\boldsymbol{\mathcal{B}}}(\hat{\boldsymbol{r}}_\alpha)\big] \tag{5.2}$$

when subject to the quantum electromagnetic field. In the following, we will use the notion Lorentz force for this force as a whole rather than distinguishing the standard Lorentz force (first two terms) from the Zeeman force (third term). Summing over all particles contained in the atom and recalling the definition (1.60) of the centre-of-mass coordinate, we have

$$m_A \ddot{\hat{\boldsymbol{r}}}_A = \hat{\boldsymbol{F}} \tag{5.3}$$

where the total Lorentz force acting on the atom is given by

$$\hat{\boldsymbol{F}} = \sum_{\alpha\in A} \hat{\boldsymbol{f}}_\alpha = \sum_{\alpha\in A}\Big\{q_\alpha \hat{\boldsymbol{\mathcal{E}}}(\hat{\boldsymbol{r}}_\alpha) + q_\alpha \mathcal{S}\Big[\dot{\hat{\boldsymbol{r}}}_\alpha \times \hat{\boldsymbol{\mathcal{B}}}(\hat{\boldsymbol{r}}_\alpha)\Big] + \nabla_\alpha\big[\hat{\boldsymbol{m}}_\alpha \cdot \hat{\boldsymbol{\mathcal{B}}}(\hat{\boldsymbol{r}}_\alpha)\big]\Big\}\,. \tag{5.4}$$

This force may be cast into a more compact form by expressing it in terms of the atomic charge and current densities. To this end, we rewrite last term by using the vector identity $\boldsymbol{a}\times(\boldsymbol{b}\times\boldsymbol{c}) = \boldsymbol{b}(\boldsymbol{a}\cdot\boldsymbol{c}) - \boldsymbol{c}(\boldsymbol{a}\cdot\boldsymbol{b})$ and exploiting the transversality (1.80) of the magnetic field,

$$\hat{\boldsymbol{F}} = \sum_{\alpha\in A} \hat{\boldsymbol{f}}_\alpha = \sum_{\alpha\in A}\Big\{q_\alpha \hat{\boldsymbol{\mathcal{E}}}(\hat{\boldsymbol{r}}_\alpha) + q_\alpha \mathcal{S}\Big[\dot{\hat{\boldsymbol{r}}}_\alpha \times \hat{\boldsymbol{\mathcal{B}}}(\hat{\boldsymbol{r}}_\alpha)\Big] + (\hat{\boldsymbol{m}}_\alpha \times \nabla_\alpha)\times \hat{\boldsymbol{\mathcal{B}}}(\hat{\boldsymbol{r}}_\alpha)\Big\}\,. \tag{5.5}$$

Recalling the definitions (1.56) and (1.57) for the atomic charge and current densities, we then find

$$\hat{\boldsymbol{F}} = \int \mathrm{d}^3 r\,\big(\hat{\rho}_A \hat{\boldsymbol{\mathcal{E}}} + \hat{\boldsymbol{j}}_A \times \hat{\boldsymbol{\mathcal{B}}}\big)\,. \tag{5.6}$$

This expression is completely analogous to the Lorentz force (1.105) used in Sect. 1.2.1 for deriving the Casimir force on a body. The only difference lies in the charge and current densities. While the Casimir force stems from the Lorentz force on the macroscopic internal charges and currents inside the body, the CP force has its origin in the Lorentz force on microscopic atomic charges and currents.

An alternative representation of the Lorentz force can be obtained by expressing the atomic charge and current densities in terms of the atomic polarisation and magnetisation according to (1.65) and (1.66):

$$\hat{\boldsymbol{F}} = -\int \mathrm{d}^3r\,(\boldsymbol{\nabla}\cdot\hat{\boldsymbol{P}}_A)\hat{\boldsymbol{\mathcal{E}}} + \int \mathrm{d}^3r\,\dot{\hat{\boldsymbol{P}}}_A\times\hat{\boldsymbol{\mathcal{B}}}$$
$$+ \int \mathrm{d}^3r\,\Big\{\boldsymbol{\nabla}\times\Big[\hat{\boldsymbol{M}}_A + \mathcal{S}\Big(\hat{\boldsymbol{P}}_A\times\dot{\hat{\boldsymbol{r}}}_A\Big)\Big]\Big\}\times\hat{\boldsymbol{\mathcal{B}}}\,. \tag{5.7}$$

Using the rule $\boldsymbol{a}\times(\boldsymbol{b}\times\boldsymbol{c}) = \boldsymbol{b}(\boldsymbol{a}\cdot\boldsymbol{c}) - \boldsymbol{c}(\boldsymbol{a}\cdot\boldsymbol{b})$, the first term can be rewritten as

$$-\int \mathrm{d}^3r\,(\boldsymbol{\nabla}\cdot\hat{\boldsymbol{P}}_A)\hat{\boldsymbol{\mathcal{E}}} = -\int \mathrm{d}^3r\,(\boldsymbol{\nabla}\hat{\boldsymbol{P}}_A)\cdot\hat{\boldsymbol{\mathcal{E}}} + \int \mathrm{d}^3r\,(\hat{\boldsymbol{\mathcal{E}}}\times\boldsymbol{\nabla})\times\hat{\boldsymbol{P}}_A$$
$$= \boldsymbol{\nabla}_A\Big[\int \mathrm{d}^3r\,\hat{\boldsymbol{P}}_A\cdot\hat{\boldsymbol{\mathcal{E}}}\Big] + \int \mathrm{d}^3r\,\hat{\boldsymbol{P}}_A\times\dot{\hat{\boldsymbol{B}}}\,. \tag{5.8}$$

To obtain the last equality, we have used the identity $\boldsymbol{\nabla}\hat{\boldsymbol{P}}_A = -\boldsymbol{\nabla}_A\hat{\boldsymbol{P}}_A$, cf. (1.63), partially integrated the second term and used the Faraday law (1.31). In a similar way, using $\boldsymbol{\nabla}\hat{\boldsymbol{M}}_A = -\boldsymbol{\nabla}_A\hat{\boldsymbol{M}}_A$ in accordance with (1.64) and exploiting the transversality (1.80) of the magnetic field, one finds

$$\int \mathrm{d}^3r\,\Big\{\boldsymbol{\nabla}\times\Big[\hat{\boldsymbol{M}}_A + \mathcal{S}\Big(\hat{\boldsymbol{P}}_A\times\dot{\hat{\boldsymbol{r}}}_A\Big)\Big]\Big\}\times\hat{\boldsymbol{\mathcal{B}}}$$
$$= -\int \mathrm{d}^3r\,\Big\{\boldsymbol{\nabla}\Big[\hat{\boldsymbol{M}}_A + \mathcal{S}\Big(\hat{\boldsymbol{P}}_A\times\dot{\hat{\boldsymbol{r}}}_A\Big)\Big]\Big\}\cdot\hat{\boldsymbol{\mathcal{B}}}$$
$$+ \int \mathrm{d}^3r\,(\hat{\boldsymbol{\mathcal{B}}}\cdot\boldsymbol{\nabla})\Big[\hat{\boldsymbol{M}}_A + \mathcal{S}\Big(\hat{\boldsymbol{P}}_A\times\dot{\hat{\boldsymbol{r}}}_A\Big)\Big]$$
$$= \boldsymbol{\nabla}_A\Big[\int \mathrm{d}^3r\,\Big(\hat{\boldsymbol{M}}_A + \hat{\boldsymbol{P}}_A\times\dot{\hat{\boldsymbol{r}}}_A\Big)\cdot\hat{\boldsymbol{\mathcal{B}}}\Big]\,. \tag{5.9}$$

with these identities, the Lorentz force reads [1]

$$\hat{\boldsymbol{F}} = \boldsymbol{\nabla}_A\Big[\int \mathrm{d}^3r\,\hat{\boldsymbol{P}}_A\cdot\hat{\boldsymbol{\mathcal{E}}} + \int \mathrm{d}^3r\,\Big(\hat{\boldsymbol{M}}_A + \hat{\boldsymbol{P}}_A\times\dot{\hat{\boldsymbol{r}}}_A\Big)\cdot\hat{\boldsymbol{\mathcal{B}}}\Big]$$
$$+ \frac{\mathrm{d}}{\mathrm{d}t}\int \mathrm{d}^3r\,\hat{\boldsymbol{P}}_A\times\hat{\boldsymbol{\mathcal{B}}}\,. \tag{5.10}$$

Finally, we need to relate the physical electromagnetic fields $\hat{\boldsymbol{\mathcal{E}}}$ and $\hat{\boldsymbol{\mathcal{B}}}$ to the fields $\hat{\boldsymbol{E}}$ and $\hat{\boldsymbol{B}}$ of the minimal coupling scheme. According to (1.78), the magnetic fields coincide, while the physical electric field differs from the minimal-coupling one by the longitudinal fields created by the charged particles. However, the latter only give rise to internal forces which do not contribute to the total atomic force due to pairwise cancellations,

$$\sum_{\alpha\in A} q_\alpha\boldsymbol{\nabla}\hat{\phi}_A(\boldsymbol{r}_\alpha) = \boldsymbol{0}\,. \tag{5.11}$$

Our result (5.4) and the subsequent manipulations hence remain valid with the replacements $\hat{\mathcal{E}} \mapsto \hat{\boldsymbol{E}}$ and $\hat{\mathcal{B}} \mapsto \hat{\boldsymbol{B}}$, and we have

$$\hat{\boldsymbol{F}} = \nabla_A \left[\int \mathrm{d}^3 r\, \hat{\boldsymbol{P}}_A \cdot \hat{\boldsymbol{E}} + \int \mathrm{d}^3 r \left(\hat{\boldsymbol{M}}_A + \hat{\boldsymbol{P}}_A \times \dot{\hat{\boldsymbol{r}}}_A \right) \cdot \hat{\boldsymbol{B}} \right] + \frac{\mathrm{d}}{\mathrm{d}t} \int \mathrm{d}^3 r\, \hat{\boldsymbol{P}}_A \times \hat{\boldsymbol{B}} \,. \tag{5.12}$$

Expanding the atomic polarisation (1.63) and magnetisation (1.64) to leading order in the relative particle coordinates $\hat{\overline{\boldsymbol{r}}}_\alpha$, the Lorentz force in long-wavelength approximation is given by

$$\hat{\boldsymbol{F}} = \nabla \left[\hat{\boldsymbol{d}} \cdot \hat{\boldsymbol{E}}(\boldsymbol{r}) + \hat{\boldsymbol{m}} \cdot \hat{\boldsymbol{B}}(\boldsymbol{r}) + \hat{\boldsymbol{d}} \times \dot{\hat{\boldsymbol{r}}}_A \cdot \hat{\boldsymbol{B}}(\boldsymbol{r}) \right] \Big|_{\boldsymbol{r} = \hat{\boldsymbol{r}}_A} + \frac{\mathrm{d}}{\mathrm{d}t} \left[\hat{\boldsymbol{d}} \times \hat{\boldsymbol{B}}(\hat{\boldsymbol{r}}_A) \right], \tag{5.13}$$

where the electric and magnetic dipole moments of the atom are given by (1.69) and (1.70). For non-magnetic atoms, the magnetic interaction can be discarded and the electric-dipole approximation

$$\hat{\boldsymbol{F}} = \nabla \left[\hat{\boldsymbol{d}} \cdot \hat{\boldsymbol{E}}(\boldsymbol{r}) + \hat{\boldsymbol{d}} \times \dot{\hat{\boldsymbol{r}}}_A \cdot \hat{\boldsymbol{B}}(\boldsymbol{r}) \right] \Big|_{\boldsymbol{r} = \hat{\boldsymbol{r}}_A} + \frac{\mathrm{d}}{\mathrm{d}t} \left[\hat{\boldsymbol{d}} \times \hat{\boldsymbol{B}}(\hat{\boldsymbol{r}}_A) \right] \tag{5.14}$$

applies.

Throughout this chapter, we will assume that the atom is sufficiently slow such that the influence of its centre-of-mass motion on the CP force can be neglected. Dispersion forces on moving atoms will be discussed in detail in Chap. 8. Discarding the velocity-dependent term, the Lorentz force simplifies to

$$\hat{\boldsymbol{F}} = \left\{ \nabla \left[\hat{\boldsymbol{d}} \cdot \hat{\boldsymbol{E}}(\boldsymbol{r}) \right] + \frac{\mathrm{d}}{\mathrm{d}t} \left[\hat{\boldsymbol{d}} \times \hat{\boldsymbol{B}}(\boldsymbol{r}) \right] \right\}_{\boldsymbol{r} = \hat{\boldsymbol{r}}_A} . \tag{5.15}$$

5.1.2 Multipolar Coupling

The electromagnetic fields $\hat{\boldsymbol{E}}'$ and $\hat{\boldsymbol{B}}'$ of the multipolar coupling scheme differ from those of the minimal coupling scheme. To obtain an expression for the Lorentz force in terms of the multipolar fields, we employ an alternative route via the Heisenberg equation. We recall from Sect. 1.2.2 that the particle positions agree in the two schemes, $\hat{\boldsymbol{r}}_\alpha = \hat{\boldsymbol{r}}'_\alpha$. As the total Hamiltonians also coincide, the same holds for time derivative of the particle positions. We can hence conclude from (1.102) that the particle velocities in the multipolar coupling scheme read

$$m_\alpha \dot{\hat{\boldsymbol{r}}}_\alpha = m_\alpha \dot{\hat{\boldsymbol{r}}}'_\alpha = \hat{\boldsymbol{p}}'_\alpha + \int \mathrm{d}^3 r\, \hat{\boldsymbol{\Xi}}'_\alpha \times \hat{\boldsymbol{B}}' \,. \tag{5.16}$$

Summing this result over all particles contained in an atom and recalling the definitions (1.60) and (1.61) for the atomic centre-of-mass position and momentum as well as that (1.91) for $\hat{\boldsymbol{\Xi}}_\alpha$, we obtain

$$m_A \dot{\hat{\boldsymbol{r}}}_A = m_A \dot{\hat{\boldsymbol{r}}}'_A = \sum_{\alpha\in A} m_\alpha \dot{\hat{\boldsymbol{r}}}'_\alpha = \hat{\boldsymbol{p}}'_A + \int \mathrm{d}^3 r\, \hat{\boldsymbol{P}}'_A \times \hat{\boldsymbol{B}}' \,. \tag{5.17}$$

The Heisenberg equation of motion now implies

$$m_A \ddot{\hat{\boldsymbol{r}}}_A = m_A \ddot{\hat{\boldsymbol{r}}}'_A = \frac{1}{\mathrm{i}\hbar}\big[m_A \dot{\hat{\boldsymbol{r}}}'_A, \hat{H}\big] = \hat{\boldsymbol{F}}' \tag{5.18}$$

where the Lorentz force assumes the form

$$\hat{\boldsymbol{F}}' = \frac{1}{\mathrm{i}\hbar}\big[\hat{\boldsymbol{p}}'_A, \hat{H}\big] + \frac{\mathrm{d}}{\mathrm{d}t}\int \mathrm{d}^3 r\, \hat{\boldsymbol{P}}'_A \times \hat{\boldsymbol{B}}' \,. \tag{5.19}$$

The different contributions to the commutator can be evaluated by recalling the multipolar Hamiltonian (1.90) as well as the commutation relations (1.62). Using the identity $\boldsymbol{\nabla}'_A \hat{\boldsymbol{P}}'_A = -\boldsymbol{\nabla} \hat{\boldsymbol{P}}'_A$, one can show that

$$\frac{1}{\mathrm{i}\hbar}\left[\hat{\boldsymbol{p}}'_A, \frac{1}{2\varepsilon_0}\int \mathrm{d}^3 r\, \hat{\boldsymbol{P}}'^2_A\right] = -\frac{1}{2\varepsilon_0}\int \mathrm{d}^3 r\, \boldsymbol{\nabla}'_A \hat{\boldsymbol{P}}'^2_A = \frac{1}{2\varepsilon_0}\int \mathrm{d}^3 r\, \boldsymbol{\nabla} \hat{\boldsymbol{P}}'^2_A = \boldsymbol{0} \,. \tag{5.20}$$

Similarly, we have ($\boldsymbol{\nabla}'_A = \boldsymbol{\nabla}_A$)

$$\frac{1}{\mathrm{i}\hbar}\left[\hat{\boldsymbol{p}}'_A, -\int \mathrm{d}^3 r\, \hat{\boldsymbol{P}}'_A \cdot \hat{\boldsymbol{E}}'\right] = \boldsymbol{\nabla}_A \int \mathrm{d}^3 r\, \hat{\boldsymbol{P}}'_A \cdot \hat{\boldsymbol{E}}' \,; \tag{5.21}$$

and by recalling the definitions (1.63), (1.64) and (1.91) and invoking the relation (5.17) above, we derive

$$\begin{aligned} &\frac{1}{\mathrm{i}\hbar}\left[\hat{\boldsymbol{p}}'_A, \sum_{\alpha\in A} \frac{1}{2m_\alpha}\left(\hat{\boldsymbol{p}}'_\alpha + \int \mathrm{d}^3 r\, \hat{\boldsymbol{\Xi}}'_\alpha \times \hat{\boldsymbol{B}}'\right)^2 - \sum_{\alpha\in A} \gamma_\alpha \hat{\boldsymbol{s}}'_\alpha \cdot \hat{\boldsymbol{B}}'(\hat{\boldsymbol{r}}'_\alpha)\right] \\ &= \boldsymbol{\nabla}_A \int \mathrm{d}^3 r \left(\hat{\boldsymbol{M}}'_A + \hat{\boldsymbol{P}}'_A \times \dot{\hat{\boldsymbol{r}}}_A\right) \cdot \hat{\boldsymbol{B}}' \,. \end{aligned} \tag{5.22}$$

Combining these results, the Lorentz force in the multipolar coupling scheme is given by [1–3]

$$\hat{\boldsymbol{F}}' = \nabla_A\left[\int \mathrm{d}^3r\, \hat{\boldsymbol{P}}'_A\cdot\hat{\boldsymbol{E}}' + \int \mathrm{d}^3r\left(\hat{\boldsymbol{M}}'_A + \hat{\boldsymbol{P}}'_A\times\dot{\hat{\boldsymbol{r}}}_A\right)\cdot\hat{\boldsymbol{B}}'\right] + \frac{\mathrm{d}}{\mathrm{d}t}\int \mathrm{d}^3r\, \hat{\boldsymbol{P}}'_A\times\hat{\boldsymbol{B}}'. \tag{5.23}$$

This is a generalisation of the free-space result for the QED Lorentz force [4, 5] to the case of magnetoelectric bodies being present.

As is obvious from comparing (5.3) and (5.18), the multipolar-coupling expression for the Lorentz force must coincide with the earlier minimal-coupling result (5.12). To verify this explicitly, we recall from Sect. 1.1.2.2 that the following quantities are invariant under a Power–Zienau–Woolley transformation: $\hat{\boldsymbol{B}}' = \hat{\boldsymbol{B}}$, $\hat{\boldsymbol{P}}'_A = \hat{\boldsymbol{P}}_A$ and $\hat{\boldsymbol{M}}'_A = \hat{\boldsymbol{M}}_A$. According to (1.52) and (1.101), the electric fields in the two schemes differ,

$$\hat{\boldsymbol{E}}' = \hat{\boldsymbol{E}} + \frac{1}{\varepsilon_0}\,\hat{\boldsymbol{P}}_A^{\perp}\,. \tag{5.24}$$

However, the relations

$$\nabla_A \int \mathrm{d}^3r\, \hat{\boldsymbol{P}}_A^{\perp 2} = -\nabla \int \mathrm{d}^3r\, \hat{\boldsymbol{P}}_A^{\perp 2} = \boldsymbol{0}\,, \tag{5.25}$$

and

$$\int \mathrm{d}^3r\, \hat{\boldsymbol{P}}_A^{\perp}\cdot\hat{\boldsymbol{P}}_A^{\parallel} = 0 \tag{5.26}$$

reveal that this difference does not contribute to the Lorentz force, so indeed we have

$$\hat{\boldsymbol{F}}' = \hat{\boldsymbol{F}}\,. \tag{5.27}$$

In close analogy to the minimal-coupling case, the Lorentz force (5.23) reduces to

$$\hat{\boldsymbol{F}}' = \nabla\left[\hat{\boldsymbol{d}}'\cdot\hat{\boldsymbol{E}}'(\boldsymbol{r}) + \hat{\boldsymbol{m}}'\cdot\hat{\boldsymbol{B}}'(\boldsymbol{r}) + \hat{\boldsymbol{d}}'\times\dot{\hat{\boldsymbol{r}}}_A\cdot\hat{\boldsymbol{B}}'(\boldsymbol{r})\right]\Big|_{\boldsymbol{r}=\hat{\boldsymbol{r}}_A} + \frac{\mathrm{d}}{\mathrm{d}t}\left[\hat{\boldsymbol{d}}'\times\hat{\boldsymbol{B}}'(\hat{\boldsymbol{r}}_A)\right] \tag{5.28}$$

when employing the long-wavelength approximation. For a non-magnetic atom, the force assumes its electric-dipole form

$$\hat{\boldsymbol{F}}' = \nabla\left[\hat{\boldsymbol{d}}'\cdot\hat{\boldsymbol{E}}'(\boldsymbol{r}) + \hat{\boldsymbol{d}}'\times\dot{\hat{\boldsymbol{r}}}_A\cdot\hat{\boldsymbol{B}}'(\boldsymbol{r})\right]\Big|_{\boldsymbol{r}=\hat{\boldsymbol{r}}_A} + \frac{\mathrm{d}}{\mathrm{d}t}\left[\hat{\boldsymbol{d}}'\times\hat{\boldsymbol{B}}'(\hat{\boldsymbol{r}}_A)\right]. \tag{5.29}$$

Neglecting the effect of centre-of-mass motion on the force, we will in the following work with the simplified expression

$$\hat{F}' = \left\{ \nabla \left[\hat{d}' \cdot \hat{E}'(r) \right] + \frac{\mathrm{d}}{\mathrm{d}t} \left[\hat{d}' \times \hat{B}'(r) \right] \right\}_{r=\hat{r}_A}. \tag{5.30}$$

5.2 Internal Atomic Dynamics

In close analogy to the Casimir force, we identify the CP force on an atom as the average Lorentz force

$$F = \left[\nabla \langle \hat{d} \cdot \hat{E}(r) \rangle + \frac{\mathrm{d}}{\mathrm{d}t} \langle \hat{d} \times \hat{B}(r) \rangle \right]_{r=r_A} \tag{5.31}$$

on its charge and current densities. While the body-assisted electromagnetic field will be assumed to be in its ground state $|\{0\}\rangle$, we will allow for the atom to be prepared in an arbitrary internal state. For an excited atom, the state of the atom–field system will then evolve in time. This is in contrast to the Casimir case where the whole system is in its stationary ground state. In order to evaluate the average force, we must first solve the coupled atom–field dynamics to obtain the time-dependent dipole and field operators in the Heisenberg picture. Employing the Born–Oppenheimer approximation, we will assume that the fast internal motion of the particles inside the atom effectively decouples from its slow centre-of-mass motion, so that we can solve the internal dynamics for given centre-or-mass position r_A. For simplicity, we will exclusively work within the multipolar coupling scheme and discard the primes distinguishing the multipolar variables from the ones of the minimal-coupling scheme.

The internal atomic dynamics in the Schrödinger picture is described by the time-dependent atomic density matrix $\hat{\sigma} = \hat{\sigma}(t)$. With the aid of the completeness relation $\sum_n |n\rangle\langle n| = \hat{I}$ ($\hat{I}$, unit operator), it can be expanded in terms of internal-energy eigenstates,

$$\hat{\sigma} = \sum_{m,n} \sigma_{mn} |m\rangle\langle n| \tag{5.32}$$

with time-dependent density matrix elements $\langle m|\hat{\sigma}|n\rangle = \sigma_{mn} = \sigma_{mn}(t)$. Note that the diagonal density-matrix elements

$$p_n(t) = \sigma_{nn}(t) \tag{5.33}$$

represent the probabilities of the atom in the respective internal state $|n\rangle$, while the off-diagonal matrix elements characterise the coherence of the internal atomic quantum state.

For easier compatibility with the above Lorentz force, we will study the internal atomic dynamics within the alternative Heisenberg frame. To that end, we introduce the time-dependent atomic flip operators $|m\rangle\langle n| = \hat{A}_{mn} = \hat{A}_{mn}(t)$. They are closely related to the density matrix, since the expansion above implies

$$\big\langle \hat{A}_{mn}(t)\big\rangle = \mathrm{tr}\big[\hat{\sigma}\hat{A}_{mn}(t)\big] = \mathrm{tr}\big[\hat{\sigma}(t)\hat{A}_{mn}\big] = \sum_{k,l} \sigma_{kl}(t)\langle l|\hat{A}_{mn}|k\rangle = \sigma_{nm}(t)\ , \tag{5.34}$$

where the second equality marks the transition from the Heisenberg to the Schrödinger picture. In particular, the expectation values of the diagonal flip operators coincide with the internal-state populations

$$\big\langle \hat{A}_{nn}(t)\big\rangle = p_n(t)\ . \tag{5.35}$$

The equal-time commutation relations of the flip operators are easily found to be

$$\big[\hat{A}_{mn}, \hat{A}_{kl}\big] = |m\rangle\langle n|k\rangle\langle l| - |k\rangle\langle l|m\rangle\langle n| = \delta_{nk}\hat{A}_{ml} - \delta_{lm}\hat{A}_{kn}\ . \tag{5.36}$$

Introducing the flip operators, the internal atomic Hamiltonian (1.92) assumes the form

$$\hat{H}_{\mathrm{A}} = \sum_n E_n \hat{A}_{nn}\ . \tag{5.37}$$

To express the atom–field coupling Hamiltonian in a similar way, we expand the electric-dipole operator by means of the completeness relation $\sum_n |n\rangle\langle n| = \hat{I}$,

$$\hat{\boldsymbol{d}} = \sum_{m,n} \boldsymbol{d}_{mn}|m\rangle\langle n| = \sum_{m,n} \boldsymbol{d}_{mn}\hat{A}_{mn}\ . \tag{5.38}$$

Substituting this expansion into (1.98) and discarding the velocity-dependent Röntgen interaction, we have

$$\hat{H}_{\mathrm{AF}} = -\sum_{m,n} \hat{A}_{mn}\boldsymbol{d}_{mn}\cdot\hat{\boldsymbol{E}}(\boldsymbol{r}_{\mathrm{A}})\ . \tag{5.39}$$

With these Hamiltonians, the Heisenberg equation of motion for the atomic flip operators can be easily evaluated. Using the commutators as given above, one finds

$$\begin{aligned}\dot{\hat{A}}_{mn} &= \frac{1}{\mathrm{i}\hbar}\big[\hat{A}_{mn}, \hat{H}\big] = \frac{1}{\mathrm{i}\hbar}\big[\hat{A}_{mn}, \hat{H}_{\mathrm{F}}\big] + \frac{1}{\mathrm{i}\hbar}\big[\hat{A}_{mn}, \hat{H}_{\mathrm{AF}}\big]\\ &= \mathrm{i}\omega_{mn}\hat{A}_{mn} + \frac{\mathrm{i}}{\hbar}\sum_k \big(\hat{A}_{mk}\boldsymbol{d}_{nk} - \hat{A}_{kn}\boldsymbol{d}_{km}\big)\cdot\hat{\boldsymbol{E}}(\boldsymbol{r}_{\mathrm{A}})\end{aligned} \tag{5.40}$$

where all operators are understood to carry a time argument t.

In order to solve this equation, we need to also consider the equation of motion for the electromagnetic field. With the expansion (1.22) for the electric-field operator, the coupling Hamiltonian (5.39) takes the form

$$\hat{H}_{\mathrm{AF}} = -\sum_{m,n}\sum_{\lambda=e,m}\int \mathrm{d}^3r \int_0^\infty \mathrm{d}\omega\, \boldsymbol{d}_{mn}\cdot\boldsymbol{G}_\lambda(\boldsymbol{r}_A,\boldsymbol{r},\omega)\cdot\hat{\boldsymbol{f}}_\lambda(\boldsymbol{r},\omega)\hat{A}_{mn} + \mathrm{H.c.} \quad (5.41)$$

Recalling the field Hamiltonian (1.93) and using the commutation relations (1.17) and (1.18), the Heisenberg equations for the fundamental fields hence read

$$\begin{aligned}\dot{\hat{\boldsymbol{f}}}_\lambda(\boldsymbol{r},\omega) &= \frac{1}{\mathrm{i}\hbar}\big[\hat{\boldsymbol{f}}_\lambda(\boldsymbol{r},\omega),\hat{H}\big] = \frac{1}{\mathrm{i}\hbar}\big[\hat{\boldsymbol{f}}_\lambda(\boldsymbol{r},\omega),\hat{H}_{\mathrm{F}}\big] + \frac{1}{\mathrm{i}\hbar}\big[\hat{\boldsymbol{f}}_\lambda(\boldsymbol{r},\omega),\hat{H}_{\mathrm{AF}}\big]\\ &= -\mathrm{i}\omega\hat{\boldsymbol{f}}_\lambda(\boldsymbol{r},\omega) + \frac{\mathrm{i}}{\hbar}\sum_{m,n}\boldsymbol{G}_\lambda^{*\mathrm{T}}(\boldsymbol{r}_A,\boldsymbol{r},\omega)\cdot\boldsymbol{d}_{mn}\hat{A}_{mn}\end{aligned} \quad (5.42)$$

where time arguments t are again not shown explicitly. The solution to this inhomogeneous linear differential equation reads

$$\begin{aligned}\hat{\boldsymbol{f}}_\lambda(\boldsymbol{r},\omega,t) &= \mathrm{e}^{-\mathrm{i}\omega(t-t_0)}\hat{\boldsymbol{f}}_\lambda(\boldsymbol{r},\omega)\\ &\quad + \frac{\mathrm{i}}{\hbar}\sum_{m,n}\int_{t_0}^{t}\mathrm{d}t'\,\mathrm{e}^{-\mathrm{i}\omega(t-t')}\boldsymbol{G}_\lambda^{*\mathrm{T}}[\boldsymbol{r}_A(t'),\boldsymbol{r},\omega]\cdot\boldsymbol{d}_{mn}\hat{A}_{mn}(t')\end{aligned} \quad (5.43)$$

where we have required the Heisenberg-picture operator $\hat{\boldsymbol{f}}_\lambda(\boldsymbol{r},\omega,t)$ to agree with its time-independent Schrödinger-picture counterpart $\hat{\boldsymbol{f}}_\lambda(\boldsymbol{r},\omega)$ at initial time t_0: $\hat{\boldsymbol{f}}_\lambda(\boldsymbol{r},\omega,t_0) = \hat{\boldsymbol{f}}_\lambda(\boldsymbol{r},\omega)$. Substituting this solution into the expansion (1.22) and invoking the integral relation (1.25) for the Green's tensor, the time-dependent electric-field operator reads

$$\begin{aligned}\hat{\underline{\boldsymbol{E}}}(\boldsymbol{r},\omega,t) &= \mathrm{e}^{-\mathrm{i}\omega(t-t_0)}\hat{\underline{\boldsymbol{E}}}(\boldsymbol{r},\omega)\\ &\quad + \frac{\mathrm{i}\mu_0}{\pi}\sum_{m,n}\omega^2\int_{t_0}^{t}\mathrm{d}t'\,\mathrm{e}^{-\mathrm{i}\omega(t-t')}\mathrm{Im}\,\boldsymbol{G}[\boldsymbol{r},\boldsymbol{r}_A(t'),\omega]\cdot\boldsymbol{d}_{mn}\hat{A}_{mn}(t')\,.\end{aligned} \quad (5.44)$$

Here, the first term is the free field, i.e., the field as it would be in the absence of the atom. The second term is the source field created by the atom.

The source field depends on the position and state of the atom at all previous times after the initial state preparation. To evaluate it more explicitly, we once more neglect the effect of atomic motion on the CP force, so that $\boldsymbol{r}_A(t) = \boldsymbol{r}_A(t') \equiv \boldsymbol{r}_A$. Next, we assume weak–atom field coupling. As shown in Sect. 6.1, this requires the field spectrum $\omega^2\mathrm{Im}\,\boldsymbol{G}(\boldsymbol{r}_A,\boldsymbol{r}_A,\omega)$ to be sufficiently flat. In particular, it may not exhibit narrow peaks in the vicinity of any atomic transition frequency. For weak coupling, we may evaluate the time integral by means of the Markov approximation. By comparison with (5.40), we assume that the dynamics of the atomic flip operators $\hat{A}_{mn}$ is dominated by oscillations with frequencies $\tilde{\omega}_{mn}$ which are yet to be determined. We expect these frequencies to deviate from the bare transition frequencies ω_{mn} due to influence of the second term on the right hand side of (5.40). Neglecting the slow non-oscillatory dynamics of the flip operators during the time interval $t_0 \le t' \le t$, we may put $\hat{A}_{mn}(t') \simeq \mathrm{e}^{\mathrm{i}\tilde{\omega}_{mn}(t'-t)}\hat{A}_{mn}(t)$, so that

$$\int_{t_0}^{t}\mathrm{d}t'\,\mathrm{e}^{-\mathrm{i}\omega(t-t')}\hat{A}_{mn}(t') \simeq \hat{A}_{mn}(t)\int_{t_0}^{t}\mathrm{d}t'\,\mathrm{e}^{-\mathrm{i}(\omega-\tilde{\omega}_{nm})(t-t')}\,. \tag{5.45}$$

Note that we have assumed $\tilde{\omega}_{mn} = -\tilde{\omega}_{nm}$, which will be confirmed by (5.65) below. In addition, we extend the lower limit of the time integral to minus infinity, so that

$$\begin{aligned}\int_{t_0}^{t}\mathrm{d}t'\,\mathrm{e}^{-\mathrm{i}(\omega-\tilde{\omega}_{nm})(t-t')} &\simeq \int_{-\infty}^{t}\mathrm{d}t'\,\mathrm{e}^{-\mathrm{i}(\omega-\tilde{\omega}_{nm})(t-t')}\\ &= \pi\delta(\omega-\tilde{\omega}_{nm}) - \mathrm{i}\,\frac{\mathcal{P}}{\omega-\tilde{\omega}_{nm}}\end{aligned} \tag{5.46}$$

and hence

$$\int_{t_0}^{t}\mathrm{d}t'\,\mathrm{e}^{-\mathrm{i}\omega(t-t')}\hat{A}_{mn}(t') \simeq \left[\pi\delta(\omega-\tilde{\omega}_{nm}) - \mathrm{i}\,\frac{\mathcal{P}}{\omega-\tilde{\omega}_{nm}}\right]\hat{A}_{mn}(t)\,. \tag{5.47}$$

This result immediately shows why the Markov approximation only applies for a sufficiently narrow field spectrum: We have effectively replaced the exact frequency-dependence of the time integral by an idealised delta function plus principal-value profile. Obviously, this is only possible for a flat field spectrum that does not resolve the deviations of the exact time integral from this profile.

Substituting our solution for the time integral, the electric field takes the form

$$\begin{aligned}\underline{\hat{\boldsymbol{E}}}(\boldsymbol{r},\omega,t) &= \mathrm{e}^{-\mathrm{i}\omega(t-t_0)}\underline{\hat{\boldsymbol{E}}}(\boldsymbol{r},\omega) + \mathrm{i}\mu_0\sum_{m,n}\left[\delta(\omega-\tilde{\omega}_{nm}) - \frac{\mathrm{i}}{\pi}\frac{\mathcal{P}}{\omega-\tilde{\omega}_{nm}}\right]\\ &\quad\times\omega^2\mathrm{Im}\,\boldsymbol{G}(\boldsymbol{r},\boldsymbol{r}_A,\omega)\cdot\boldsymbol{d}_{mn}\hat{A}_{mn}(t)\,.\end{aligned} \tag{5.48}$$

The field at time t now only depends of the value of the atomic flip operators at the same time t. Physically, the Markov approximation thus means that the field has no memory of the internal atomic state at earlier times.

Having solved the atom-dependent field dynamics, we return our attention to the internal equation of motion for the atom. As a preparation, we note that the total field operator commutes with the atomic flip operators at equal times. We may hence rearrange the product in (5.40) without changing the result:

$$\dot{\hat{A}}_{mn} = \mathrm{i}\omega_{mn}\hat{A}_{mn} + \frac{\mathrm{i}}{\hbar}\sum_k \int_0^\infty \mathrm{d}\omega \big[\big(\hat{A}_{mk}\boldsymbol{d}_{nk} - \hat{A}_{kn}\boldsymbol{d}_{km}\big)\cdot\hat{\underline{\boldsymbol{E}}}(\boldsymbol{r}_A,\omega) + \hat{\underline{\boldsymbol{E}}}^\dagger(\boldsymbol{r}_A,\omega)\cdot\big(\boldsymbol{d}_{nk}\hat{A}_{mk} - \boldsymbol{d}_{km}\hat{A}_{kn}\big)\big] . \quad (5.49)$$

Substituting our solution (5.48) for the electric field and evaluating operator products according to

$$\hat{A}_{mn}(t)\hat{A}_{kl}(t) = \delta_{nk}\hat{A}_{ml}(t) , \quad (5.50)$$

we find a closed equation of motion for the atomic flip operators:

$$\begin{aligned}\dot{\hat{A}}_{mn}(t) &= \mathrm{i}\omega_{mn}\hat{A}_{mn}(t)\\ &+ \frac{\mathrm{i}}{\hbar}\sum_k \int_0^\infty \mathrm{d}\omega \big\{\mathrm{e}^{-\mathrm{i}\omega(t-t_0)}\big[\hat{A}_{mk}(t)\boldsymbol{d}_{nk} - \hat{A}_{kn}(t)\boldsymbol{d}_{km}\big]\cdot\hat{\underline{\boldsymbol{E}}}(\boldsymbol{r}_A,\omega)\\ &+ \mathrm{e}^{\mathrm{i}\omega(t-t_0)}\hat{\underline{\boldsymbol{E}}}^\dagger(\boldsymbol{r}_A,\omega)\cdot\big[\boldsymbol{d}_{nk}\hat{A}_{mk}(t) - \boldsymbol{d}_{km}\hat{A}_{kn}(t)\big]\big\}\\ &- \sum_{k,l}\big[\boldsymbol{d}_{nk}\cdot\boldsymbol{C}_{kl}\hat{A}_{ml}(t) - \boldsymbol{d}_{km}\cdot\boldsymbol{C}_{nl}\hat{A}_{kl}(t)\big]\\ &+ \sum_{k,l}\big[\boldsymbol{d}_{nk}\cdot\boldsymbol{C}^*_{ml}\hat{A}_{lk}(t) - \boldsymbol{d}_{km}\cdot\boldsymbol{C}^*_{kl}\hat{A}_{ln}(t)\big] \end{aligned} \quad (5.51)$$

with coefficients

$$\begin{aligned}\boldsymbol{C}_{mn} = \boldsymbol{C}_{mn}(\boldsymbol{r}_A) &= \frac{\mu_0}{\hbar}\Theta(\tilde{\omega}_{nm})\tilde{\omega}_{nm}^2 \mathrm{Im}\,\boldsymbol{G}(\boldsymbol{r}_A,\boldsymbol{r}_A,\tilde{\omega}_{nm})\cdot\boldsymbol{d}_{mn}\\ &- \frac{\mathrm{i}\mu_0}{\pi\hbar}\mathcal{P}\int_0^\infty \frac{\mathrm{d}\omega}{\omega-\tilde{\omega}_{nm}}\omega^2 \mathrm{Im}\,\boldsymbol{G}(\boldsymbol{r}_A,\boldsymbol{r}_A,\omega)\cdot\boldsymbol{d}_{mn} .\end{aligned} \quad (5.52)$$

As a result of our rearrangement (5.49), this equation appears in normal ordering with field annihilation operators $\hat{\underline{\boldsymbol{E}}}$ to the right and creation operators $\hat{\underline{\boldsymbol{E}}}^\dagger$ to the left. This ordering will simplify our calculation considerably.

Next, we take expectation values of (5.51). We assume that the electromagnetic field is prepared in its ground state at initial time t_0, so that its density matrix reads $\hat{\rho} = \hat{\rho}(t_0) = |\{0\}\rangle\langle\{0\}|$. The definition (1.19) of the ground state together with the field expansion (1.22) implies that

$$\hat{\underline{\boldsymbol{E}}}(\boldsymbol{r},\omega)|\{0\}\rangle = \boldsymbol{0} \quad \forall\,\boldsymbol{r},\omega . \quad (5.53)$$

As a benefit of the normal ordering used, the free electric field does hence not contribute to the dynamics of the quantum-averaged flip operators,

$$\big\langle\dot{\hat{A}}_{mn}(t)\big\rangle = \mathrm{i}\omega_{mn}\big\langle\hat{A}_{mn}(t)\big\rangle + \sum_{k,l}\big[\boldsymbol{d}_{km}\cdot\boldsymbol{C}_{nl}\big\langle\hat{A}_{kl}(t)\big\rangle + \boldsymbol{d}_{nk}\cdot\boldsymbol{C}^*_{ml}\big\langle\hat{A}_{lk}(t)\big\rangle$$
$$- \boldsymbol{d}_{nk}\cdot\boldsymbol{C}_{kl}\big\langle\hat{A}_{ml}(t)\big\rangle - \boldsymbol{d}_{km}\cdot\boldsymbol{C}^*_{kl}\big\langle\hat{A}_{ln}(t)\big\rangle\big]\,. \tag{5.54}$$

In order to decouple this set of linear differential equations, we assume that the atom is free of quasi-degenerate transitions. Exact degeneracies $\omega_{mn} = \omega_{m'n'}$ naturally occur in most atoms when the states in each pair $|n\rangle$, $|n'\rangle$ and $|m\rangle$, $|m'\rangle$ belong to common manifolds of degenerate energy eigenstates, e.g., states that only differ in the z-component of the electronic orbital angular momentum. Quasi-degeneracies are transitions $\omega_{mn} \simeq \omega_{kl}$ where no two states belong to the same manifold. In addition, we assume that the free atom is unpolarised in each of its energy eigenstates, $\boldsymbol{d}_{nn} = \boldsymbol{0}$, and that states of a degenerate manifold are not connected by electric-dipole transitions, $\boldsymbol{d}_{nn'} = \boldsymbol{0}$. Both of these conditions are guaranteed by atomic selection rules.

Under the conditions above, the fast-oscillating off-diagonal flip operators effectively decouple from the non-oscillating diagonal ones as well as from each other. For $m \neq n$, we hence retain only those terms on the right hand side of the equation which are also proportional to $\langle\hat{A}_{mn}\rangle$. Using the fact that $\boldsymbol{d}_{nn} = \boldsymbol{0}$, we find

$$\big\langle\dot{\hat{A}}_{mn}(t)\big\rangle = \Big[\mathrm{i}\omega_{mn} - \sum_{k}(\boldsymbol{d}_{nk}\cdot\boldsymbol{C}_{kn} + \boldsymbol{d}_{km}\cdot\boldsymbol{C}^*_{km})\Big]\big\langle\hat{A}_{mn}(t)\big\rangle\,. \tag{5.55}$$

Note that contributions of the type $\boldsymbol{d}_{m'm}\cdot\boldsymbol{C}_{nn'}\langle\hat{A}_{m'n'}\rangle$ from exact degeneracies vanish due to $\boldsymbol{d}_{nn'} = \boldsymbol{0}$, whereas terms of the type $\boldsymbol{d}_{nk}\cdot\boldsymbol{C}_{kn'}\langle\hat{A}_{mn'}\rangle$ are excluded by atomic selection rules.

The diagonal flip operators are non-oscillating and mutually coupled. Retaining only diagonal terms on the right hand side of (5.54), we find

$$\big\langle\dot{\hat{A}}_{nn}(t)\big\rangle = -\sum_{k}(\boldsymbol{d}_{nk}\cdot\boldsymbol{C}_{kn} + \boldsymbol{d}_{kn}\cdot\boldsymbol{C}^*_{kn})\big\langle\hat{A}_{nn}(t)\big\rangle$$
$$+ \sum_{k}(\boldsymbol{d}_{kn}\cdot\boldsymbol{C}_{nk} + \boldsymbol{d}_{nk}\cdot\boldsymbol{C}^*_{nk})\big\langle\hat{A}_{kk}(t)\big\rangle\,. \tag{5.56}$$

Once more, terms of the type $\boldsymbol{d}_{nk}\cdot\boldsymbol{C}_{kn'}\langle\hat{A}_{nn'}\rangle$ or $\boldsymbol{d}_{kn}\cdot\boldsymbol{C}_{nk'}\langle\hat{A}_{kk'}\rangle$ are excluded by atomic selection rules.

Real and imaginary parts of the coefficients in the above equations can be taken according to

$$\sum_k \boldsymbol{d}_{nk}\cdot\boldsymbol{C}_{kn} = \tfrac{1}{2}\sum_{k<n}\Gamma_{nk} + \mathrm{i}\sum_k \delta\omega_{nk} = \tfrac{1}{2}\Gamma_n + \mathrm{i}\delta\omega_n\,, \tag{5.57}$$

$$\sum_k \boldsymbol{d}_{kn}\cdot\boldsymbol{C}^*_{kn} = \sum_{k<n}\tfrac{1}{2}\Gamma_{nk} - \mathrm{i}\sum_k \delta\omega_{nk} = \tfrac{1}{2}\Gamma_n - \mathrm{i}\delta\omega_n \tag{5.58}$$

where we have introduced

$$\delta\omega_n = \sum_k \delta\omega_{nk}\,, \tag{5.59}$$

$$\Gamma_n = \sum_{k<n}\Gamma_{nk} \tag{5.60}$$

with

$$\delta\omega_{nk} = \delta\omega_{nk}(\boldsymbol{r}_A) = -\frac{\mu_0}{\pi\hbar}\mathcal{P}\int_0^\infty \frac{\mathrm{d}\omega}{\omega-\tilde{\omega}_{nk}}\,\omega^2 \boldsymbol{d}_{nk}\cdot\mathrm{Im}\,\boldsymbol{\mathsf{G}}^{(1)}(\boldsymbol{r}_A,\boldsymbol{r}_A,\omega)\cdot\boldsymbol{d}_{kn}\,, \tag{5.61}$$

$$\Gamma_{nk} = \Gamma_{nk}(\boldsymbol{r}_A) = \frac{2\mu_0}{\hbar}\,\tilde{\omega}^2_{nk}\boldsymbol{d}_{nk}\cdot\mathrm{Im}\,\boldsymbol{\mathsf{G}}(\boldsymbol{r}_A,\boldsymbol{r}_A,\tilde{\omega}_{nk})\cdot\boldsymbol{d}_{kn} \tag{5.62}$$

and noted that the Green's tensor is symmetric, cf. (A.4). With these definitions, the equations of motion for the expectation values of the atomic flip operators take the final form

$$\langle\dot{\hat{A}}_{nn}(t)\rangle = -\Gamma_n\langle\hat{A}_{nn}(t)\rangle + \sum_{k>n}\Gamma_{kn}\langle\hat{A}_{kk}(t)\rangle\,, \tag{5.63}$$

$$\langle\dot{\hat{A}}_{mn}(t)\rangle = \left[\mathrm{i}\tilde{\omega}_{mn} - \tfrac{1}{2}(\Gamma_m+\Gamma_n)\right]\langle\hat{A}_{mn}(t)\rangle \quad \text{for } m\neq n \tag{5.64}$$

where we have identified the frequencies $\tilde{\omega}_{mn}$ as

$$\tilde{\omega}_{mn} = \omega_{mn} + \delta\omega_m - \delta\omega_n\,. \tag{5.65}$$

For the evaluation of the CP force, we will also require two-time correlation functions of the atomic flip operators. They can be obtained by means of the quantum regression theorem. To that end, we integrate the equation of motion for the expectation values of the off-diagonal flip operators,

$$\langle\hat{A}_{mn}(t)\rangle = \mathrm{e}^{[\mathrm{i}\tilde{\omega}_{mn}-(\Gamma_m+\Gamma_n)/2](t-t_0)}\langle\hat{A}_{mn}\rangle \quad \text{for } m\neq n\,. \tag{5.66}$$

It follows that expectation values at different times are related according to

$$\langle\hat{A}_{mn}(t)\rangle = f(t-t')\langle\hat{A}_{mn}(t')\rangle \quad \text{for } t\geq t' \tag{5.67}$$

where the function

$$f_{mn}(t-t') = \mathrm{e}^{[\mathrm{i}\tilde{\omega}_{mn}-(\Gamma_m+\Gamma_n)/2](t-t')} \tag{5.68}$$

only depends on the time difference. The quantum regression theorem or Onsager–Lax regression theorem states that whenever this is the case, two-time correlation functions are given by [6–8]

$$\langle \hat{A}_{mn}(t)\hat{A}_{kl}(t')\rangle = f_{mn}(t-t')\langle \hat{A}_{mn}(t')\hat{A}_{kl}(t')\rangle \quad \text{for } t \geq t' . \tag{5.69}$$

Recalling (5.50), we hence have

$$\langle \hat{A}_{mn}(t)\hat{A}_{kl}(t')\rangle = \mathrm{e}^{[\mathrm{i}\tilde{\omega}_{mn}-(\Gamma_m+\Gamma_n)/2](t-t')}\delta_{nk}\langle \hat{A}_{ml}(t')\rangle \quad \text{for } t \geq t' . \tag{5.70}$$

The applicability of the quantum regression theorem is a consequence of the Markov approximation, according to which the electric field is insensitive to the influence of the atom at earlier times.

After this little digression, we return to the dynamics (5.63) and (5.66) of single atomic flip operators. The time-dependence of the atomic density matrix elements follows immediately by virtue of the relations (5.34) and (5.35),

$$\dot{p}_n(t) = -\Gamma_n p_n(t) + \sum_{k>n} \Gamma_{kn} p_k(t) , \tag{5.71}$$

$$\sigma_{mn}(t) = \mathrm{e}^{[-\mathrm{i}\tilde{\omega}_{mn}-(\Gamma_m+\Gamma_n)/2](t-t_0)}\sigma_{mn} \quad \text{for } m \neq n . \tag{5.72}$$

This internal dynamics of an atom, governed by the spontaneous and irreversible emission of real photons, is known as spontaneous decay. Due to spontaneous decay, the population p_n of a given state $|n\rangle$ is reduced via to downward transitions to lower lying states, while gaining via to population transfer from the states above. Note that the decay-induced dynamics is probability-conserving, (5.60) implies that

$$\frac{\mathrm{d}}{\mathrm{d}t}\sum_n p_n(t) = 0 . \tag{5.73}$$

An atom initially excited an energy eigenstate, $\hat{\sigma}(t_0) = |n\rangle\langle n|$, will decay to an incoherent superposition of energy eigenstates,

$$\hat{\sigma}(t) = \sum_{k\leq n} p_k(t)|k\rangle\langle k| \quad \text{for } t \geq t_0 . \tag{5.74}$$

In the long-time limit, the atom will reach its ground state $\hat{\sigma}(t\to\infty) = |0\rangle\langle 0|$ from which no further decay is possible since $\Gamma_0 = 0$.

The transition rates (5.62) depend on the respective dipole matrix elements and the imaginary part of Green's tensor taken at the atomic transitions frequencies. The total decay rate (5.60) of an excited state is obtained by summing the transition rates to all lower lying states. The lifetime of an excited state $T_1 = 1/\Gamma_n$ is commonly known as the longitudinal time or T_1-time. It sets the time scale for the incoherent population transfer or energy relaxation.

In free space, (A.26) shows that

$$\mathrm{Im}\,\boldsymbol{G}^{(0)}(\boldsymbol{r},\boldsymbol{r},\omega) = \frac{\omega}{6\pi c}\,\boldsymbol{I}\,, \tag{5.75}$$

so the transition rates take the forms

$$\Gamma_{nk} = \frac{\omega_{nk}^3 |\boldsymbol{d}_{nk}|^2}{3\pi\varepsilon_0 \hbar c^3}\,. \tag{5.76}$$

They depend on the internal structure of the atom and are commonly known as the Einstein A-coefficients, cf. Sect. 7.3.1 below. The total decay rate reads

$$\Gamma_n = \frac{1}{3\pi\varepsilon_0 \hbar c^3} \sum_{k<n} \omega_{nk}^3 |\boldsymbol{d}_{nk}|^2\,. \tag{5.77}$$

The off-diagonal density matrix elements are only non-vanishing if the atom is initially prepared in a coherent superposition of energy eigenstates. For instance, consider an atom prepared in a qubit state $(|0\rangle+|1\rangle)/\sqrt{2}$ whose initial density matrix reads $\hat{\sigma}(t_0) = \frac{1}{2}(|0\rangle\langle 0| + |0\rangle\langle 1| + |1\rangle\langle 0| + |1\rangle\langle 1|)$. According to (5.72), they undergo damped oscillations. The associated damping time $T_2 = 2/(\Gamma_n + \Gamma_k)$ is known as the transverse time or T_2-time. It sets the timescale for the loss of decoherence or phase relaxation of an internal atomic state due to spontaneous decay. For a coherent superposition of an excited atomic state with the ground state, we have $T_2 = 2T_1$. This relation between the longitudinal and transverse time is characteristic for the energy and phase relaxation induced by spontaneous decay. Other loss mechanisms may induce pure phase relaxation, leading to $T_2 < 2T_1$ [9].

The oscillations of the off-diagonal atomic density matrix elements are governed by the atomic transition frequencies. According to (5.65), these frequencies acquire shifts (5.59) due to the emission and absorption of photons. The shift (5.61) of an energy level E_n due to transitions to a state $|k\rangle$ depends on the respective atomic transition frequency and dipole matrix element as well as the electromagnetic Green's tensor. Note that we have separated the Green's tensor into its bulk and scattering parts according to (1.111). We have assumed the Lamb shift associated with the free-space Green's tensor to be already included in the bare transition frequencies ω_{mn} by making the replacement $\boldsymbol{G} \mapsto \boldsymbol{G}^{(1)}$ (see the discussion in Sect. 4.1 of Vol. I). The remaining frequency shifts are hence entirely due to the present bodies surrounding the atom. Using the integration contour of Fig. 4.1, they can be cast into the form

$$\delta\omega_{nk} = -\frac{\mu_0}{\hbar}\sum_{k<n}\tilde{\omega}_{nk}^2 \boldsymbol{d}_{nk}\cdot\mathrm{Re}\,\boldsymbol{G}^{(1)}(\boldsymbol{r}_A,\boldsymbol{r}_A,\tilde{\omega}_{nk})\cdot\boldsymbol{d}_{kn} + \frac{\mu_0}{\pi\hbar}\sum_k\int_0^\infty \mathrm{d}\xi\,\frac{\tilde{\omega}_{kn}\xi^2}{\tilde{\omega}_{kn}^2+\xi^2}\boldsymbol{d}_{nk}\cdot\boldsymbol{G}^{(1)}(\boldsymbol{r}_A,\boldsymbol{r}_A,\mathrm{i}\xi)\cdot\boldsymbol{d}_{kn}\,, \qquad (5.78)$$

with the two terms being due to virtual and real photons, respectively. Note that this equation for the frequency shifts depends in turn on the shifted transition frequencies. It is hence an implicit equation from which the shifts have to be determined as a self-consistent solution, cf. Sect. 5.5 below.

In the perturbative limit $\delta\omega_m, \delta\omega_n \ll \omega_{mn}$, the frequency shifts are given by the explicit equation

$$\delta\omega_{nk} = -\frac{\mu_0}{\hbar}\sum_{k<n}\omega_{nk}^2 \boldsymbol{d}_{nk}\cdot\mathrm{Re}\,\boldsymbol{G}^{(1)}(\boldsymbol{r}_A,\boldsymbol{r}_A,\omega_{nk})\cdot\boldsymbol{d}_{kn} + \frac{\mu_0}{\pi\hbar}\sum_k\int_0^\infty \mathrm{d}\xi\,\frac{\omega_{kn}\xi^2}{\omega_{kn}^2+\xi^2}\boldsymbol{d}_{nk}\cdot\boldsymbol{G}^{(1)}(\boldsymbol{r}_A,\boldsymbol{r}_A,\mathrm{i}\xi)\cdot\boldsymbol{d}_{kn}\,. \qquad (5.79)$$

The associated shift $\Delta E_n = \hbar\delta\omega_n = \sum_k \hbar\delta\omega_{nk}$ of an atomic energy level agrees with the CP potential (4.40)–(4.42) found in Sect. 4.1.2.

Apart from being relevant for atoms prepared in coherent superposition states, the off-diagonal atomic density matrix elements also govern the average dipole moment of the atom. Using the expansion (5.38) and the property $\boldsymbol{d}_{nn} = \boldsymbol{0}$, it is given by

$$\langle\hat{\boldsymbol{d}}(t)\rangle = \sum_{m\neq n}\boldsymbol{d}_{mn}\langle\hat{A}_{mn}(t)\rangle = \sum_{m\neq n}\boldsymbol{d}_{mn}\sigma_{nm}(t)\,. \qquad (5.80)$$

The average dipole moment obviously vanishes when the atom is initially prepared in an incoherent superposition of energy eigenstates. An atom in an energy eigenstate $|n\rangle$ can acquire a non-vanishing dipole moment when it is driven by to an applied electric field. The relation between the induced dipole moment and the electric field takes a particularly simple form in Fourier space,

$$\langle\hat{\underline{\boldsymbol{d}}}(\omega)\rangle_n = \boldsymbol{\alpha}_n(\omega)\cdot\underline{\boldsymbol{E}}(\boldsymbol{r}_A,\omega)\,. \qquad (5.81)$$

The relevant proportionality constant $\boldsymbol{\alpha}_n(\omega)$ is the polarisability of the atom in state $|n\rangle$.

5.3 Atomic Polarisability

As we have seen in Chap. 4, the non-resonant CP potential is governed by the atomic polarisability which depends on the atomic transitions. To analyse it in more detail, we need to solve the internal dynamics of a driven atom and find the time-dependent average dipole moment (5.80). We model the applied electric field by assuming the quantum electric field to be prepared in a coherent state $|\{\underline{\boldsymbol{E}}(\boldsymbol{r},\omega)\}\rangle$,

$$\hat{\underline{\boldsymbol{E}}}(\boldsymbol{r},\omega)|\{\underline{\boldsymbol{E}}(\boldsymbol{r},\omega)\}\rangle = \underline{\boldsymbol{E}}(\boldsymbol{r},\omega)|\{\underline{\boldsymbol{E}}(\boldsymbol{r},\omega)\}\rangle \quad \forall \boldsymbol{r},\omega\,. \tag{5.82}$$

The equations of motion (5.51) for the atomic flip operators then take the form

$$\begin{aligned}\dot{\hat{A}}_{mn}(t) &= \mathrm{i}\omega_{mn}\hat{A}_{mn}(t)\\ &\quad+\frac{\mathrm{i}}{\hbar}\sum_k\int_0^\infty \mathrm{d}\omega\big\{\mathrm{e}^{-\mathrm{i}\omega(t-t_0)}\big[\hat{A}_{mk}(t)\boldsymbol{d}_{nk}-\hat{A}_{kn}(t)\boldsymbol{d}_{km}\big]\cdot\underline{\boldsymbol{E}}(\boldsymbol{r}_A,\omega)\\ &\qquad+\mathrm{e}^{\mathrm{i}\omega(t-t_0)}\underline{\boldsymbol{E}}^*(\boldsymbol{r}_A,\omega)\cdot\big[\boldsymbol{d}_{nk}\hat{A}_{mk}(t)-\boldsymbol{d}_{km}\hat{A}_{kn}(t)\big]\big\}\\ &\quad-\sum_{k,l}\big[\boldsymbol{d}_{nk}\cdot\boldsymbol{C}_{kl}\hat{A}_{ml}(t)-\boldsymbol{d}_{km}\cdot\boldsymbol{C}_{nl}\hat{A}_{kl}(t)\big]\\ &\quad+\sum_{k,l}\big[\boldsymbol{d}_{nk}\cdot\boldsymbol{C}^*_{ml}\hat{A}_{lk}(t)-\boldsymbol{d}_{km}\cdot\boldsymbol{C}^*_{kl}\hat{A}_{ln}(t)\big]\end{aligned} \tag{5.83}$$

After taking expectation values and decomposing the coefficients into real and imaginary parts according to (5.57) and (5.58), the off-diagonal flip operators are found to be governed by

$$\begin{aligned}\big\langle\dot{\hat{A}}_{mn}(t)\big\rangle &= \big[\mathrm{i}\tilde{\omega}_{mn}-\tfrac{1}{2}(\Gamma_m+\Gamma_n)\big]\big\langle\hat{A}_{mn}(t)\big\rangle\\ &\quad+\frac{\mathrm{i}}{\hbar}\sum_k\int_0^\infty \mathrm{d}\omega\big\{\mathrm{e}^{-\mathrm{i}\omega(t-t_0)}\underline{\boldsymbol{E}}(\boldsymbol{r}_A,\omega)\cdot\big[\boldsymbol{d}_{nk}\big\langle\hat{A}_{mk}(t)\big\rangle-\boldsymbol{d}_{km}\big\langle\hat{A}_{kn}(t)\big\rangle\big]\\ &\qquad+\mathrm{e}^{\mathrm{i}\omega(t-t_0)}\underline{\boldsymbol{E}}^*(\boldsymbol{r}_A,\omega)\cdot\big[\boldsymbol{d}_{nk}\big\langle\hat{A}_{mk}(t)\big\rangle-\boldsymbol{d}_{km}\big\langle\hat{A}_{kn}(t)\big\rangle\big]\big\}\,.\end{aligned} \tag{5.84}$$

This inhomogeneous linear differential equation can be easily integrated. Using the Fourier relation

$$\boldsymbol{E}(\boldsymbol{r},t) = \int_0^\infty \mathrm{d}\omega\big[\mathrm{e}^{-\mathrm{i}\omega t}\underline{\boldsymbol{E}}(\boldsymbol{r},\omega)+\mathrm{e}^{\mathrm{i}\omega t}\underline{\boldsymbol{E}}^*(\boldsymbol{r},\omega)\big]\,, \tag{5.85}$$

the result can be given in the form

$$\begin{aligned}\langle\hat{A}_{mn}(t)\rangle &= \mathrm{e}^{[\mathrm{i}\tilde{\omega}_{mn}-(\Gamma_m+\Gamma_n)/2](t-t_0)}\langle\hat{A}_{mn}\rangle \\ &\quad+\frac{\mathrm{i}}{\hbar}\sum_k\int_{t_0}^{t}\mathrm{d}t'\mathrm{e}^{[\mathrm{i}\tilde{\omega}_{mn}-(\Gamma_m+\Gamma_n)/2](t-t')} \\ &\quad\times\left[\langle\hat{A}_{mk}(t')\rangle\boldsymbol{d}_{nk}-\langle\hat{A}_{kn}(t')\rangle\boldsymbol{d}_{km}\right]\cdot\boldsymbol{E}(\boldsymbol{r}_A,t')\,. \end{aligned}\tag{5.86}$$

Note that the dynamics of a the flip operators for a driven atom does not meet the requirements of the quantum regression theorem.

We can now evaluate the average dipole moment (5.80) for an atom in an energy eigenstate $|n\rangle$ using our result (5.86). We restrict our attention to a time scale $t \ll 1/\Gamma_n$ where the atom effectively remains in its initial state $|n\rangle$, so that $\langle\hat{A}_{kl}(t')\rangle \simeq \langle\hat{A}_{kl}(t)\rangle \simeq \delta_{kn}\delta_{ln}$. Using the property $\boldsymbol{d}_{nn} = \boldsymbol{0}$, we then find

$$\begin{aligned}\langle\hat{\boldsymbol{d}}(t)\rangle_n &= \sum_{k\neq n}\left[\boldsymbol{d}_{nk}\langle\hat{A}_{nk}(t)\rangle_n+\boldsymbol{d}_{kn}\langle\hat{A}_{kn}(t)\rangle_n\right] \\ &= \frac{\mathrm{i}}{\hbar}\sum_k\int_{-\infty}^{t}\mathrm{d}t'\,\left\{\mathrm{e}^{[\mathrm{i}\tilde{\omega}_{nk}-(\Gamma_m+\Gamma_n)/2](t-t')}\boldsymbol{d}_{nk}\boldsymbol{d}_{kn}\right. \\ &\quad\left.-\,\mathrm{e}^{[\mathrm{i}\tilde{\omega}_{kn}-(\Gamma_m+\Gamma_n)/2](t-t')}\boldsymbol{d}_{kn}\boldsymbol{d}_{nk}\right\}\cdot\boldsymbol{E}(\boldsymbol{r}_A,t')\,. \end{aligned}\tag{5.87}$$

Note that we have applied the Markov approximation by letting $t_0 \to -\infty$.

To identify the atomic polarisability, we need to take Fourier components

$$\begin{aligned}\langle\underline{\hat{\boldsymbol{d}}}(\omega)\rangle_n &= \frac{1}{2\pi}\int_{-\infty}^{\infty}\mathrm{d}t\,\mathrm{e}^{\mathrm{i}\omega t}\langle\hat{\boldsymbol{d}}(t)\rangle_n \\ &= \frac{\mathrm{i}}{\hbar}\sum_k\frac{1}{2\pi}\int_{-\infty}^{\infty}\mathrm{d}t\,\mathrm{e}^{\mathrm{i}\omega t}\int_{-\infty}^{t}\mathrm{d}t'\,\left\{\mathrm{e}^{[\mathrm{i}\tilde{\omega}_{nk}-(\Gamma_m+\Gamma_n)/2](t-t')}\boldsymbol{d}_{nk}\boldsymbol{d}_{kn}\right. \\ &\quad\left.-\,\mathrm{e}^{[\mathrm{i}\tilde{\omega}_{kn}-(\Gamma_m+\Gamma_n)/2](t-t')}\boldsymbol{d}_{kn}\boldsymbol{d}_{nk}\right\}\cdot\boldsymbol{E}(\boldsymbol{r}_A,t')\,. \end{aligned}\tag{5.88}$$

Substituting the Fourier decomposition (5.85) of the electric field, the t'-integral can be easily performed. Evaluating the t-integral according to

$$\frac{1}{2\pi}\int_{-\infty}^{\infty}\mathrm{d}t\,\mathrm{e}^{\mathrm{i}(\omega-\omega')t} = \delta(\omega-\omega')\,,\tag{5.89}$$

we find

$$\langle \hat{\underline{\boldsymbol{d}}}(\omega)\rangle_n = \frac{1}{\hbar}\sum_k \left[\frac{\boldsymbol{d}_{nk}\boldsymbol{d}_{kn}}{\tilde{\omega}_{kn} - \omega - \frac{\mathrm{i}}{2}(\Gamma_n + \Gamma_k)} + \frac{\boldsymbol{d}_{kn}\boldsymbol{d}_{nk}}{\tilde{\omega}_{kn} + \omega + \frac{\mathrm{i}}{2}(\Gamma_n + \Gamma_k)}\right]\cdot \underline{\boldsymbol{E}}(\boldsymbol{r}_A, \omega)\,. \tag{5.90}$$

By comparison with (5.81), the polarisability of an atom in an energy eigenstate $|n\rangle$ is given by the dispersion formula

$$\begin{aligned}\boldsymbol{\alpha}_n(\omega) &= \boldsymbol{\alpha}_n(\boldsymbol{r}_A, \omega)\\ &= \frac{1}{\hbar}\sum_k \left[\frac{\boldsymbol{d}_{nk}\boldsymbol{d}_{kn}}{\tilde{\omega}_{kn} - \omega - \frac{\mathrm{i}}{2}(\Gamma_n + \Gamma_k)} + \frac{\boldsymbol{d}_{kn}\boldsymbol{d}_{nk}}{\tilde{\omega}_{kn} + \omega + \frac{\mathrm{i}}{2}(\Gamma_n + \Gamma_k)}\right].\end{aligned} \tag{5.91}$$

The atomic polarisability exhibits resonance lines centred around the shifted atomic transition frequencies $\tilde{\omega}_{mn}$ with widths $\Gamma_m + \Gamma_n$. The rates of spontaneous decay Γ_n may hence be interpreted as a finite widths of the associated energy levels $|n\rangle$. Note that conventions of quantum field theory suggest the level width to enter the non-resonant, second term of the polarisability with an opposite sign [10], which has led to some controversies in the past [11–14]. The issue was resolved by Milonni and Boyd [15]. Using a dynamical derivation similar to ours, they showed that the definition (5.81) of the atomic polarisability leads to the above correct signs for the level widths. These signs are necessary to ensure the convergence of the t'-integral in (5.88) and guarantee the Schwarz reflection principle

$$\boldsymbol{\alpha}_n^*(\omega) = \boldsymbol{\alpha}_n(-\omega^*)\,. \tag{5.92}$$

Recalling that the line shifts and widths depend on the position of the atom, we note that the atomic polarisability is also position-dependent, in general. Furthermore, the shifts and widths may induce an anisotropy of the atomic polarisability. To see this, consider an atom in an isotropic state $|n\rangle$ such that the relation (4.16) holds. Even so, we cannot in general use this relation to simplify the polarisability (5.91) since the shifts and widths may vary within manifolds of energy-degenerate states, $\Gamma_{k'} \neq \Gamma_{k''}$, $\delta\omega_{k'} \neq \delta\omega_{k''}$ in spite of $k', k'' \in \{\overline{k}\}$. In other words, the atomic polarisability may be anisotropic even when the state of the atom is isotropic; the isotropy of the atom is broken by the influence of its anisotropic environment. This complication does not arise in free space where $\delta\omega_k = 0$ and $\Gamma_{k'} = \Gamma_{k''}$ whenever $k', k'' \in \{\overline{k}\}$, cf. (5.77). Exploiting the relation (4.16), we then find that the polarisability of an atom in an isotropic state is indeed isotropic, $\boldsymbol{\alpha}_n(\omega) = \alpha_n(\omega)\boldsymbol{I}$ with

$$\alpha_n(\omega) = \frac{2}{3\hbar}\sum_k \frac{\omega_{kn}|\boldsymbol{d}_{nk}|^2}{\omega_{kn}^2 - [\omega + \frac{\mathrm{i}}{2}(\Gamma_n + \Gamma_k)]^2}\,. \tag{5.93}$$

In the perturbative limit $|\delta\omega_m|, |\delta\omega_n|, \Gamma_m, \Gamma_n \ll \omega_{mn}$, the polarisability reduces to (4.46) as introduced in Sect. 4.1.2. The perturbative polarisability is unaffected by the atomic environment. It is hence always isotropic for an atom in an isotropic state, just like the free-space polarisability.

5.4 Casimir–Polder Force

Having solved the coupled atom–field dynamics, we can now evaluate the average Lorentz force (5.31). Using the expansion (5.38) of the dipole operator and the field expansions (1.22) and (1.26), it takes the form

$$\begin{aligned}\boldsymbol{F}(t) = \sum_{m,n} \int_0^\infty \mathrm{d}\omega \Big[\nabla\big\langle \hat{A}_{mn}(t)\boldsymbol{d}_{mn}\cdot\hat{\underline{\boldsymbol{E}}}(\boldsymbol{r},\omega,t)\big\rangle \\ + \nabla\big\langle \hat{\underline{\boldsymbol{E}}}^\dagger(\boldsymbol{r},\omega,t)\cdot\boldsymbol{d}_{mn}\hat{A}_{mn}(t)\big\rangle\Big]_{\boldsymbol{r}=\boldsymbol{r}_A} \\ + \frac{\mathrm{d}}{\mathrm{d}t}\sum_{m,n}\int_0^\infty \mathrm{d}\omega\Big[\big\langle \hat{A}_{mn}(t)\boldsymbol{d}_{mn}\times\hat{\underline{\boldsymbol{B}}}(\boldsymbol{r}_A,\omega,t)\big\rangle \\ - \big\langle \hat{\underline{\boldsymbol{B}}}^\dagger(\boldsymbol{r}_A,\omega,t)\times\boldsymbol{d}_{mn}\hat{A}_{mn}(t)\big\rangle\Big]\,. \end{aligned} \tag{5.94}$$

For convenience, we have again arranged the operator products in normal ordering. Assuming the atom–field coupling to be weak, we may use our solution (5.44) for the time-dependent electric field. The corresponding solution for the magnetic field

$$\begin{aligned}\hat{\underline{\boldsymbol{B}}}(\boldsymbol{r},\omega,t) = \mathrm{e}^{-\mathrm{i}\omega(t-t_0)}\hat{\underline{\boldsymbol{B}}}(\boldsymbol{r},\omega) \\ + \frac{\mu_0}{\pi}\sum_{m,n}\omega\int_{t_0}^{t}\mathrm{d}t'\,\mathrm{e}^{-\mathrm{i}\omega(t-t')}\nabla\times\mathrm{Im}\,\boldsymbol{\mathsf{G}}(\boldsymbol{r},\boldsymbol{r}_A,\omega)\cdot\boldsymbol{d}_{mn}\hat{A}_{mn}(t')\end{aligned} \tag{5.95}$$

follows immediately by means of (1.8). We assume the electromagnetic field to be prepared in its ground state at initial time, $\hat{\rho} = \hat{\rho}(t_0) = |\{0\}\rangle\langle\{0\}|$. We then have (5.53) for the electric field; the corresponding relation

$$\hat{\underline{\boldsymbol{B}}}(\boldsymbol{r},\omega)|\{0\}\rangle = \boldsymbol{0} \quad \forall\,\boldsymbol{r},\omega \tag{5.96}$$

for the magnetic field follows from the definition (1.19) of the ground state together with the field expansion (1.26). By virtue of the normal ordering used, the free electric and magnetic fields do hence not contribute to the average Lorentz force and we find

$$
\begin{aligned}
\boldsymbol{F}(t) &= \frac{\mathrm{i}\mu_0}{\pi} \sum_{m,n,k,l} \int_0^\infty \mathrm{d}\omega\, \omega^2 \boldsymbol{\nabla} \boldsymbol{d}_{mn} \cdot \operatorname{Im} \boldsymbol{G}(\boldsymbol{r}_A, \boldsymbol{r}_A, \omega) \cdot \boldsymbol{d}_{kl} \\
&\quad \times \int_{t_0}^{t} \mathrm{d}t' \Big[\mathrm{e}^{-\mathrm{i}\omega(t-t')} \big\langle \hat{A}_{mn}(t) \hat{A}_{kl}(t') \big\rangle - \mathrm{e}^{\mathrm{i}\omega(t-t')} \big\langle \hat{A}_{kl}(t') \hat{A}_{mn}(t) \big\rangle \Big] \\
&\quad + \frac{\mu_0}{\pi} \sum_{m,n,k,l} \int_0^\infty \mathrm{d}\omega\, \omega \boldsymbol{d}_{mn} \times [\boldsymbol{\nabla} \times \operatorname{Im} \boldsymbol{G}(\boldsymbol{r}_A, \boldsymbol{r}_A, \omega)] \cdot \boldsymbol{d}_{kl} \\
&\quad \times \frac{\mathrm{d}}{\mathrm{d}t} \int_{t_0}^{t} \mathrm{d}t' \Big[\mathrm{e}^{-\mathrm{i}\omega(t-t')} \big\langle \hat{A}_{mn}(t) \hat{A}_{kl}(t') \big\rangle + \mathrm{e}^{\mathrm{i}\omega(t-t')} \big\langle \hat{A}_{kl}(t') \hat{A}_{mn}(t) \big\rangle \Big] \\
&= \frac{\mathrm{i}\mu_0}{\pi} \sum_{m,n,k,l} \int_0^\infty \mathrm{d}\omega\, \omega^2 \boldsymbol{\nabla} \boldsymbol{d}_{mn} \cdot \operatorname{Im} \boldsymbol{G}(\boldsymbol{r}_A, \boldsymbol{r}_A, \omega) \cdot \boldsymbol{d}_{kl} \\
&\quad \times \int_{t_0}^{t} \mathrm{d}t' \mathrm{e}^{-\mathrm{i}\omega(t-t')} \big\langle \hat{A}_{mn}(t) \hat{A}_{kl}(t') \big\rangle \\
&\quad + \frac{\mu_0}{\pi} \sum_{m,n,k,l} \int_0^\infty \mathrm{d}\omega\, \omega \boldsymbol{d}_{mn} \times [\boldsymbol{\nabla} \times \operatorname{Im} \boldsymbol{G}(\boldsymbol{r}_A, \boldsymbol{r}_A, \omega)] \cdot \boldsymbol{d}_{kl} \\
&\quad \times \frac{\mathrm{d}}{\mathrm{d}t} \int_{t_0}^{t} \mathrm{d}t' \mathrm{e}^{-\mathrm{i}\omega(t-t')} \big\langle \hat{A}_{mn}(t) \hat{A}_{kl}(t') \big\rangle + \text{C.c.}
\end{aligned}
\tag{5.97}
$$

where we use the notation $\boldsymbol{\nabla} \boldsymbol{G}(\boldsymbol{r}_A, \boldsymbol{r}_A, \omega) \equiv \boldsymbol{\nabla} \boldsymbol{G}(\boldsymbol{r}, \boldsymbol{r}_A, \omega)|_{\boldsymbol{r}=\boldsymbol{r}_A}$.

Due to the property $\boldsymbol{d}_{nn} = \boldsymbol{0}$, atomic two-time correlation functions are only needed for the off-diagonal flip operators. They have been derived in the previous section by means of the quantum regression theorem. Substituting the result (5.70) into the expression above, the average Lorentz force reads

$$
\begin{aligned}
\boldsymbol{F}(t) &= \frac{\mathrm{i}\mu_0}{\pi} \sum_{m,n,k} \int_0^\infty \mathrm{d}\omega\, \omega^2 \boldsymbol{\nabla} \boldsymbol{d}_{mk} \cdot \operatorname{Im} \boldsymbol{G}(\boldsymbol{r}_A, \boldsymbol{r}_A, \omega) \cdot \boldsymbol{d}_{kn} \\
&\quad \times \int_{t_0}^{t} \mathrm{d}t'\, \mathrm{e}^{[-\mathrm{i}(\omega-\tilde{\omega}_{mk})-(\Gamma_m+\Gamma_k)/2](t-t')} \big\langle \hat{A}_{mn}(t') \big\rangle + \text{C.c.}
\end{aligned}
$$

$$+\frac{\mu_0}{\pi}\sum_{m,n,k}\int_0^\infty \mathrm{d}\omega\,\omega\boldsymbol{d}_{mk}\times[\nabla\times\mathrm{Im}\,\boldsymbol{G}(\boldsymbol{r}_A,\boldsymbol{r}_A,\omega)]\cdot\boldsymbol{d}_{kn}$$
$$\times\frac{\mathrm{d}}{\mathrm{d}t}\int_{t_0}^{t}\mathrm{d}t'\,\mathrm{e}^{[-\mathrm{i}(\omega-\tilde{\omega}_{mk})-(\Gamma_m+\Gamma_k)/2](t-t')}\big\langle\hat{A}_{mn}(t')\big\rangle+\mathrm{C.c.}\qquad(5.98)$$

The time integrals can be evaluated with the aid of the Markov approximation. With $\hat{A}_{mn}(t')\simeq\mathrm{e}^{-\mathrm{i}\tilde{\omega}_{mn}(t-t')}\hat{A}_{mn}(t)$ and $t_0\to-\infty$, we find

$$\int_{t_0}^{t}\mathrm{d}t'\,\mathrm{e}^{[-\mathrm{i}(\omega-\tilde{\omega}_{mk})-(\Gamma_m+\Gamma_k)/2](t-t')}\big\langle\hat{A}_{mn}(t')\big\rangle$$
$$\simeq\big\langle\hat{A}_{mn}(t)\big\rangle\int_{-\infty}^{t}\mathrm{d}t'\,\mathrm{e}^{[-\mathrm{i}(\omega-\tilde{\omega}_{nk})-(\Gamma_m+\Gamma_k)/2](t-t')}$$
$$=-\frac{\mathrm{i}\big\langle\hat{A}_{mn}(t)\big\rangle}{\omega-\tilde{\omega}_{nk}-\frac{\mathrm{i}}{2}(\Gamma_m+\Gamma_k)}\qquad(5.99)$$

and similarly

$$\frac{\mathrm{d}}{\mathrm{d}t}\int_{t_0}^{t}\mathrm{d}t'\,\mathrm{e}^{[-\mathrm{i}(\omega-\tilde{\omega}_{mk})-(\Gamma_m+\Gamma_k)/2](t-t')}\big\langle\hat{A}_{mn}(t')\big\rangle$$
$$=\big\langle\hat{A}_{mn}(t)\big\rangle+\big[-\mathrm{i}(\omega-\tilde{\omega}_{mk})-\tfrac{\mathrm{i}}{2}(\Gamma_m+\Gamma_k)\big]$$
$$\times\int_{t_0}^{t}\mathrm{d}t'\,\mathrm{e}^{[-\mathrm{i}(\omega-\tilde{\omega}_{mk})-(\Gamma_m+\Gamma_k)/2](t-t')}\big\langle\hat{A}_{mn}(t')\big\rangle$$
$$\simeq\big\langle\hat{A}_{mn}(t)\big\rangle+\big\langle\hat{A}_{mn}(t)\big\rangle\big[\mathrm{i}(\tilde{\omega}_{mk}-\omega)-\tfrac{\mathrm{i}}{2}(\Gamma_m+\Gamma_k)\big]$$
$$\times\int_{-\infty}^{t}\mathrm{d}t'\,\mathrm{e}^{[-\mathrm{i}(\omega-\tilde{\omega}_{nk})-(\Gamma_m+\Gamma_k)/2](t-t')}$$
$$=\frac{\tilde{\omega}_{mn}\big\langle\hat{A}_{mn}(t)\big\rangle}{\omega-\tilde{\omega}_{nk}-\frac{\mathrm{i}}{2}(\Gamma_m+\Gamma_k)}.\qquad(5.100)$$

Note that the time derivative has been evaluated prior to applying the Markov approximation.

The final result for the CP force can be obtained by combining (5.98)–(5.100). Let us first consider the case where the atom is prepared in an incoherent superposition of energy eigenstates. As seen from (5.63) and (5.64), it will remain in an incoherent

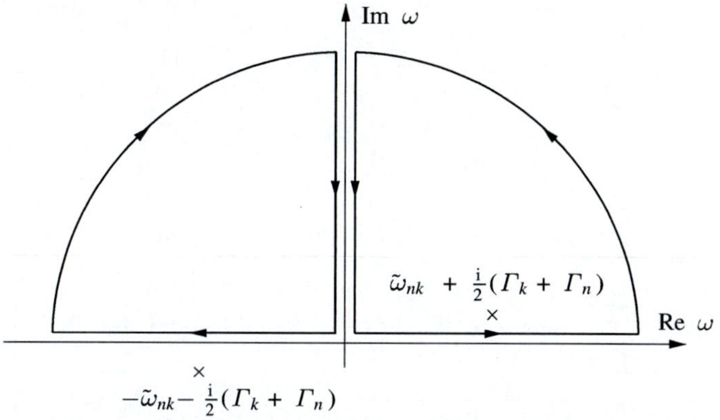

Fig. 5.1 Integration contours used for transforming real-frequency integrals into ones along the positive imaginary axis plus contributions from the poles

state for all times, so that $\langle \hat{A}_{mn}(t)\rangle = 0$ for $m \neq n$. Due to $\tilde{\omega}_{mm} = 0$, the contribution from (5.100) then vanishes. With $\langle \hat{A}_{nn}(t)\rangle = p_n(t)$, the CP force can be given as [1–3, 16]

$$\boldsymbol{F}(\boldsymbol{r}_A, t) = \sum_n p_n(t)\boldsymbol{F}_n(\boldsymbol{r}_A) \tag{5.101}$$

with

$$\boldsymbol{F}_n(\boldsymbol{r}_A) = \frac{\mu_0}{\pi}\sum_k \int_0^\infty \mathrm{d}\omega\, \omega^2 \frac{\boldsymbol{\nabla}\boldsymbol{d}_{nk}\cdot \mathrm{Im}\,\boldsymbol{G}^{(1)}(\boldsymbol{r}_A, \boldsymbol{r}_A, \omega)\cdot \boldsymbol{d}_{kn}}{\omega - \tilde{\omega}_{nk} - \frac{\mathrm{i}}{2}(\Gamma_n + \Gamma_k)} + \text{C.c.} \tag{5.102}$$

The total force on an atom in an incoherent superposition state is hence a weighted sum over force components associated with the populated states. The dynamics of the force is governed by that of the probabilities $p_n(t)$ while the position-dependence is contained in the force components $\boldsymbol{F}_n(\boldsymbol{r}_A)$.

Let us first discuss the force components in more detail. Note that we have discarded the self-force associated with the free-space Green's tensor $\boldsymbol{G}^{(0)}$ by making the replacement $\boldsymbol{G} \mapsto \boldsymbol{G}^{(1)}$. In order to decompose the force into its non-resonant components from virtual photons and off-resonant components due to real photons, we write $\mathrm{Im}\,\boldsymbol{G} = (\boldsymbol{G} - \boldsymbol{G}^*)/(2\mathrm{i})$ and make use of the Schwarz reflection principle (A.3). We then apply the integration contour depicted in Fig. 5.1 to rotate the integrals over the positive and negative real frequency axes to the positive imaginary axis. Note that in contrast to the contour of Fig. 4.1 used in the time-independent perturbative calculation of Chap. 4, the poles are now situated away from the real axis. As a consequence, we pick up the full residuum from one of them rather than half-contributions from both poles. We obtain

$$\boldsymbol{F}_n(\boldsymbol{r}_A) = \boldsymbol{F}_n^{\text{nres}}(\boldsymbol{r}_A) + \boldsymbol{F}_n^{\text{res}}(\boldsymbol{r}_A) \tag{5.103}$$

with

$$\boldsymbol{F}_n^{\text{nres}}(\boldsymbol{r}_A) = -\frac{\hbar\mu_0}{2\pi}\int_0^\infty \mathrm{d}\xi\,\xi^2 \boldsymbol{\nabla}\mathrm{tr}\big\{\big[\boldsymbol{\alpha}_n^{\mathrm{T}}(\mathrm{i}\xi) + \boldsymbol{\alpha}_n^{\mathrm{T}}(-\mathrm{i}\xi)\big]\cdot\boldsymbol{G}^{(1)}(\boldsymbol{r}_A, \boldsymbol{r}_A, \mathrm{i}\xi)\big\}\,, \tag{5.104}$$

$$\boldsymbol{F}_n^{\text{res}}(\boldsymbol{r}_A) = \mu_0 \sum_{k<n} \Omega_{nk}^2 \boldsymbol{\nabla}\boldsymbol{d}_{nk}\cdot\boldsymbol{G}^{(1)}(\boldsymbol{r}_A, \boldsymbol{r}_A, \Omega_{nk})\cdot\boldsymbol{d}_{kn} + \text{C.c.} \tag{5.105}$$

where we have defined the complex transition frequencies

$$\Omega_{nk} = \Omega_{nk}(\boldsymbol{r}_A) = \tilde{\omega}_{nk} + \tfrac{\mathrm{i}}{2}(\Gamma_n + \Gamma_k) \tag{5.106}$$

and recalled the definition (5.91) of the atomic polarisability. We finally assume the atomic Hamiltonian to be time-reversal invariant, so that the dipole-matrix elements are real, $\boldsymbol{d}_{kn} = \boldsymbol{d}_{nk}$. The Green's tensor then only enters in conjunction with the symmetric tensor $(\boldsymbol{d}_{kn}\boldsymbol{d}_{nk})^{\mathrm{T}} = \boldsymbol{d}_{nk}\boldsymbol{d}_{kn} = \boldsymbol{d}_{kn}\boldsymbol{d}_{nk}$. Exploiting Onsager reciprocity (A.4), the relation

$$\begin{aligned}\boldsymbol{\nabla}\boldsymbol{G}^{(1)}(\boldsymbol{r}, \boldsymbol{r}, \omega) &= \boldsymbol{\nabla}\boldsymbol{G}^{(1)}(\boldsymbol{r}, \boldsymbol{r}', \omega)\big|_{\boldsymbol{r}'=\boldsymbol{r}} + \boldsymbol{\nabla}\boldsymbol{G}^{(1)}(\boldsymbol{r}', \boldsymbol{r}, \omega)\big|_{\boldsymbol{r}'=\boldsymbol{r}}\\ &= \boldsymbol{\nabla}\boldsymbol{G}^{(1)}(\boldsymbol{r}, \boldsymbol{r}', \omega)\big|_{\boldsymbol{r}'=\boldsymbol{r}} + \boldsymbol{\nabla}\boldsymbol{G}^{(1)\mathrm{T}}(\boldsymbol{r}, \boldsymbol{r}', \omega)\big|_{\boldsymbol{r}'=\boldsymbol{r}}\end{aligned} \tag{5.107}$$

then shows that the replacement $\boldsymbol{\nabla}\boldsymbol{G}^{(1)}(\boldsymbol{r}_A, \boldsymbol{r}_A, \omega) \mapsto \frac{1}{2}\boldsymbol{\nabla}\boldsymbol{G}^{(1)}(\boldsymbol{r}, \boldsymbol{r}, \omega)|_{\boldsymbol{r}=\boldsymbol{r}_A}$ applies, resulting in

$$\begin{aligned}\boldsymbol{F}_n^{\text{nres}}(\boldsymbol{r}_A) = -\frac{\hbar\mu_0}{4\pi}\int_0^\infty \mathrm{d}\xi\,\xi^2\\ \times \boldsymbol{\nabla}\mathrm{tr}\big\{[\boldsymbol{\alpha}_n(\mathrm{i}\xi) + \boldsymbol{\alpha}_n(-\mathrm{i}\xi)]\cdot\boldsymbol{G}^{(1)}(\boldsymbol{r}, \boldsymbol{r}, \mathrm{i}\xi)\big\}\big|_{\boldsymbol{r}=\boldsymbol{r}_A}\,,\end{aligned} \tag{5.108}$$

$$\boldsymbol{F}_n^{\text{res}}(\boldsymbol{r}_A) = \frac{\mu_0}{2}\sum_{k<n} \Omega_{nk}^2 \boldsymbol{\nabla}\boldsymbol{d}_{nk}\cdot\boldsymbol{G}^{(1)}(\boldsymbol{r}, \boldsymbol{r}, \Omega_{nk})\cdot\boldsymbol{d}_{kn}\big|_{\boldsymbol{r}=\boldsymbol{r}_A} + \text{C.c.} \tag{5.109}$$

The non-resonant and resonant force components depend on the shifted and broadened transition frequencies. As discussed in the previous section, the shifts and widths are position-dependent and they may induce an anisotropy of the atomic polarisability. In addition, the finite level widths have the effect that the polarisability is complex valued even at purely imaginary frequencies. The force components hence depend on the real combination $\frac{1}{2}[\boldsymbol{\alpha}_n(\mathrm{i}\xi) + \boldsymbol{\alpha}_n(-\mathrm{i}\xi)]$ rather than $\boldsymbol{\alpha}_n(\mathrm{i}\xi)$ alone. This is in contrast to our findings (4.41) on the basis of leading-order perturbation theory. Note that the gradients in (5.108) and (5.109) above only act on the position arguments of the Green's tensor and not on the position-dependent shifts, widths and polarisability, so the CP force is not a conservative force in general.

As discussed in the previous Sect. 5.2, the frequency shifts and broadenings typically induce an anisotropy of the atomic polarisability even if the atom is in an isotropic state. This complication does not arise if the atom is either placed in an isotropic environment ($\delta\omega_{k'} = \delta\omega_{k''}$ and $\Gamma_{k'} = \Gamma_{k''}$ whenever $k', k'' \in \{\overline{k}\}$) or sufficiently far from any body ($\delta\omega_k = 0$ and $\Gamma_{k'} = \Gamma_{k''}$ whenever $k', k'' \in \{\overline{k}\}$). In these cases, the polarisability of an atom in an isotropic state takes the form (5.93) and the force components above simplify to

$$\boldsymbol{F}_n^{\text{nres}}(\boldsymbol{r}_A) = -\frac{\hbar\mu_0}{4\pi}\int_0^\infty \mathrm{d}\xi\,\xi^2[\alpha_n(\mathrm{i}\xi)+\alpha_n(-\mathrm{i}\xi)]\boldsymbol{\nabla}_A \mathrm{tr}\,\boldsymbol{G}^{(1)}(\boldsymbol{r}_A,\boldsymbol{r}_A,\mathrm{i}\xi)\,, \tag{5.110}$$

$$\boldsymbol{F}_n^{\text{res}}(\boldsymbol{r}_A) = \frac{\mu_0}{6}\sum_{k<n}\Omega_{nk}^2|\boldsymbol{d}_{nk}|^2\boldsymbol{\nabla}\mathrm{tr}\,\boldsymbol{G}^{(1)}(\boldsymbol{r},\boldsymbol{r},\Omega_{nk})\big|_{\boldsymbol{r}=\boldsymbol{r}_A} + \text{C.c.} \tag{5.111}$$

In the perturbative limit $|\delta\omega_n|, |\delta\omega_k|, \Gamma_n, \Gamma_k \ll \omega_{nk}$, we have $\boldsymbol{\alpha}_n(-\mathrm{i}\xi) = \boldsymbol{\alpha}_n(\mathrm{i}\xi)$ and the force components reduce to our result (4.40)–(4.42) together with (1.119). In this limit, the CP force is always conservative and it simplifies as above for an atom in an isotropic state. Casimir and Polder's concept of identifying the CP force with the gradient of the atom–field coupling energy is hence only valid within leading-order perturbation theory. The influence of body-induced frequency shifts and broadenings is a higher-order correction. Their impact on the magnitude of the non-resonant and resonant forces will be studied via an example in the next section.

Even in the case of negligible shifts and widths, time-independent perturbation theory can only provide snapshots of the CP force at given times. Let us follow the dynamics of the CP force on an atom initially prepared in an excited energy eigenstate $|n\rangle$, $p_n(t_0) = 1$. The dynamics of the state populations $p_n(t)$ is governed by (5.71). For times that are short with respect to the relevant decay rate Γ_n, the CP force (5.101) is given by

$$\boldsymbol{F}(\boldsymbol{r}_A, t) \simeq \boldsymbol{F}(\boldsymbol{r}_A, t_0) = \boldsymbol{F}_n(\boldsymbol{r}_A) \quad \text{for } (t-t_0)\Gamma_n \ll 1\,. \tag{5.112}$$

On this time-scale, time-independent descriptions may be used. As time progresses, the lower lying levels will become populated, resulting in a superposition force

$$\boldsymbol{F}(\boldsymbol{r}_A, t) = \sum_{k\le n} p_k(t)\boldsymbol{F}_k(\boldsymbol{r}_A) \tag{5.113}$$

for intermediate times. In the long-time limit, the atom will decay to its ground state $p_k(t\to\infty) = \delta_{k0}$ and the atom will be subject to the ground-state force

$$\boldsymbol{F}(\boldsymbol{r}_A, t\to\infty) = \boldsymbol{F}_0(\boldsymbol{r}_A)\,. \tag{5.114}$$

Table 5.1 Different interpretations of the ground-state CP potential in multipolar coupling, depending on the chosen operator ordering

Ordering	Normal	Symmetric	Anti-normal
Relevant fluctuations	Atomic	Field	Atomic+field

Our dynamical calculation of the CP force implies a simple physical interpretation of the force: the quantum fluctuations of the electric dipole moment lead to the emission of an electric source field (5.44) which then acts on the atom itself, leading to a force (5.94). This process is known as radiation reaction. In our picture, the quantum fluctuations of the electric field, known as vacuum fluctuations do not contribute to the force, recall (5.53). This is a consequence of the chosen normal ordering [17]. Analysing the physical origin of the CP potential for different operator orderings, Milonni and Shih showed that the ground-state potential must be entirely attributed to atomic fluctuations for normal ordering; is entirely due to field fluctuations for symmetric ordering; and is partly due to both for anti-normal ordering [18, 19], cf. Table 5.1.

The problem was later analysed in more detail for ground-state vs excited atoms in front of a perfectly conducting plate using symmetric ordering [20], cf. Table 5.2.

Working within the minimal coupling scheme (as opposed to the multipolar scheme employed by us), it was concluded that the nonretarded potential is entirely due to atomic fluctuations. In the retarded regime, the non-resonant potential was entirely attributed to field fluctuations, while the resonant potential was found to be equally due to atomic and field fluctuations. The preference for symmetric ordering was motivated by the fact that for this ordering the contributions from atomic and field fluctuations are averages of hermitian operators, i.e., physical observables [21]. However, since these two components can never be observed separately, the chosen ordering and hence interpretation remains a matter of taste. We have chosen the normal ordering simply for calculational convenience.

After this little digression, let us turn our attention an atom which is initially prepared in a coherent superposition of energy eigenstates. In this case, the off-diagonal atomic density matrix elements become relevant and both (5.99) and (5.100) contribute to the force (5.98). Recalling (5.34), we find

$$\boldsymbol{F}(\boldsymbol{r}_A, t) = \sum_{m,n} \sigma_{mn}(t) \boldsymbol{F}_{mn}(\boldsymbol{r}_A) \tag{5.115}$$

with coherent force components

$$\boldsymbol{F}_{mn}(\boldsymbol{r}_A) = \frac{\mu_0}{\pi} \sum_k \int_0^\infty \mathrm{d}\omega\, \omega^2 \left[\frac{\nabla \boldsymbol{d}_{nk} \cdot \mathrm{Im}\, \boldsymbol{G}^{(1)}(\boldsymbol{r}_A, \boldsymbol{r}_A, \omega) \cdot \boldsymbol{d}_{km}}{\omega - \tilde{\omega}_{mk} - \frac{\mathrm{i}}{2}(\Gamma_n + \Gamma_k)} \right.$$
$$\left. + \frac{\nabla \boldsymbol{d}_{km} \cdot \mathrm{Im}\, \boldsymbol{G}^{(1)}(\boldsymbol{r}_A, \boldsymbol{r}_A, \omega) \cdot \boldsymbol{d}_{nk}}{\omega - \tilde{\omega}_{nk} + \frac{\mathrm{i}}{2}(\Gamma_m + \Gamma_k)} \right]$$

Table 5.2 Fluctuations dominating the CP potential of ground-state versus excited atoms in minimal coupling for symmetric ordering, depending on the distance regime

Distance	Retarded	Nonretarded
Ground-state atom	Field	Atomic
Excited atom	Atomic+field	Atomic

$$-\frac{\mu_0}{\pi}\sum_k\int_0^\infty \mathrm{d}\omega\,\omega\tilde{\omega}_{mn}\left\{\frac{\boldsymbol{d}_{nk}\times\left[\nabla\times\mathrm{Im}\,\boldsymbol{G}^{(1)}(\boldsymbol{r}_A,\boldsymbol{r}_A,\omega)\right]\cdot\boldsymbol{d}_{km}}{\omega-\tilde{\omega}_{mk}-\frac{\mathrm{i}}{2}(\Gamma_n+\Gamma_k)}\right.$$
$$\left.-\frac{\boldsymbol{d}_{km}\times\left[\nabla\times\mathrm{Im}\,\boldsymbol{G}^{(1)}(\boldsymbol{r}_A,\boldsymbol{r}_A,\omega)\right]\cdot\boldsymbol{d}_{nk}}{\omega-\tilde{\omega}_{nk}+\frac{\mathrm{i}}{2}(\Gamma_m+\Gamma_k)}\right\}. \tag{5.116}$$

They can be decomposed into contributions from virtual and real photon by means of the integration contour from Fig. 5.1. We find

$$\boldsymbol{F}_{mn}(\boldsymbol{r}_A)=\boldsymbol{F}^{\mathrm{nres}}_{mn}(\boldsymbol{r}_A)+\boldsymbol{F}^{\mathrm{res}}_{mn}(\boldsymbol{r}_A) \tag{5.117}$$

with non-resonant forces

$$\boldsymbol{F}^{\mathrm{nres}}_{mn}(\boldsymbol{r}_A)=-\frac{\hbar\mu_0}{2\pi}\int_0^\infty \mathrm{d}\xi\,\xi^2\nabla\mathrm{tr}\left\{\left[\boldsymbol{\alpha}^{\mathrm{T}}_{mn}(\mathrm{i}\xi)+\boldsymbol{\alpha}^{\mathrm{T}}_{n}(-\mathrm{i}\xi)\right]\cdot\boldsymbol{G}^{(1)}(\boldsymbol{r}_A,\boldsymbol{r}_A,\mathrm{i}\xi)\right\}$$
$$-\frac{\hbar\mu_0\tilde{\omega}_{mn}}{2\pi\mathrm{i}}\int_0^\infty \mathrm{d}\xi\,\xi\mathrm{tr}\left\{\left[\boldsymbol{\alpha}^{\mathrm{T}}_{mn}(\mathrm{i}\xi)-\boldsymbol{\alpha}^{\mathrm{T}}_{mn}(-\mathrm{i}\xi)\right]\times\left[\nabla\times\boldsymbol{G}^{(1)}(\boldsymbol{r}_A,\boldsymbol{r}_A,\mathrm{i}\xi)\right]\right\} \tag{5.118}$$

and resonant forces

$$\boldsymbol{F}^{\mathrm{res}}_{mn}(\boldsymbol{r}_A)=\mu_0\sum_{k<m}\Omega^2_{mnk}\nabla\boldsymbol{d}_{nk}\cdot\boldsymbol{G}^{(1)}(\boldsymbol{r}_A,\boldsymbol{r}_A,\Omega_{mnk})\cdot\boldsymbol{d}_{km}$$
$$+\mu_0\sum_{k<n}\Omega^{*2}_{nmk}\nabla\boldsymbol{d}_{km}\cdot\boldsymbol{G}^{(1)}(\boldsymbol{r}_A,\boldsymbol{r}_A,-\Omega^*_{nmk})\cdot\boldsymbol{d}_{nk}$$
$$-\mu_0\tilde{\omega}_{mn}\sum_{k<m}\Omega_{mnk}\boldsymbol{d}_{nk}\times\left[\nabla\times\boldsymbol{G}^{(1)}(\boldsymbol{r}_A,\boldsymbol{r}_A,\Omega_{mnk})\cdot\boldsymbol{d}_{km}\right]$$
$$+\mu_0\tilde{\omega}_{mn}\sum_{k<n}\Omega^*_{nmk}\boldsymbol{d}_{km}\times\left[\nabla\times\boldsymbol{G}^{(1)}(\boldsymbol{r}_A,\boldsymbol{r}_A,-\Omega^*_{nmk})\cdot\boldsymbol{d}_{nk}\right]. \tag{5.119}$$

Here, we have defined the off-diagonal polarisability

$$\boldsymbol{\alpha}_{mn}(\omega) = \boldsymbol{\alpha}_{mn}(\boldsymbol{r}_A, \omega)$$
$$= \frac{1}{\hbar}\sum_k \left[\frac{\boldsymbol{d}_{nk}\boldsymbol{d}_{km}}{\tilde{\omega}_{km}-\omega-\frac{\mathrm{i}}{2}(\Gamma_n+\Gamma_k)} + \frac{\boldsymbol{d}_{km}\boldsymbol{d}_{nk}}{\tilde{\omega}_{kn}+\omega+\frac{\mathrm{i}}{2}(\Gamma_m+\Gamma_k)}\right] \tag{5.120}$$

as well as

$$\Omega_{mnk} = \Omega_{mnk}(\boldsymbol{r}_A) = \tilde{\omega}_{mk} + \tfrac{\mathrm{i}}{2}(\Gamma_n + \Gamma_k)\,, \tag{5.121}$$

which generalise (5.91) and (5.106). In contrast to the diagonal forces $\boldsymbol{F}_{nn}$, the off-diagonal force components contain contributions from both electric and magnetic fields. The contribution from the magnetic field exhibits an entirely different vector structure. Coherent forces associated with off-diagonal density matrix elements (5.72) oscillate extremely rapidly around a zero mean. The oscillation frequency being given by the respective atomic transition frequency $\tilde{\omega}_{mn}$, it is very difficult to resolve them experimentally.

5.5 Excited Atom in Front of a Plate

To illustrate the impact of level shifts and widths on the CP force, let us study the simple example of two-level atom at nonretarded distance $z_A \ll c/(\omega_{10}\sqrt{\varepsilon})$ from a semi-infinite dielectric half space. We describe the permittivity of the half space by a single-resonance Drude–Lorentz model

$$\varepsilon(\omega) = 1 + \frac{\omega_\mathrm{P}^2}{\omega_\mathrm{T}^2 - \omega^2 - \mathrm{i}\omega\gamma} \tag{5.122}$$

with plasma frequency ω_P, transverse resonance frequency ω_T and damping constant γ. As seen from (A.48) in App. A.3.2, the nonretarded Green's tensor of such a half space reads

$$\boldsymbol{G}^{(1)}(\boldsymbol{r}, \boldsymbol{r}, \omega) = \frac{c^2}{32\pi\omega^2 z^3}\,\frac{\varepsilon(\omega) - 1}{\varepsilon(\omega) + 1}\begin{pmatrix}1&0&0\\0&1&0\\0&0&2\end{pmatrix}. \tag{5.123}$$

It exhibits a surface-plasmon resonance at $\omega_\mathrm{S} = \sqrt{\omega_\mathrm{T}^2 + \omega_\mathrm{P}^2/2}$, where the denominator $\varepsilon(\omega) + 1$ becomes very small.

We begin by determining the shifted transition frequency $\tilde{\omega}_{10} = \omega_{10} + \delta\omega$ of the two-level atom. According to (5.65), the frequency shift $\delta\omega = \delta\omega_1 - \delta\omega_0$ has contributions from the ground- and excited-state level shifts. While the ground-state shift is purely non-resonant, the excited-state shift exhibits both resonant and non-resonant contribution, as given by the two terms in (5.78). In the vicinity of the surface-plasmon frequency, the resonant contribution contained in the excited-state level shift dominates over the non-resonant ones. Substituting the nonretarded Green's tensor into the resonant term in (5.78) and recalling (4.104), we find [1]

$$\delta\omega(z_A) = \delta\omega_1(z_A) = -\frac{\boldsymbol{d}_{01}^2 + (\boldsymbol{d}_{01}\cdot\boldsymbol{e}_z)^2}{32\pi\varepsilon_0\hbar z_A^3}\,\frac{|\varepsilon[\omega_{10}+\delta\omega(z_A)]|^2-1}{|\varepsilon[\omega_{10}+\delta\omega(z_A)]+1|^2} \tag{5.124}$$

where the transition-dipole matrix element has been assumed to be real. With the permittivity being given as above, this is a fifth-order polynomial equation for $\delta\omega$.

In a perturbative approximation, we neglect $\delta\omega$ on the right hand side of the equation. The resulting frequency shift is displayed in Fig. 5.2(i) for two different distances z_A. The frequency shift shows a typical dispersion profile centred around the surface-plasmon frequency $\omega_S \simeq 1.13\omega_T$, with the shift being positive for atomic frequencies above the surface-plasmon resonance and negative below the resonance. Due to its $1/z_A^3$-dependence, the profile becomes more pronounced as the atom approaches the surface. We also display the exact results for the frequency shift, obtained by solving (5.124) numerically. We note that the profile of the exact shift is considerably steeper in the vicinity of the resonance. This is due to a positive feedback effect: for instance, bare frequencies slightly above the resonance lead to a positive perturbative shift, driving the true frequency further away from the resonance.

The decay rate $\Gamma = \Gamma_1$ of the excited state can be found by substituting the nonretarded Green's tensor (5.123) into (5.62) and recalling (4.98) [1]:

$$\Gamma(z_A) = \frac{\boldsymbol{d}_{01}^2 + (\boldsymbol{d}_{01}\cdot\boldsymbol{e}_z)^2}{8\pi\varepsilon_0\hbar z_A^3}\,\frac{\operatorname{Im}\varepsilon(\tilde{\omega}_{10})}{|\varepsilon(\tilde{\omega}_{10})+1|^2}\,. \tag{5.125}$$

In a perturbative approximation, we replace $\tilde{\omega}_{10}$ with the bare frequency. As shown in Fig. 5.2(ii), the perturbative decay rate exhibits a typical absorption line profile. We also display the exact decay rate obtained by taking the frequency shift into account. The exact absorption line is considerably more narrow than the perturbative line. This is the case because the exact atomic transition frequency is always further away from the resonance than the bare one.

Note that the decay rate is proportional to the imaginary part of the permittivity. It can hence be strongly reduced by using a superconducting surface with $\operatorname{Im}\varepsilon \ll \operatorname{Re}\varepsilon$. This effect has been studied in detail for the closely related case of magnetic transitions, i.e., spin flips [22]. The strong sensitivity of the decay rate to superconducting surfaces is in stark contract with the behaviour of CP potential, which does not change very much when replacing an ordinary metal with a superconductor.

With these preparations at hand, we can determine the CP force acting on the excited atom. We begin with the dominant resonant contribution. Substituting the Green's tensor (5.123) into the force expression (5.109), one obtains [1–3, 16]

$$\boldsymbol{F}_1^{\text{res}}(z_A) = F_1^{\text{res}}(z_A)\boldsymbol{e}_z = -\frac{3\big[\boldsymbol{d}_{01}^2 + (\boldsymbol{d}_{01}\cdot\boldsymbol{e}_z)^2\big]}{32\pi\varepsilon_0 z_A^4}\,\frac{|\varepsilon(\Omega_{10})|^2-1}{|\varepsilon(\Omega_{10})+1|^2}\,\boldsymbol{e}_z \tag{5.126}$$

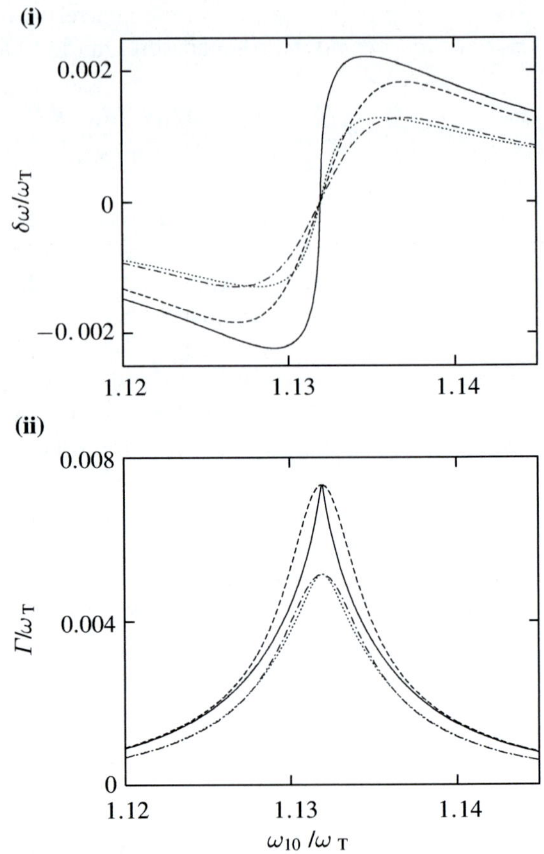

Fig. 5.2 Exact (*solid* and *dotted lines*) and perturbative (*dashed* and *dot-dashed lines*) **(i)** frequency shift and **(ii)** decay rate for a two-level atom at distances $z_A = 0.0075\lambda_T$ (*solid* and *dashed lines*) or $z_A = 0.009\lambda_T$ (*dotted* and *dot-dashed lines*) from a dielectric half space as a function of the bare atomic transition frequency with parameters $\lambda_T = 2\pi c/\omega_T$, $\omega_P/\omega_T = 0.75$, $\gamma/\omega_T = 0.01$, $\omega_T^2 \boldsymbol{d}_{10}^2/(3\pi\hbar\varepsilon_0 c^3) = 10^{-7}$. The atomic dipole moment is perpendicular to the surface

where the complex transition frequency (5.106) reads

$$\Omega_{10} = \tilde{\omega}_{10} + \tfrac{\mathrm{i}}{2}\Gamma = \omega_{10} + \delta\omega + \tfrac{\mathrm{i}}{2}\Gamma \,. \tag{5.127}$$

To illustrate the impact of the surface-induced level shift and width on the force, we again begin with the perturbative approximation $\Omega_{10} \simeq \omega_{10}$. As shown in Fig. 5.3, the resonant force exhibits a dispersion profile centred around the surface-plasmon frequency. The force is attractive for atomic frequencies below ω_S and repulsive for frequencies greater than ω_S. Note that this corresponds to the two cases $|\varepsilon(\omega_{10})| > 1$ and $|\varepsilon(\omega_{10})| < 1$ as discussed in Sect. 4.2.2. For comparison, we show the separate and combined effects of frequency shift and width on the force. The frequency shift leads to a narrowing of the dispersion profile due to the above mentioned positive feedback effect. The level width induces a broadening of the profile, accompanied by a reduction. This can be easily understood by noting that the Drude–Lorentz permittivity (5.122) at complex transition frequency reads

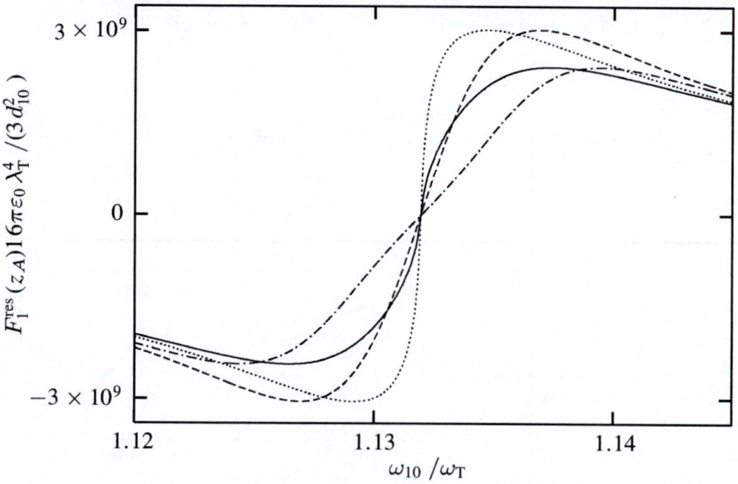

Fig. 5.3 Resonant CP force on an excited two-level atom at distance $z_A = 0.0075\lambda_T$ from a dielectric half space as a function of the bare atomic transition frequency (*solid line*), with parameters as in Fig. 5.2. For comparison, we also show the perturbative result without level shift and width (*dashed lines*) and the separate effects of shift (*dotted lines*) and width (*dash-dotted lines*)

$$\varepsilon(\Omega_{10}) = 1 + \frac{\omega_P^2}{\omega_T^2 - \tilde{\omega}_{10}^2 - i(\gamma + \Gamma)\tilde{\omega}_{10}} \quad \text{for } \gamma, \Gamma \ll \omega_T\,. \tag{5.128}$$

The width of the exact profile is hence given by the sum $\gamma + \Gamma$ of the atomic and medium line widths rather than the medium line width alone. The figure shows that the width-induced broadening of the dispersion force profile almost cancels with the shift-induced narrowing. As a result, the exact CP force including the full shifted and broadened atomic transition frequency has almost the same profile as the perturbative one, but with a considerably reduced force.

The non-resonant force component (5.108) can also be calculated with the aid of the Green's tensor (5.123). Using the explicit form (5.91) for the polarisability, we find [1–3, 16]

$$\begin{aligned} \boldsymbol{F}_1^{\text{nres}}(z_A) = F_1^{\text{nres}}(z_A)\boldsymbol{e}_z = \frac{3\big[\boldsymbol{d}_{01}^2 + (\boldsymbol{d}_{01}\cdot\boldsymbol{e}_z)^2\big]}{32\pi^2\varepsilon_0 z_A^4} \int_0^\infty \mathrm{d}\xi\, \frac{\varepsilon(i\xi) - 1}{\varepsilon(i\xi) + 1} \\ \times \frac{\tilde{\omega}_{10}}{\tilde{\omega}_{10}^2 + \big(\xi + \frac{1}{2}\Gamma\big)^2} \frac{\tilde{\omega}_{10}^2 + \xi^2 + \frac{1}{4}\Gamma^2}{\tilde{\omega}_{10}^2 + \big(\xi - \frac{1}{2}\Gamma\big)^2}\,\boldsymbol{e}_z\,. \end{aligned} \tag{5.129}$$

We note that the influence of the level width on the non-resonant force is very weak, the leading-order dependence being $O(\Gamma^2)$. This can be understood from the fact that the non-resonant force is due to virtual emission and reabsorption processes, which are only weakly affected by decay-induced broadening. Formally, the linear

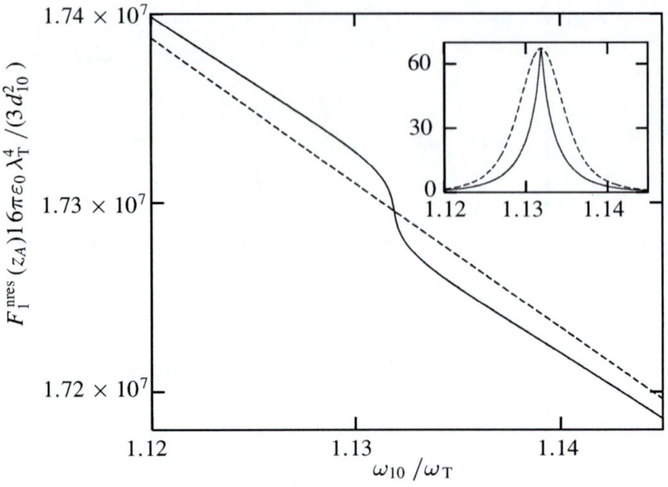

Fig. 5.4 Non-resonant CP force on an excited two-level atom at distance $z_A = 0.0075\lambda_\mathrm{T}$ from dielectric half space as a function of the bare atomic transition frequency (*solid line*). The assumptions of Fig. 5.2 apply. For comparison, we also show the perturbative result without level shift and width (*dashed lines*). The inset displays the difference between the force with and without consideration of broadening (*solid lines*) and the same difference is displayed when ignoring the shifting (*dashed lines*)

term in Γ is absent, because the atomic polarisability enters the non-resonant force components (5.108) only as a combination $\frac{1}{2}[\boldsymbol{\alpha}_n(\mathrm{i}\xi) + \boldsymbol{\alpha}_n(-\mathrm{i}\xi)]$.

The effects of level shift and width on the non-resonant force are illustrated in Fig. 5.4. The non-resonant force in the perturbative approximation is purely repulsive for downward transitions, in agreement with our findings from Sect. 4.2.2. In contrast to the resonant force, it shows a very weak dependence on the atomic transition frequency. The frequency shifting has the effect of slightly increasing the force for $\omega_{10} < \omega_\mathrm{S}$ or slightly decreasing it for $\omega_{10} > \omega_\mathrm{S}$. The effect of the level width is so weak that it is not visible in the displayed curves. Only by plotting the difference between the results with and without broadening, a slight reduction of the force becomes visible in the vicinity of ω_S where Γ is largest. Note that for a two-level atom, we have $\boldsymbol{F}_0(z_A) = \boldsymbol{F}_0^{\mathrm{nres}}(z_A) = -\boldsymbol{F}_1^{\mathrm{nres}}(z_A)$. In this case, our discussion for the non-resonant excited-state force hence also applies to the ground-state force.

The results of these Sects. 5.4 and 5.5 have well illustrated the validity limits of the time-independent leading-order perturbative description to the CP force as presented in the previous Chap. 4. We have seen that due to the decay-induced dynamics of the CP force, the static results are valid only for atoms in their stationary ground-state or on time scales which are short compared to the life times associated with spontaneous decay. In addition, the perturbative approach neglects the effect of level shifts and widths on the force. They may strongly affect the dominate resonant force on an excited atom. In particular, line broadening limits a possible resonant enhancement of the CP force.

References

1. S.Y. Buhmann, L. Knöll, D.G. Welsch, D.T. Ho, Phys. Rev. A **70**(5), 052117 (2004)
2. S.Y. Buhmann, D.T. Ho, T. Kampf, D.G. Welsch, Eur. Phys. J. D **35**(1), 15 (2005)
3. S.Y. Buhmann, D.G. Welsch, Prog. Quantum Electron. **31**(2), 51 (2007)
4. C. Baxter, M. Babiker, R. Loudon, Phys. Rev. A **47**(2), 1278 (1993)
5. V.E. Lembessis, M. Babiker, C. Baxter, R. Loudon, Phys. Rev. A **48**(2), 1594 (1993)
6. L. Onsager, Phys. Rev. **37**(4), 405 (1931)
7. M. Lax, Phys. Rev. **129**(5), 2342 (1963)
8. M.O. Scully, M.S. Zubairy, *Quantum Optics* (Cambridge University Press, Cambridge, 1997)
9. W. Vogel, D.G. Welsch, *Quantum Optics*, 3rd edn. (Wiley-VCH, Berlin, 2006)
10. D.L. Andrews, S. Naguleswaran, G.E. Stedman, Phys. Rev. A **57**(6), 4925 (1998)
11. A.D. Buckingham, P. Fischer, Phys. Rev. A **61**(3), 035801 (2000)
12. G.E. Stedman, S. Naguleswaran, D.L. Andrews, L.C. Dávila Romero, Phys. Rev. A **63**(4), 047801 (2001)
13. A.D. Buckingham, P. Fischer, Phys. Rev. A **63**(4), 047802 (2001)
14. D.L. Andrews, L.C. Dávila-Romero, G.E. Stedman, Phys. Rev. A **67**(5), 055801 (2003)
15. P.W. Milonni, R.W. Boyd, Phys. Rev. A **69**(2), 023814 (2004)
16. S.Y. Buhmann, D.T. Ho, T. Kampf, L. Knöll, D.G. Welsch, Opt. Spectrosc. (USSR) **99**(3), 466 (2005)
17. P.W. Milonni, Phys. Rev. A **25**(5), 1315 (1982)
18. P.W. Milonni, M.L. Shih, Phys. Rev. A **45**(7), 4241 (1992)
19. P.W. Milonni, *The Quantum Vacuum* (Academic Press, New York, 1994)
20. D. Meschede, W. Jhe, E.A. Hinds, Phys. Rev. A **41**(3), 1587 (1990)
21. J. Dalibard, J. Dupont-Roc, C. Cohen-Tanoudji, J. Phys. (Paris) **43**(11), 1617 (1982)
22. S. Scheel, P.K. Rekdal, P.L. Knight, E.A. Hinds, Phys. Rev. A **72**(4), 042901 (2005)

Chapter 6
Casimir–Polder Forces in Cavity Quantum Electrodynamics

The interaction of an excited atom with the quantised electromagnetic field can be strongly enhanced in confined, resonator-like geometries such as planar, spherical or cylindrical cavities formed of highly reflecting mirrors. The structure of the field inside such a cavity and its interaction with atoms is the subject of cavity QED. In this chapter, we use basic concepts of cavity QED to investigate the phenomenon of strong atom–field coupling and its impact on the CP force. For simplicity, we will exclusively work within the multipolar coupling scheme, without using primes to indicate multipolar variables.

We begin with a time-independent analysis that generalises Casimir and Polder's method as employed in Chap. 4 beyond the perturbative regime. Introducing the Jaynes–Cummings model, we will show that the strongly coupled atom–field system may be characterised by dressed atomic states. The CP potential on an atom in such a state is governed by the vacuum Rabi frequency.

Next, we study the dynamics of CP force for strong atom–field coupling, generalising the approach of Chap. 5. We will demonstrate that strong coupling facilitates a continuous and reversible emission and reabsorption of a single photon by an excited atom, leading to periodic Rabi oscillations of the atomic internal state. The CP force on a strongly coupled atom exhibits similar periodic oscillations.

6.1 Static Theory

The field inside a cavity typically forms standing-wave modes of certain discrete frequencies ω_ν. The near-resonant coupling of such a mode to the downward transition ω_{mn} of an excited atom can lead to a strongly enhanced atom–field interaction. In particular, the energy shift can become comparable to the atomic-transition and mode-excitation energies. In this strong coupling regime, perturbation theory no longer applies. Instead, the relevant strong atom–field interaction has to be treated exactly.

S. Y. Buhmann, *Dispersion Forces II*, Springer Tracts in Modern Physics 248,
DOI: 10.1007/978-3-642-32466-6_6,

6.1.1 Jaynes–Cummings Model

To develop an exact approach to strong atom–field coupling, we need to focus on the strongly interacting part of the atom–field system. In doing so, we will arrive at the Jaynes–Cummings model which describes the coupling of a single atomic transition to a single cavity mode.

To model the atom, we consider only those two levels $|1\rangle$ and $|0\rangle$ whose transition is strongly coupled to a resonator mode ν. For such a two-level atom, the Hamiltonian (1.92) simplifies to

$$\hat{H}_A = E_0|0\rangle\langle 0| + E_1|1\rangle\langle 1| \tag{6.1}$$

for a given centre-of-mass position $\boldsymbol{r}_A$. By means of the reduced completeness relation $|0\rangle\langle 0| + |1\rangle\langle 1| = \hat{I}$, the atomic Hamiltonian may be written in the alternative form

$$\hat{H}_A = \tfrac{1}{2}\,\hbar\omega_{10}\hat{\sigma}_z + \tfrac{1}{2}\,(E_0 + E_1)\hat{I} \tag{6.2}$$

with $\hat{\sigma}_z = |1\rangle\langle 1| - |0\rangle\langle 0|$ being one of the three Pauli operators. One commonly discards the last term, which is a state-independent constant, so that

$$\hat{H}_A = \tfrac{1}{2}\hbar\omega_{10}\hat{\sigma}_z\,. \tag{6.3}$$

The expansion (5.38) of the electric-dipole operator simplifies to

$$\hat{\boldsymbol{d}} = \boldsymbol{d}_{01}\hat{\sigma} + \boldsymbol{d}_{10}\hat{\sigma}^\dagger \tag{6.4}$$

for a two-level atom, with $\hat{\sigma} = |0\rangle\langle 1|$ being the Pauli lowering operator. The atom–field interaction (4.36) hence reads

$$\hat{H}_{\mathrm{AF}} = -\big(\boldsymbol{d}_{01}\hat{\sigma} + \boldsymbol{d}_{10}\hat{\sigma}^\dagger\big)\cdot\hat{\boldsymbol{E}}(\boldsymbol{r}_A)\,. \tag{6.5}$$

Next, we need to adapt our description of the electromagnetic field to the strong-coupling problem. With the field expansion (1.22), the interaction Hamiltonian takes the form

$$\hat{H}_{\mathrm{AF}} = -\int_0^\infty \mathrm{d}\omega \sum_{\lambda=e,m}\int \mathrm{d}^3r\,\big(\boldsymbol{d}_{01}\hat{\sigma} + \boldsymbol{d}_{10}\hat{\sigma}^\dagger\big)\cdot\boldsymbol{G}_\lambda(\boldsymbol{r}_A,\boldsymbol{r},\omega)\cdot\hat{\boldsymbol{f}}_\lambda(\boldsymbol{r},\omega) + \mathrm{H.c.} \tag{6.6}$$

This suggests the introduction of photon annihilation and creation operators $\hat{a}(\boldsymbol{r},\omega)$ and $\hat{a}^\dagger(\boldsymbol{r},\omega)$ according to

$$\hat{a}(\boldsymbol{r},\omega) = -\frac{1}{\hbar g(\boldsymbol{r},\omega)}\sum_{\lambda=e,m}\int \mathrm{d}^3r'\,\boldsymbol{d}_{10}\cdot\boldsymbol{G}_\lambda(\boldsymbol{r},\boldsymbol{r}',\omega)\cdot\hat{\boldsymbol{f}}_\lambda(\boldsymbol{r}',\omega) \tag{6.7}$$

where the normalisation

$$g(\boldsymbol{r},\omega)=\sqrt{\frac{\mu_0}{\pi\hbar}\,\omega^2\boldsymbol{d}_{10}\cdot\mathrm{Im}\,\boldsymbol{G}(\boldsymbol{r},\boldsymbol{r},\omega)\cdot\boldsymbol{d}_{01}} \tag{6.8}$$

is needed to ensure the canonical commutation relations (6.11) and (6.18) as given below. With this definition, the interaction Hamiltonian takes the much simpler form

$$\hat{H}_{\mathrm{AF}}=\int_0^\infty \mathrm{d}\omega\,\hbar g(\boldsymbol{r}_A,\omega)\big[\hat{a}(\boldsymbol{r}_A,\omega)+\hat{a}^\dagger(\boldsymbol{r}_A,\omega)\big]\big(\hat{\sigma}+\hat{\sigma}^\dagger\big)\,, \tag{6.9}$$

showing that $g(\boldsymbol{r},\omega)$ can be interpreted as an atom–field coupling strength. We have assumed the dipole matrix elements to be real, $\boldsymbol{d}_{01}=\boldsymbol{d}_{10}$.

According to the definitions above, the commutation relations of $\hat{a}$ and $\hat{a}^\dagger$ follow from those of $\hat{\boldsymbol{f}}_\lambda$ and $\hat{\boldsymbol{f}}^\dagger_\lambda$, recall (1.17) and (1.18). Using the integral relation (1.25), one finds

$$\big[\hat{a}(\boldsymbol{r},\omega),\hat{a}(\boldsymbol{r},\omega')\big]=\big[\hat{a}^\dagger(\boldsymbol{r},\omega)\,,\hat{a}^\dagger(\boldsymbol{r},\omega')\big]=0\,, \tag{6.10}$$

$$\big[\hat{a}(\boldsymbol{r},\omega),\hat{a}^\dagger(\boldsymbol{r}',\omega')\big]=\frac{g(\boldsymbol{r},\boldsymbol{r}',\omega)}{g(\boldsymbol{r},\omega)g(\boldsymbol{r}',\omega)}\,\delta(\omega-\omega') \tag{6.11}$$

with

$$g(\boldsymbol{r},\boldsymbol{r}',\omega)=\frac{\mu_0}{\hbar\pi}\omega^2\boldsymbol{d}_{10}\cdot\mathrm{Im}\boldsymbol{G}(\boldsymbol{r},\boldsymbol{r}',\omega)\cdot\boldsymbol{d}_{01}\,. \tag{6.12}$$

Just like $\hat{\boldsymbol{f}}_\lambda$, the operator $\hat{a}$ has the meaning of an annihilation operator: Recalling the definition (1.19) of the ground-state of the body-assisted electromagnetic field, our construction immediately implies

$$\hat{a}(\boldsymbol{r},\omega)|\{0\}\rangle=0\quad\forall\boldsymbol{r},\omega\,. \tag{6.13}$$

The creation operators $\hat{a}^\dagger(\boldsymbol{r},\omega)$ can be used to define single-quantum Fock states

$$|1(\boldsymbol{r},\omega)\rangle=\hat{a}^\dagger(\boldsymbol{r},\omega)|\{0\}\rangle\,. \tag{6.14}$$

The above commutation relations show that these states are orthogonal with respect to frequency, but not with respect to position,

$$\langle 1(\boldsymbol{r},\omega)|1(\boldsymbol{r}',\omega')\rangle=\frac{g(\boldsymbol{r},\boldsymbol{r}',\omega)}{g(\boldsymbol{r},\omega)g(\boldsymbol{r}',\omega)}\,\delta(\omega-\omega')\,. \tag{6.15}$$

They are eigenstates of the field Hamiltonian (1.93) carrying one quantum of energy $\hbar\omega$,

$$\hat{H}_{\mathrm{F}}|1(\boldsymbol{r},\omega)\rangle = \hbar\omega|1(\boldsymbol{r},\omega)\rangle \ . \tag{6.16}$$

This can be seen from the commutation relation

$$\left[\hat{H}_{\mathrm{F}},\hat{a}^{\dagger}(\mathbf{r},\omega)\right] = \hbar\omega\hat{a}^{\dagger}(\mathbf{r},\omega) \ , \tag{6.17}$$

which is a consequence our definition of $\hat{a}$ and $\hat{a}^{\dagger}$ and the fundamental commutation relations (1.17) and (1.18). Physically, the states $|1(\boldsymbol{r},\omega)\rangle$ represent the single photon resonantly emitted by our excited two-level atom at position $\boldsymbol{r}$. Their non-orthogonality reflects the fact that photons emitted at one position $\boldsymbol{r}$ have a non-vanishing probability of being reabsorbed by an atom at a different position $\boldsymbol{r}'$.

For an atom at fixed centre-of-mass position $\boldsymbol{r}_A$, we have photon creation and annihilation operators $\hat{a}(\boldsymbol{r}_A,\omega)\equiv\hat{a}(\omega)$ and $\hat{a}^{\dagger}(\boldsymbol{r}_A,\omega)\equiv\hat{a}^{\dagger}(\omega)$ with commutation relations

$$\left[\hat{a}(\omega),\hat{a}(\omega')\right] = \left[\hat{a}^{\dagger}(\omega),\hat{a}^{\dagger}(\omega')\right] = 0 \ , \quad \left[\hat{a}(\omega),\hat{a}^{\dagger}(\omega')\right] = \delta(\omega-\omega') \ , \tag{6.18}$$

so that the associated single photon states $|1(\boldsymbol{r}_A,\omega)\rangle\equiv|1(\omega)\rangle$ are orthogonal,

$$\langle 1(\omega)|1(\omega')\rangle = \delta(\omega-\omega') \ . \tag{6.19}$$

Finally, we restrict our attention to a single cavity mode ν which has a Lorentzian profile with mid-frequency ω_ν and width γ_ν,

$$g_\nu^2(\boldsymbol{r},\omega) = g_\nu^2(\boldsymbol{r})\,\frac{\frac{1}{4}\gamma_\nu^2}{(\omega-\omega_\nu)^2+\frac{1}{4}\gamma_\nu^2} \ . \tag{6.20}$$

We apply the single-mode approximation by assuming that the mode dominates the field spectrum in its vicinity,

$$g^2(\boldsymbol{r},\omega) \simeq g_\nu^2(\boldsymbol{r},\omega) \quad \text{for } |\omega-\omega_\nu| \le \Delta\omega/2 \tag{6.21}$$

($\Delta\omega$: distance between two neighbouring modes; $\gamma_\nu \ll \Delta\omega$). The field spectrum is sketched in Fig. 6.1. In order to realise strong coupling, we require the mode to be very narrow, $\gamma_\nu \ll \omega_\nu$. Introducing the Q-factor of the cavity $Q=\omega_\nu/\gamma_\nu$, this is equivalent to assuming a high-Q cavity.

Creation and annihilation operators for photons of the single mode ν can be introduced according to

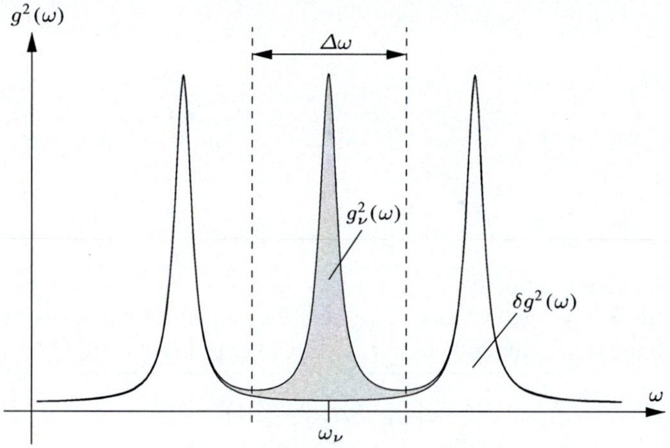

Fig. 6.1 Single-mode approximation for the field spectrum inside a cavity

$$\hat{a} = \sqrt{\frac{\gamma_\nu}{2\pi}} \int_{\omega_\nu-\Delta\omega/2}^{\omega_\nu+\Delta\omega/2} \mathrm{d}\omega \, \frac{\hat{a}(\omega)}{\sqrt{(\omega-\omega_\nu)^2+\frac{1}{4}\gamma_\nu^2}} \,. \tag{6.22}$$

Their commutation relations follow from those of $\hat{a}(\omega)$ and $\hat{a}^\dagger(\omega)$ as given by (6.18) above. For a sufficiently narrow mode with $\gamma_\nu \ll \Delta\omega < \omega_\nu$, we have

$$\int_{\omega_\nu-\Delta\omega/2}^{\omega_\nu+\Delta\omega/2} \frac{\mathrm{d}\omega}{(\omega-\omega_\nu)^2+\frac{1}{4}\gamma_\nu^2} \simeq \int_{-\infty}^{\infty} \frac{\mathrm{d}\omega}{(\omega-\omega_\nu)^2+\frac{1}{4}\gamma_\nu^2} = \frac{2\pi}{\gamma_\nu} \,, \tag{6.23}$$

so that

$$\big[\hat{a},\hat{a}\big] = \big[\hat{a}^\dagger,\hat{a}^\dagger\big] = 0 \,, \quad \big[\hat{a},\hat{a}^\dagger\big] = 1 \,. \tag{6.24}$$

As an immediate consequence, the associated single-photon state

$$|1_\nu\rangle = \hat{a}^\dagger|\{0\}\rangle \tag{6.25}$$

is normalised to unity,

$$\langle 1_\nu|1_\nu\rangle = 1 \,. \tag{6.26}$$

Substituting (6.20) and (6.22) into (6.9), the interaction Hamiltonian in single-mode approximation ($\int_0^\infty \mathrm{d}\omega \mapsto \int_{\omega_\nu-\Delta\omega/2}^{\omega_\nu+\Delta\omega/2} \mathrm{d}\omega$) takes the simple form

$$\hat{H}_{\mathrm{AF}} = \tfrac{1}{2}\hbar\Omega_{\mathrm{R}}\big(\hat{a}+\hat{a}^\dagger\big)\big(\hat{\sigma}+\hat{\sigma}^\dagger\big) \tag{6.27}$$

where

$$\Omega_\mathrm{R} = \Omega_\mathrm{R}(\boldsymbol{r}_A) = \sqrt{2\pi\gamma_\nu g^2(\boldsymbol{r}_A,\omega_\nu)} = \sqrt{\frac{2\mu_0\gamma_\nu\omega_\nu^2}{\hbar}\boldsymbol{d}_{10}\cdot\mathrm{Im}\boldsymbol{G}(\boldsymbol{r}_A,\boldsymbol{r}_A,\omega_\nu)\cdot\boldsymbol{d}_{01}} \tag{6.28}$$

is the vacuum Rabi frequency.

For a sufficiently narrow mode $\gamma_\nu \ll \Delta\omega$, the commutator of the field Hamiltonian (1.93) with the single-mode operator (6.22) may approximately be given as

$$\left[\hat{H}_\mathrm{F},\hat{a}\right] = -\hbar\omega_\nu\hat{a}\ . \tag{6.29}$$

This commutator is faithfully reproduced by the simpler single-mode Hamiltonian

$$\hat{H}_\mathrm{F} = \hbar\omega_\nu\hat{a}^\dagger\hat{a}\ . \tag{6.30}$$

Combining our results (6.3), (6.27) and (6.30), we obtain the Hamiltonian of the Jaynes–Cummings model [1]

$$\hat{H} = \hbar\omega_\nu\hat{a}^\dagger\hat{a} + \tfrac{1}{2}\hbar\omega_{10}\hat{\sigma}_z + \tfrac{1}{2}\hbar\Omega_\mathrm{R}\left(\hat{a}+\hat{a}^\dagger\right)\left(\hat{\sigma}+\hat{\sigma}^\dagger\right). \tag{6.31}$$

It describes the idealised interaction of a single cavity mode with a single atomic transition. The level scheme of the James–Cummings model is depicted in Fig. 6.2. In the figure, we have introduced the atom–field detuning

$$\Delta = \omega_\nu - \omega_{10}\ . \tag{6.32}$$

Fig. 6.2 James–Cummings model

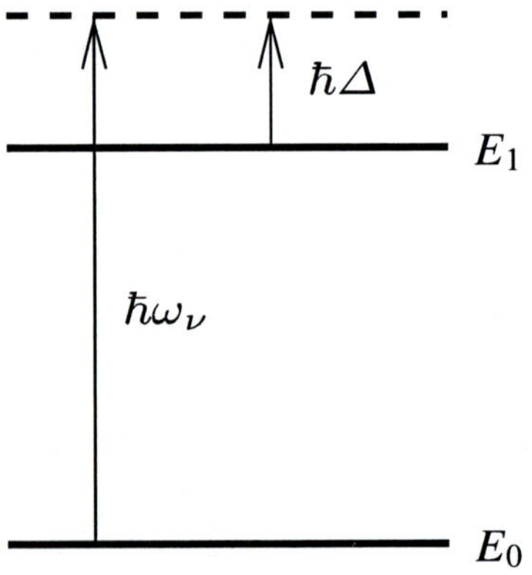

We are only interested in the resonant, energy-conserving interaction processes where the atom either makes a downward transition while emitting a photon ($\hat{a}^\dagger\hat{\sigma}$) or it reabsorbs the photon while making an upward transition ($\hat{a}\hat{\sigma}^\dagger$). Discarding the other two, purely non-resonant interaction terms, (5.133) approximates to

$$\hat{H}_{\mathrm{AF}} = \tfrac{1}{2}\hbar\Omega_{\mathrm{R}}\left(\hat{a}\hat{\sigma}^\dagger + \hat{a}^\dagger\hat{\sigma}\right) \tag{6.33}$$

and the Jaynes–Cummings Hamiltonian reads

$$\hat{H} = \hbar\omega_\nu\hat{a}^\dagger\hat{a} + \tfrac{1}{2}\hbar\omega_{10}\hat{\sigma}_z + \tfrac{1}{2}\hbar\Omega_{\mathrm{R}}\left(\hat{a}\hat{\sigma}^\dagger + \hat{a}^\dagger\hat{\sigma}\right). \tag{6.34}$$

The neglect of the non-resonant interactions is known as the rotating-wave approximation. The name stems from the fact that the uncoupled systems carry a periodic time dependence in the Heisenberg picture, $\hat{a}(t) = \hat{a}\mathrm{e}^{-\mathrm{i}\omega_\nu(t-t_0)}$ and $\hat{\sigma}(t) = \hat{\sigma}\mathrm{e}^{-\mathrm{i}\omega_{10}(t-t_0)}$. We have retained the terms $\hat{\sigma}(t)\hat{a}^\dagger(t) = \hat{\sigma}\hat{a}^\dagger\mathrm{e}^{-\mathrm{i}(\omega_{10}-\omega_\nu)(t-t_0)}$ and $\hat{\sigma}^\dagger(t)\hat{a}(t) = \hat{\sigma}^\dagger\hat{a}\mathrm{e}^{\mathrm{i}(\omega_{10}-\omega_\nu)(t-t_0)}$ whose phase is slowly rotating. In the interaction picture, this means that the electromagnetic wave is co-rotating with the atom; the two counter-rotating terms have been discarded.

6.1.2 Casimir–Polder Potential

Generalising Casimir and Polder's approach, the CP potential of a strongly coupled excited atom inside a cavity can be obtained as follows: We start from the uncoupled state $|1\rangle|\{0\}\rangle$ of an excited atom with the field being in its ground state and calculate the position-dependent energy shift due to the atom–field interaction (6.33).

The coupling Hamiltonian in rotating-wave approximation induces transitions to the state $|0\rangle|1_\nu\rangle$ where the atom is in its ground state and one single-mode photon is present. On the subspace spanned by the two states $|1\rangle|\{0\}\rangle$ and $|0\rangle|1_\nu\rangle$, the Jaynes–Cummings Hamiltonian can be given in the matrix form

$$\hat{H} = \begin{pmatrix} \frac{1}{2}\hbar\omega_{10} & \frac{1}{2}\hbar\Omega_{\mathrm{R}} \\ \frac{1}{2}\hbar\Omega_{\mathrm{R}} & \hbar\omega_\nu - \frac{1}{2}\hbar\omega_{10} \end{pmatrix}. \tag{6.35}$$

This simple two-dimensional matrix can be diagonalised in a straightforward manner, yielding the two eigenenergies

$$E_\pm = \tfrac{1}{2}\hbar\omega_\nu \pm \tfrac{1}{2}\hbar\Omega \tag{6.36}$$

with

$$\Omega = \Omega(\boldsymbol{r}_A) = \sqrt{\Omega_{\mathrm{R}}^2(\boldsymbol{r}_A) + \Delta^2} \tag{6.37}$$

being the generalised Rabi frequency. Using the relation [2]

$$\frac{1}{\sqrt{1+\cot^2(\alpha)}} = \sin(\alpha)\,, \quad \alpha \in [0,\pi] \tag{6.38}$$

the associated eigenstates can be given as

$$|+\rangle = \cos\theta_{\mathrm{c}}|1\rangle|\{0\}\rangle + \sin\theta_{\mathrm{c}}|0\rangle|1_\nu\rangle\,, \tag{6.39}$$

$$|-\rangle = -\sin\theta_{\mathrm{c}}|1\rangle|\{0\}\rangle + \cos\theta_{\mathrm{c}}|0\rangle|1_\nu\rangle\,, \tag{6.40}$$

where the coupling angle $\theta_{\mathrm{c}} = \theta_{\mathrm{c}}(\boldsymbol{r}_A)$ is defined according to

$$\tan(2\theta_{\mathrm{c}}) = -\frac{\Omega_{\mathrm{R}}}{\Delta}\,, \quad \theta_{\mathrm{c}} \in [0,\pi/2]\,. \tag{6.41}$$

The states $|\pm\rangle$ are known as dressed atomic states [3]. When the system is in a strongly coupled dressed state, the excitation is shared between the atom and the cavity mode.

According to Casimir and Polder, the CP potential is the position-dependent part of the energy shift. For an atom–field system prepared in one of the dressed states $|+\rangle$ or $|-\rangle$, the CP potential is given by [4]

$$\begin{aligned} U_\pm(\boldsymbol{r}_A) &= \pm\tfrac{1}{2}\hbar\Omega(\boldsymbol{r}_A) \\ &= \pm\tfrac{1}{2}\sqrt{2\hbar\mu_0\gamma_\nu\omega_\nu^2\boldsymbol{d}_{10}\cdot\mathrm{Im}\,\boldsymbol{G}(\boldsymbol{r}_A,\boldsymbol{r}_A,\omega_\nu)\cdot\boldsymbol{d}_{10} + \hbar^2\Delta^2}\,, \end{aligned} \tag{6.42}$$

from which the associated CP force

$$\boldsymbol{F}_\pm(\boldsymbol{r}_A) = -\boldsymbol{\nabla}_A U_\pm(\boldsymbol{r}_A) \tag{6.43}$$

can be obtained. A result of this kind was first obtained in [5, 6] for an atom in a one-dimensional, perfectly conducting cavity. To interpret this result, we recall (1.28) to see that the ground-state fluctuations of the electric field due to the mode ν can be given as

$$\left\langle[\Delta\hat{\boldsymbol{E}}(\boldsymbol{r})]^2\right\rangle_\nu = \frac{\hbar\mu_0}{\pi}\int\limits_{\omega_\nu-\Delta\omega/2}^{\omega_\nu+\Delta\omega/2} \mathrm{d}\omega\,\omega^2\mathrm{tr}[\mathrm{Im}\boldsymbol{G}(\boldsymbol{r},\boldsymbol{r},\omega)]\,. \tag{6.44}$$

Using the mode profile (6.20) and carrying out the integral according to (6.23), we find

$$\left\langle[\Delta\hat{\boldsymbol{E}}(\boldsymbol{r})]^2\right\rangle_\nu = \tfrac{1}{2}\hbar\mu_0\gamma_\nu\omega_\nu^2\mathrm{tr}[\mathrm{Im}\boldsymbol{G}(\boldsymbol{r},\boldsymbol{r},\omega_\nu)]\,. \tag{6.45}$$

A comparison shows that the resonant strong-coupling CP potential is essentially determined by the ground-state fluctuations of the electric field at the position of the atom. For a system prepared in the state $|+\rangle$ the atom is repelled from regions of high field fluctuations, while for the state $|-\rangle$, it is attracted towards these regions.

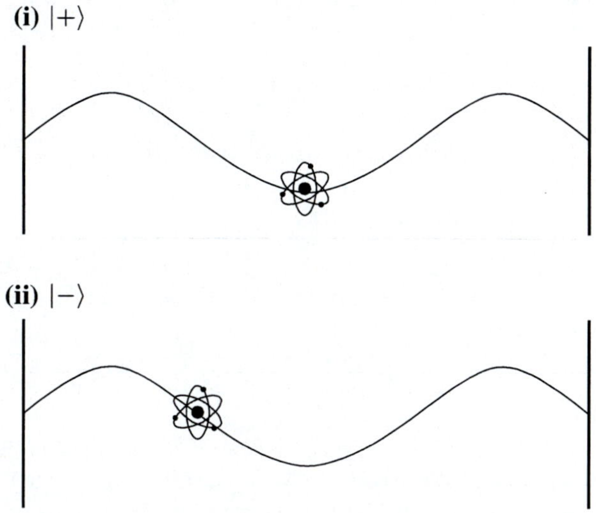

Fig. 6.3 An atom strongly interacting with a cavity mode, with the system being prepared in state **(i)** $|+\rangle$ or **(ii)** $|-\rangle$

A simple example is shown in Fig. 6.3 where we have sketched the behaviour of an atom interacting with a standing wave of an ideal planar cavity. For state $|+\rangle$, the atom is drawn towards the nodes of the wave while for the state $|+\rangle$, it is attracted towards the antinodes of the wave.

The results of perturbation theory can be recovered in the weak-coupling limit where the vacuum Rabi frequency is much smaller than the atom–field detuning $\Omega_{\mathrm{R}} \ll |\Delta|$. In this case, the coupling angle (6.41) approaches $\theta_c = \pi/2$ for positive detuning ($\Delta > 0$) and $\theta_c = 0$ for negative detuning ($\Delta > 0$). Consequently, the dressed states (6.39) and (6.40) approximate to

$$|+\rangle = \begin{cases} |0\rangle|1_\nu\rangle & \text{for } \Delta > 0\,, \\ -|1\rangle|\{0\}\rangle & \text{for } \Delta < 0\,, \end{cases} \tag{6.46}$$

$$|-\rangle = \begin{cases} |1\rangle|\{0\}\rangle & \text{for } \Delta > 0\,, \\ |0\rangle|1_\nu\rangle & \text{for } \Delta < 0\,. \end{cases} \tag{6.47}$$

The CP potentials (6.42) can be approximated by expanding the square root. With the state identifications above, we find

$$U_1(\boldsymbol{r}_A) = U_\pm(\boldsymbol{r}_A) = -\frac{\hbar\Omega_{\mathrm{R}}^2(\boldsymbol{r}_A)}{4\Delta} = -\frac{\hbar\Omega_{\mathrm{R}}^2(\boldsymbol{r}_A)}{4(\omega_\nu - \omega_{10})}\,. \tag{6.48}$$

This result can be cast into a more familiar form by employing the Kramers–Kronig relation for the response function $\omega^2 \boldsymbol{G}^{(1)}(\boldsymbol{r}_A, \boldsymbol{r}_A, \omega)$,

$$\omega^2 \mathrm{Re}\boldsymbol{G}^{(1)}(\boldsymbol{r}_A, \boldsymbol{r}_A, \omega) = \frac{1}{\pi}\mathcal{P}\int_{-\infty}^{\infty} \frac{\mathrm{d}\omega'}{\omega' - \omega}\omega'^2 \mathrm{Im}\boldsymbol{G}^{(1)}(\boldsymbol{r}_A, \boldsymbol{r}_A, \omega') \,. \tag{6.49}$$

For a sufficiently narrow mode with shape (6.20), it leads to

$$\begin{aligned}
&\mu_0\omega^2 \boldsymbol{d}_{10}\cdot\mathrm{Re}\boldsymbol{G}^{(1)}(\boldsymbol{r}_A, \boldsymbol{r}_A, \omega)\cdot\boldsymbol{d}_{01} \\
&= \frac{\mu_0}{\pi}\omega_\nu^2 \boldsymbol{d}_{10}\cdot\mathrm{Im}\boldsymbol{G}^{(1)}(\boldsymbol{r}_A, \boldsymbol{r}_A, \omega_\nu)\cdot\boldsymbol{d}_{01}\mathcal{P}\int_{-\infty}^{\infty} \frac{\mathrm{d}\omega'}{\omega' - \omega}\frac{\gamma_\nu^2/4}{(\omega'-\omega_\nu)^2 + \frac{1}{4}\gamma_\nu^2} \\
&= \frac{\hbar\Omega_\mathrm{R}^2(\boldsymbol{r}_A)}{4(\omega_\nu - \omega)} \,,
\end{aligned} \tag{6.50}$$

where we have noted that $\mathrm{Im}\boldsymbol{G}^{(1)} = \mathrm{Im}\boldsymbol{G}$ in the single-mode approximation and the frequency integral has been carried out by means of (6.23). Combining (6.48) with (6.50), we find that the CP potential reads

$$U_1(\boldsymbol{r}_A) = -\mu_0\omega_{10}^2 \boldsymbol{d}_{10}\cdot\mathrm{Re}\,\boldsymbol{G}^{(1)}(\boldsymbol{r}_A, \boldsymbol{r}_A, \omega_{10})\cdot\boldsymbol{d}_{01} \,. \tag{6.51}$$

This agrees with our perturbative result (4.42) for the resonant CP potential of an excited two-level atom. The absence of the non-resonant potential is due to the rotating-wave approximation made. The Jaynes–Cummings model is hence able to give an exact account of the resonant interaction of a two-level atom with a single cavity mode, which reduces to the perturbative result in the weak-coupling limit.

The opposite limit of strong atom–field coupling $\Omega_\mathrm{R} \gg |\Delta|$ is realised in particular for exact resonance, $\omega_{10} = \omega_\nu$. In this case, the coupling angle reads $\theta_\mathrm{c} = \pi/4$, so the states (6.39) and (6.40) are given by

$$|\pm\rangle = \frac{1}{\sqrt{2}}\big(\pm|1\rangle|\{0\}\rangle + |0\rangle|1_\nu\rangle\big) \,. \tag{6.52}$$

The excitation is hence evenly distributed between the atom and the cavity mode. The associated CP potentials read

$$U_\pm(\boldsymbol{r}_A) = \pm\tfrac{1}{2}\sqrt{2\hbar\mu_0\gamma_\nu\omega_\nu^2 \boldsymbol{d}_{10}\cdot\mathrm{Im}\boldsymbol{G}(\boldsymbol{r}_A, \boldsymbol{r}_A, \omega_\nu)\cdot\boldsymbol{d}_{10}} \,. \tag{6.53}$$

Let us next consider a situation where the system is not prepared in one of the eigenstates $|\pm\rangle$, but in a more general state

$$|\theta\rangle = \cos\theta|1\rangle|\{0\}\rangle + \sin\theta|0\rangle|1_\nu\rangle \,, \quad \theta \in [0, \pi] \,, \tag{6.54}$$

which includes the special cases

$$|\theta{=}0\rangle = |1\rangle|\{0\}\rangle\,, \quad |\theta{=}\theta_c\rangle = |+\rangle\,, \quad |\theta{=}\theta_c{+}\pi/2\rangle = |-\rangle\,. \tag{6.55}$$

The CP potential for this state is the weighted average of the two eigenstate potentials. Combining (6.39), (6.40) and (6.42), we find

$$\begin{aligned} U_\theta(\boldsymbol{r}_\mathrm{A}) &= |\langle\theta|+\rangle|^2 U_+(\boldsymbol{r}_A) + |\langle\theta|-\rangle|^2 U_-(\boldsymbol{r}_A) \\ &= \cos^2(\theta-\theta_\mathrm{c})U_+(\boldsymbol{r}_A) + \sin^2(\theta-\theta_\mathrm{c})U_-(\boldsymbol{r}_A) \\ &= \tfrac{1}{2}\cos[2(\theta-\theta_\mathrm{c})]\hbar\Omega(\boldsymbol{r}_A)\,. \end{aligned} \tag{6.56}$$

Care has to be taken when evaluating the CP force (1.119), since the overlap functions $|\langle\theta|+\rangle|^2$ and $|\langle\theta|-\rangle|^2$ carry a position-dependence via the coupling angle,

$$\boldsymbol{F}_\theta(\boldsymbol{r}_A) = -\frac{1}{2}\cos[2(\theta-\theta_\mathrm{c})]\hbar\nabla_A\Omega(\boldsymbol{r}_A) - \sin[2(\theta-\theta_\mathrm{c})]\hbar\Omega(\boldsymbol{r}_A)\nabla_A\theta_\mathrm{c}(\boldsymbol{r}_A)\,. \tag{6.57}$$

To evaluate the second gradient, we take the derivative of the definition (6.41) and make use of the identities

$$\frac{1}{\sqrt{1+\tan^2(\alpha)}} = \begin{cases} \cos(\alpha) & \text{for } \alpha\in[0,\pi/2]\,, \\ -\cos(\alpha) & \text{for } \alpha\in[\pi/2,\pi]\,, \end{cases} \tag{6.58}$$

to find

$$\nabla_A\theta_\mathrm{c}(\boldsymbol{r}_A) = -\frac{\cos^2(2\theta_\mathrm{c})\nabla_A\Omega_\mathrm{R}(\boldsymbol{r}_A)}{2\Delta}\,. \tag{6.59}$$

The CP force thus reads

$$\begin{aligned} \boldsymbol{F}_\theta(\boldsymbol{r}_A) &= -\tfrac{1}{2}\cos[2(\theta-\theta_\mathrm{c})]\hbar\nabla_A\Omega(\boldsymbol{r}_A) \\ &\quad + \tfrac{1}{2}\sin[2(\theta-\theta_\mathrm{c})]\cos^2(2\theta_\mathrm{c})\frac{\Omega(\boldsymbol{r}_A)}{\Delta}\hbar\nabla_A\Omega_\mathrm{R}(\boldsymbol{r}_A)\,. \end{aligned} \tag{6.60}$$

To bring this result into a more compact form, we make use of the relations

$$\Omega(\boldsymbol{r}_A) = \frac{\Omega_\mathrm{R}(\boldsymbol{r}_A)}{\sin(2\theta_\mathrm{c})}\,, \tag{6.61}$$

$$\nabla_A\Omega(\boldsymbol{r}_A) = \sin(2\theta_\mathrm{c})\nabla_A\Omega_\mathrm{R}(\boldsymbol{r}_A) \tag{6.62}$$

which follow directly from the definitions (6.37) and (6.41) together with the identity (6.38). Substituting these results and using the definition (6.41) once more, the CP force on an atom in a superposition state (6.54) can be given as [4]

$$\boldsymbol{F}_\theta(\boldsymbol{r}_\mathrm{A}) = -\tfrac{1}{2}\big\{\cos[2(\theta-\theta_\mathrm{c})] + \cot(2\theta_\mathrm{c})\sin[2(\theta-\theta_\mathrm{c})]\big\}\hbar\nabla_A\Omega(\boldsymbol{r}_A)\,. \tag{6.63}$$

Let us consider some special cases. When the system is prepared in one of its eigenstates $|\pm\rangle$, (6.55) shows that $\theta - \theta_c = 0, \pi/2$ and we recover the previous result (6.43) with (6.42). For the strictly uncoupled state $|1\rangle|\{0\}\rangle$, we have $\theta = 0$ and the CP force vanishes. This is due to cancellations of contributions from U_+ and U_-. This result is not in contradiction with our non-vanishing perturbative result (6.51) where the system is in fact in its a weakly coupled eigenstate rather than $|1\rangle|\{0\}\rangle$.

The strong-coupling CP potential hence sensitively depends on the system's state. Two kinds of states can be prepared in a particularly simple way. Firstly, the atom may enter the cavity in its ground state and is then excited by an external laser, such that the system ends up in the uncoupled state $|1\rangle|\{0\}\rangle$. Alternatively, the atom may be excited outside the cavity where $\Omega_R \ll |\Delta|$. According to (6.46) and (6.47), the systems state $|1\rangle|\{0\}\rangle$ hence coincides with one of the two eigenstates $|\pm\rangle$, depending on the sign of the detuning. If the atom then enters the cavity sufficiently slowly so that the adiabatic approximation holds, the system will remain in this eigenstate. As a result, the atom is in the cavity, with the system being prepared in one of the states $|\pm\rangle$.

6.2 Dynamical Approach

As shown in Chap. 5, a time-independent analysis of the CP force is incomplete with two respects: It fails to reproduce the dynamics of the force; and it does not account for the body-induced frequency shifts and broadenings. To overcome these deficiencies in the case of strong coupling, we generalise the dynamical approach presented in Chap. 5. We first solve the strongly coupled atom–field dynamics and then evaluate the average Lorentz force to find the CP force.

6.2.1 Internal Atomic Dynamics

To account for the body-induced shift and broadening of the atomic transition frequency, we need to include the interaction of the atom with the full spectrum of the electromagnetic field. Improving the single-mode approximation (6.21), we separate the field spectrum (6.8) into two components

$$g^2(\boldsymbol{r}, \omega) = g_\nu^2(\boldsymbol{r}, \omega) + \delta g^2(\boldsymbol{r}, \omega) \,. \tag{6.64}$$

We assume that the single-mode contribution $g_\nu^2(\boldsymbol{r}, \omega)$ has a narrow Lorentzian profile (6.20) and that the underlying field continuum $\delta g^2(\boldsymbol{r}, \omega)$ is flat in the vicinity of the atomic transition frequency and hence only weakly coupled to the atom. The two contributions are illustrated in Fig. 6.1. The dynamics of this improved single-mode approximation is not adequately described by the Jaynes–Cummings Hamiltonian

(6.34). Instead, by combining (6.1), (6.9) and (6.17) and employing the rotating-wave approximation, we have

$$\hat{H} = \int_0^\infty \mathrm{d}\omega\, \hbar\omega \hat{a}^\dagger(\omega)\hat{a}(\omega) + E_0|0\rangle\langle 0| + E_1|1\rangle\langle 1| + \int_0^\infty \mathrm{d}\omega\, \hbar g(\boldsymbol{r}_A,\omega)\big[\hat{a}(\omega)|1\rangle\langle 0| + \hat{a}^\dagger(\omega)|0\rangle\langle 1|\big]\,. \tag{6.65}$$

The atom–field dynamics for strong coupling can most conveniently be solved in the Schrödinger picture. We assume that the atom is initially excited, with the field being in its vacuum state $|\psi(t_0)\rangle = |1\rangle|\{0\}\rangle$. The state of the system at later times can thus be given in the form

$$|\psi(t)\rangle = \psi_1(t)|1\rangle|\{0\}\rangle + \int_0^\infty \mathrm{d}\omega\, \psi_0(\omega,t)|0\rangle|1(\omega)\rangle \tag{6.66}$$

where the coefficients obey the normalisation

$$|\psi_1(t)|^2 + \int_0^\infty \mathrm{d}\omega\, |\psi_0(\omega,t)|^2 = 1 \tag{6.67}$$

and fulfil the initial conditions

$$\psi_1(t_0) = 1\,,\quad \psi_0(\omega,t_0) = 0\,. \tag{6.68}$$

The Schrödinger equation

$$\mathrm{i}\hbar\frac{\partial}{\partial t}|\psi(t)\rangle = \hat{H}|\psi(t)\rangle \tag{6.69}$$

is hence equivalent to the set of equations

$$\dot{\psi}_1(t) = -\mathrm{i}\frac{E_1}{\hbar}\psi_1(t) - \mathrm{i}\int_0^\infty \mathrm{d}\omega\, g(\boldsymbol{r}_A,\omega)\psi_0(\omega,t)\,, \tag{6.70}$$

$$\dot{\psi}_0(\omega,t) = -\mathrm{i}\left(\frac{E_0}{\hbar} + \omega\right)\psi_0(\omega,t) - \mathrm{i}g(\boldsymbol{r}_A,\omega)\psi_1(t)\,. \tag{6.71}$$

The formal solution to the second equation together with the initial condition above is given by

$$\psi_0(\omega, t) = -\mathrm{i}g(\boldsymbol{r}_A, \omega) \int\limits_{t_0}^{t} \mathrm{d}t' \, \mathrm{e}^{-\mathrm{i}(E_0/\hbar+\omega)(t-t')} \psi_1(t') \,. \tag{6.72}$$

Substituting it into the first equation, we obtain

$$\dot{\psi}_1(t) = -\mathrm{i}\frac{E_1}{\hbar}\psi_1(t) - \int\limits_0^\infty \mathrm{d}\omega \, g^2(\boldsymbol{r}_A, \omega) \int\limits_{t_0}^{t} \mathrm{d}t' \, \mathrm{e}^{-\mathrm{i}(E_0/\hbar+\omega)(t-t')} \psi_1(t') \,. \tag{6.73}$$

To solve this equation, we first need to evaluate the frequency integrals. To that end, we apply the decomposition (6.64) of the field spectrum:

$$\begin{aligned} \dot{\psi}_1(t) = &-\mathrm{i}\frac{E_1}{\hbar}\psi_1(t) - \int\limits_0^\infty \mathrm{d}\omega \, \delta g^2(\boldsymbol{r}_A, \omega) \int\limits_{t_0}^{t} \mathrm{d}t' \, \mathrm{e}^{-\mathrm{i}(E_0/\hbar+\omega)(t-t')} \psi_1(t') \\ &- \int\limits_0^\infty \mathrm{d}\omega \, g_\nu^2(\boldsymbol{r}_A, \omega) \int\limits_{t_0}^{t} \mathrm{d}\tau \, \mathrm{e}^{-\mathrm{i}(E_0/\hbar+\omega)(t-t')} \psi_1(t') \,. \end{aligned} \tag{6.74}$$

We assume that the field continuum is assumed to be sufficiently flat, so that its contribution can be evaluated by means of the Markov approximation. Writing $\psi_1(t') = \mathrm{e}^{-\mathrm{i}(\tilde{E}_1/\hbar)(t'-t)}\psi_1(t)$ and extending the lower limit of the time integral to minus infinity, we find

$$\begin{aligned} &\int\limits_0^\infty \mathrm{d}\omega \, \delta g^2(\boldsymbol{r}_A, \omega) \int\limits_{t_0}^{t} \mathrm{d}t' \, \mathrm{e}^{-\mathrm{i}(E_0/\hbar+\omega)(t-t')} \psi_1(t') \\ &\quad \simeq \psi_1(t) \int\limits_0^\infty \mathrm{d}\omega \, \delta g^2(\boldsymbol{r}_A, \omega) \int\limits_{-\infty}^{t} \mathrm{d}t' \, \mathrm{e}^{-\mathrm{i}(\omega-\tilde{\omega}_{10})(t-t')/\hbar} \\ &\quad = \left[\pi\delta g^2(\boldsymbol{r}_A, \tilde{\omega}_{10}) - \mathrm{i}\mathcal{P}\int\limits_0^\infty \frac{\mathrm{d}\omega}{\omega-\tilde{\omega}_{10}} \, \delta g^2(\boldsymbol{r}_A, \omega)\right]\psi_1(t) \\ &\quad = \left(\mathrm{i}\delta\omega_1' + \frac{1}{2}\Gamma_1'\right)\psi_1(t) \end{aligned} \tag{6.75}$$

where we have defined $\tilde{\omega}_{10} = (\tilde{E}_1 - E_0)/\hbar$ and used (5.46). To obtain the last line, we have combined the definitions (6.8), (6.20), (6.28) and (6.64) for the field spectrum; used the integral

$$\mathcal{P}\int_0^\infty \frac{d\omega}{\omega-\tilde{\omega}_{10}}\frac{1}{(\omega-\omega_\nu)^2+\frac{1}{4}\gamma_\nu^2} = \frac{2\pi}{\gamma_\nu}\frac{\omega_\nu-\tilde{\omega}_{10}}{(\tilde{\omega}_{10}-\omega_\nu)^2+\frac{1}{4}\gamma_\nu^2} \quad \text{for } \gamma_\nu \ll \omega_\nu\,; \tag{6.76}$$

and introduced

$$\delta\omega_1' = \delta\omega_1'(\boldsymbol{r}_A) = -\frac{\mu_0}{\pi\hbar}\mathcal{P}\int_0^\infty \frac{d\omega}{\omega-\tilde{\omega}_{10}}\omega^2 \boldsymbol{d}_{10}\cdot\mathrm{Im}\boldsymbol{G}^{(1)}(\boldsymbol{r}_A,\boldsymbol{r}_A,\omega)\cdot\boldsymbol{d}_{01} + \Omega_\mathrm{R}^2\frac{\frac{1}{4}\Delta}{\Delta^2+\frac{1}{4}\gamma_\nu^2}\,, \tag{6.77}$$

$$\Gamma_1' = \Gamma_1'(\boldsymbol{r}_A) = \frac{2\mu_0}{\hbar}\tilde{\omega}_{10}^2\boldsymbol{d}_{01}\cdot\mathrm{Im}\boldsymbol{G}(\boldsymbol{r}_A,\boldsymbol{r}_A,\tilde{\omega}_{10})\cdot\boldsymbol{d}_{10} - \Omega_\mathrm{R}^2\frac{\frac{1}{4}\gamma_\nu}{\Delta^2+\frac{1}{4}\gamma_\nu^2}\,. \tag{6.78}$$

Note that the atom–field detuning

$$\Delta = \Delta(\boldsymbol{r}_A) = \omega_\nu - \tilde{\omega}_{10} \tag{6.79}$$

is now defined with respect to the shifted atomic transition frequency.

The last term in (6.74) is the contribution from the mode ν which is assumed to be very narrow, $\gamma_\nu \ll \omega_\nu$. The Markov approximation is hence not applicable. Instead, exploit the Lorentzian profile (6.20) of the mode and perform the frequency integral by means of the relation

$$\int_0^\infty d\omega\,\frac{e^{-i\omega x}}{(\omega-\omega_\nu)^2+\gamma_\nu^2/4} = \frac{2\pi}{\gamma_\nu}\,e^{-i\omega_\nu x-\gamma_\nu|x|/2} \quad \text{for } \gamma_\nu/2 \ll \omega_\nu\,. \tag{6.80}$$

Recalling the definition (6.28) of the Rabi frequency, we find

$$\int_0^\infty d\omega g_\nu^2(\boldsymbol{r}_A,\omega)\int_{t_0}^t dt' e^{-i(E_0/\hbar+\omega)(t-t')}\psi_1(t') = \frac{\Omega_\mathrm{R}^2}{4}\int_{t_0}^t dt'\, e^{[-i(E_0/\hbar+\omega_\nu)-\gamma_\nu/2](t-t')}\psi_1(t')\,. \tag{6.81}$$

Having evaluated the contributions from both the field continuum and the single mode, the equation of motion for $\psi_1(t)$ reads

$$\dot{\psi}_1(t) = \left(-\mathrm{i}\frac{\tilde{E}_1}{\hbar} - \frac{\Gamma_1'}{2}\right)\psi_1(t) - \frac{\Omega_\mathrm{R}^2}{4}\int\limits_{t_0}^{t} \mathrm{d}t' \mathrm{e}^{[-\mathrm{i}(E_0/\hbar+\omega_\nu)-\gamma_\nu/2](t-t')}\psi_1(t') \quad (6.82)$$

where we have identified $\tilde{E}_1 = E_1 + \hbar\delta\omega_1'$, i.e.,

$$\tilde{\omega}_{10} = \omega_{10} + \delta\omega_1' \,. \quad (6.83)$$

In other words, $\delta\omega_1'$ as given by (6.77) is the continuum-induced frequency shift of the excited atomic level. To simplify the equation for $\psi_1(t)$, we write the solution in the form

$$\psi_1(t) = \mathrm{e}^{(-\mathrm{i}\tilde{E}_1/\hbar - \Gamma_1'/2)(t-t_0)}\phi_1(t) \quad (6.84)$$

where according to (6.82), $\phi_1(t)$ obeys the integro-differential equation

$$\dot{\phi}_1(t) = -\frac{\Omega_\mathrm{R}^2}{4}\int\limits_{t_0}^{t} \mathrm{d}t' \mathrm{e}^{[-\mathrm{i}\Delta-(\gamma_\nu-\Gamma_1')/2](t-t')}\phi_1(t') \,. \quad (6.85)$$

Differentiating with respect to t, it is found to be equivalent to the second-order differential equation

$$\ddot{\phi}_1(t) + \left[\mathrm{i}\Delta + \frac{1}{2}(\gamma_\nu - \Gamma_1')\right]\dot{\phi}_1(t) + \frac{1}{4}\Omega_\mathrm{R}^2\phi_1(t) = 0 \quad (6.86)$$

The associated initial conditions

$$\phi_1(t_0) = 1 \,, \quad \dot{\phi}_1(t_0) = 0 \quad (6.87)$$

follow from (6.68), (6.84) and (6.85). The solution reads

$$\phi_1(t) = \mathrm{e}^{\Omega_+(t-t_0)}c_+ + \mathrm{e}^{\Omega_-(t-t_0)}c_- \,, \quad (6.88)$$

with complex frequencies

$$\Omega_\pm = \Omega_\pm(\boldsymbol{r}_A) = -\frac{1}{2}\left[\mathrm{i}\Delta + \frac{1}{2}(\gamma_\nu - \Gamma_1')\right] \mp \frac{1}{2}\sqrt{\left[\mathrm{i}\Delta + \frac{1}{2}(\gamma_\nu - \Gamma_1')\right]^2 - \Omega_\mathrm{R}^2} \quad (6.89)$$

and coefficients

$$c_\pm = c_\pm(\boldsymbol{r}_A) = \frac{\Omega_\mp}{\Omega_\mp - \Omega_\pm} \,. \quad (6.90)$$

Combining (6.88) with (6.84) we finally obtain [4, 7]

$$\psi_1(t) = \mathrm{e}^{(-\mathrm{i}\tilde{E}_1/\hbar - \Gamma_1'/2 + \Omega_+)(t-t_0)} c_+ + \mathrm{e}^{(-\mathrm{i}\tilde{E}_1/\hbar - \Gamma_1'/2 + \Omega_-)(t-t_0)} c_- \,. \tag{6.91}$$

Having completely solved the atom–field dynamics, let us first make contact with the weak-coupling results obtained in the previous Chap. 5. Weak atom–field coupling is realised if the mode ν is very broad, $\gamma_\nu \gg 2\Omega_\mathrm{R}$, or far detuned from the atomic transition frequency, $|\Delta| \gg 2\Omega_\mathrm{R}^2/\gamma_\nu$. In both cases, the first term under the square root in (6.89) is much larger than the second one and a Taylor expansion yields

$$\Omega_+ = -\mathrm{i}\Delta - \frac{1}{2}\left(\gamma_\nu - \Gamma_1'\right), \tag{6.92}$$

$$\begin{aligned}\Omega_- &= \frac{\mathrm{i}\Omega_\mathrm{R}^2}{4}\frac{\Delta}{\Delta^2 + \frac{1}{4}\gamma_\nu^2} - \frac{\Omega_\mathrm{R}^2}{8}\frac{\gamma_\nu}{\Delta^2 + \frac{1}{4}\gamma_\nu^2}\\ &= -\mathrm{i}(\delta\omega_1 - \delta\omega_1') - \frac{1}{2}(\Gamma_1 - \Gamma_1')\,.\end{aligned} \tag{6.93}$$

The coefficients (6.90) approximate to $c_+ = 0$, $c_- = 1$. Combining the shift (6.77) and rate (6.78) due to the field continuum with the contributions from Ω_-, (6.91) reduces to

$$\psi_1(t) = \mathrm{e}^{[-\mathrm{i}\tilde{E}_1/\hbar - \Gamma_1/2)(t-t_0)} \tag{6.94}$$

with $\tilde{E}_1 = E_1 + \hbar\delta\omega_1$. Here,

$$\begin{aligned}\delta\omega_1 &= \delta\omega_1(\boldsymbol{r}_A)\\ &= -\frac{\mu_0}{\pi\hbar}\mathcal{P}\int_0^\infty \frac{\mathrm{d}\omega}{\omega - \tilde{\omega}_{10}}\omega^2 \boldsymbol{d}_{10}\cdot\mathrm{Im}\boldsymbol{G}^{(1)}(\boldsymbol{r}_A, \boldsymbol{r}_A, \omega)\cdot\boldsymbol{d}_{01}\end{aligned} \tag{6.95}$$

$$\Gamma_1 = \Gamma_1(\boldsymbol{r}_A) = \frac{2\mu_0}{\hbar}\tilde{\omega}_{10}^2 \boldsymbol{d}_{01}\cdot\mathrm{Im}\boldsymbol{G}(\boldsymbol{r}_A, \boldsymbol{r}_A, \tilde{\omega}_{10})\cdot\boldsymbol{d}_{10} \tag{6.96}$$

with

$$\tilde{\omega}_{10} = \omega_{10} + \delta\omega_1 \tag{6.97}$$

are the shift and width of the excited level associated with full field spectrum (6.64), including both the mode ν and the continuum. We have thus reproduced (5.59)–(5.62) as obtained in Chap. 5. Note however that the lower-level shift is absent from $\tilde{\omega}_{10}$ as a result of the rotating-wave approximation made. Our calculation has revealed the conditions under which the Markov approximation applies. It can be used if the field spectrum has either no resonance at the atomic transition frequency or only a sufficiently broad one.

Let us turn our attention to the opposite strong-coupling regime which is realised if the mode ν is both sufficiently narrow, $\gamma_\nu \ll 2\Omega_\mathrm{R}$, and near-resonant with the atomic transition, $|\Delta| \ll 2\Omega_\mathrm{R}^2/\gamma_\nu$. In this case, we can neglect the terms $\mathrm{i}\Delta(\gamma_\nu - \Gamma_1')$

and $(\gamma_\nu - \Gamma_1')/4$ to find

$$\Omega_\pm = -\frac{1}{2}\left[\mathrm{i}\Delta + \frac{1}{2}(\gamma_\nu - \Gamma_1')\right] \mp \frac{1}{2}\Omega\,. \tag{6.98}$$

with the generalised Rabi frequency Ω being given as in (6.37). Recall however, that the detuning (6.79) is defined with respect to the shifted atomic transition frequency. The coefficients (6.90) reduce to

$$c_\pm = \frac{\Omega \mp \Delta}{2\Omega} = \begin{cases} \cos^2\theta_\mathrm{c}\,, \\ \sin^2\theta_\mathrm{c} \end{cases} \tag{6.99}$$

where we have introduced the coupling angle in the form (6.41) and made use of (6.58). Substituting these results into (6.91) and recalling the definition (6.78) of Γ_1', we find

$$\psi_1(t) = \mathrm{e}^{-\gamma(t-t_0)/2}\left[\mathrm{e}^{-\mathrm{i}E_+(t-t_0)/\hbar}\cos^2\theta_\mathrm{c} + \mathrm{e}^{-\mathrm{i}E_-(t-t_0)/\hbar}\sin^2\theta_\mathrm{c}\right]. \tag{6.100}$$

Here,

$$\gamma = \gamma(\boldsymbol{r}_A) = \frac{1}{2}(\gamma_\nu + \Gamma_1') \tag{6.101}$$

is the total damping rate and

$$E_\pm = E_\pm(\boldsymbol{r}_A) = \tfrac{1}{2}\left(E_0 + \tilde{E}_1 + \hbar\omega_\nu\right) \pm \tfrac{1}{2}\hbar\Omega \tag{6.102}$$

with $\tilde{E}_1 = E_1 + \hbar\delta\omega_1'$ are the eigenenergies of the system.

In the case of exact resonance, $\tilde{\omega}_{10} = \omega_\nu$, the coupling angle assumes the value $\theta_\mathrm{c} = \pi/4$ and $\psi_1(t)$ simplifies to

$$\psi_1(t) = \frac{\mathrm{e}^{-\gamma(t-t_0)/2}}{2}\left[\mathrm{e}^{-\mathrm{i}E_+(t-t_0)/\hbar} + \mathrm{e}^{-\mathrm{i}E_-(t-t_0)/\hbar}\right]. \tag{6.103}$$

Using the explicit form of the eigenenergies, it can be given in the alternative form

$$\psi_1(t) = \mathrm{e}^{[-\mathrm{i}(E_0+\tilde{E}_1+\hbar\omega_\nu)/\hbar-\gamma](t-t_0)/2}\cos[\Omega(t-t_0)/2]\,. \tag{6.104}$$

With the state $|\psi(t)\rangle$ of the system being given by (6.66), the excited-state probability $p_1(t) = |\langle 1|\psi(t)\rangle|^2 = |\psi_1(t)|^2$ of the atom is given by

$$p_1(t) = \mathrm{e}^{-\gamma(t-t_0)}\cos^2[\Omega(t-t_0)/2]\,. \tag{6.105}$$

The population of the excited atomic state hence undergoes damped Rabi oscillations [8] during which it is continuously excited and de-excited. The associated

Rabi frequency is affected by the shift $\delta\omega_1'$ of the excited level induced by the field continuum.

To follow the whereabouts of the atomic excitation energy during the Rabi oscillations, let us turn our attention to the field amplitude $\psi_0(\omega, t)$. Substituting our solution (6.104) for $\psi_1(t)$, it is given by

$$\psi_0(\omega, t) = -\mathrm{i}g(\boldsymbol{r}_A, \omega) \int_{t_0}^{t} \mathrm{d}t' \, \mathrm{e}^{-\mathrm{i}(E_0/\hbar+\omega)(t-t')} \times \mathrm{e}^{[-\mathrm{i}(E_0+\tilde{E}_1+\hbar\omega_\nu)/\hbar-\gamma](t'-t_0)/2} \cos[\Omega(t'-t_0)/2] \,. \tag{6.106}$$

To find the amplitude of the single-mode excited state (6.25), we calculate the projection $\psi_{1_\nu}(t) = \langle 1_\nu | \psi(t) \rangle$,

$$\psi_{1_\nu}(t) = \sqrt{\frac{\gamma_\nu}{2\pi}} \int\limits_{\omega_\nu-\Delta\omega/2}^{\omega_\nu+\Delta\omega/2} \mathrm{d}\omega \, \frac{\psi_0(\omega, t)}{\sqrt{(\omega-\omega_\nu)^2 + \frac{1}{4}\gamma_\nu^2}} \,. \tag{6.107}$$

Combining (6.106) and (6.107), we neglect the contribution from the field continuum and evaluate the frequency integral by means of (6.20), (6.28) and (6.80),

$$\psi_{1_\nu}(t) = -\mathrm{i}\frac{\Omega_\mathrm{R}}{2} \int\limits_{t_0}^{t} \mathrm{d}t' \, \mathrm{e}^{[-\mathrm{i}(E_0/\hbar+\omega_\nu)-\gamma_\nu/2](t-t')} \times \mathrm{e}^{[-\mathrm{i}(E_0+\tilde{E}_1+\hbar\omega_\nu)/\hbar-\gamma](t'-t_0)/2} \cos[\Omega(t'-t_0)/2] \,. \tag{6.108}$$

In the strong-coupling limit, the time integral leads to

$$\psi_{1_\nu}(t) = -\mathrm{i}\mathrm{e}^{[-\mathrm{i}(E_0+\tilde{E}_1+\hbar\omega_\nu)/\hbar-\gamma](t-t_0)/2} \sin[\Omega(t-t_0)/2] \,. \tag{6.109}$$

The probability $p_{1_\nu}(t) = |\langle 1_\nu | \psi(t) \rangle|^2 = |\psi_{1_\nu}(t)|^2$ of a single-mode photon being present in the cavity is hence given by

$$p_{1_\nu}(t) = \mathrm{e}^{-\gamma(t-t_0)} \sin^2[\Omega(t-t_0)/2] \,. \tag{6.110}$$

Comparing this with $p_1(t)$ as given above, we conclude that the excitation initially present in the atom is continuously exchanged between the atom and the cavity mode. In other words, the atom reversibly emits and reabsorbs a single photon ν. Adding (6.105) and (6.110), we find

$$p_1(t) + p_{1_\nu}(t) = \mathrm{e}^{-\gamma(t-t_0)} \,. \tag{6.111}$$

This shows that the excitation is slowly lost during the Rabi oscillations. According to (6.101), two processes contribute to this energy dissipation: the atom may emit a photon into the field continuum which is not confined to the cavity and hence is not reabsorbed. Alternatively, a single-mode photon may leak out of the cavity due to imperfect reflection of the cavity walls. The two loss processes are effective at alternating times during the Rabi oscillations. The total damping rate is hence the average of the atomic decay rate Γ_1' and the photonic decay rate γ_ν.

The dynamical analysis including atomic and cavity losses has led to a more refined definition of the strong-coupling regime. Strong coupling is realised whenever Rabi oscillations may be observed. As seen, this requires the mode to be narrow, $\gamma_\nu \ll 2\Omega_\mathrm{R}$, and near-resonant with the atomic transition, $|\Delta| \ll 2\Omega_\mathrm{R}^2/\gamma_\nu$. These conditions allow for a whole range of possible coupling angles. Recall that the simple static analysis of the previous section as based on the Jaynes–Cummings model had suggested a much more reduced notion of strong coupling: With the width of the resonance being neglected, we had placed a much more severe limit $|\Delta| \ll \Omega_\mathrm{R}$ on the detuning, corresponding to $\theta_\mathrm{c} \simeq \pi/4$.

Finally, let us consider the case where the system is initially prepared in a superposition (6.54) of atomic and field excitations, $|\psi(t_0)\rangle = |\theta\rangle$. The state of the system can still be given in the form (6.66), but with modified initial conditions (6.68). Recalling the definitions (6.22) and (6.25) of $|1_\nu\rangle$, we now have

$$\psi_1(t_0) = \cos\theta\,, \quad \psi_0(\omega,t_0) = \sqrt{\frac{\gamma_\nu}{2\pi}}\,\frac{\sin\theta}{\sqrt{(\omega-\omega_\nu)^2+\frac{1}{4}\gamma_\nu^2}}\,. \tag{6.112}$$

With these initial conditions, the solution to the field equation (6.71) is given by

$$\psi_0(\omega,t) = \sqrt{\frac{\gamma_\nu}{2\pi}}\,\frac{\sin\theta\mathrm{e}^{-\mathrm{i}(E_0/\hbar+\omega)(t-t_0)}}{\sqrt{(\omega-\omega_\nu)^2+\frac{1}{4}\gamma_\nu^2}} - \mathrm{i}g(\boldsymbol{r}_A,\omega)\int_{t_0}^{t}\mathrm{d}t'\,\mathrm{e}^{-\mathrm{i}(E_0/\hbar+\omega)(t-t')}\psi_1(t')\,. \tag{6.113}$$

We substitute this result back into the atomic equation (6.70). Evaluating the contribution from the first term in (6.112) via the single-mode approximation $g^2(\boldsymbol{r}_A,\omega) \simeq g_\nu^2(\boldsymbol{r}_A,\omega)$ by combining (6.20), (6.28) and (6.80), we obtain

$$\dot{\psi}_1(t) = -\mathrm{i}\frac{E_1}{\hbar}\psi_1(t) - \mathrm{i}\frac{\Omega_\mathrm{R}}{2}\sin\theta\mathrm{e}^{[-\mathrm{i}(E_0/\hbar+\omega_\nu)-\gamma_\nu/2](t-t_0)} - \int_0^\infty\mathrm{d}\omega\,g^2(\boldsymbol{r}_A,\omega)\int_{t_0}^{t}\mathrm{d}t'\,\mathrm{e}^{-\mathrm{i}(E_0/\hbar+\omega)(t-t')}\psi_1(t')\,. \tag{6.114}$$

We now follow exactly the same steps as before. Decomposing the field spectrum into its single-mode and continuum contributions, we find

$$\dot{\psi}_1(t) = \left(-\mathrm{i}\frac{\tilde{E}_1}{\hbar} - \frac{\Gamma_1'}{2}\right)\psi_1(t) - \mathrm{i}\,\frac{\Omega_\mathrm{R}}{2}\sin\theta \mathrm{e}^{[-\mathrm{i}(E_0/\hbar+\omega_\nu)-\gamma_\nu/2](t-t_0)}$$
$$- \frac{\Omega_\mathrm{R}^2}{4}\int_{t_0}^{t} \mathrm{d}t'\, \mathrm{e}^{[-\mathrm{i}(E_0/\hbar+\omega_\nu)-\gamma_\nu/2](t-t')}\psi_1(t')\,. \tag{6.115}$$

Writing $\psi_1(t)$ in the form (6.84), we obtain the integro-differential equation

$$\dot{\phi}_1(t) = -\mathrm{i}\frac{\Omega_\mathrm{R}}{2}\sin\theta \mathrm{e}^{[-\mathrm{i}\Delta-(\gamma_\nu-\Gamma_1')/2](t-t_0)}$$
$$- \frac{\Omega_\mathrm{R}^2}{4}\int_{t_0}^{t} \mathrm{d}t'\, \mathrm{e}^{[-\mathrm{i}\Delta-(\gamma_\nu-\Gamma_1')/2](t-t')}\phi_1(t')\,. \tag{6.116}$$

By taking the time derivative, we recover the same second-order differential equation (6.86) as before, but with generalised initial conditions

$$\phi_1(t_0) = \cos\theta\,, \qquad \dot{\phi}_1(t_0) = -\frac{\mathrm{i}}{2}\Omega_\mathrm{R}\sin\theta\,. \tag{6.117}$$

It is solved by (6.88) with (6.89) as before, but with modified coefficients

$$c_\pm = c_\pm(\boldsymbol{r}_A) = \frac{\Omega_\mp\cos\theta + \frac{\mathrm{i}}{2}\Omega_\mathrm{R}\sin\theta}{\Omega_\mp - \Omega_\pm}\,. \tag{6.118}$$

Consequently, $\psi_1(t)$ is again given by (6.91), but with the modified coefficients.

In the strong-coupling limit, the approximation (6.98) applies and the coefficients reduce to

$$c_\pm = \frac{\left(\Omega \mp \Delta\right)\cos\theta \pm \Omega_\mathrm{R}\sin\theta}{2\Omega}\,. \tag{6.119}$$

Introducing the coupling angle in the form (6.41) and using the relations (6.38) and (6.58), they can alternatively be given as

$$c_+ = \cos^2\theta_c\cos\theta + \sin\theta_c\cos\theta_c\sin\theta = \cos\theta_\mathrm{c}\cos(\theta-\theta_\mathrm{c})\,, \tag{6.120}$$
$$c_- = \sin^2\theta_c\cos\theta - \sin\theta_c\cos\theta_c\sin\theta = -\sin\theta_\mathrm{c}\sin(\theta-\theta_\mathrm{c})\,. \tag{6.121}$$

With these approximations, the excited-state amplitude simplifies to

$$\psi_1(t) = \mathrm{e}^{[-\mathrm{i}(E_0+\tilde{E}_1+\hbar\omega_\nu)/\hbar-\gamma](t-t_0)/2}\Big[\cos\theta_\mathrm{c}\cos(\theta-\theta_\mathrm{c})\mathrm{e}^{-\mathrm{i}\Omega(t-t_0)}$$
$$- \sin\theta_\mathrm{c}\sin(\theta-\theta_\mathrm{c})\mathrm{e}^{\mathrm{i}\Omega(t-t_0)}\Big]\,. \tag{6.122}$$

In the case of exact resonance, $\tilde{\omega}_{10} = \omega_\nu$, the coupling angle reads $\theta_\mathrm{c} = \pi/4$, so that

$$\times \frac{\mathrm{d}}{\mathrm{d}t}\big[\psi_1^*(t)\psi_0(\omega,t) - \psi_1(t)\psi_0^*(\omega,t)\big]. \qquad (6.126)$$

Using to our solution (6.72) for the coupled atom–field dynamics, we may eliminate $\psi_0^*(\omega,t)$:

$$\begin{aligned}\boldsymbol{F}(t) = &-\frac{\mathrm{i}\mu_0}{\pi}\int_0^\infty \mathrm{d}\omega\,\omega^2 \nabla \boldsymbol{d}_{10}\cdot \mathrm{Im}\boldsymbol{G}(\boldsymbol{r}_A,\boldsymbol{r}_A,\omega)\cdot \boldsymbol{d}_{01}\\ &\times \int_{t_0}^{t} \mathrm{d}t'\Big[\mathrm{e}^{\mathrm{i}(E_0/\hbar+\omega)(t-t')}\psi_1(t)\psi_1^*(t') - \mathrm{e}^{-\mathrm{i}(E_0/\hbar+\omega)(t-t')}\psi_1^*(t)\psi_1(t')\Big]\\ &+\frac{\mu_0}{\pi}\int_0^\infty \mathrm{d}\omega\,\omega \boldsymbol{d}_{10}\times\big[\nabla\times \mathrm{Im}\boldsymbol{G}(\boldsymbol{r}_A,\boldsymbol{r}_A,\omega)\cdot\boldsymbol{d}_{01}\big]\\ &\times\frac{\mathrm{d}}{\mathrm{d}t}\int_{t_0}^{t} \mathrm{d}t'\Big[\mathrm{e}^{\mathrm{i}(E_0/\hbar+\omega)(t-t')}\psi_1(t)\psi_1^*(t') + \mathrm{e}^{-\mathrm{i}(E_0/\hbar+\omega)(t-t')}\psi_1^*(t)\psi_1(t')\Big].\end{aligned}$$

(6.127)

Finally, by substituting (6.91) for $\psi_1(t)$ and carrying out the time integral, the CP force on an initially excited atom reads

$$\begin{aligned}\boldsymbol{F}_1(\boldsymbol{r}_A,t) = &\frac{2\mu_0}{\pi}\int_0^\infty \mathrm{d}\omega\,\omega^2 \nabla \boldsymbol{d}_{10}\cdot \mathrm{Im}\,\boldsymbol{G}^{(1)}(\boldsymbol{r}_A,\boldsymbol{r}_A,\omega)\cdot\boldsymbol{d}_{01}\mathrm{Re}\,s(\omega,t-t_0)\\ &-\frac{2\mu_0}{\pi}\int_0^\infty \mathrm{d}\omega\,\omega \boldsymbol{d}_{10}\times\big[\nabla\times\mathrm{Im}\boldsymbol{G}^{(1)}(\boldsymbol{r}_A,\boldsymbol{r}_A,\omega)\cdot\boldsymbol{d}_{01}\big]\frac{\mathrm{d}}{\mathrm{d}t}\mathrm{Im}\,s(\omega,t-t_0)\end{aligned}$$

(6.128)

with

$$\begin{aligned}s(\omega,t) &= s(\boldsymbol{r}_A,\omega,t)\\ &= \sum_{p,q=\pm}\frac{\mathrm{e}^{(-\Gamma_1'+\Omega_p+\Omega_q^*)t} - \mathrm{e}^{[\mathrm{i}(\omega-\tilde{\omega}_{10})-\Gamma_1'/2+\Omega_p]t}}{\omega-\tilde{\omega}_{10}-\frac{\mathrm{i}}{2}\Gamma_1'+\mathrm{i}\Omega_q^*}c_p c_q^*. \qquad (6.129)\end{aligned}$$

This is the exact force on a two-level atom valid for both weak and strong atom–field coupling. Note that we have discarded the self-force associated with the free-space Green's tensor $\boldsymbol{G}^{(0)}$ by making the replacement $\boldsymbol{G}\mapsto\boldsymbol{G}^{(1)}$.

To make contact with the results of the previous Chap. 5, let us first consider the weak-coupling limit where the approximations (6.92), (6.93) and $c_+ = 0$, $c_- = 1$

hold. Neglecting the rapidly oscillating second terms of the numerators in (6.129), we recover the result of the Markov approximation

$$s(\omega,t)=\frac{\mathrm{e}^{-\Gamma_1 t}}{\omega-\tilde{\omega}_{10}-\frac{\mathrm{i}}{2}\Gamma_1}\,,\quad \frac{\mathrm{d}}{\mathrm{d}t}s(\omega,t)=0\,. \tag{6.130}$$

with $\tilde{\omega}_{10}$ containing the full shift according to (6.97). The second result has been obtained by recalling that the time derivative has to be performed before carrying out the integral. The CP force hence reduces to

$$\boldsymbol{F}_1(\boldsymbol{r}_A,t)=\mathrm{e}^{-\Gamma_1(t-t_0)}\boldsymbol{F}_1(\boldsymbol{r}_A)\,, \tag{6.131}$$

$$\boldsymbol{F}_1(\boldsymbol{r}_A)=\frac{\mu_0}{\pi}\int_0^\infty \mathrm{d}\omega\,\omega^2\frac{\nabla\boldsymbol{d}_{10}\cdot\mathrm{Im}\boldsymbol{G}(\boldsymbol{r},\boldsymbol{r}_A,\omega)\cdot\boldsymbol{d}_{01}\big|_{\boldsymbol{r}=\boldsymbol{r}_A}}{\omega-\tilde{\omega}_{10}-\frac{\mathrm{i}}{2}\Gamma_1}+\mathrm{C.c.} \tag{6.132}$$

This is in agreement with (5.101) and (5.102) as found in Chap. 5 on the basis of the Markov approximation.

In the strong-coupling regime of a sufficiently narrow cavity mode ($\gamma_\nu\le 2\Omega_\mathrm{R}$) that is near-resonant with the atomic transition ($|\Delta|\ll 2\Omega_\mathrm{R}^2/\gamma_\nu$), the CP force is dominated by the single-mode contribution. In close analogy with (6.20) and (6.21), we may hence approximate

$$\omega^2\mathrm{Im}\boldsymbol{G}^{(1)}(\boldsymbol{r},\boldsymbol{r}',\omega)\simeq\omega_\nu^2\mathrm{Im}\boldsymbol{G}^{(1)}(\boldsymbol{r},\boldsymbol{r}',\omega_\nu)\frac{\frac{1}{4}\gamma_\nu^2}{(\omega-\omega_\nu)^2+\frac{1}{4}\gamma_\nu^2}\,. \tag{6.133}$$

The frequency integral can then be carried out by means of (6.80), resulting in

$$\begin{aligned}\boldsymbol{F}_1(\boldsymbol{r}_A,t)&=\mu_0\gamma_\nu\omega_\nu^2\nabla\boldsymbol{d}_{10}\cdot\mathrm{Im}\,\boldsymbol{G}^{(1)}(\boldsymbol{r}_A,\boldsymbol{r}_A,\omega_\nu)\cdot\boldsymbol{d}_{01}\mathrm{Re}\,s(t-t_0)\\&\quad-\mu_0\gamma_\nu\omega_\nu\boldsymbol{d}_{10}\times\left[\nabla\times\mathrm{Im}\,\boldsymbol{G}^{(1)}(\boldsymbol{r}_A,\boldsymbol{r}_A,\omega_\nu)\cdot\boldsymbol{d}_{01}\right]\frac{\mathrm{d}}{\mathrm{d}t}\,\mathrm{Im}\,s(t-t_0)\,.\end{aligned} \tag{6.134}$$

with

$$s(t)=s(\boldsymbol{r}_A,t)=\sum_{p,q=\pm}\frac{\mathrm{e}^{(-\Gamma_1'+\Omega_p+\Omega_q^*)t}-\mathrm{e}^{(\mathrm{i}\Delta-\gamma+\Omega_p)t}}{\Delta+\frac{\mathrm{i}}{2}(\gamma_\nu-\Gamma_1')+\mathrm{i}\Omega_q^*}\,c_pc_q^*\,. \tag{6.135}$$

With the strong-coupling approximations (6.98) and (6.99), $s(t)$ further simplifies to

$$s(t)=\frac{\mathrm{e}^{-\gamma t}}{\Omega}\left[\sin^2\theta_\mathrm{c}-\cos^2\theta_\mathrm{c}+\mathrm{e}^{-\mathrm{i}\Omega t}\cos^2\theta_\mathrm{c}-\mathrm{e}^{\mathrm{i}\Omega t}\sin^2\theta_\mathrm{c}\right], \tag{6.136}$$

so that

$$\operatorname{Re} s(t) = -\frac{\mathrm{e}^{-\gamma t}}{\Omega} \cos(2\theta_\mathrm{c})[1 - \cos(\Omega t)] , \tag{6.137}$$

$$\frac{\mathrm{d}}{\mathrm{d}t} \operatorname{Im} s(t) = -\mathrm{e}^{-\gamma t} \cos(\Omega t) . \tag{6.138}$$

The CP force for strong atom–field coupling is hence given by [4]

$$\begin{aligned} \boldsymbol{F}_1(\boldsymbol{r}_A, t) = &-\frac{\mu_0\gamma_\nu}{\Omega}\omega_\nu^2 \boldsymbol{\nabla} \boldsymbol{d}_{10} \cdot \operatorname{Im} \boldsymbol{G}^{(1)}(\boldsymbol{r}_A, \boldsymbol{r}_A, \omega_\nu) \cdot \boldsymbol{d}_{01} \\ &\times \cos(2\theta_\mathrm{c}) \mathrm{e}^{-\gamma(t-t_0)}\{1 - \cos[\Omega(t - t_0)]\} \\ &+ \mu_0\gamma_\nu\omega_\nu \boldsymbol{d}_{10} \times \left[\boldsymbol{\nabla} \times \operatorname{Im} \boldsymbol{G}^{(1)}(\boldsymbol{r}_A, \boldsymbol{r}_A, \omega_\nu) \cdot \boldsymbol{d}_{01}\right] \\ &\times \mathrm{e}^{-\gamma(t-t_0)} \cos[\Omega(t - t_0)] . \end{aligned} \tag{6.139}$$

It has two components which arise due to the action of the electric and magnetic field on the atom. The electric force has a constant sign while its amplitude undergoes damped Rabi oscillations. The magnetic force exhibits Rabi oscillations during which its sign changes periodically. Its magnitude is roughly Ω/ω_ν times that of the electric force. It is hence only relevant in the recently considered superstrong-coupling regime [9] where Ω/ω_ν might be comparable to or even greater than unity.

In the ordinary strong-coupling regime, the magnetic force is negligible. To simplify the electric force, we exploit the symmetry (A.4) of the Green's tensor to make the replacement $\boldsymbol{\nabla} \boldsymbol{G}^{(1)}(\boldsymbol{r}_A, \boldsymbol{r}_A, \omega) \mapsto \frac{1}{2}\boldsymbol{\nabla} \boldsymbol{G}^{(1)}(\boldsymbol{r}, \boldsymbol{r}, \omega)|_{\boldsymbol{r}=\boldsymbol{r}_A}$ and recall the definitions (6.28) and (6.37) of the vacuum Rabi frequency Ω_R and its generalised counterpart Ω. Making use of the relations (6.61) and (6.62), we may write

$$\begin{aligned} \frac{\mu_0\gamma_\nu}{\Omega}\omega_\nu^2 \boldsymbol{\nabla} \boldsymbol{d}_{10} \cdot \operatorname{Im} \boldsymbol{G}^{(1)}(\boldsymbol{r}_A, \boldsymbol{r}_A, \omega_\nu) \cdot \boldsymbol{d}_{01} &= \frac{\mu_0\hbar\boldsymbol{\nabla}_A\Omega_\mathrm{R}^2}{4\Omega} \\ &= \frac{1}{2}\hbar\boldsymbol{\nabla}\sqrt{\Omega_\mathrm{R}^2(\boldsymbol{r}) + \Delta^2}\Big|_{\boldsymbol{r}=\boldsymbol{r}_A} . \end{aligned} \tag{6.140}$$

The CP force can hence be given as [4, 7]

$$\boldsymbol{F}_1(\boldsymbol{r}_A, t) = \mathrm{e}^{-\gamma(t-t_0)} \cos(2\theta_\mathrm{c})\{1 - \cos[\Omega(t - t_0)]\}\boldsymbol{F}_+(\boldsymbol{r}_A) \tag{6.141}$$

with

$$\boldsymbol{F}_+(\boldsymbol{r}_A) = -\tfrac{1}{2}\hbar\boldsymbol{\nabla}\sqrt{\Omega_\mathrm{R}^2(\boldsymbol{r}) + \Delta^2}\Big|_{\boldsymbol{r}=\boldsymbol{r}_A} . \tag{6.142}$$

Like the weak-coupling result, the force factorises into a function of time and a force component that carries all the main spatial dependence. The force component resembles the static result (6.43) with (6.42), but the detuning now contains the body-induced frequency shift $\delta\omega_1'$. Since the gradient in (6.142) only acts on the position argument of the vacuum Rabi-frequency, the CP force is not a conservative force in general. Recall from Sect. 5.4 that the body-induced frequency shift has similar consequences for weak atom–field coupling. In the case of exact resonance $\tilde{\omega}_{10} = \omega_\nu$, we have $\theta_c = \pi/4$ and the CP force on an initially excited atom vanishes at all times. This is in agreement with the result of static treatment as discussed below (6.63) in Sect. 6.1.2. For finite detuning, a vanishing force at initial time eventually evolves into a finite force whose amplitude exhibits Rabi oscillations.

To conclude the section, let us consider the CP force for the more general initial state $|\psi(t_0)\rangle = |\theta\rangle$ as given by (6.54). In this case, $\psi_0(\omega, t)$ is given by (6.113) which we substitute into (6.126). Focussing our attention to the force associated with the mode ν, we make use of the single-mode approximation (6.20) with (6.21) and evaluate the frequency integrals according to (6.80). Recalling (6.28), we find

$$\begin{aligned}
\boldsymbol{F}_\theta(\boldsymbol{r}_A, t) = {} & \mu_0\gamma_\nu\omega_\nu^2 \nabla \boldsymbol{d}_{10}\cdot \mathrm{Im}\,\boldsymbol{G}^{(1)}(\boldsymbol{r}_A, \boldsymbol{r}_A, \omega_\nu)\cdot \boldsymbol{d}_{01} \\
& \times \mathrm{Re}\left[s(t-t_0) - \frac{2}{\Omega_\mathrm{R}}\sin\theta\, q(t-t_0)\right] \\
& - \mu_0\gamma_\nu\omega_\nu \boldsymbol{d}_{10} \times \left[\nabla \times \mathrm{Im}\,\boldsymbol{G}^{(1)}(\boldsymbol{r}_A, \boldsymbol{r}_A, \omega_\nu)\cdot \boldsymbol{d}_{01}\right] \\
& \times \frac{\mathrm{d}}{\mathrm{d}t}\,\mathrm{Im}\left[s(t-t_0) - \frac{2}{\Omega_\mathrm{R}}\sin\theta\, q(t-t_0)\right].
\end{aligned} \tag{6.143}$$

with

$$q(t) = q(\boldsymbol{r}_A, t) = \mathrm{e}^{[\mathrm{i}\Delta - (\gamma_\nu + \Gamma_1')/2]t}\left(c_+\mathrm{e}^{\Omega_+ t} + c_-\mathrm{e}^{\Omega_- t}\right). \tag{6.144}$$

Note that $s(\omega, t)$ is still given by (6.135), but with coefficients $c_\pm$ according to (6.118). Using the strong-coupling approximations (6.98), (6.120) and (6.121) and recalling (6.61), one finds

$$\begin{aligned}
s(t) - \frac{2}{\Omega_\mathrm{R}}\sin\theta\, q(t) = {} & \frac{\mathrm{e}^{-\gamma t}}{\Omega}\big[\sin^2(\theta-\theta_c) - \cos^2(\theta-\theta_c) \\
& + \left(\mathrm{e}^{-\mathrm{i}\Omega t}\cot\theta_c - \mathrm{e}^{\mathrm{i}\Omega t}\tan\theta_c\right)\sin(\theta-\theta_c)\cos(\theta-\theta_c)\big]
\end{aligned} \tag{6.145}$$

and hence

$$\begin{aligned}
& \mathrm{Re}\left[s(t-t_0) - \frac{2}{\Omega_\mathrm{R}}\sin\theta q(t-t_0)\right] \\
& = -\frac{\mathrm{e}^{-\gamma t}}{\Omega}\{\cos[2(\theta-\theta_c)] + \cot(2\theta_c)\sin[2(\theta-\theta_c)]\cos(\Omega t)\},
\end{aligned} \tag{6.146}$$

$$\frac{\mathrm{d}}{\mathrm{d}t}\mathrm{Im}\left[s(t-t_0)-\frac{2}{\Omega_\mathrm{R}}\sin\theta q(t-t_0)\right]=-\mathrm{e}^{-\gamma t}\frac{\sin[2(\theta-\theta_\mathrm{c})]}{\sin(2\theta_\mathrm{c})}\cos(\Omega t)\}\,. \tag{6.147}$$

The strong-coupling force for an initial state $|\theta\rangle$ follows immediately [4],

$$\begin{aligned}\boldsymbol{F}_\theta(\boldsymbol{r}_A,t)=&-\frac{\mu_0\gamma_\nu}{\Omega}\omega_\nu^2\nabla\boldsymbol{d}_{10}\cdot\mathrm{Im}\boldsymbol{G}^{(1)}(\boldsymbol{r}_A,\boldsymbol{r}_A,\omega_\nu)\cdot\boldsymbol{d}_{01}\mathrm{e}^{-\gamma(t-t_0)}\\&\times\{\cos[2(\theta-\theta_\mathrm{c})]+\cot(2\theta_\mathrm{c})\sin[2(\theta-\theta_\mathrm{c})]\cos[\Omega(t-t_0)]\}\\&+\mu_0\gamma_\nu\omega_\nu\boldsymbol{d}_{10}\times\left[\nabla\times\mathrm{Im}\,\boldsymbol{G}^{(1)}(\boldsymbol{r}_A,\boldsymbol{r}_A,\omega_\nu)\cdot\boldsymbol{d}_{01}\right]\\&\times\mathrm{e}^{-\gamma(t-t_0)}\frac{\sin[2(\theta-\theta_\mathrm{c})]}{\sin(2\theta_\mathrm{c})}\cos[\Omega(t-t_0)]\,.\end{aligned} \tag{6.148}$$

Just like the force on an initially excited atom, it contains of electric and magnetic contributions. The magnetic contribution is again oscillating around a zero mean. It is smaller than the electric-field contribution by a factor Ω/ω_ν and only observable in the superstrong-coupling regime. For ordinary strong coupling, the force is dominated by the electric-field contribution. Using the relation (6.140) as found above, it can be given in the form [4]

$$\begin{aligned}\boldsymbol{F}_\theta(\boldsymbol{r}_A,t)=&\,\mathrm{e}^{-\gamma(t-t_0)}\{\cos[2(\theta-\theta_\mathrm{c})]\\&+\cot(2\theta_\mathrm{c})\sin[2(\theta-\theta_\mathrm{c})]\cos[\Omega(t-t_0)]\}\boldsymbol{F}_+(\boldsymbol{r}_A)\,.\end{aligned} \tag{6.149}$$

The force at initial time

$$\boldsymbol{F}_\theta(\boldsymbol{r}_A,t)=\{\cos[2(\theta-\theta_\mathrm{c})]+\cot(2\theta_\mathrm{c})\sin[2(\theta-\theta_\mathrm{c})]\}\boldsymbol{F}_+(\boldsymbol{r}_A) \tag{6.150}$$

reduces to our static result (6.63) when neglecting the frequency shift $\delta\omega_1'$.

The dynamics of the strong-coupling force for different initial states $|\theta\rangle$ is shown in Fig. 6.4. We observe that the curves may be grouped into pairs of curves with opposite signs. The members of each pair lie at opposite points in the polar diagram for 2θ. Recalling the definition (6.54) of $|\theta\rangle$, we see that the relative distribution of the excitation between the atom and the field mode is also exactly opposite for the two members of a pair. The roles of atom and field are hence exchanged, photon emissions processes are replaced with absorptions and vice versa. The opposite signs of the resulting force is then plausible since the resonant CP force can be interpreted as a consequence of photon recoil.

The figure further reveals that there are two extremes of behaviour: For the initial states with $2\theta=2\theta_\mathrm{c}$ (a_1), $2\theta_\mathrm{c}+\pi$ (a_3), the force shows no oscillations and is purely

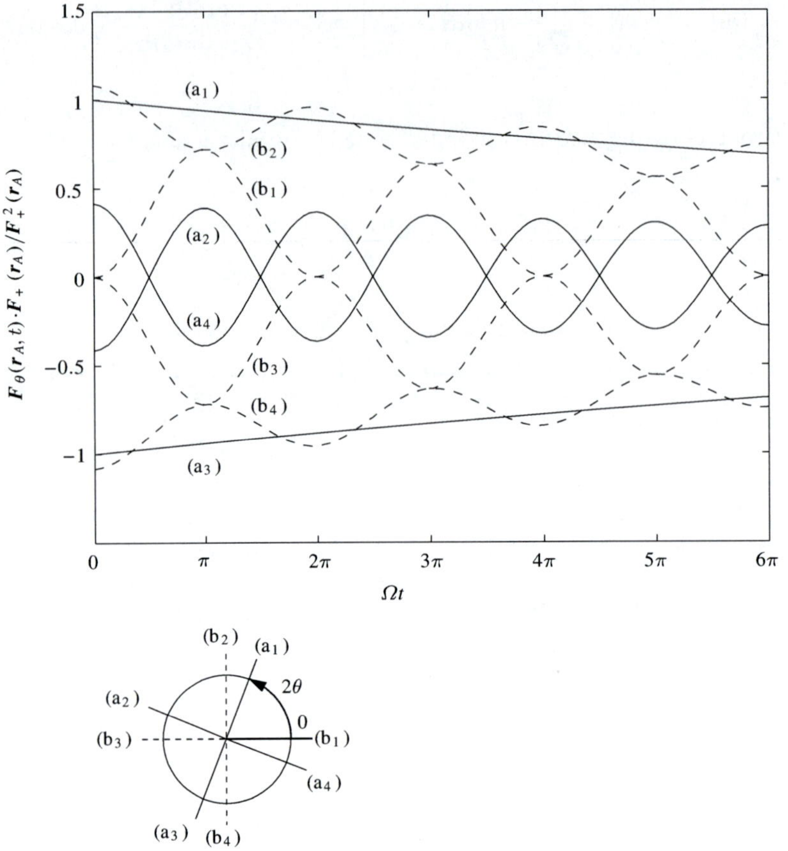

Fig. 6.4 Dynamics of the strong-coupling CP force for different initial states $|\theta\rangle$: $2\theta = 2\theta_c$ (a_1), $2\theta_c + \pi/2$ (a_2), $2\theta_c + \pi$ (a_3), $2\theta_c + 3\pi/2$ (a_4), 0 (b_1), $\pi/2$ (b_2), π (b_3), $3\pi/2$ (b_4). The coupling angle and damping rate are held constant at $2\theta_c = 3\pi/8$ and $\gamma = 0.05\Omega$. The polar diagram indicates the angles 2θ for the various curves

exponentially damped as a function of time. These initial states are the dressed states $|\pm\rangle$, i.e., the quasi-stationary approximate eigenstates of the system. On the other extreme, the initial states $2\theta = 2\theta_c + \pi/2$ (a_2), $2\theta_c + 3\pi/2$ (a_4) correspond to equal-weight superpositions of the eigenstates. They lead to Rabi oscillations of maximal amplitude around a zero mean value. For other values of θ (b_1)–(b_4), the temporal behaviour of the force is a combination of oscillating and non-oscillating components. Note that $2\theta = 0$ (b_1) corresponds to the initial state $|1\rangle|\{0\}\rangle$ of a purely excited atom. As discussed below (6.142), only the magnitude of the force oscillates in time in this case, leaving the sign invariant.

References

1. E.T. Jaynes, F.W. Cummings, Proc. IEEE **51**(1), 89 (1963)
2. I.N. Bronstein, K.A. Semendjajew, G. Musiol, H. Mühlig, *Taschenbuch der Mathematik* (Harry Deutsch, Frankfurt am Main, 1995)
3. C. Cohen-Tannoudji, S. Reynaud, J. Phys. B: At. Mol. Opt. Phys. **10**(3), 345 (1977)
4. S.Y. Buhmann, D.G. Welsch, Phys. Rev. A **77**(1), 012110 (2008)
5. S. Haroche, M. Brune, J.M. Raimond, Europhys. Lett. **14**(1), 19 (1991)
6. B.G. Englert, J. Schwinger, A.O. Barut, M.O. Scully, Europhys. Lett. **14**(1), 25 (1991)
7. S.Y. Buhmann, D.G. Welsch, Prog. Quantum Electron. **31**(2), 51 (2007)
8. I.I. Rabi, Phys. Rev. **51**(8), 652 (1937)
9. D. Meiser, P. Meystre, Phys. Rev. A **74**(6), 065801 (2006)

Chapter 7
Thermal Casimir–Polder Forces

When calculating dispersion forces, we have mostly assumed the body-assisted field to be in its ground state. Dispersion interactions of ground-state atoms and bodies are then pure quantum effects. They are due to the ground-state fluctuations of the electromagnetic field, in other words: virtual photons. As seen in the previous three Chaps. 4–6, the situation may drastically change when considering excited atoms. They can emit real photons, leading to resonant forces which exhibit an oscillatory spatial behaviour and a non-trivial dynamics.

In this chapter, we are going to complement the analysis by considering excited fields. In particular, we will allow for a uniform finite temperature of the bodies and the electromagnetic field. The fluctuations of the electromagnetic field will then exhibit a thermal, classical component in addition to the ground-state, quantum contribution. In other words, real thermal photons will be present in addition to the virtual ones. To study their influence on the CP interaction, we will follow the two approaches developed in Chaps. 4 and 5. We first calculate the CP potential for a given atomic state and environment temperature using perturbation theory. To account for the temporal evolution of the force, we use our alternative approach by solving the coupled atom–field dynamics and evaluating the average Lorentz force. To illustrate the closely interrelated dependences of the CP interaction on position, time, temperature, atomic and material properties, we study the example of a single atom or molecule interacting with a plate.

For simplicity, we will work within the multipolar coupling scheme throughout this chapter, dropping the primes that indicate multipolar variables.

7.1 Static Theory

Following Casimir and Polder's famous concept as laid out in Chap. 4, the potential can be derived from the energy shift induced by the atom–field interaction. In order to define such an energy shift, the uncoupled system has to be prepared in an energy

S. Y. Buhmann, *Dispersion Forces II*, Springer Tracts in Modern Physics 248,
DOI: 10.1007/978-3-642-32466-6_7, © Springer-Verlag Berlin Heidelberg 2012

eigenstate. A thermal state of the electromagnetic field of uniform temperature T as given by (1.38) is not an eigenstate of the Hamiltonian $\hat{H}_{\mathrm{F}}$. Instead, it is an incoherent superposition of energy eigenstates $|\psi\rangle$ with probabilities p_ψ, as described by a density matrix

$$\hat{\rho} = \sum_{\psi} p_\psi |\psi\rangle\langle\psi| \,. \tag{7.1}$$

To extend the notion of an energy shift to such a mixed initial state, we introduce the average energy shift

$$\langle \Delta E\rangle = \sum_{\psi} p_\psi \langle\psi|\hat{H}_{\mathrm{AF}}|\psi\rangle + \sum_{\psi} p_\psi \sum_{I\neq\psi} \frac{\langle\psi|\hat{H}_{\mathrm{AF}}|I\rangle\langle I|\hat{H}_{\mathrm{AF}}|\psi\rangle}{E_\psi - E_I} + \ldots\,, \tag{7.2}$$

which as an generalisation of (1.120)–(1.122). Inserting the completeness relation $\sum_\phi |\phi\rangle\langle\phi| = \hat{I}$ and using the above definition of the density matrix, the average energy shift can be given in the more compact form

$$\langle \Delta E\rangle = \big\langle \hat{H}_{\mathrm{AF}}\big\rangle + \left\langle \sum_I \frac{\hat{H}_{\mathrm{AF}}|I\rangle\langle I|\hat{H}_{\mathrm{AF}}}{\hat{H}_{\mathrm{A}} + \hat{H}_{\mathrm{F}} - E_I} \right\rangle + \ldots \tag{7.3}$$

where $\langle\ldots\rangle = \mathrm{tr}(\ldots\hat{\rho})$ and we have assumed that $\langle\psi|\hat{H}_{\mathrm{AF}}|\psi\rangle = 0$.

We will calculate this energy shift for an atom in an incoherent superposition of internal-energy eigenstates,

$$\hat{\sigma} = \sum_n p_n |n\rangle\langle n| \tag{7.4}$$

with the field being in a thermal state $\hat{\rho}_T$ of uniform temperature T. The uncoupled atom–field system hence reads $\hat{\rho} = \hat{\sigma} \otimes \hat{\rho}_T$. The thermal field state (1.38) can be expanded in a basis of Fock states (1.37),

$$\hat{\rho}_T = \frac{1}{Z}\Bigg[|\{0\}\rangle\langle\{0\}| + \sum_{j=1}^{\infty} \sum_{\lambda_1\ldots\lambda_j=e,m} \int \mathrm{d}^3 r_1 \cdots \int \mathrm{d}^3 r_j \int_0^\infty \mathrm{d}\omega_1 \cdots \int_0^\infty \mathrm{d}\omega_j$$
$$\times\, \mathrm{e}^{-\hbar(\omega_1 + \cdots + \omega_j)/(k_{\mathrm{B}}T)}$$
$$\times\, |\mathbf{1}_{\lambda_1}(\boldsymbol{r}_1,\omega_1)\ldots\mathbf{1}_{\lambda_j}(\boldsymbol{r}_j,\omega_j)\rangle\langle\mathbf{1}_{\lambda_1}(\boldsymbol{r}_1,\omega_1)\ldots\mathbf{1}_{\lambda_j}(\boldsymbol{r}_j,\omega_j)| \Bigg] \tag{7.5}$$

where

$$Z = \mathrm{tr}\Bigg[|\{0\}\rangle\langle\{0\}| + \sum_{j=1}^{\infty} \sum_{\lambda_1\ldots\lambda_j=e,m} \int \mathrm{d}^3 r_1 \cdots \int \mathrm{d}^3 r_j \int_0^\infty \mathrm{d}\omega_1 \cdots \int_0^\infty \mathrm{d}\omega_j$$

$$\times\, \mathrm{e}^{-\hbar(\omega_1+\cdots+\omega_j)/(k_B T)}$$

$$\times\, |\mathbf{1}_{\lambda_1}(\boldsymbol{r}_1,\omega_1)\ldots\mathbf{1}_{\lambda_j}(\boldsymbol{r}_j,\omega_j)\rangle\langle\mathbf{1}_{\lambda_1}(\boldsymbol{r}_1,\omega_1)\ldots\mathbf{1}_{\lambda_j}(\boldsymbol{r}_j,\omega_j)|\Bigg] \tag{7.6}$$

is the partition function.

The multipolar coupling Hamiltonian (4.36) for a non-magnetic atom in electric-dipole approximation is linear in the field variables. With the thermal density matrix being diagonal, the first-order contribution $\langle\hat{H}_{\mathrm{AF}}\rangle$ to the energy shift vanishes and we need to calculate the second-order contribution. The total density matrix is a superposition of states $|n\rangle|\mathbf{1}_{\lambda_1}(\boldsymbol{r}_1,\omega_1)\ldots\mathbf{1}_{\lambda_j}(\boldsymbol{r}_j,\omega_j)\rangle$ with j photons. Non vanishing contributions to the energy shift are hence due to intermediate states which contain either $j+1$ photons, $|I\rangle = |k\rangle|\mathbf{1}_{\lambda'_1}(\boldsymbol{r}'_1,\omega'_1)\ldots\mathbf{1}_{\lambda'_{j+1}}(\boldsymbol{r}'_{j+1},\omega'_{j+1})\rangle$, or $j-1$ photons, $|I\rangle = |k\rangle|\mathbf{1}_{\lambda'_1}(\boldsymbol{r}'_1,\omega'_1)\ldots\mathbf{1}_{\lambda'_{j-1}}(\boldsymbol{r}'_{j-1},\omega'_{j-1})\rangle$. In the former case, the energy denominator reads $E_\psi - E_I = E_n - (E_k + \hbar\omega) = -\hbar(\omega - \omega_{nk})$, in the latter case, we have $E_\psi - E_I = E_n + \hbar\omega - E_k = \hbar(\omega + \omega_{nk})$. Here, ω is the frequency of the photon that has been added or subtracted with respect to the initial state. The required matrix elements of the interaction Hamiltonian can be calculated by recalling the field expansion (1.22) and the Fock-state definition (1.37). One finds

$$\begin{aligned}
&\langle k|\langle\mathbf{1}_{\lambda'_1}(\boldsymbol{r}'_1,\omega'_1)\ldots\mathbf{1}_{\lambda'_{j+1}}(\boldsymbol{r}'_{j+1},\omega'_{j+1})|\,-\hat{\boldsymbol{d}}\cdot\hat{\boldsymbol{E}}(\boldsymbol{r}_A)|\mathbf{1}_{\lambda_1}(\boldsymbol{r}_1,\omega_1)\ldots\mathbf{1}_{\lambda_j}(\boldsymbol{r}_j,\omega_j)\rangle|n\rangle\\
&= -\sqrt{j+1}\sum_{\lambda=e,m}\int \mathrm{d}^3r\int_0^\infty \mathrm{d}\omega\big[\boldsymbol{d}_{kn}\cdot\boldsymbol{G}^*_\lambda(\boldsymbol{r}_A,\boldsymbol{r},\omega)\big]_i\\
&\quad\times\langle\mathbf{1}_{\lambda'_1}(\boldsymbol{r}'_1,\omega'_1)\ldots\mathbf{1}_{\lambda'_{j+1}}(\boldsymbol{r}'_{j+1},\omega'_{j+1})|1_{\lambda,i}(\boldsymbol{r},\omega)\mathbf{1}_{\lambda_1}(\boldsymbol{r}_1,\omega_1)\ldots\mathbf{1}_{\lambda_j}(\boldsymbol{r}_j,\omega_j)\rangle
\end{aligned} \tag{7.7}$$

and

$$\begin{aligned}
&\langle k|\langle\mathbf{1}_{\lambda'_1}(\boldsymbol{r}'_1,\omega'_1)\ldots\mathbf{1}_{\lambda'_{j-1}}(\boldsymbol{r}'_{j-1},\omega'_{j-1})|\,-\hat{\boldsymbol{d}}\cdot\hat{\boldsymbol{E}}(\boldsymbol{r}_A)|\mathbf{1}_{\lambda_1}(\boldsymbol{r}_1,\omega_1)\ldots\mathbf{1}_{\lambda_j}(\boldsymbol{r}_j,\omega_j)\rangle|n\rangle\\
&= -\sqrt{j}\sum_{\lambda=e,m}\int \mathrm{d}^3r\int_0^\infty \mathrm{d}\omega\big[\boldsymbol{d}_{kn}\cdot\boldsymbol{G}_\lambda(\boldsymbol{r}_A,\boldsymbol{r},\omega)\big]_i\\
&\quad\times\langle\mathbf{1}_{\lambda'_1}(\boldsymbol{r}'_1,\omega'_1)\ldots\mathbf{1}_{\lambda'_{j-1}}(\boldsymbol{r}'_{j-1},\omega'_{j-1})1_{\lambda,i}(\boldsymbol{r},\omega)|\mathbf{1}_{\lambda_1}(\boldsymbol{r}_1,\omega_1)\ldots\mathbf{1}_{\lambda_j}(\boldsymbol{r}_j,\omega_j)\rangle\,.
\end{aligned} \tag{7.8}$$

We substitute these results into the second-order energy shift. Using the commutation relations (1.17) and (1.18), scalar products of Fock states reduce to sums over products of delta functions. We carry out all sums over λ and integrals over $\boldsymbol{r}$ and ω. Using the integral relation (1.25) and noting that all trivial integrals are compensated for by respective contributions from the partition function, we find

$$\langle \Delta E \rangle = \langle \Delta_2 E \rangle = \frac{\mu_0}{\pi} \sum_n p_n \sum_k \mathcal{P} \int_0^\infty \mathrm{d}\omega\, \omega^2 \boldsymbol{d}_{nk} \cdot \mathrm{Im}\boldsymbol{G}(\boldsymbol{r}_A, \boldsymbol{r}_A, \omega) \boldsymbol{d}_{kn}$$
$$\times \left[\frac{\sum_{j=0}^\infty j \mathrm{e}^{-j\hbar\omega/(k_B T)}}{\sum_{l=0}^\infty \mathrm{e}^{-l\hbar\omega/(k_B T)}} \frac{1}{\omega + \omega_{nk}} - \frac{\sum_{j=0}^\infty (j+1) \mathrm{e}^{-j\hbar\omega/(k_B T)}}{\sum_{l=0}^\infty \mathrm{e}^{-l\hbar\omega/(k_B T)}} \frac{1}{\omega - \omega_{nk}} \right]. \tag{7.9}$$

The geometric sums over j can easily be carried out, resulting in

$$\langle \Delta E \rangle = \frac{\mu_0}{\pi} \sum_n p_n \sum_k \mathcal{P} \int_0^\infty \mathrm{d}\omega\, \omega^2 \left[\frac{n(\omega)}{\omega + \omega_{nk}} - \frac{n(\omega) + 1}{\omega - \omega_{nk}} \right]$$
$$\times \boldsymbol{d}_{nk} \cdot \mathrm{Im}\boldsymbol{G}(\boldsymbol{r}_A, \boldsymbol{r}_A, \omega) \cdot \boldsymbol{d}_{kn} \tag{7.10}$$

where

$$n(\omega) = \frac{1}{\mathrm{e}^{\hbar\omega/(k_B T)} - 1} \tag{7.11}$$

is the average thermal photon number in accordance with Bose–Einstein statistics.

To extract the CP potential, we decompose the Green's tensor into its bulk and scattering parts. We discard the position-independent infinite self-energy associated with the bulk part by making the replacement $\boldsymbol{G} \mapsto \boldsymbol{G}^{(1)}$. We proceed by writing $\mathrm{Im}\,\boldsymbol{G} = (\boldsymbol{G} - \boldsymbol{G}^*)/(2\mathrm{i})$ and using the Schwarz reflection principle (A.3), as well as the identity

$$n(-\omega) = -[n(\omega) + 1] \tag{7.12}$$

to obtain

$$U(\boldsymbol{r}_A) = \frac{\mu_0}{2\pi \mathrm{i}} \sum_n p_n \sum_k \mathcal{P} \int_{-\infty}^\infty \mathrm{d}\omega\, \omega^2 \left[\frac{n(\omega)}{\omega + \omega_{nk}} - \frac{n(\omega) + 1}{\omega - \omega_{nk}} \right]$$
$$\times \boldsymbol{d}_{nk} \cdot \boldsymbol{G}(\boldsymbol{r}_A, \boldsymbol{r}_A, \omega) \cdot \boldsymbol{d}_{kn}\,. \tag{7.13}$$

The result can be cast into a more explicit form by means of contour-integral techniques. As a preparation, we note that the integrand in (7.13) has poles at $\omega = \pm\omega_{nk}$. In addition, the thermal photon number $n(\omega)$ exhibits poles in the upper half plane for purely imaginary frequencies $\omega = \mathrm{i}\xi_j$,

$$\xi_j = \frac{2\pi k_B T}{\hbar} j\,, \quad j = 0, 1, 2, \ldots \tag{7.14}$$

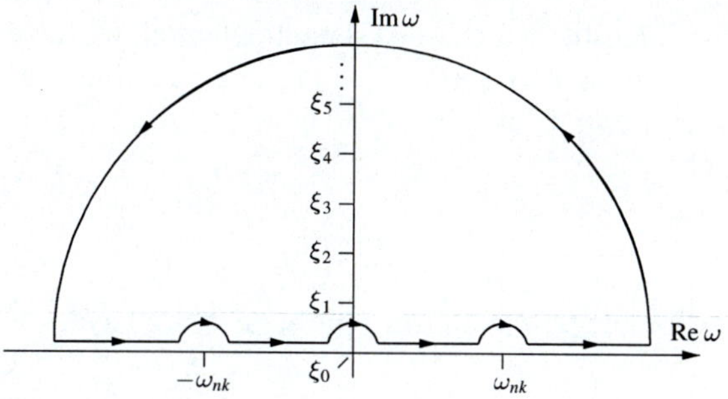

Fig. 7.1 Integration contour used for transforming the real-frequency CP integral into a Matsubara sum plus contributions from the atomic poles

which are commonly known as the Matsubara frequencies [1]. In the vicinity of the Matsubara frequencies, we have

$$n(\omega) \simeq \frac{k_B T}{\hbar} \frac{1}{\omega - \mathrm{i}\xi_j} . \tag{7.15}$$

Having determined the poles of the integrand in the upper half of the complex frequency plane, we apply the integration contour depicted in Fig. 7.1. It transforms the principal-value integral along the real frequency axis into integrals along an infinite semi-circle plus integrals along infinitesimally small semi-circles around the poles at $\omega = \pm\omega_{nk}$ and $\omega = \mathrm{i}\xi_0 = 0$ plus residues from all other Matsubara poles ξ_j $(j > 0)$, which are enclosed by the contour. The integral along the infinite semi-circle vanishes and the thermal CP potential can be given as [2–4]

$$U(\boldsymbol{r}_A) = \sum_n p_n U_n(\boldsymbol{r}_A) , \tag{7.16}$$

$$U_n(\boldsymbol{r}_A) = U_n^{\mathrm{nres}}(\boldsymbol{r}_A) + U_n^{\mathrm{res}}(\boldsymbol{r}_A) \tag{7.17}$$

with non-resonant potentials

$$U_n{}^{\mathrm{nres}}(\boldsymbol{r}_A) = \mu_0 k_B T {\sum_{j=0}^{\infty}}' \xi_j^2 \mathrm{tr}\big[\boldsymbol{\alpha}_n(\mathrm{i}\xi_j)\cdot \boldsymbol{G}^{(1)}(\boldsymbol{r}_A, \boldsymbol{r}_A, \mathrm{i}\xi_j)\big] \tag{7.18}$$

and resonant potentials

$$\begin{aligned} U_n^{\mathrm{res}}(\boldsymbol{r}_A) = &-\mu_0 \sum_{k<n} [n(\omega_{nk}) + 1]\omega_{nk}^2 \boldsymbol{d}_{nk}\cdot \mathrm{Re}\, \boldsymbol{G}^{(1)}(\boldsymbol{r}_A, \boldsymbol{r}_A, \omega_{nk})\cdot \boldsymbol{d}_{kn} \\ &+ \mu_0 \sum_{k>n} n(\omega_{kn})\omega_{kn}^2 \boldsymbol{d}_{nk}\cdot \mathrm{Re}\, \boldsymbol{G}^{(1)}(\boldsymbol{r}_A, \boldsymbol{r}_A, \omega_{kn})\cdot \boldsymbol{d}_{kn} . \end{aligned} \tag{7.19}$$

To obtain the non-resonant potential, we have recalled the definition (4.43) of the atomic polarisability. The prime at the Matsubara sum indicates that the $j = 0$ term carries half-weight,

$$\sum_{j=0}^{\infty}{}' f_j = \frac{1}{2} f_0 + \sum_{j=1}^{\infty} f_j \,. \tag{7.20}$$

For an atom in an isotropic state $\hat{\sigma}$, we have

$$\sum_{n'\in\{\overline{n}\}} \sum_{k'\in\{\overline{k}\}} \boldsymbol{d}_{n'k'}\boldsymbol{d}_{k'n'} = \frac{1}{3} \sum_{n'\in\{\overline{n}\}} \sum_{k'\in\{\overline{k}\}} |\boldsymbol{d}_{n'k'}|^2 \boldsymbol{I} \tag{7.21}$$

in close analogy with (4.16), so the potentials reduce to

$$U_n^{\mathrm{nres}}(\boldsymbol{r}_A) = \mu_0 k_{\mathrm{B}} T \sum_{j=0}^{\infty}{}' \xi_j^2 \alpha_n(\mathrm{i}\xi_j) \mathrm{tr}\, \boldsymbol{G}^{(1)}(\boldsymbol{r}_A, \boldsymbol{r}_A, \mathrm{i}\xi_j) \tag{7.22}$$

and

$$\begin{aligned} U_n^{\mathrm{res}}(\boldsymbol{r}_A) = &-\frac{\mu_0}{3} \sum_{k<n} [n(\omega_{nk}) + 1] \omega_{nk}^2 |\boldsymbol{d}_{nk}|^2 \mathrm{tr}\big[\mathrm{Re}\, \boldsymbol{G}^{(1)}(\boldsymbol{r}_A, \boldsymbol{r}_A, \omega_{nk})\big] \\ &+ \frac{\mu_0}{3} \sum_{k>n} n(\omega_{kn}) \omega_{kn}^2 |\boldsymbol{d}_{nk}|^2 \mathrm{tr}\big[\mathrm{Re}\, \boldsymbol{G}^{(1)}(\boldsymbol{r}_A, \boldsymbol{r}_A, \omega_{kn})\big] \,, \end{aligned} \tag{7.23}$$

with the isotropic polarisability being given by (4.46). The results (7.16)–(7.19) were originally found on the basis of linear-response theory [5].

Comparing the thermal CP potential with the result (4.40)–(4.42) for zero temperature, we note that in both cases the potential contains a non-resonant component due to virtual photons that depends on the Green's tensor at purely imaginary frequencies as well as a resonant part due to real photons that depends on the Green's tensor at the real atomic transition frequencies. However, the details of these components have changed drastically when going from zero to finite temperature. For the non-resonant potential, a continuous integral at zero temperature has been replaced by a discrete Matsubara sum at finite temperature:

$$\frac{\hbar}{2\pi} \int_0^{\infty} \mathrm{d}\xi\, f(\mathrm{i}\xi) \mapsto k_{\mathrm{B}} T \sum_{j=0}^{\infty}{}' f(\mathrm{i}\xi_j) \,. \tag{7.24}$$

This replacement rule can be used to obtain non-resonant CP potentials from known zero-temperature results.

The resonant part of the potential at zero temperature was entirely due to spontaneous emission and hence associated with downward transitions only. At finite temperature, these potential components are enhanced by stimulated emission caused by the thermal photons, leading to a factor $n(\omega_{nk})+1$. In addition, new potential components due to the absorption of thermal photons have appeared. They are proportional to $n(\omega_{kn})$ and carry a different sign. The generalisation from zero-temperature to thermal resonant potentials can thus be implemented via the replacement rule

$$\sum_{k<n} f(\omega_{nk}) \mapsto \sum_{k<n}[n(\omega_{nk})+1]f(\omega_{nk}) - \sum_{k>n} n(\omega_{kn})f(\omega_{kn})\,. \qquad (7.25)$$

At zero temperature, resonant potential components only appear for excited atoms. At finite temperature, they are more generally present whenever the atom is out of equilibrium with its environment. This is true, in particular, for a ground-state atom. To be at equilibrium with its environment, the atom must itself be in a thermal state at temperature T,

$$\hat{\sigma} = \hat{\sigma}_T = \frac{\mathrm{e}^{-\hat{H}_A/(k_\mathrm{B}T)}}{\mathrm{tr}\big[\mathrm{e}^{-\hat{H}_A/(k_\mathrm{B}T)}\big]}\,, \qquad (7.26)$$

so that

$$p_n = \frac{\mathrm{e}^{-E_n/(k_\mathrm{B}T)}}{\sum_k \mathrm{e}^{-E_k/(k_\mathrm{B}T)}}\,. \qquad (7.27)$$

Using the relations

$$p_n = \mathrm{e}^{-\hbar\omega_{nk}/(k_\mathrm{B}T)}\,p_k\,, \qquad (7.28)$$

$$n(\omega)+1 = \mathrm{e}^{\hbar\omega/(k_\mathrm{B}T)}n(\omega)\,, \qquad (7.29)$$

we find that the resonant potentials associated with upward and downward transitions mutually cancel,

$$\begin{aligned} U^{\mathrm{res}}(\boldsymbol{r}_A) &= \sum_n p_n U_n^{\mathrm{res}}(\boldsymbol{r}_A) \\ &= -\mu_0 \sum_n p_n \sum_{k<n}[n(\omega_{nk})+1]\omega_{nk}^2 \boldsymbol{d}_{nk}\cdot\mathrm{Re}\,\boldsymbol{G}^{(1)}(\boldsymbol{r}_A,\boldsymbol{r}_A,\omega_{nk})\cdot\boldsymbol{d}_{kn} \\ &\quad + \mu_0 \sum_n p_n \sum_{k>n} n(\omega_{kn})\omega_{kn}^2 \boldsymbol{d}_{nk}\cdot\mathrm{Re}\,\boldsymbol{G}^{(1)}(\boldsymbol{r}_A,\boldsymbol{r}_A,\omega_{kn})\cdot\boldsymbol{d}_{kn} \\ &= -\mu_0 \sum_n \sum_{k<n} \big\{p_n[n(\omega_{nk})+1] - p_k n(\omega_{nk})\big\}\omega_{nk}^2 \\ &\quad \times \boldsymbol{d}_{nk}\cdot\mathrm{Re}\,\boldsymbol{G}^{(1)}(\boldsymbol{r}_A,\boldsymbol{r}_A,\omega_{nk})\cdot\boldsymbol{d}_{kn} = 0\,. \end{aligned} \qquad (7.30)$$

At thermal equilibrium, the CP potential is hence entirely non-resonant. Introducing the polarisability of an atom at temperature T as

$$\alpha_T(\omega) = \sum_n p_n \alpha_n(\omega) , \tag{7.31}$$

the equilibrium potential reads

$$U(\boldsymbol{r}_A) = \mu_0 k_B T \sum_{j=0}^{\infty}{}' \xi_j^2 \alpha_T(\mathrm{i}\xi_j) \mathrm{tr}\, \boldsymbol{G}^{(1)}(\boldsymbol{r}_A, \boldsymbol{r}_A, \mathrm{i}\xi_j) . \tag{7.32}$$

This result was first obtained using linear-response theory [6, 7].

Let us discuss the behaviour of the thermal CP potential for low and high temperatures. We begin with the non-resonant potential (7.18) which is given in terms of a Matsubara sum over positive imaginary frequencies. According to (7.14), the temperature determines how densely spaced the terms of this sum are. The frequency-dependence of the terms is governed by the atom-body separations and the characteristic atomic and medium resonance frequencies. As a result, the temperature and spatial dependences of the non-resonant potential are strongly intertwined. To quantify this, we make use of the Abel–Plana formula [8, 9]

$$\sum_{j=0}^{\infty}{}' f(j) = \int_0^{\infty} \mathrm{d}x\, f(x) + \mathrm{i} \int_0^{\infty} \mathrm{d}y\, \frac{f(\mathrm{i}y) - f(-\mathrm{i}y)}{\mathrm{e}^{2\pi y} - 1} . \tag{7.33}$$

Applying it to (7.18) and using the Schwarz reflection principle (A.4) for the Green's tensor together with the corresponding property

$$\boldsymbol{\alpha}_n^*(\omega) = \boldsymbol{\alpha}_n(-\omega^*) \tag{7.34}$$

of the atomic polarisability, we find

$$U_n^{\mathrm{nres}}(\boldsymbol{r}_A) = \frac{\hbar\mu_0}{2\pi} \int_0^{\infty} \mathrm{d}\xi\, \xi^2 \mathrm{tr}\big[\boldsymbol{\alpha}_n(\mathrm{i}\xi) \cdot \boldsymbol{G}^{(1)}(\boldsymbol{r}_A, \boldsymbol{r}_A, \mathrm{i}\xi)\big] - \frac{\mu_0\hbar}{\pi} \int_0^{\infty} \mathrm{d}\omega\, \omega^2 \frac{\mathrm{Im}\,\mathrm{tr}\big[\boldsymbol{\alpha}_n(\omega) \cdot \boldsymbol{G}^{(1)}(\boldsymbol{r}_A, \boldsymbol{r}_A, \omega)\big]}{\mathrm{e}^{\hbar\omega/k_B T} - 1} . \tag{7.35}$$

To estimate the magnitude of the second term on the right hand side, we recall the Born expansion (4.26) with (4.27). It shows that $\boldsymbol{G}^{(1)}(\boldsymbol{r}_A, \boldsymbol{r}_A, \omega)$ is proportional to $\mathrm{e}^{(|\boldsymbol{r}_A - \boldsymbol{s}_1| + |\boldsymbol{s}_K - \boldsymbol{r}_A|)\omega/c}$ and becomes rapidly oscillating for frequencies $\omega > c/r$ (r: atom–body distances), effectively limiting the range of the integral. For low temperatures $k_B T \ll \hbar c/r_+$ (r_+: maximum of all atom–body distances), the argument of $\mathrm{e}^{\hbar\omega/k_B T}$ becomes very large for the major part of the integral, $\hbar\omega/k_B T \simeq \hbar c/(k_B T r_+) \gg 1$. The second term in (7.35) is hence small, so we recover the zero-temperature result (4.41) as found in Chap. 4. Roughly speaking, one may say that in the low-temperature limit, the Matsubara sum governing the thermal CP potential becomes so densely spaced that it approaches the integral of the zero-temperature result.

Note that the threshold temperature $\hbar c/(k_{\mathrm{B}} r_+)$ for the onset of the low-temperature limit depends on the atom–body separation. Physically, it is the temperature of thermal radiation whose wavelength is equal to the atom–body distance. As a result of the intertwined position- and temperature-dependences, it is not possible to globally define a threshold temperature for the non-resonant CP potential; the low-temperature limit is a non-uniform limit [10, 11]. To stress its position-dependence, we will refer to this limit as the geometric low-temperature limit.

Conversely, the non-uniformity implies that for a given temperature, the low-temperature limit is a good approximation only for distances that are smaller than the thermal wavelength $\lambda_T = hc/(k_{\mathrm{B}} T)$. At room temperature ($T = 300\,\mathrm{K}$), the thermal wavelength is $\lambda_T = 48.0\,\mu\mathrm{m}$. The zero-temperature results presented in the previous chapters of this book as well as in Vol. I have to be considered with this condition in mind; this is particularly true for all retarded, large-distance limits. We will elaborate this point further in Sect. 7.2 below when studying the example of an atom in front of a plate.

In the opposite geometric high-temperature limit $k_{\mathrm{B}} T \gg \hbar c/r_-$ (r_-: minimum of all atom–body distances), the Matsubara sum becomes very coarse-grained. The Born expansion (4.26) with (4.27) shows that the terms $j > 0$ are proportional to $\mathrm{e}^{-(|\boldsymbol{r}_A - \boldsymbol{s}_1| + |\boldsymbol{s}_K - \boldsymbol{r}_A|)\xi_j/c}$. The argument of this exponential takes large negative values, $(|\boldsymbol{r}_A - \boldsymbol{s}_1| + |\boldsymbol{s}_K - \boldsymbol{r}_A|)\xi_j/c > r_- \xi_j/c > j k_{\mathrm{B}} T r_-/(\hbar c) \gg 1$, so all terms $j > 0$ are exponentially small. Only the $j = 0$ term contributes, and the non-resonant potential simplifies to

$$U_n^{\mathrm{nres}}(\boldsymbol{r}_A) = -\frac{k_{\mathrm{B}} T}{2\varepsilon_0} \operatorname{tr}\left\{ \boldsymbol{\alpha}_n \cdot \left[\frac{\omega^2}{c^2} \boldsymbol{G}^{(1)}(\boldsymbol{r}_A, \boldsymbol{r}_A, \omega) \right]_{\omega=0} \right\}. \tag{7.36}$$

Here, $\boldsymbol{\alpha}_n \equiv \boldsymbol{\alpha}_n(0)$ is the static polarisability. Note that the non-resonant potential for high temperatures is proportional to $k_{\mathrm{B}} T$ rather than $\hbar$ and may hence be identified as a classical effect. The geometric high-temperature limit is again non-uniform, because the threshold temperature depends on the atom–body separation.

The situation is a lot simpler for the resonant CP potential, whose temperature-dependence is separate from its position-dependence. We can hence define a spectroscopic low-temperature limit $k_{\mathrm{B}} T \ll \hbar\omega_-$ (ω_-: minimum of the relevant atomic transition frequencies). In this limit, the thermal photon number (7.11) becomes much smaller than unity, $n(\omega_{nk}), n(\omega_{kn}) \ll 1$, so the resonant thermal potential (7.19) reduces to the corresponding zero-temperature result (4.42). The opposite, spectroscopic high-temperature limit is given by $k_{\mathrm{B}} T \gg \hbar\omega_+$ (ω_+: maximum of the relevant atomic transition frequencies). The photon number (7.11) then approaches

$$-[n(\omega_{nk}) + 1] = n(\omega_{kn}) = \frac{k_{\mathrm{B}} T}{\hbar\omega_{kn}} \tag{7.37}$$

and the resonant potential (7.19) simplifies to

$$U_n^{\mathrm{res}}(\boldsymbol{r}_A) = -\frac{\mu_0 k_\mathrm{B} T}{\hbar} \sum_k \omega_{nk} \boldsymbol{d}_{nk} \cdot \mathrm{Re}\, \boldsymbol{G}^{(1)}(\boldsymbol{r}_A, \boldsymbol{r}_A, |\omega_{nk}|) \cdot \boldsymbol{d}_{kn} \,. \tag{7.38}$$

The geometric and spectroscopic low- and high-temperature limits are relatively independent of each other. The former constrains the temperature via the distances, whereas the latter compares it to transition frequencies. As the one exception to this rule, we note that the spectroscopic high-temperature limit prevents the non-resonant potential from ever exhibiting the geometric low-temperature behaviour presented above. To understand this constraint, we note that the conditions $\hbar\omega_+ \ll k_\mathrm{B}T$ and $k_\mathrm{B}T \ll \hbar c/r_+$ of a spectroscopically high but geometrically low temperature imply that the potential is nonretarded, $r_+ \ll c/\omega_+$. The nonretarded limit requires a separate limiting procedure which we will demonstrate in the following for perfectly conducting bodies.

We begin by considering the non-resonant potential (7.18) and recall the Born expansion (4.26) with (4.27) of the Green's tensor. In the nonretarded limit, the polarisability limits the Matsubara sum to terms with $|\boldsymbol{r}_A - \boldsymbol{s}_1|\xi_j/c, |\boldsymbol{s}_K - \boldsymbol{r}_A|\xi_j/c \ll 1$. The Green's tensor is hence well approximated by its electrostatic equivalent (4.29) and the nonretarded potential reads

$$U_n^{\mathrm{nres}}(\boldsymbol{r}_A) = \frac{k_\mathrm{B} T}{\varepsilon_0} {\sum_{j=0}^{\infty}}' \mathrm{tr}\big[\boldsymbol{\alpha}_n(\mathrm{i}\xi_j) \cdot \boldsymbol{G}^{(1)}(\boldsymbol{r}_A, \boldsymbol{r}_A)\big] \,. \tag{7.39}$$

Recalling the definition (4.43) of the atomic polarisability, the Matsubara sum can be performed according to

$$ {\sum_{j=0}^{\infty}}' \frac{1}{a^2 + j^2} = \frac{\pi}{2a} \coth(\pi a) \,. \tag{7.40}$$

Using the identity

$$\coth[\hbar\omega/(2k_\mathrm{B}T)] = 2n(\omega) + 1 \,, \tag{7.41}$$

which follows immediately from the definition of the thermal photon number, the result can be given in the form

$$U_n^{\mathrm{nres}}(\boldsymbol{r}_A) = -\frac{1}{\varepsilon_0} \sum_k \big[n(\omega_{kn}) + \tfrac{1}{2}\big] \boldsymbol{d}_{nk} \cdot \boldsymbol{G}^{(1)}(\boldsymbol{r}_A, \boldsymbol{r}_A) \cdot \boldsymbol{d}_{kn} \,. \tag{7.42}$$

Similarly, we may replace the Green's tensor featuring in the resonant CP potential (7.19) with its purely real electrostatic counterpart. Recalling the relation (7.12), we find

$$U_n^{\mathrm{res}}(\boldsymbol{r}_A) = \frac{1}{\varepsilon_0} \sum_k n(\omega_{kn}) \boldsymbol{d}_{nk} \cdot \mathbf{G}^{(1)}(\boldsymbol{r}_A, \boldsymbol{r}_A) \cdot \boldsymbol{d}_{kn} \,. \tag{7.43}$$

Combining the non-resonant and resonant results, the terms proportional to the photon number cancel. Invoking the completeness relation $\sum_k \boldsymbol{d}_{nk} \boldsymbol{d}_{kn} = \langle \hat{\boldsymbol{d}} \hat{\boldsymbol{d}} \rangle_n$, we obtain a total potential [12]

$$U_n(\boldsymbol{r}_A) = -\frac{\langle \hat{\boldsymbol{d}} \cdot \mathbf{G}^{(1)}(\boldsymbol{r}_A, \boldsymbol{r}_A) \cdot \hat{\boldsymbol{d}} \rangle_n}{2\varepsilon_0} \,. \tag{7.44}$$

The thermal CP is hence independent of T for an atom in an energy eigenstate at nonretarded distance from perfectly conducting bodies. As it must, this temperature-invariant result agrees with our zero-temperature potential (4.34) found in Sect. 4.1. Under the stated conditions, the zero-temperature potential is hence universally valid for all temperatures, despite the fact that the non-resonant and resonant parts of the thermal CP potential may strongly vary with temperature. This result is particularly relevant for particles whose dominant transition frequencies are small, such that the nonretarded limit extends over a large distance range. Examples include polar molecules with their low-frequency rotational and vibrational transitions or highly excited Rydberg atoms with their densely spaced spectrum of neighbouring states. At such small frequencies, the behaviour of real metals is very well approximated by the perfect conductor, so that our idealised assumptions are well justified. We will demonstrate this in more detail when studying examples in the following Sect. 7.2.

The demonstrated temperature-invariance refers to temperature-dependences induced by the photons only. The nonretarded CP potential near perfectly conducting bodies will depend on temperature if the atom is not in an energy eigenstate, but in a thermal state (7.26). Using the eigenstate result above, the total potential (7.16) in this case reads

$$U(\boldsymbol{r}_A) = -\frac{\langle \hat{\boldsymbol{d}} \cdot \mathbf{G}^{(1)}(\boldsymbol{r}_A, \boldsymbol{r}_A) \cdot \hat{\boldsymbol{d}} \rangle_T}{2\varepsilon_0} \,. \tag{7.45}$$

The thermal average of the atomic dipole fluctuations depends on temperature and thus results in a T-dependent potential. As a rule of thumb, the fluctuating atomic dipole moment becomes larger for larger temperatures, leading to a stronger potential.

7.2 Atom or Molecule in Front of a Plate

Thermal CP forces have been studied for a variety of geometries, including spheres [13], cylinders [13–16] and planar [17, 18], cylindrical [13, 19, 20] or spherical cavities [13]. To illuminate the intertwined position- and temperature dependences of the thermal CP potential, we concentrate on the simplest example of an atom in front of a plate. We begin with the case of a perfectly conducting plate, identifying different asymptotic regimes for the distance dependence and confirming the general temperature-invariance of the nonretarded interaction as formulated above. We will then study the more realistic case of a metal plate in order to evaluate the reliability or shortcomings of the idealised perfect-conductor results.

As a starting point, we recall the scattering Green's tensor of a plate of arbitrary reflectivity as given in Sect. 4.2. Using the expression (4.60) for the Green's tensor at imaginary frequencies, the non-resonant CP potential at finite temperature T reads [21]

$$U_n^{\mathrm{nres}}(z_A) = \frac{\mu_0 k_{\mathrm{B}} T}{4\pi} {\sum_{j=0}^{\infty}}' \xi_j^2 \int_{\xi_j/c}^{\infty} \mathrm{d}\kappa^{\perp}\, \mathrm{e}^{-2\kappa^{\perp} z_A} \biggl\{ \alpha_n^{\parallel}(\mathrm{i}\xi_j) r_s - \biggl[\frac{\kappa^{\perp 2} c^2}{\xi_j^2} \alpha_n^{\parallel}(\mathrm{i}\xi_j) + \biggl(\frac{\kappa^{\perp 2} c^2}{\xi_j^2} - 1 \biggr) \alpha_n^{\perp}(\mathrm{i}\xi_j) \biggr] r_p \biggr\} \tag{7.46}$$

with $\alpha^{\parallel} = \frac{1}{2}(\alpha_{xx} + \alpha_{yy})$ and $\alpha^{\perp} = \alpha_{zz}$. The Green's tensor (4.58) at real frequencies naturally separates into two components associated with propagating vs evanescent waves. The resonant CP potential (7.19) at finite temperature separates accordingly, where

$$U_n^{\mathrm{prop}}(z_A) = \frac{\mu_0}{8\pi} \biggl\{ \sum_{k<n} [n(\omega_{nk}) + 1] - \sum_{k>n} n(\omega_{kn}) \biggr\} \omega_{nk}^2 \times \int_0^{\omega_{nk}/c} \mathrm{d}k^{\perp} \biggl\{ |\mathbf{d}_{nk}^{\parallel}|^2 \mathrm{Im}\bigl(\mathrm{e}^{2\mathrm{i}k^{\perp} z_A} r_s\bigr) - \biggl[\frac{k^{\perp 2} c^2}{\omega_{nk}^2} |\mathbf{d}_{nk}^{\parallel}|^2 + 2 \biggl(\frac{k^{\perp 2} c^2}{\omega_{nk}^2} - 1 \biggr) |\mathbf{d}_{nk}^{\perp}|^2 \biggr] \mathrm{Im}\bigl(\mathrm{e}^{2\mathrm{i}k^{\perp} z_A} r_p\bigr) \biggr\} \tag{7.47}$$

is the contribution due to real, propagating photons and [21]

$$U_n^{\mathrm{evan}}(z_A) = -\frac{\mu_0}{8\pi} \biggl\{ \sum_{k<n} [n(\omega_{nk}) + 1] - \sum_{k>n} n(\omega_{kn}) \biggr\} \omega_{nk}^2 \times \int_0^{\infty} \mathrm{d}\kappa^{\perp}\, \mathrm{e}^{-2\kappa^{\perp} z_A} \biggl\{ |\mathbf{d}_{nk}^{\parallel}|^2 \mathrm{Re}(r_s) + \biggl[\frac{\kappa^{\perp 2} c^2}{\omega_{nk}^2} |\mathbf{d}_{nk}^{\parallel}|^2 + 2 \biggl(\frac{\kappa^{\perp 2} c^2}{\omega_{nk}^2} + 1 \biggr) |\mathbf{d}_{nk}^{\perp}|^2 \biggr] \mathrm{Re}(r_p) \biggr\} \tag{7.48}$$

is due to real, evanescent photons. The total potential is the sum (4.61) of these three components. Note that rather than computing the three potentials explicitly, we could have inferred them from the zero-temperature results (4.62)–(4.64) by virtue of the replacement rules (7.24) and (7.25).

7.2.1 Perfectly Conducting Plate

For a perfectly conducting plate, the reflection coefficients are independent of frequency and given by $r_s = -1$ and $r_p = -1$, cf. Sect. 3.3.1 in Vol. I.

Thermal Equilibrium

With these reflection coefficients, the non-resonant potential (7.46) can be evaluated explicitly by carrying out the $\kappa^{\perp}$-integral [21],

$$U_n^{\mathrm{nres}}(z_A) = -\frac{k_{\mathrm{B}}T}{16\pi\varepsilon_0 z_A^3}\sum_{j=0}^{\infty}{}' \mathrm{e}^{-2\xi_j z_A/c}\left[\alpha_n^{\parallel}(\mathrm{i}\xi_j)\left(1+2\,\frac{\xi_j z_A}{c}+4\,\frac{\xi_j^2 z_A^2}{c^2}\right)\right.$$
$$\left.+\,\alpha_n^{\perp}(\mathrm{i}\xi_j)\left(1+2\,\frac{\xi_j z_A}{c}\right)\right]. \tag{7.49}$$

The non-resonant thermal CP potential in front of a perfectly conducting plate has been derived by a variety of methods, including linear-response theory [6] and normal-mode QED in free space [22].

The intertwined dependences of the thermal CP potential on the three parameters temperature, distance and atomic transition frequencies can now be seen very explicitly. The temperature governs the spacing of the Matsubara sum while distance and transition frequencies compete in determining the effective summation range. Asymptotic results can be obtained whenever pairs of parameters take extreme values in comparison to each other. As listed in Table 7.1, three types of limits can be distinguished, depending on which two parameters are compared to one another. By comparing the distance with the atomic transition frequencies, we define the retarded limit $z_A\omega_-/c \gg 1$ (ω_-: minimum of the relevant atomic transition frequencies) and the nonretarded limit $z_A\omega_+/c \ll 1$ (ω_+: maximum of the relevant atomic transition frequencies). Relating temperature and transition frequencies, we identify the spectroscopic low-temperature regime $k_{\mathrm{B}}T \ll \hbar\omega_-$ and the opposite spectroscopic high-temperature regime $k_{\mathrm{B}}T \gg \hbar\omega_+$. And finally, a comparison of distance and temperature leads to the notions of a geometric low-temperature limit $k_{\mathrm{B}}T \ll \hbar c/z_A$ and its high-temperature counterpart $k_{\mathrm{B}}T \gg \hbar c/z_A$. Two out of the eight possible combinations of limiting conditions are logically inconsistent: It is impossible to simultaneously realise the retarded, spectroscopic high-temperature and geometric

Table 7.1 Limiting cases for the thermal CP potential

Limit	Condition	z	T
Retarded	$z_A\omega_-/c \gg 1$	$z_\omega \ll z_A$	$T_z \ll T_\omega$
Nonretarded	$z_A\omega_+/c \ll 1$	$z_A \ll z_\omega$	$T_\omega \ll T_z$
Spectroscopic low-temperature	$k_\mathrm{B}T \ll \hbar\omega_-$	$z_\omega \ll z_T$	$T \ll T_\omega$
Spectroscopic high-temperature	$k_\mathrm{B}T \gg \hbar\omega_+$	$z_T \ll z_\omega$	$T_\omega \ll T$
Geometric low-temperature	$k_\mathrm{B}T \ll \hbar c/z_A$	$z_A \ll z_T$	$T \ll T_z$
Geometric high-temperature	$k_\mathrm{B}T \gg \hbar c/z_A$	$z_T \ll z_A$	$T_z \ll T$

We have introduced characteristic spectroscopic and thermal lengths $z_\omega = c/\omega_\pm$ and $z_T = \hbar c/(k_\mathrm{B}T)$ as well as characteristic geometric and spectroscopic temperatures $T_z = \hbar c/(z_A k_\mathrm{B})$ and $T_\omega = \hbar\omega_\pm/k_\mathrm{B}$.

low-temperature limits; the same is true for a combination of the nonretarded, spectroscopic low-temperature and geometric high-temperature limits.

When discussing the distance-dependence of the CP potential for a given temperature, it is useful to introduce characteristic spectroscopic and thermal lengths $z_\omega = c/\omega_\pm$ and $z_T = \hbar c/(k_\mathrm{B}T)$. They are measures of the wavelengths of the relevant atomic transitions and the thermal radiation, respectively. As shown in Table 7.1, the spectroscopic low- and high-temperature limits define the relation between the two characteristic length scales. For a given temperature, the retarded vs nonretarded limits compare the distance z_A with the spectroscopic length z_ω while the geometric low- vs high-temperature limits relate z_A and z_T.

In a similar way, one may introduce spectroscopic temperatures $T_\omega = \hbar\omega_\pm/k_\mathrm{B}$, which denote the temperature required to noticeably populate all or at least one of the excited atomic levels; and a geometric temperature $T_z = \hbar c/(z_A k_\mathrm{B})$, i.e., the temperature of radiation whose wavelength is of the order z_A. They are useful when analysing the temperature-dependence of the CP potential for fixed atomic position. From this point of view, the retarded and nonretarded limits set the two characteristic temperatures in proportion to one another, cf. Table 7.1. The spectroscopic and geometric low- or high-temperature limits then relate the given environment temperature to the spectroscopic and geometric temperatures, respectively.

We first discuss the distance-dependence of the potential for a given temperature, distinguishing two cases. We begin with the spectroscopic low-temperature limit $k_\mathrm{B}T \ll \hbar\omega_-$ where $z_\omega \ll z_T$. It is realised for most ground-state atoms in a room-temperature environment, because typical transition frequencies are of the order of several 10^{15} rad/s, much larger than $\omega_T = k_\mathrm{B}T/\hbar = 3.93 \times 10^{13}$ rad/s ($T = 300$ K). One may identify three asymptotic distance regimes for the non-resonant potential. For very large distances $z_A \gg z_T$ with $z_T = \hbar c/(k_\mathrm{B}T)$, the geometric high-temperature limit applies, the argument of the exponential $\mathrm{e}^{-2\xi_j z_A/c}$ in (7.49) taking large negative values, $2z_A\xi_j/c > jk_\mathrm{B}Tz_A/(\hbar c) = jz_A/z_T \gg 1$ ($j > 0$). Only the $j = 0$ term contributes, and the non-resonant potential simplifies to

$$U_n^\mathrm{nres}(z_A) = -\frac{k_\mathrm{B}T\left(\alpha_n^\parallel + \alpha_n^\perp\right)}{32\pi\varepsilon_0 z_A^3}\,. \tag{7.50}$$

with $\alpha_n^{\parallel} \equiv \alpha_n^{\parallel}(0)$, $\alpha_n^{\perp} \equiv \alpha_n^{\perp}(0)$. This result is a special case of (7.36) and may be obtained more directly by using the electrostatic Green's tensor (4.87) of the plate.

For smaller distances $z_A \ll z_T$, the geometric low-temperature limit is is realised. The Abel-Plana formula (7.33) then shows that the Matsubara sum then becomes so densely spaced that it is well approximated by an integral. We recover the zero-temperature result (4.69). Due to our assumption $z_\omega \ll z_T$, the geometric low-temperature regime is compatible with both the retarded and nonretarded limits. For intermediate distances $z_\omega \ll z_A \ll z_T$, the retarded zero-temperature result (4.70) is valid,

$$U_n^{\text{nres}}(z_A) = -\frac{\hbar c\left(2\alpha_n^{\parallel} + \alpha_n^{\perp}\right)}{32\pi^2\varepsilon_0 z_A^4}\,. \tag{7.51}$$

For even smaller distances $z_A \ll z_\omega$, the nonretarded zero-temperature potential (4.73) applies,

$$U_n^{\text{nres}}(z_A) = -\frac{1}{64\pi\varepsilon_0 z_A^3}\sum_k \operatorname{sgn}(\omega_{kn})\left(|\boldsymbol{d}_{nk}^{\parallel}|^2 + 2|\boldsymbol{d}_{nk}^{\perp}|^2\right). \tag{7.52}$$

On the contrary, the spectroscopic high-temperature limit $k_B T \gg \hbar\omega_+$ means that $z_T \ll z_\omega$. This limit applies to particles dominated by low-frequency transitions, such as polar molecules or Rydberg atoms. For large distances $z_A \gg z_T$, the geometric high-temperature limit leads to (7.50), as before. For smaller distances $z_A \ll z_T$, the condition $z_T \ll z_\omega$ automatically implies a nonretarded limit $z_A \ll z_\omega$. The atomic polarisability then limits the Matsubara sum to terms with $\xi_j z_A/c \le \omega_+ z_A/c \ll 1$, so (7.49) simplifies to

$$U_n^{\text{nres}}(z_A) = -\frac{k_B T}{16\pi\varepsilon_0 z_A^3}{\sum_{j=0}^{\infty}}'\left[\alpha_n^{\parallel}(\mathrm{i}\xi_j) + \alpha_n^{\perp}(\mathrm{i}\xi_j)\right]. \tag{7.53}$$

Recalling the definition of the polarisability (4.15), the Matsubara sum can be carried out according to (7.40),

$${\sum_{j=0}^{\infty}}'\alpha_n^{\parallel}(\mathrm{i}\xi_j) = \frac{1}{2k_B T}\sum_k\left[n(\omega_{kn}) + \tfrac{1}{2}\right]|\boldsymbol{d}_{nk}^{\parallel}|^2 \simeq \tfrac{1}{2}\alpha_n^{\parallel}\,, \tag{7.54}$$

$${\sum_{j=0}^{\infty}}'\alpha_n^{\perp}(\mathrm{i}\xi_j) = \frac{1}{k_B T}\sum_k\left[n(\omega_{kn}) + \tfrac{1}{2}\right]|\boldsymbol{d}_{nk}^{\perp}|^2 \simeq \tfrac{1}{2}\alpha_n^{\perp}\,, \tag{7.55}$$

where the second equalities are valid for $k_B T \gg \hbar\omega_+$. With these results, the thermal CP potential (7.53) again assumes the form (7.50). In the spectroscopic high-temperature regime, this distance law hence holds globally for all distances.

The CP potential of a particle at thermal equilibrium with its environment is entirely non-resonant. It can hence immediately obtained from our previous results by evaluating the thermal average (7.16) with (7.27). Using (7.49), the equilibrium potential reads [21]

$$U(z_A) = -\frac{k_\mathrm{B}T}{16\pi\varepsilon_0 z_A^3}\sum_{j=0}^{\infty}{}' \mathrm{e}^{-2\xi_j z_A/c}\biggl[\alpha_T^\parallel(\mathrm{i}\xi_j)\biggl(1+2\frac{\xi_j z_A}{c}+4\frac{\xi_j^2 z_A^2}{c^2}\biggr) + \alpha_T^\perp(\mathrm{i}\xi_j)\biggl(1+2\frac{\xi_j z_A}{c}\biggr)\biggr], \tag{7.56}$$

with the polarisability at temperature T being given by (7.31). In the spectroscopic low-temperature regime, which typically applies to atoms at room temperature, the particle is essentially in its ground state, $p_n \simeq \delta_{n0}$. The results (7.50)–(7.52) then imply that

$$U(z_A) = -\frac{k_\mathrm{B}T\left(\alpha_0^\parallel + \alpha_0^\perp\right)}{32\pi\varepsilon_0 z_A^3} \tag{7.57}$$

for large distances $z_A \gg z_T$,

$$U(z_A) = -\frac{\hbar c\left(2\alpha_0^\parallel + \alpha_0^\perp\right)}{32\pi^2\varepsilon_0 z_A^4} \tag{7.58}$$

for intermediate distances $z_\omega \ll z_A \ll z_T$ and

$$U(z_A) = -\frac{\langle \hat{\boldsymbol{d}}^{\parallel 2} + 2\hat{\boldsymbol{d}}^{\perp 2}\rangle_0}{64\pi\varepsilon_0 z_A^3} \tag{7.59}$$

for small distances $z_A \ll z_\omega$. The potential is always attractive. When moving away from the plate, it first decreases as $1/z_A^3$, then falls off more rapidly as $1/z_A^4$ before returning to a $1/z_A^3$ asymptote at very large distances. This behaviour with its three asymptotic regimes is illustrated in Fig. 7.2(i). Note that the coefficient of the large-distance power law is smaller than that of the small-distance asymptote by a factor of the order of $k_\mathrm{B}T/(\hbar\omega_-)$. This is a result of the more rapid decay of the potential for intermediate distances.

In the spectroscopic high-temperature regime, the equilibrium potential is governed by (7.50) for all distances, so that

$$U(z_A) = -\frac{k_\mathrm{B}T\left(\alpha_T^\parallel + \alpha_T^\perp\right)}{32\pi\varepsilon_0 z_A^3}\,. \tag{7.60}$$

Using the relation (7.28) for the probabilities, the spectroscopic high-temperature limit can be performed according to

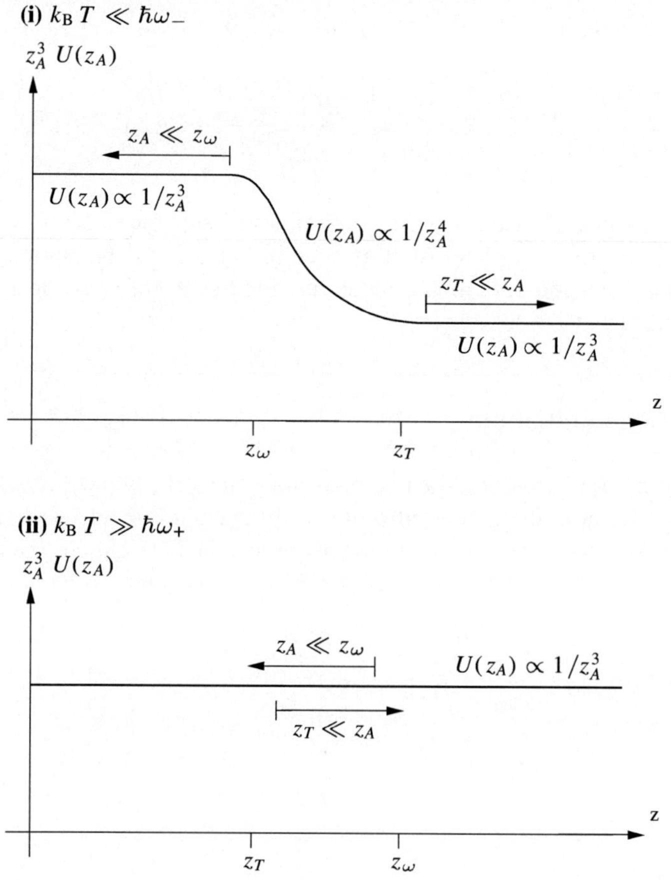

Fig. 7.2 Schematic illustration of the distance-dependence of the CP potential of an atom at thermal equilibrium with a perfectly conducting plate in the spectroscopic (**i**) high-temperature and (**ii**) low-temperature limits

$$k_B T \alpha_T^{\parallel} = k_B T \sum_{n,k} p_n \frac{|\boldsymbol{d}_{nk}^{\parallel}|^2}{\hbar\omega_{kn}} = \frac{k_B T}{2} \sum_{n,k} (p_n - p_k) \frac{|\boldsymbol{d}_{nk}^{\parallel}|^2}{\hbar\omega_{kn}} \simeq \tfrac{1}{2} \sum_{n,k} p_n |\boldsymbol{d}_{nk}^{\parallel}|^2 , \tag{7.61}$$

$$k_B T \alpha_T^{\perp} = 2k_B T \sum_{n,k} p_n \frac{|\boldsymbol{d}_{nk}^{\perp}|^2}{\hbar\omega_{kn}} = k_B T \sum_{n,k} (p_n - p_k) \frac{|\boldsymbol{d}_{nk}^{\perp}|^2}{\hbar\omega_{kn}} \simeq \sum_{n,k} p_n |\boldsymbol{d}_{nk}^{\perp}|^2 , \tag{7.62}$$

so the potential reads

$$U(z_A) = -\frac{\langle \hat{\boldsymbol{d}}^{\|2} + 2\hat{\boldsymbol{d}}^{\perp 2}\rangle_T}{64\pi\varepsilon_0 z_A^3}\,. \tag{7.63}$$

This result is a special case of (7.45), as can be seen using the Green's tensor (4.87) of the perfectly conducting plate. As illustrated in Fig. 7.2(ii), the nonretarded regime overlaps with the geometric high-temperature regime, resulting in the global asymptotic distance law (7.63).

Thermal Non-Equilibrium

We next turn our attention to the two resonant potentials, which are relevant for an atom which is not at thermal equilibrium with its environment. With the reflection coefficients $r_s = -1$ and $r_p = +1$ of the perfectly conducting plate, the propagating-wave potential (7.47) can be easily integrated to give [21]

$$\begin{aligned} U_n^{\mathrm{prop}}(z_A) = &-\frac{1}{32\pi\varepsilon_0 z_A^3}\Big\{\sum_{k<n}[n(\omega_{nk})+1] - \sum_{k>n} n(\omega_{kn})\Big\} \\ &\times \big\{|\boldsymbol{d}_{nk}^{\|}|^2\big[\cos(2x) + 2x\sin(2x) - 4x^2\cos(2x) - 1 + 2x^2\big] \\ &+ 2|\boldsymbol{d}_{nk}^{\perp}|^2\big[\cos(2x) + 2x\sin(2x) - 1 - 2x^2\big]\big\}_{x=\omega_{nk}z_A/c}\,. \end{aligned} \tag{7.64}$$

We see that the spatial dependence of the potential decouples completely from its temperature-dependence. Temperature enters only via the thermal photon number to determine the amplitudes of the potentials associated with the various atomic transitions. The potential reduces to

$$\begin{aligned} U_n^{\mathrm{prop}}(z_A) = &\frac{\mu_0}{16\pi z_A}\Big\{\sum_{k<n}[n(\omega_{nk})+1] - \sum_{k>n} n(\omega_{kn})\Big\}\omega_{nk}^2 \\ &\times\Big\{|\boldsymbol{d}_{nk}^{\|}|^2\Big[2\cos\Big(\frac{2\omega_{nk}z_A}{c}\Big) - 1\Big] + 2|\boldsymbol{d}_{nk}^{\perp}|^2\Big\} \end{aligned} \tag{7.65}$$

in the retarded limit and vanishes in the nonretarded limit,

$$U_n^{\mathrm{prop}}(z_A) = 0\,. \tag{7.66}$$

The resonant potential due to evanescent photons can be treated in a very similar way. Using the reflection coefficients for the perfectly conducting plate, the potential (7.48) is found to be [21]

$$U_n^{\mathrm{evan}}(z_A) = -\frac{1}{32\pi\varepsilon_0 z_A^3}\left\{\sum_{k<n}[n(\omega_{nk})+1] - \sum_{k>n} n(\omega_{kn})\right\} \times \left[|\boldsymbol{d}_{nk}^{\parallel}|^2\left(1 - 2\frac{\omega_{nk}^2 z_A^2}{c^2}\right) + 2|\boldsymbol{d}_{nk}^{\perp}|^2\left(1 + 2\frac{\omega_{nk}^2 z_A^2}{c^2}\right)\right]. \tag{7.67}$$

Its asymptotes in the retarded and nonretarded limits read

$$U_n^{\mathrm{evan}}(z_A) = \frac{\mu_0}{16\pi z_A}\left\{\sum_{k<n}[n(\omega_{nk})+1] - \sum_{k>n} n(\omega_{kn})\right\} \omega_{nk}^2\left(|\boldsymbol{d}_{nk}^{\parallel}|^2 - 2|\boldsymbol{d}_{nk}^{\perp}|^2\right) \tag{7.68}$$

and

$$U_n^{\mathrm{evan}}(z_A) = -\frac{1}{32\pi\varepsilon_0 z_A^3}\left\{\sum_{k<n}[n(\omega_{nk})+1] - \sum_{k>n} n(\omega_{kn})\right\}\left(|\boldsymbol{d}_{nk}^{\parallel}|^2 + 2|\boldsymbol{d}_{nk}^{\perp}|^2\right), \tag{7.69}$$

respectively.

Combining the obtained results, we can now find the total thermal CP potential of an atom in an eigenstate. For instance, in the retarded limit, the non-resonant potential with its $1/z_A^3$ or $1/z_A^4$ asymptote becomes negligible to leading order in $1/z_A$. The propagating-wave potential (7.65) combines with the evanescent-wave contribution (4.80) to give a spatially oscillating potential

$$U_n(z_A) = \frac{\mu_0}{8\pi z_A}\left\{\sum_{k<n}[n(\omega_{nk})+1] - \sum_{k>n} n(\omega_{kn})\right\} \omega_{nk}^2\, |\boldsymbol{d}_{nk}^{\parallel}|^2 \cos\left(\frac{2\omega_{nk} z_A}{c}\right). \tag{7.70}$$

In contrast to the zero-temperature case, upward and downward atomic transitions contribute to this potential. An oscillating potential is present even for an atom or molecule in its ground state. The amplitudes of the contributions are governed by the thermal photon number; they increase for increasing temperature.

Temperature-Dependence

For large atomic transition frequencies, the oscillating potential may become difficult to resolve and observe. For this reason, let us also discuss the non-oscillating, next-to-leading order retarded potential. It only contains non-resonant contributions and those from evanescent waves; and it exhibits an interesting dependence on temperature. To see this, we make use of the spectroscopic and geometric temperatures T_ω and T_z as introduced in Table 7.1 where $T_z \ll T_\omega$ in the retarded limit. We begin with very low temperatures such that the geometric low-temperature limit holds, $T \ll T_z$. These conditions are equivalent to $z_\omega \ll z_A \ll z_T$, so the non-resonant potential is given by (7.51). The next-to-leading contribution from the evanescent-wave potential (7.67) in the retarded limit reads

$$U_n^{\text{evan}}(z_A) = -\frac{1}{32\pi\varepsilon_0 z_A^3}\sum_{k<n}\left(|\boldsymbol{d}_{nk}^{\parallel}|^2 + 2|\boldsymbol{d}_{nk}^{\perp}|^2\right). \tag{7.71}$$

Here, we have exploited the inequality $T \ll T_z \ll T_\omega$ which is automatically implied in the retarded limit, so that $n(|\omega_{nk}|) \ll 1$. In other words, a geometric low-temperature limit implies a spectroscopic one in the retarded limit. Adding the non-resonant and evanescent-wave contributions, we find a temperature-independent potential [21]

$$U_n(z_A) = -\frac{1}{32\pi\varepsilon_0 z_A^3}\sum_{k<n}\left(|\boldsymbol{d}_{nk}^{\parallel}|^2 + 2|\boldsymbol{d}_{nk}^{\perp}|^2\right) - \frac{\hbar c\left(2\alpha_n^{\parallel} + \alpha_n^{\perp}\right)}{32\pi^2\varepsilon_0 z_A^4}. \tag{7.72}$$

in the spectroscopic low-temperature limit $T \ll T_z$.

For intermediate temperatures $T_z \ll T \ll T_\omega$ which are geometrically large but spectroscopically small, the next-to-leading evanescent-wave potential is still given by (7.71). The condition $T \gg T_z$ being equivalent to $z_A \gg z_T$, the non-resonant potential is now given by (7.50). The total potential [21]

$$U_n(z_A) = -\frac{1}{32\pi\varepsilon_0 z_A^3}\sum_{k<n}\left(|\boldsymbol{d}_{nk}^{\parallel}|^2 + 2|\boldsymbol{d}_{nk}^{\perp}|^2\right) - \frac{k_B T\left(\alpha_n^{\parallel} + \alpha_n^{\perp}\right)}{32\pi\varepsilon_0 z_A^3} \tag{7.73}$$

thus varies linearly with temperature in an intermediate range $T_z \ll T \ll T_\omega$.

For spectroscopically large temperatures $T \gg T_\omega$, the next-to-leading contribution from the evanescent-wave potential approximates to

$$U_n^{\text{evan}}(z_A) = \frac{1}{32\pi\varepsilon_0 z_A^3}\sum_{k}\left(\frac{k_B T}{\hbar\omega_{kn}} - \frac{1}{2}\right)\left(|\boldsymbol{d}_{nk}^{\parallel}|^2 + 2|\boldsymbol{d}_{nk}^{\perp}|^2\right) \tag{7.74}$$

where we have used (7.12) to write

$$n(\omega_{nk}) + 1 = -n(\omega_{kn}) \simeq -\left(\frac{k_B T}{\hbar\omega_{kn}} - \frac{1}{2}\right). \tag{7.75}$$

It partially cancels with the non-resonant potential (7.50). Invoking the completeness relation, the total potential reads [21]

$$U_n(z_A) = -\frac{\langle \hat{\boldsymbol{d}}^{\parallel 2} + 2\hat{\boldsymbol{d}}^{\perp 2}\rangle_n}{64\pi\varepsilon_0 z_A^3} \tag{7.76}$$

in the spectroscopic high-temperature limit $T \gg T_\omega$. Rather remarkably, the potential saturates for large T due to cancellations between resonant and evanescent-wave contributions. Furthermore, this retarded high-temperature potential agrees with the nonretarded zero-temperature result (4.85) when ignoring the spatially oscillating contribution.

The temperature-dependence of the non-oscillating CP potential in the retarded limit is illustrated in Fig. 7.3(i) [21]. We see that the potential is temperature-independent in the geometric low-temperature regime; varies linearly with temperature for intermediate temperatures; and saturates to another constant value in the spectroscopic high-temperature regime. This behaviour can best be observed for atoms in their ground states or low-lying excited states where the retarded regime sets in for very small distances.

At nonretarded distances $T_\omega \ll T_z$, the geometric low-temperature limit overlaps with the spectroscopic high-temperature limit. As we will see, this leads to a potential which is independent of T over the entire range of possible temperatures. For the geometric low-temperature regime $T \ll T_z$, we have either $z_A \ll z_\omega \ll z_T$ or $z_A \ll z_T \ll z_\omega$, so the non-resonant CP potential is given by (7.52) or (7.53). We note that the former result is just a limiting case of the latter more general one. Performing the Matsubara sum by means of (7.54) and (7.55), we find

$$U_n^{\mathrm{nres}}(z_A) = -\frac{1}{32\pi\varepsilon_0 z_A^3}\sum_k \left[n(\omega_{kn}) + \tfrac{1}{2}\right]\left(|\boldsymbol{d}_{nk}^{\parallel}|^2 + 2|\boldsymbol{d}_{nk}^{\perp}|^2\right). \tag{7.77}$$

Using the property (7.12) of the thermal photon number, the evanescent-wave potential (7.69) can be written in a similar form

$$U_n^{\mathrm{evan}}(z_A) = \frac{1}{32\pi\varepsilon_0 z_A^3}\sum_k n(\omega_{kn})\left(|\boldsymbol{d}_{nk}^{\parallel}|^2 + 2|\boldsymbol{d}_{nk}^{\perp}|^2\right). \tag{7.78}$$

The terms depending on the thermal photon number cancel, leaving a temperature-independent total potential (7.76) in the geometric low-temperature regime $T \ll T_z$.

For spectroscopically high temperatures $T \gg T_\omega$, we have either $z_A \ll z_T \ll z_\omega$ or $z_T \ll z_A \ll z_\omega$. In both cases, the non-resonant potential is given by (7.50), recall the discussions above and below (7.53). Invoking the atomic polarisability (4.15), it can be cast into the form

$$U_n^{\mathrm{nres}}(z_A) = -\frac{1}{32\pi\varepsilon_0 z_A^3}\sum_k \frac{k_{\mathrm{B}}T}{\hbar\omega_{kn}}\left(|\boldsymbol{d}_{nk}^{\parallel}|^2 + 2|\boldsymbol{d}_{nk}^{\perp}|^2\right). \tag{7.79}$$

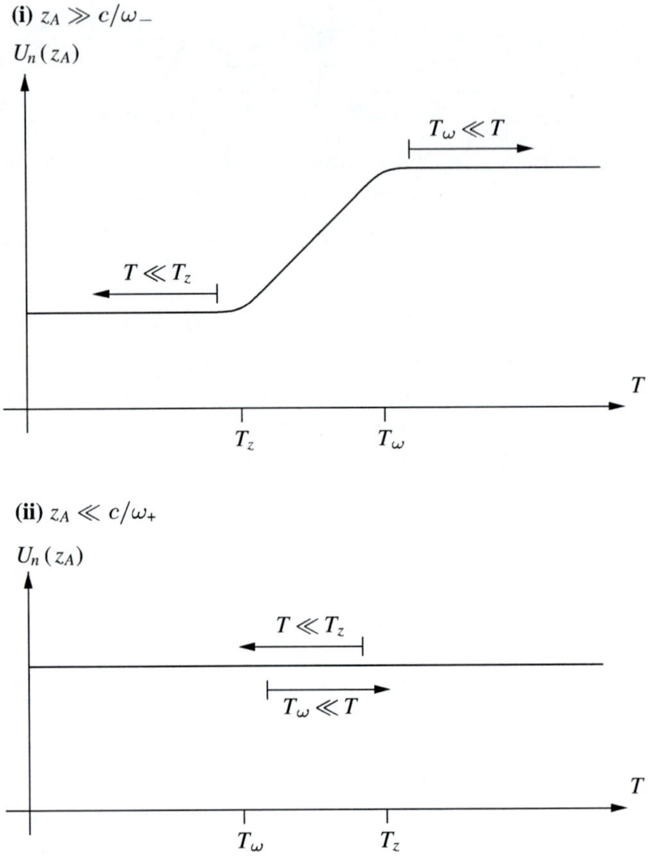

Fig. 7.3 Schematic illustration of the temperature-dependence of the non-oscillating CP potential of an atom in front of perfectly conducting plate in the **(i)** retarded and **(ii)** nonretarded limits

By using the relation (7.12) for the photon number once more, the evanescent-wave potential (7.69) reads

$$U_n^{\text{evan}}(z_A) = \frac{1}{32\pi\varepsilon_0 z_A^3}\sum_k \left(\frac{k_B T}{\hbar\omega_{kn}} - \frac{1}{2}\right)\left(|\boldsymbol{d}_{nk}^{\parallel}|^2 + 2|\boldsymbol{d}_{nk}^{\perp}|^2\right) \tag{7.80}$$

in the spectroscopic high-temperature limit. Combining the two, the contributions linear in T cancel and the total CP potential is again given by (7.76).

In the nonretarded limit, the temperature-independent result (7.76) is thus valid both in the geometric low-temperature and the spectroscopic high-temperature regimes. As shown in Fig. 7.3(ii), these two regimes overlap, so that the potential is independent of temperature across the entire temperature scale [21]. Rather remarkably, the zero-temperature result hence holds globally for arbitrary temperatures. The temperature invariance may best be observed for atoms in highly excited Rydberg states or polar molecules. Their potentials are dominated by long-wavelength transitions, so that the nonretarded regime extends over a large distance range. Note that the temperature invariance of the nonretarded CP potential for a perfectly conducting plate is a special case of the invariance (7.44) for perfect conductors of arbitrary shapes as demonstrated in the previous section.

We have just discussed the temperature-dependence vs invariance for a fixed atomic position, while in the beginning of the section, we have analysed the position-dependence for a given temperature. These are complementary aspects of the thermal CP potential, which is a non-factorisable function of the two parameters position and temperature. In this sense, the temperature-invariance shown in Fig. 7.3(ii) and the global power law for the low-temperature position-dependence found in Fig. 7.2(ii) are two sides of the same coin.

7.2.2 *Half Space*

Considering the thermal CP potential of an atom or molecule near a perfectly conducting plate, we have found surprising results. The equilibrium potential of a particle with dominant long-wavelength transitions can be governed by a single global power law. Furthermore, the CP potential of a particle in an energy eigenstate may become entirely independent of temperature due to mutual cancellations of non-resonant and evanescent-wave contributions.

In order to judge to what extend these effects can be observed under realistic conditions, let us next consider a semi-infinite electric half space of permittivity $\varepsilon(\omega)$. Note that plates of finite thickness have also been studied [14, 23, 24].

Thermal Equilibrium

Using the reflection coefficients (A.41) and (A.42) of an electric half space as given in Appendix A.3.2 with $\mu(\omega) \equiv 1$, the non-resonant CP potential (7.46) reads [21]

$$U_n^{\mathrm{nres}}(z_A) = \frac{\mu_0 k_\mathrm{B} T}{4\pi} {\sum_{j=0}^{\infty}}' \xi_j^2 \int_{\xi_j/c}^{\infty} \mathrm{d}\kappa^\perp \, \mathrm{e}^{-2\kappa^\perp z_A} \left\{ \alpha_n^\parallel(\mathrm{i}\xi_j) \frac{\kappa^\perp - \kappa_1^\perp}{\kappa^\perp + \kappa_1^\perp} + \left[\left(1 - \frac{\kappa^{\perp 2} c^2}{\xi_j^2} \right) \alpha_n^\perp(\mathrm{i}\xi_j) - \frac{\kappa^{\perp 2} c^2}{\xi_j^2} \alpha_n^\parallel(\mathrm{i}\xi_j) \right] \frac{\varepsilon(\mathrm{i}\xi_j)\kappa^\perp - \kappa_1^\perp}{\varepsilon(\mathrm{i}\xi_j)\kappa^\perp + \kappa_1^\perp} \right\} \tag{7.81}$$

with

$$\kappa_1^\perp = \sqrt{\kappa^{\perp 2} + [\varepsilon(\mathrm{i}\xi_j) - 1] \frac{\xi_j^2}{c^2}} \, . \tag{7.82}$$

Asymptotic results can be derived in close analogy with the perfect-conductor case. In the spectroscopic low-temperature limit $k_B T \ll \hbar\omega_-$, we have $z_\omega \ll z_T$ and three distance regimes may be distinguished. For large distances $z_A \gg z_T$, the geometric high-temperature limit applies and the Matsubara is dominated by the $j = 0$ term,

$$U_n^{\text{nres}}(z_A) = -\frac{k_B T\left(\alpha_n^{\parallel} + \alpha_n^{\perp}\right)}{32\pi\varepsilon_0 z_A^3}\frac{\varepsilon - 1}{\varepsilon + 1} \tag{7.83}$$

where $\varepsilon \equiv \varepsilon(0)$. In the opposite geometric low-temperature regime $z \ll z_T$, the Matsubara sum is well approximated by an integral and recover the zero-temperature results (4.91) and (4.92). We hence have

$$U_n^{\text{nres}}(z_A) = -\frac{3\hbar c}{64\pi^2\varepsilon_0 z_A^4}\int_1^\infty \mathrm{d}v\left[\left(\frac{\alpha_n^{\parallel} + \alpha_n^{\perp}}{v^2} - \frac{\alpha_n^{\perp}}{v^4}\right)\frac{\varepsilon v - \sqrt{\varepsilon - 1 + v^2}}{\varepsilon v + \sqrt{\varepsilon - 1 + v^2}} - \frac{\alpha_n^{\parallel}}{v^4}\frac{v - \sqrt{\varepsilon - 1 + v^2}}{v + \sqrt{\varepsilon - 1 + v^2}}\right] \tag{7.84}$$

for intermediate distances $z_\omega \ll z_A \ll z_T$ and

$$U_n^{\text{nres}}(z_A) = -\frac{\hbar}{32\pi^2\varepsilon_0 z_A^3}\int_0^\infty \mathrm{d}\xi\left[\alpha_n^{\parallel}(\mathrm{i}\xi) + \alpha_n^{\perp}(\mathrm{i}\xi)\right]\frac{\varepsilon(\mathrm{i}\xi) - 1}{\varepsilon(\mathrm{i}\xi) + 1} \tag{7.85}$$

for small, nonretarded distances $z_A \ll z_\omega$.

In the spectroscopic high-temperature limit $k_B T \gg \hbar\omega_+$ with $z_T \ll z_\omega$, one can show that a single global power law applies for all distances. As before, the geometric high-temperature limit (7.83) applies for large distances $z_A \gg z_T$. In the opposite, geometric low-temperature limit $z_A \ll z_T$, the nonretarded limit automatically applies, $z_A \ll z_T \ll z_\omega$. With the approximation $\xi_j z_A/c \le \omega_+ z_A/c \ll 1$, (7.81) then simplifies to

$$U_n^{\text{nres}}(z_A) = -\frac{k_B T}{16\pi\varepsilon_0 z_A^3}{\sum_{j=0}^{\infty}}'\left[\alpha_n^{\parallel}(\mathrm{i}\xi_j) + \alpha_n^{\perp}(\mathrm{i}\xi_j)\right]\frac{\varepsilon(\mathrm{i}\xi_j) - 1}{\varepsilon(\mathrm{i}\xi_j) + 1}\,. \tag{7.86}$$

The factor $[\alpha_n(\mathrm{i}\xi_j) + \alpha_n^{\perp}(\mathrm{i}\xi_j)]$ is proportional to $1/[1 + (\xi_1/\omega_{kn})^2 j^2]$ with a large coefficient $\xi_1/|\omega_{kn}| \ge 2\pi k_B T/(\hbar\omega_+) \gg 1$. Again, the Matsubara sum is hence effectively limited to its $j = 0$ term and the potential given by (7.83).

In close similarity to the zero-temperature case as discussed in Sect. 4.2.2, the nonresonant potential for a metal plate is well approximated by the perfect-conductor model for large distances: The potentials (7.83) and (7.84) agree with their perfect-conductor counterparts (7.50) and (7.51) for a metal with $\varepsilon(0) = \infty$. One exception to this general rule of thumb was noted by Boström and Sernelius [25]: The

Casimir force between two metal plates is not well approximated by the respective perfect-conductor result for very large distances. The origin of this discrepancy is the contribution from the reflection coefficient for s-polarised waves

$$r_s = \frac{\kappa^\perp - \kappa_1^\perp}{\kappa^\perp + \kappa_1^\perp} = \frac{\kappa^\perp - \sqrt{\kappa^{\perp 2} + [\varepsilon(\mathrm{i}\xi) - 1]\xi^2/c^2}}{\kappa^\perp + \sqrt{\kappa^{\perp 2} + [\varepsilon(\mathrm{i}\xi) - 1]\xi^2/c^2}} \,. \tag{7.87}$$

In the geometric high-temperature limit, the Casimir force is dominated by the contribution from $\xi = 0$. However, the limit $|\omega| \to 0$ does not commute with the perfect-conductor limit. If we perform the perfect-conductor limit $\varepsilon(\mathrm{i}\xi) \to \infty$ first, we find the frequency-independent result $r_s = -1$. On the other hand, for a real metal, we have $\varepsilon(\omega) \propto 1/\omega$ in the limit $|\omega| \to 0$, so that $r_s \to 0$. The perfect-conductor model with $r_s = -1$ will thus predict Casimir force which is larger than the actual result for a metal, where s-polarised waves do not contribute in the geometric high-temperature limit. This complication does not arise for the CP potential, because it is dominated by contributions from p-polarised waves.

Summing (7.81) in accordance with (7.16), the entirely non-resonant potential of an atom or molecule at thermal equilibrium with its environment reads

$$U(z_A) = \frac{\mu_0 k_\mathrm{B} T}{4\pi} {\sum_{j=0}^{\infty}}' \xi_j^2 \int_{\xi_j/c}^{\infty} \mathrm{d}\kappa^\perp \, \mathrm{e}^{-2\kappa^\perp z_A} \left\{ \alpha_T^\parallel(\mathrm{i}\xi_j) \frac{\kappa^\perp - \kappa_1^\perp}{\kappa^\perp + \kappa_1^\perp} \right.$$
$$\left. + \left[\left(1 - \frac{\kappa^{\perp 2} c^2}{\xi_j^2}\right) \alpha_T^\perp(\mathrm{i}\xi_j) - \frac{\kappa^{\perp 2} c^2}{\xi_j^2} \alpha_T^\parallel(\mathrm{i}\xi_j) \right] \frac{\varepsilon(\mathrm{i}\xi_j)\kappa^\perp - \kappa_1^\perp}{\varepsilon(\mathrm{i}\xi_j)\kappa^\perp + \kappa_1^\perp} \right\} . \tag{7.88}$$

This is the famous Lifshitz result [26, 27]. He inferred it from the thermal Casimir energy of two plates by assuming one of the plates to consists of a dilute gas of atoms. The Lifshitz derivation was later extended to a magnetoelectric plate [28]. Linear-response theory has been used as an alternative method [6].

Limiting cases can easily be obtained from the above. In the spectroscopic low-temperature regime, we have

$$U(z_A) = -\frac{k_\mathrm{B} T \left(\alpha_0^\parallel + \alpha_0^\perp\right)}{32\pi\varepsilon_0 z_A^3} \frac{\varepsilon - 1}{\varepsilon + 1} \tag{7.89}$$

for large distances $z_A \gg z_T$,

$$U(z_A) = -\frac{3\hbar c}{64\pi^2\varepsilon_0 z_A^4}\int\limits_1^\infty \mathrm{d}v \left[\left(\frac{\alpha_0^{\parallel}+\alpha_0^{\perp}}{v^2}-\frac{\alpha_0^{\perp}}{v^4}\right)\frac{\varepsilon v-\sqrt{\varepsilon-1+v^2}}{\varepsilon v+\sqrt{\varepsilon-1+v^2}} - \frac{\alpha_0^{\parallel}}{v^4}\frac{v-\sqrt{\varepsilon-1+v^2}}{v+\sqrt{\varepsilon-1+v^2}}\right] \tag{7.90}$$

for intermediate distances $z_\omega \ll z_A \ll z_T$ and

$$U(z_A) = -\frac{\hbar}{32\pi^2\varepsilon_0 z_A^3}\int\limits_0^\infty \mathrm{d}\xi \left[\alpha_0^{\parallel}(\mathrm{i}\xi)+\alpha_0^{\perp}(\mathrm{i}\xi)\right]\frac{\varepsilon(\mathrm{i}\xi)-1}{\varepsilon(\mathrm{i}\xi)+1} \tag{7.91}$$

for small distances $z_A \ll z_\omega$. For spectroscopically high temperatures, (7.83) with (7.61) and (7.62) leads to

$$U(z_A) = -\frac{\langle \hat{\boldsymbol{d}}^{\parallel 2}+2\hat{\boldsymbol{d}}^{\perp 2}\rangle_T}{64\pi\varepsilon_0 z_A^3}\frac{\varepsilon-1}{\varepsilon+1} \tag{7.92}$$

for all distances. The equilibrium potential near a purely electric half space is thus always attractive, where three asymptotic regimes may be distinguished for low temperatures and a single power law governs the interaction for high temperatures.

As in the zero-temperature case, we note that the perfect-conductor results are a good approximation for metallic half spaces at large distances, where low-frequency contributions are dominant. At smaller distances, the CP potential for a real metal is generally smaller than the perfect-conductor prediction. This difference is relevant for the nonretarded low-temperature potential (7.91). On the contrary, the spectroscopic high-temperature result (7.92) for a metal plate agrees with the perfect-conductor prediction at all distances. Here, high temperatures, rather than large distances, cause the potential to be dominated by low-frequency contributions.

Thermal Non-Equilibrium

When the atom is not at thermal equilibrium with its environment, we need to include the propagating- and evanescent-wave contributions. With the reflection coefficients (A.41) and (A.42) from App. A.3.2, the propagating-wave potential (7.47) reads

$$U_n^{\mathrm{prop}}(z_A) = \frac{\mu_0}{8\pi}\left\{\sum_{k<n}[n(\omega_{nk})+1] - \sum_{k>n} n(\omega_{kn})\right\}\omega_{nk}^2$$
$$\times \int_0^{|\omega_{nk}|/c} \mathrm{d}k^\perp \left\{ |\boldsymbol{d}_{nk}^\parallel|^2 \mathrm{Im}\left[\mathrm{e}^{2\mathrm{i}k^\perp z_A} \frac{k^\perp - k_1^\perp}{k^\perp + k_1^\perp}\right] - \left[\frac{k^{\perp 2}c^2}{\omega_{nk}^2} |\boldsymbol{d}_{nk}^\parallel|^2 \right.\right.$$
$$\left.\left. + 2\left(\frac{k^{\perp 2}c^2}{\omega_{nk}^2} - 1\right)|\boldsymbol{d}_{nk}^\perp|^2\right] \mathrm{Im}\left[\mathrm{e}^{2\mathrm{i}k^\perp z_A} \frac{\varepsilon(|\omega_{nk}|)k^\perp - k_1^\perp}{\varepsilon(|\omega_{nk}|)k^\perp + k_1^\perp}\right]\right\} \tag{7.93}$$

with

$$k_1^\perp = \sqrt{[\varepsilon(|\omega_{nk}|) - 1]\,\frac{\omega_{nk}^2}{c^2} + k^{\perp 2}}\,, \qquad \mathrm{Im}\,k_1^\perp > 0\,. \tag{7.94}$$

In the retarded limit $z_A \gg c/|\omega_{nk}|$, we may apply the estimates $k^\perp \simeq |\omega_{nk}|/c$ and $k_1^\perp \simeq \sqrt{\varepsilon(|\omega_{nk}|)}|\omega_{nk}|/c$ to find

$$U_n^{\mathrm{prop}}(z_A) = \frac{\mu_0}{16\pi z_A}\left\{\sum_{k<n}[n(\omega_{nk})+1] - \sum_{k>n} n(\omega_{kn})\right\}\omega_{nk}^2$$
$$\times \left\{ 2|\boldsymbol{d}_{nk}^\perp|^2 \mathrm{Re}\left[\frac{\sqrt{\varepsilon(|\omega_{nk}|)} - 1}{\sqrt{|\varepsilon(\omega_{nk}|)} + 1}\right]\right.$$
$$\left. + |\boldsymbol{d}_{nk}^\parallel|^2 \mathrm{Re}\left[\left(2\mathrm{e}^{2\mathrm{i}|\omega_{nk}|z_A/c} - 1\right)\frac{\sqrt{\varepsilon(|\omega_{nk}|)} - 1}{\sqrt{\varepsilon(|\omega_{nk}|)} + 1}\right]\right\}, \tag{7.95}$$

cf. the derivation of (4.97). In the nonretarded limit $z_A \ll c/|\omega_{nk}|$, the approximations $\mathrm{e}^{2\mathrm{i}k^\perp z_A} \simeq 1$ and $k_1^\perp \simeq k^\perp$ lead to

$$U_n^{\mathrm{prop}}(z_A) = \frac{\mu_0}{12\pi c}\left\{\sum_{k<n}[n(\omega_{nk})+1] - \sum_{k>n} n(\omega_{kn})\right\}|\omega_{nk}|^3$$
$$\times \left(4|\boldsymbol{d}_{nk}^\perp|^2 - |\boldsymbol{d}_{nk}^\parallel|^2\right)\frac{\mathrm{Im}\varepsilon(|\omega_{nk}|)}{|\varepsilon(|\omega_{nk}|) + 1|^2}\Bigg] . \tag{7.96}$$

The evanescent-wave potential (7.48) reads [21]

$$U_n^{\text{evan}}(z_A) = -\frac{\mu_0}{8\pi}\left\{\sum_{k<n}[n(\omega_{nk})+1] - \sum_{k>n} n(\omega_{kn})\right\}\omega_{nk}^2$$
$$\times \int_0^\infty \mathrm{d}\kappa^\perp\, \mathrm{e}^{-2\kappa^\perp z_A}\left\{|\boldsymbol{d}_{nk}^\parallel|^2 \mathrm{Re}\left[\frac{\kappa^\perp - \kappa_1^\perp}{\kappa^\perp + \kappa_1^\perp}\right] + \left[\frac{\kappa^{\perp 2}c^2}{\omega_{nk}^2}|\boldsymbol{d}_{nk}^\parallel|^2\right.\right.$$
$$\left.\left. + 2\left(\frac{\kappa^{\perp 2}c^2}{\omega_{nk}^2}+1\right)|\boldsymbol{d}_{nk}^\perp|^2\right]\mathrm{Re}\left[\frac{\varepsilon(|\omega_{nk}|)\kappa^\perp - \kappa_1^\perp}{\varepsilon(|\omega_{nk}|)\kappa^\perp + \kappa_1^\perp}\right]\right\} \quad (7.97)$$

for an electric half space with

$$\kappa_1^\perp = \sqrt{\kappa^{\perp 2} - [\varepsilon(|\omega_{nk}|) - 1]\frac{\omega_{nk}^2}{c^2}}\,, \qquad \mathrm{Re}\,\kappa_1^\perp > 0\,. \quad (7.98)$$

In the retarded limit, the estimates $\kappa^\perp \simeq |\omega_{nk}|/c$ and $\kappa_1^\perp \simeq \sqrt{\varepsilon(|\omega_{nk}|)}|\omega_{nk}|/c$ result in the asymptote

$$U_n^{\text{evan}}(z_A) = \frac{\mu_0}{16\pi z_A}\left\{\sum_{k<n}[n(\omega_{nk})+1] - \sum_{k>n} n(\omega_{kn})\right\}\omega_{nk}^2$$
$$\times \left(|\boldsymbol{d}_{nk}^\parallel|^2 - 2|\boldsymbol{d}_{nk}^\perp|^2\right)\mathrm{Re}\left[\frac{\sqrt{\varepsilon(|\omega_{nk}|)} - 1}{\sqrt{\varepsilon(|\omega_{nk}|)} + 1}\right]. \quad (7.99)$$

In the nonretarded limit, the evanescent-wave potential reduces to ($\kappa_1^\perp \simeq \kappa^\perp$)

$$U_n^{\text{evan}}(z_A) = -\frac{1}{32\pi\varepsilon_0 z_A^3}\left\{\sum_{k<n}[n(\omega_{nk})+1] - \sum_{k>n} n(\omega_{kn})\right\}$$
$$\times \left(|\boldsymbol{d}_{nk}^\parallel|^2 + 2|\boldsymbol{d}_{nk}^\perp|^2\right)\frac{|\varepsilon(|\omega_{nk}|)|^2 - 1}{|\varepsilon(|\omega_{nk}|) + 1|^2}\,. \quad (7.100)$$

Note that for both propagating- and evanescent-wave potentials, the temperature-dependence is separate from the position dependence. In the spectroscopic low-temperature limit, the estimate $n(|\omega_{nk}|) \ll 1$ leads to the zero-temperature results as given in Sect. 4.2.2. For spectroscopically high temperatures, the resonant potentials above become proportional to $k_B T/(\hbar|\omega_{nk}|)$.

The total thermal CP potential for an atom or a molecule in an energy eigenstate is the sum of the non-resonant, propagating- and evanescent-wave contributions. In the retarded limit, the result is dominated by the resonant terms (7.95) and (7.99) which combine to a spatially oscillating result [5, 29]

$$U_n(z_A) = \frac{\mu_0}{8\pi z_A}\left\{\sum_{k<n}[n(\omega_{nk})+1] - \sum_{k>n} n(\omega_{kn})\right\}\omega_{nk}^2|\boldsymbol{d}_{nk}^{\parallel}|^2 \times \mathrm{Re}\left[\mathrm{e}^{2\mathrm{i}|\omega_{nk}|z_A/c}\,\frac{\sqrt{\varepsilon(|\omega_{nk}|)}-1}{\sqrt{\varepsilon(|\omega_{nk}|)}+1}\right]. \tag{7.101}$$

In the nonretarded regime, the non-resonant contribution dominates the potential together with that from evanescent waves (7.100). The former can be obtained from (7.81) by making the approximation $\kappa_1^{\perp} \simeq \kappa^{\perp}$, performing the $\kappa^{\perp}$-integral and retaining only leading orders in $z_A\xi_j/c$. The result is given by (7.86), it contains (7.83) and (7.85) as special cases of spectroscopically high or low temperatures. The total nonretarded potential [5, 29]

$$\begin{aligned} U_n(z_A) = &-\frac{k_\mathrm{B}T}{16\pi\varepsilon_0 z_A^3}\sum_{j=0}^{\infty}{}'\left[\alpha_n^{\parallel}(\mathrm{i}\xi_j)+\alpha_n^{\perp}(\mathrm{i}\xi_j)\right]\frac{\varepsilon(\mathrm{i}\xi_j)-1}{\varepsilon(\mathrm{i}\xi_j)+1} \\ &-\frac{1}{32\pi\varepsilon_0 z_A^3}\left\{\sum_{k<n}[n(\omega_{nk})+1] - \sum_{k>n} n(\omega_{kn})\right\} \\ &\times\left(|\boldsymbol{d}_{nk}^{\parallel}|^2 + 2|\boldsymbol{d}_{nk}^{\perp}|^2\right)\frac{|\varepsilon(|\omega_{nk}|)|^2-1}{|\varepsilon(|\omega_{nk}|)+1|^2} \end{aligned} \tag{7.102}$$

is hence governed by a $1/z_A^3$ power law. For a non-dispersive dielectric, these results have first been found on the basis of normal-mode QED [30].

As an example, we consider a ground-state LiH molecule near an Au surface at room temperature ($T = 300$ K). The potential of such a polar molecule is dominated by rotational transitions of very low frequency ($\omega_{10} = 2.79 \times 10^{12}$ rad/s [31]). As a consequence, spatial oscillations of the thermal CP potential can be easily resolved and the thermal energy is relatively large with respect to the molecular excitation energy, $k_\mathrm{B}T/(\hbar\omega_{10}) = 14$. The first manifold of excited rotational states is triply degenerate. Labelling these states as $|1_1\rangle, |1_2\rangle, |1_3\rangle$, we have

$$\boldsymbol{d}_{1_10} = \frac{d_{01}}{\sqrt{6}}\begin{pmatrix}1\\ \mathrm{i}\\ 0\end{pmatrix}, \quad \boldsymbol{d}_{1_20} = \frac{d_{01}}{\sqrt{3}}\begin{pmatrix}0\\ 0\\ 1\end{pmatrix}, \quad \boldsymbol{d}_{1_30} = \frac{d_{01}}{\sqrt{6}}\begin{pmatrix}-1\\ \mathrm{i}\\ 0\end{pmatrix} \tag{7.103}$$

with $d_{10} = 1.96 \times 10^{-29}$ Cm [31], cf. (B.16)–(B.18) in App. B. The permittivity of Au can be given by a Drude model (4.93) with parameters $\omega_\mathrm{P} = 1.37 \times 10^{16}$ rad/s and $\gamma = 5.32 \times 10^{13}$ rad/s [32]. Figure 7.4 displays the thermal CP potential of the ground-state LiH molecule in front of an Au surface as obtained from a numerical integration of (7.81), (7.93) and (7.97) [29]. We also show the separate non-resonant, evanescent-wave and propagating-wave contributions. Their behaviour is in agreement with our analytical analysis. The non-resonant potential is monotonous and

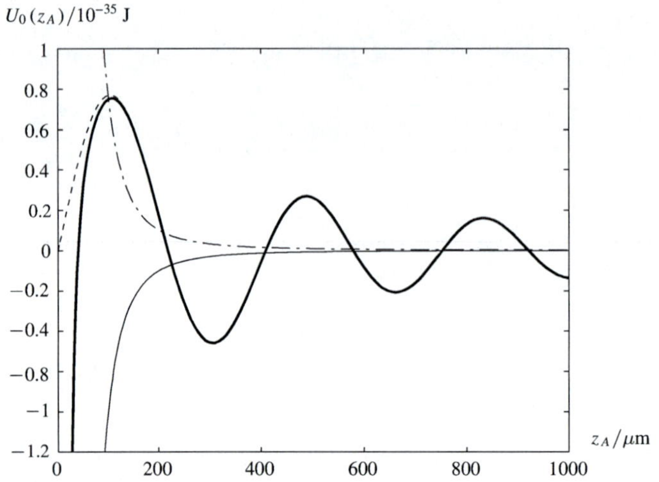

Fig. 7.4 Thermal CP potential of a ground-state LiH molecule in front of an Au half space at $T = 300$ K (*thick line*). The non-resonant (*thin solid line*), evanescent-wave (*dotted line*) and propagating-wave contributions (*dashed line*) are shown separately

attractive as expected from (7.83). The evanescent-wave potential is monotonous and repulsive as anticipated from (7.100), note that $|\varepsilon(\omega_{10})| \gg 1$ for a metal. The propagating-wave potential is spatially oscillating for large distances, cf. (7.95), and takes a finite value (7.96) on the surface. When combining the three components, we see that the spatial oscillations due to propagating waves dominate at large distances, with the oscillation period being $\lambda_{10}/2 = \pi c/\omega_{10} = 340\,\mu\text{m}$. Note that spatially oscillating resonant potentials are enhanced inside planar [17, 18] or cylindrical cavities [19, 20] whose dimensions match $\lambda_{10}/2$. At short distances, the non-resonant and evanescent-wave contributions strongly cancel each other, leaving an attractive potential due to the slightly stronger non-resonant contribution.

The observed behaviour of the thermal CP potential is very generic. By changing the half-space material to a dielectric, one can reduce the amplitude or modify the phase of the spatial oscillations. One may further change the sign of the evanescent-wave contribution by choosing a material with $|\varepsilon(\omega_{10})| < 1$, thus greatly enhancing the attractive short-range potential. The polar molecules NH, OH, OD, NaCs and KCs similarly exhibit a dominant rotational transition. We may expect their thermal CP potentials near a metal half space to mainly differ by the period of the spatial oscillations. Finally, we note that the potential as displayed in Fig. 7.4 is due to an upward transition. For downward transitions as relevant for excited molecules, the signs of all three contributions to the thermal CP potential are reversed. However, in this case the evanescent-wave potential is slightly larger than the non-resonant one, to that the total potential remains attractive at short distances.

Let us next compare the ground-state potential with the potential at thermal equilibrium (7.88). This time, we place a YbF molecule in front of the Au half space. Like LiH, this polar molecule exhibits rotational transitions to a triply degenerate excited

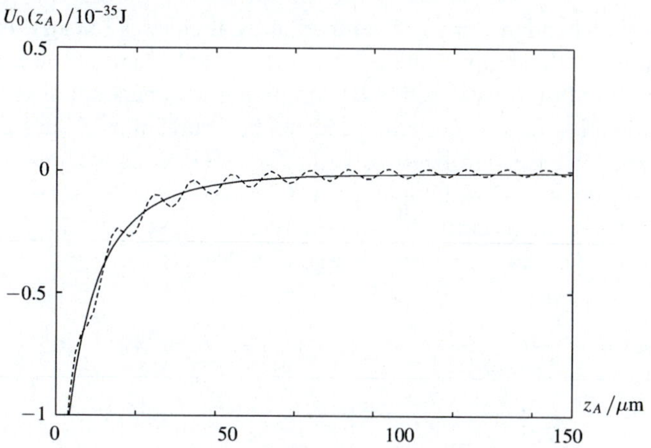

Fig. 7.5 CP potential of a YbF molecule at equilibrium with an Au half space at $T = 300$ K (*solid line*). For comparison, we also show the potential for the molecule in its ground state (*dashed line*)

manifold ($\omega_{10} = 9.05 \times 10^{10}$ rad/s, $d_{10} = 1.31 \times 10^{-29}$ Cm). In addition, vibrational excitations also contribute. The triply degenerate manifold of vibrationally excited states $|2_1\rangle$, $|2_2\rangle$, $|2_3\rangle$ is similar in structure to the manifold of rotational excitations, but with a considerably larger excitation energy ($\omega_{20} = 9.54 \times 10^{13}$ rad/s, $d_{20} = 8.60 \times 10^{-31}$ Cm) [31]. Note that the molecules CaF, BaF, LiRb, NaRb and LiCs have a similar level structure.

In Fig. 7.5, we display the CP potential of a YbF molecule for both the ground-state and thermal equilibrium cases [29]. The ground-state potential exhibits short-wavelength oscillations at large distances. They are due to vibrational transitions and have a period $\lambda_{20}/2 = \pi c/\omega_{20} = 10\,\mu$m. The period $\lambda_{10}/2 = \pi c/\omega_{10} = 10$ mm of oscillations due to rotational transitions is so large that they are not visible on the displayed distance range. Instead, rotational transitions with their longer wavelength and hence longer range dominate the ground-state potential at short distances.

The potential at thermal equilibrium is monotonous and purely attractive. Strikingly, the spatial oscillations of the non-equilibrium, ground-state potential are centred around the non-oscillating equilibrium potential. The attractive short-distance potentials agree closely in the two cases. The effect of thermal equilibrium therefore is to simply remove the spatial oscillations of the thermal CP potential in favour of a monotonous result of the same overall magnitude.

Temperature-Dependence

Let us finally investigate the temperature-dependence of the potential. In particular, we want to analyse which of our results for the perfectly conducting plate remain qualitatively valid for a plate consisting of a metal. We had found that when ignoring the spatial oscillations, the perfect-conductor potential in the retarded limit is temperature-independent for small temperatures, then rises linearly and saturates to a flat asymptote as the temperature is further increased. On the contrary, the potential in the nonretarded limit was seen to be completely independent of temperature.

Analytic results for a metal half space can be obtained in close analogy with the perfect-conductor case. We begin with the retarded limit where $T_z \ll T_\omega$ in the retarded limit. For very low temperatures $T \ll T_z$, the conditions $z_\omega \ll z_A \ll z_T$ imply a non-resonant potential (7.84). The relevant next-to-leading retarded contribution from the evanescent-wave potential (7.97) reads

$$U_n^{\text{evan}}(z_A) = -\frac{1}{32\pi\varepsilon_0 z_A^3}\left\{\sum_{k<n}[n(\omega_{nk})+1] - \sum_{k>n} n(\omega_{kn})\right\}\left(|\boldsymbol{d}_{nk}^{\parallel}|^2 + 2|\boldsymbol{d}_{nk}^{\perp}|^2\right) \times \operatorname{Re}\left[\frac{\sqrt{\varepsilon(\omega_{nk})}-1}{\sqrt{\varepsilon(\omega_{nk})}+1}\right]. \tag{7.104}$$

Adding the two results and exploiting $T \ll T_z \ll T_\omega$ to discard terms proportional to the photon number, we find

$$U_n(z_A) = -\frac{1}{32\pi\varepsilon_0 z_A^3}\sum_{k<n}\left(|\boldsymbol{d}_{nk}^{\parallel}|^2 + 2|\boldsymbol{d}_{nk}^{\perp}|^2\right)\operatorname{Re}\left[\frac{\sqrt{\varepsilon(\omega_{nk})}-1}{\sqrt{\varepsilon(\omega_{nk})}+1}\right] - \frac{3\hbar c}{64\pi^2\varepsilon_0 z_A^4}\int_1^\infty \mathrm{d}v\left[\left(\frac{\alpha_n^{\parallel}+\alpha_n^{\perp}}{v^2} - \frac{\alpha_n^{\perp}}{v^4}\right)\frac{\varepsilon v - \sqrt{\varepsilon-1+v^2}}{\varepsilon v + \sqrt{\varepsilon-1+v^2}} - \frac{\alpha_n^{\parallel}}{v^4}\,\frac{v-\sqrt{\varepsilon-1+v^2}}{v+\sqrt{\varepsilon-1+v^2}}\right]. \tag{7.105}$$

For intermediate temperatures $T_z \ll T \ll T_\omega$, we have $z_A \gg z_T$, so the non-resonant potential is given by (7.83). Combining it with the low-temperature form of the evanescent-wave potential, the non-oscillating potential assumes the form

$$U_n(z_A) = -\frac{1}{32\pi\varepsilon_0 z_A^3}\sum_{k<n}\left(|\boldsymbol{d}_{nk}^{\parallel}|^2 + 2|\boldsymbol{d}_{nk}^{\perp}|^2\right)\operatorname{Re}\left[\frac{\sqrt{\varepsilon(\omega_{nk})}-1}{\sqrt{\varepsilon(\omega_{nk})}+1}\right] - \frac{k_B T\left(\alpha_n^{\parallel}+\alpha_n^{\perp}\right)}{32\pi\varepsilon_0 z_A^3}\,\frac{\varepsilon-1}{\varepsilon+1}. \tag{7.106}$$

For very large temperatures $T \gg T_\omega$, the evanescent-wave potential above can be simplified using (7.75). Adding the non-resonant contribution (7.83), we have

$$U_n(z_A) = -\frac{1}{64\pi\varepsilon_0 z_A^3}\sum_k (|\boldsymbol{d}_{nk}^{\parallel}|^2 + 2|\boldsymbol{d}_{nk}^{\perp}|^2)\mathrm{Re}\left[\frac{\sqrt{\varepsilon(|\omega_{kn}|)}-1}{\sqrt{\varepsilon(|\omega_{kn}|)}+1}\right]$$
$$-\frac{k_B T}{32\pi\varepsilon_0 z_A^3}\sum_k \frac{|\boldsymbol{d}_{nk}^{\parallel}|^2+2|\boldsymbol{d}_{nk}^{\perp}|^2}{\hbar\omega_{kn}}\left\{1-\mathrm{Re}\left[\frac{\sqrt{\varepsilon(|\omega_{kn}|)}-1}{\sqrt{\varepsilon(|\omega_{kn}|)}+1}\right]\right\}. \tag{7.107}$$

At nonretarded distances $T_\omega \ll T_z$, we only need to distinguish two regimes. For spectroscopically small temperatures $T \ll T_\omega$, we have $z_A \ll z_\omega \ll z_T$, so the non-resonant potential is given by (7.85). Adding the nonretarded evanescent-wave potential (7.100), one has [5, 29]

$$U_n(z_A) = -\frac{\hbar}{32\pi^2\varepsilon_0 z_A^3}\int_0^\infty \mathrm{d}\xi\,\left[\alpha_n^{\parallel}(\mathrm{i}\xi)+\alpha_n^{\perp}(\mathrm{i}\xi)\right]\frac{\varepsilon(\mathrm{i}\xi)-1}{\varepsilon(\mathrm{i}\xi)+1}$$
$$-\frac{1}{32\pi\varepsilon_0 z_A^3}\sum_{k<n}(|\boldsymbol{d}_{nk}^{\parallel}|^2+2|\boldsymbol{d}_{nk}^{\perp}|^2)\frac{|\varepsilon(|\omega_{nk}|)|^2-1}{|\varepsilon(|\omega_{nk}|)+1|^2}. \tag{7.108}$$

For spectroscopically large temperatures $T_\omega \ll T$, we have either $z_A \ll z_T \ll z_\omega$ or $z_T \ll z_A$. In both cases, the non-resonant potential is given by (7.83), see the discussion around (7.86). Performing the high-temperature limit of the nonretarded evanescent-wave potential (7.100) according to (7.75), we find

$$U_n(z_A) = -\frac{1}{64\pi\varepsilon_0 z_A^3}\sum_k (|\boldsymbol{d}_{nk}^{\parallel}|^2 + 2|\boldsymbol{d}_{nk}^{\perp}|^2)\frac{|\varepsilon(|\omega_{nk}|)|^2-1}{|\varepsilon(|\omega_{nk}|)+1|^2}$$
$$-\frac{k_B T}{32\pi\varepsilon_0 z_A^3}\sum_k \frac{|\boldsymbol{d}_{nk}^{\parallel}|^2+2|\boldsymbol{d}_{nk}^{\perp}|^2}{\hbar\omega_{kn}}\left[1-\frac{|\varepsilon(|\omega_{nk}|)|^2-1}{|\varepsilon(|\omega_{nk}|)+1|^2}\right]. \tag{7.109}$$

Let us apply these results to a metallic half space and compare with the predictions from the perfect-conductor case. In the retarded limit, the non-resonant contribution to the potential exactly coincides with its perfect-conductor counterpart, whereas the evanescent-wave contribution is only very poorly represented by a perfect conductor, because the latter requires $|\sqrt{\varepsilon(|\omega_{kn}|)}| \gg 1$. As a consequence, the potential for a real metal is constant for low temperatures and rises linearly for intermediate temperatures, but never saturates in the high-temperature limit. This is due to an imperfect cancellation of the temperature-dependent terms in (7.107). On the contrary, the potential in the retarded limit is quite well represented by the perfect-conductor result provided that $\varepsilon(\mathrm{i}|\omega_{nk}|)$, $|\varepsilon(|\omega_{nk}|)| \gg 1$. In this case, the two asymptotes above coincide, making the nonretarded potential for a metal half space temperature-independent across the entire range.

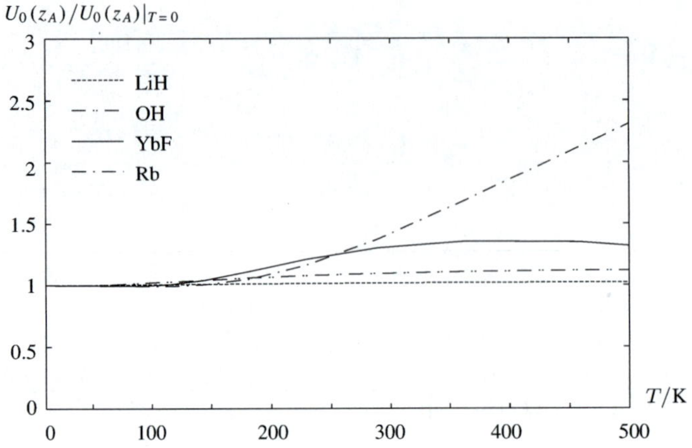

Fig. 7.6 Temperature-dependence of the CP potential of various ground-state atoms and molecules at distance $z_A = 5\mu\text{m}$ from an Au half space

To demonstrate the temperature-dependence, we display the non-oscillating part of the CP potential of different ground-state atoms and molecules for a fixed distance $z_A = 5\mu\text{m}$ from an Au half space. As seen in Fig. 7.6, the potentials for all species are virtually constant for temperatures below 100K $\approx 0.2T_z$ [21]. The potential of LiH is dominated by long-wavelength rotational transitions and hence strongly nonretarded, $z_A\omega_{10}/c = 0.046$. It is therefore temperature-independent across the entire displayed range. The rotational transition frequency of OH is slightly larger ($\omega_{10} = 1.58 \times 10^{13}$ rad/s [31]), so that the potential is only weakly nonretarded with $z_A\omega_{10}/c = 0.26$. As a consequence, we notice a slight increase of the potential as the temperature is increased beyond 150 K. The potential for YbF is dominated by a vibrational transition of even larger frequency ($\omega_{10} = 9.54 \times 10^{13}$ rad/s [31]), making it weakly retarded with $z_A\omega_{10}/c = 1.6$. The interaction increases by 30 % as the temperature is raised from zero to 500 K. Finally, the CP potential of ^{87}Rb is dominated by an $5^2S_{1/2} \to 5^2P_{3/2}$ transition ($\omega_{10} = 2.41 \times 10^{15}$ rad/s [33]) with $z_A\omega_{10}/c = 40$ and hence strongly retarded. As a result, the potential increases linearly with temperature for $T \gtrsim 200$ K. With $T_\omega \approx 18.000$ K, the potential does not saturate within the displayed range.

The temperature-invariance of the CP potential at nonretarded distances is a result of strong cancellations between non-resonant and evanescent-wave contributions. To demonstrate this, Fig. 7.7 displays these two components separately for a ground-state LiH molecule in front of an Au half space at different temperatures [21]. We see that both the attractive non-resonant potential and the repulsive evanescent-wave contribution strongly increase as the temperature is increased. However, the resulting total potential remains constant within the displayed nonretarded distance range ($z_\omega = 100\,\mu\text{m}$).

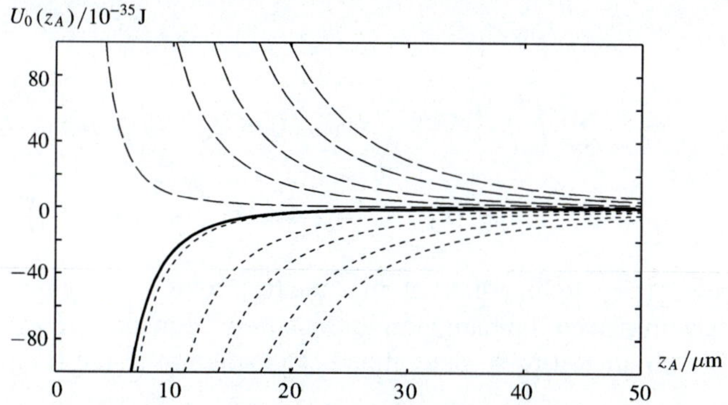

Fig. 7.7 Thermal CP potential of a ground-state LiH molecule in front of a Au surface. We show the total potential (*solid line*) as well as its evanescent (*dashed*) and non-resonant (*dotted*) contributions for temperatures $T = 10\,\mathrm{K}, 50\,\mathrm{K}, 100\,\mathrm{K}, 200\,\mathrm{K}, 300\,\mathrm{K}$ (*left to right*)

7.3 Dynamical Approach

As seen in Chap. 5, resonant forces acting on excited atoms have a limited lifetime. They disappear as the atom settles back into its ground state while spontaneously emitting a real photon. The situation is very similar at finite temperature. We have already seen that resonant force components are not present when the atom is at thermal equilibrium with its environment. An atom initially prepared in an arbitrary state will absorb thermal photons and undergo spontaneous or stimulated emission. These dynamical processes will eventually lead the atom to its equilibrium state. As in the zero-temperature case, there will be no resonant forces in the long-time limit.

To see this explicitly, we will once more employ our dynamical approach to the CP force. To that end, we first solve the atom–field dynamics in the presence of thermal photons and then evaluate the average Lorentz force to obtain the time-dependent CP force.

7.3.1 Internal Atomic Dynamics

For weak atom–field coupling, the internal atomic dynamics is governed by (5.51) as derived in Sect. 5.2. Thermal photons enter these equations via the free-field operators $\underline{\hat{\boldsymbol{E}}}, \underline{\hat{\boldsymbol{E}}}^{\dagger}$. Since the thermal density matrix of the electromagnetic field (7.1) is diagonal, they start to contribute at second order in these operators. Their influence can be made explicit by iterating. To that end, we formally solve the truncated equation (5.51),

$$
\begin{aligned}
\dot{\hat{A}}_{mn}(t) &= \mathrm{i}\tilde{\omega}_{mn}\hat{A}_{mn}(t)\\
&+\frac{\mathrm{i}}{\hbar}\sum_k\int_0^\infty \mathrm{d}\omega\big\{\mathrm{e}^{-\mathrm{i}\omega(t-t_0)}\big[\hat{A}_{mk}(t)\boldsymbol{d}_{nk}-\hat{A}_{kn}(t)\boldsymbol{d}_{km}\big]\cdot\hat{\underline{\boldsymbol{E}}}(\boldsymbol{r}_A,\omega)\\
&+\mathrm{e}^{\mathrm{i}\omega(t-t_0)}\hat{\underline{\boldsymbol{E}}}^\dagger(\boldsymbol{r}_A,\omega)\cdot\big[\boldsymbol{d}_{nk}\hat{A}_{mk}(t)-\boldsymbol{d}_{km}\hat{A}_{kn}(t)\big]\big\}\,, \qquad (7.110)
\end{aligned}
$$

which expresses the coupling of the atom to the free fields. To obtain a self-consistent result, we have replaced the bare atomic transition frequencies ω_{mn} by the shifted ones $\tilde{\omega}_{mn}$, which are yet to be determined. The solution to the truncated equation reads

$$
\begin{aligned}
\hat{A}_{mn}(t) &= \mathrm{e}^{\mathrm{i}\tilde{\omega}_{mn}(t-t_0)}\hat{A}_{mn}+\frac{\mathrm{i}}{\hbar}\sum_k\int_0^\infty \mathrm{d}\omega\int_{t_0}^t \mathrm{d}t'\,\mathrm{e}^{\mathrm{i}\tilde{\omega}_{mn}(t-t')}\\
&\times\big\{\mathrm{e}^{-\mathrm{i}\omega(t'-t_0)}\big[\hat{A}_{mk}(t')\boldsymbol{d}_{nk}-\hat{A}_{kn}(t')\boldsymbol{d}_{km}\big]\cdot\hat{\underline{\boldsymbol{E}}}(\boldsymbol{r}_A,\omega)\\
&+\mathrm{e}^{\mathrm{i}\omega(t'-t_0)}\hat{\underline{\boldsymbol{E}}}^\dagger(\boldsymbol{r}_A,\omega)\cdot\big[\boldsymbol{d}_{nk}\hat{A}_{mk}(t')-\boldsymbol{d}_{km}\hat{A}_{kn}(t')\big]\big\}\,. \qquad (7.111)
\end{aligned}
$$

Returning to (5.51) and replacing the atomic operators in the second and third rows by their iterative solution (7.111), we find the required self-consistent equation for the internal atomic dynamics

$$
\begin{aligned}
\dot{\hat{A}}_{mn}(t) &= \mathrm{i}\omega_{mn}\hat{A}_{mn}(t)-\frac{1}{\hbar^2}\sum_{k,l}\int_0^\infty \mathrm{d}\omega\int_0^\infty \mathrm{d}\omega'\int_{t_0}^t \mathrm{d}t'\\
&\times\Big(\mathrm{e}^{\mathrm{i}\omega(t'-t_0)-\mathrm{i}\omega'(t-t_0)}\\
&\times\big\{\mathrm{e}^{\mathrm{i}\tilde{\omega}_{mk}(t-t')}\big[\hat{A}_{ml}(t')\boldsymbol{d}_{kl}-\hat{A}_{lk}(t')\boldsymbol{d}_{lm}\big]\cdot\hat{\underline{\boldsymbol{E}}}^\dagger(\boldsymbol{r}_A,\omega)\hat{\underline{\boldsymbol{E}}}(\boldsymbol{r}_A,\omega')\cdot\boldsymbol{d}_{nk}\\
&+\mathrm{e}^{\mathrm{i}\tilde{\omega}_{kn}(t-t')}\big[\hat{A}_{kl}(t')\boldsymbol{d}_{nl}-\hat{A}_{ln}(t')\boldsymbol{d}_{lk}\big]\cdot\hat{\underline{\boldsymbol{E}}}^\dagger(\boldsymbol{r}_A,\omega)\hat{\underline{\boldsymbol{E}}}(\boldsymbol{r}_A,\omega')\cdot\boldsymbol{d}_{km}\big\}\\
&+\mathrm{e}^{\mathrm{i}\omega(t-t_0)-\mathrm{i}\omega'(t'-t_0)}\\
&\times\big\{\mathrm{e}^{\mathrm{i}\tilde{\omega}_{mk}(t-t')}\boldsymbol{d}_{nk}\cdot\hat{\underline{\boldsymbol{E}}}^\dagger(\boldsymbol{r}_A,\omega)\hat{\underline{\boldsymbol{E}}}(\boldsymbol{r}_A,\omega')\cdot\big[\hat{A}_{ml}(t')\boldsymbol{d}_{kl}-\hat{A}_{lk}(t')\boldsymbol{d}_{lm}\big]\\
&+\mathrm{e}^{\mathrm{i}\tilde{\omega}_{kn}(t-t')}\boldsymbol{d}_{km}\cdot\hat{\underline{\boldsymbol{E}}}^\dagger(\boldsymbol{r}_A,\omega)\hat{\underline{\boldsymbol{E}}}(\boldsymbol{r}_A,\omega')\cdot\big[\hat{A}_{kl}(t')\boldsymbol{d}_{nl}-\hat{A}_{ln}(t')\boldsymbol{d}_{lk}\big]\big\}\Big)\\
&-\sum_{k,l}\big[\boldsymbol{d}_{nk}\cdot\boldsymbol{C}_{kl}\hat{A}_{ml}(t)-\boldsymbol{d}_{km}\cdot\boldsymbol{C}_{nl}\hat{A}_{kl}(t)\big]\\
&+\sum_{k,l}\big[\boldsymbol{d}_{nk}\cdot\boldsymbol{C}^*_{ml}\hat{A}_{lk}(t)-\boldsymbol{d}_{km}\cdot\boldsymbol{C}^*_{kl}\hat{A}_{ln}(t)\big] \qquad (7.112)
\end{aligned}
$$

which is quadratic in the free-field operators and linear in the coefficients $\boldsymbol{C}_{mn}$. We have used the fact that the operators in (7.111) commute at this level of approximation. In addition, we have discarded terms proportional to $\hat{\underline{\boldsymbol{E}}}$, $\hat{\underline{\boldsymbol{E}}}^\dagger$, $\hat{\underline{\boldsymbol{E}}}\,\hat{\underline{\boldsymbol{E}}}$ and $\hat{\underline{\boldsymbol{E}}}^\dagger\hat{\underline{\boldsymbol{E}}}^\dagger$, which do not contribute for a thermal state.

We next take expectation values, assuming the electromagnetic field to be prepared in a thermal state at initial time, $\hat{\rho} = \hat{\rho}(t_0) = \hat{\rho}_T$. Averages of the free fields can then be evaluated by using the field expansion (1.22) and the thermal averages (1.39) of the fundamental fields,

$$\langle \underline{\hat{\boldsymbol{E}}}^{\dagger}(\boldsymbol{r}, \omega)\underline{\hat{\boldsymbol{E}}}(\boldsymbol{r}', \omega')\rangle_T = \frac{\hbar\mu_0}{\pi} n(\omega)\omega^2 \mathrm{Im}\boldsymbol{G}(\boldsymbol{r}, \boldsymbol{r}', \omega)\delta(\omega - \omega') \,. \tag{7.113}$$

With this result, the above equation leads to

$$\begin{aligned}
\langle \dot{\hat{A}}_{mn}(t)\rangle &= \mathrm{i}\omega_{mn}\langle \hat{A}_{mn}(t)\rangle \\
&\quad - \frac{\mu_0}{\pi\hbar}\sum_{k,l}\int_0^{\infty} \mathrm{d}\omega\, \omega^2 n(\omega)\int_{t_0}^{t} \mathrm{d}t' \left[\mathrm{e}^{-\mathrm{i}\omega(t-t')} + \mathrm{e}^{\mathrm{i}\omega(t-t')}\right] \\
&\quad \times \Big\{\mathrm{e}^{\mathrm{i}\tilde{\omega}_{mk}(t-t')}\boldsymbol{d}_{nk}\cdot \mathrm{Im}\boldsymbol{G}(\boldsymbol{r}_A, \boldsymbol{r}_A, \omega)\cdot\left[\boldsymbol{d}_{kl}\langle \hat{A}_{ml}(t')\rangle - \boldsymbol{d}_{lm}\langle \hat{A}_{lk}(t')\rangle\right] \\
&\quad\quad + \mathrm{e}^{\mathrm{i}\tilde{\omega}_{kn}(t-t')}\boldsymbol{d}_{km}\cdot \mathrm{Im}\boldsymbol{G}(\boldsymbol{r}_A, \boldsymbol{r}_A, \omega)\cdot\left[\boldsymbol{d}_{nl}\langle \hat{A}_{kl}(t')\rangle - \boldsymbol{d}_{lk}\langle \hat{A}_{ln}(t')\rangle\right]\Big\} \\
&\quad - \sum_{k,l}\left[\boldsymbol{d}_{nk}\cdot \boldsymbol{C}_{kl}\langle \hat{A}_{ml}(t)\rangle - \boldsymbol{d}_{km}\cdot \boldsymbol{C}_{nl}\langle \hat{A}_{kl}(t)\rangle\right] \\
&\quad + \sum_{k,l}\left[\boldsymbol{d}_{nk}\cdot \boldsymbol{C}^{*}_{ml}\langle \hat{A}_{lk}(t)\rangle - \boldsymbol{d}_{km}\cdot \boldsymbol{C}^{*}_{kl}\langle \hat{A}_{ln}(t)\rangle\right]
\end{aligned} \tag{7.114}$$

where we have used the fact that the Green's tensor is symmetric. The time integrals can be evaluated via the Markov equation as described by (5.45)–(5.47) in Sect. 5.2. Combining the results with the terms on the last two rows of (7.114) and recalling the definitions (5.52), the equations for the internal atomic dynamics can be written in exactly the same form as (5.54) as in the zero-temperature case. However, the coefficients $\boldsymbol{C}_{mn}$ in these equations now take the values

$$\begin{aligned}
\boldsymbol{C}_{mn} &= \boldsymbol{C}_{mn}(\boldsymbol{r}_A) \\
&= \frac{\mu_0}{\hbar}\Theta(\tilde{\omega}_{nm})[n(\tilde{\omega}_{nm}) + 1]\tilde{\omega}_{nm}^2 \mathrm{Im}\,\boldsymbol{G}(\boldsymbol{r}_A, \boldsymbol{r}_A, \tilde{\omega}_{nm})\cdot\boldsymbol{d}_{mn} \\
&\quad + \frac{\mu_0}{\hbar}\Theta(\tilde{\omega}_{mn})n(\tilde{\omega}_{mn})\tilde{\omega}_{mn}^2 \mathrm{Im}\,\boldsymbol{G}(\boldsymbol{r}_A, \boldsymbol{r}_A, \tilde{\omega}_{mn})\cdot\boldsymbol{d}_{mn} \\
&\quad - \frac{\mathrm{i}\mu_0}{\pi\hbar}\mathcal{P}\int_0^{\infty} \frac{\mathrm{d}\omega}{\omega - \tilde{\omega}_{nm}}[n(\omega) + 1]\omega^2 \mathrm{Im}\,\boldsymbol{G}(\boldsymbol{r}_A, \boldsymbol{r}_A, \omega)\cdot\boldsymbol{d}_{mn} \\
&\quad + \frac{\mathrm{i}\mu_0}{\pi\hbar}\mathcal{P}\int_0^{\infty} \frac{\mathrm{d}\omega}{\omega - \tilde{\omega}_{mn}} n(\omega)\omega^2 \mathrm{Im}\,\boldsymbol{G}(\boldsymbol{r}_A, \boldsymbol{r}_A, \omega)\cdot\boldsymbol{d}_{mn}
\end{aligned} \tag{7.115}$$

where the new contributions proportional to $n(\omega)$ are due to the thermal photons.

We next follow the similar steps as in the zero-temperature case, cf. (5.55)–(5.64). Assuming the atom to be free of quasi-degenerate transitions, the off-diagonal atomic flip operators decouple from each other as well as from the diagonal ones. Taking real and imaginary parts of the coefficients, we find

$$\big\langle \dot{\hat{A}}_{nn}(t)\big\rangle = -\Gamma_n\big\langle \hat{A}_{nn}(t)\big\rangle + \sum_k \Gamma_{kn}\big\langle \hat{A}_{kk}(t)\big\rangle\,, \tag{7.116}$$

$$\big\langle \dot{\hat{A}}_{mn}(t)\big\rangle = \big[\mathrm{i}\tilde{\omega}_{mn} - \tfrac{1}{2}(\Gamma_m+\Gamma_n)\big]\big\langle \hat{A}_{mn}(t)\big\rangle \quad (m\neq n) \tag{7.117}$$

where the identification (5.65) has been made. The atomic frequency shifts and transition rates are now given by

$$\delta\omega_n = \sum_k \delta\omega_{nk}\,, \tag{7.118}$$

$$\Gamma_n = \sum_k \Gamma_{nk} \tag{7.119}$$

with

$$\begin{aligned}\delta\omega_{nk} &= \delta\omega_{nk}(\boldsymbol{r}_A)\\ &= -\frac{\mu_0}{\pi\hbar}\mathcal{P}\int_0^\infty \frac{\mathrm{d}\omega}{\omega-\tilde{\omega}_{nk}}\,\omega^2\boldsymbol{d}_{nk}\cdot\mathrm{Im}\,\mathbf{G}^{(1)}(\boldsymbol{r}_A,\boldsymbol{r}_A,\omega)\cdot\boldsymbol{d}_{kn}\\ &\quad+\frac{\mu_0}{\pi\hbar}\mathcal{P}\int_0^\infty \mathrm{d}\omega\,\omega^2\left[\frac{n(\omega)}{\omega+\tilde{\omega}_{nk}}-\frac{n(\omega)}{\omega-\tilde{\omega}_{nk}}\right]\\ &\quad\times\boldsymbol{d}_{nk}\cdot\mathrm{Im}\,\mathbf{G}(\boldsymbol{r}_A,\boldsymbol{r}_A,\omega)\cdot\boldsymbol{d}_{kn}\,,\end{aligned} \tag{7.120}$$

$$\begin{aligned}\Gamma_{nk} &= \Gamma_{nk}(\boldsymbol{r}_A)\\ &= \begin{cases}\dfrac{2\mu_0}{\hbar}\,\tilde{\omega}_{nk}^2[n(\tilde{\omega}_{nk})+1]\boldsymbol{d}_{nk}\cdot\mathrm{Im}\,\mathbf{G}(\boldsymbol{r}_A,\boldsymbol{r}_A,\tilde{\omega}_{nk})\cdot\boldsymbol{d}_{kn} & \text{for } k<n\,,\\[2ex] \dfrac{2\mu_0}{\hbar}\,\tilde{\omega}_{kn}^2 n(\tilde{\omega}_{kn})\boldsymbol{d}_{nk}\cdot\mathrm{Im}\,\mathbf{G}(\boldsymbol{r}_A,\boldsymbol{r}_A,\tilde{\omega}_{kn})\cdot\boldsymbol{d}_{kn} & \text{for } k>n\,.\end{cases}\end{aligned} \tag{7.121}$$

Using the relation (5.35), the dynamics of the atomic density matrix elements immediately follows from (7.116):

$$\dot{p}_n(t) = -\Gamma_n p_n(t) + \sum_k \Gamma_{kn} p_k(t)\,. \tag{7.122}$$

In contrast to the zero-temperature case, it is governed by three processes. As before, an atom in a given state $|n\rangle$ can spontaneously emit a real photon while making a downward transition to a state $|k\rangle$ $(k < n)$. In addition, thermal photons can trigger stimulated decay of the atom, again accompanied by the emission of a real photon. The corresponding downward transition rate (7.121) is proportional to $n(\tilde{\omega}_{nk}) + 1$, where stimulated and spontaneous emission are represented by the first and second terms in the square brackets. Thirdly, the atom can absorb a thermal photon while making an upward transition $(k > n)$. The respective transition rate is proportional to the thermal photon number $n(\tilde{\omega}_{kn})$.

As a result of spontaneous and induced upward and downward transitions, the internal-state population of the atom is redistributed as time progresses. By summing (7.122) over all states and using (7.119), we can verify that this process is probability-conserving, so that (5.73) holds. An atom or molecule initially prepared in an energy eigenstate, $\hat{\sigma}(t_0) = |n\rangle\langle n|$, will eventually evolve into a mixed state

$$\hat{\sigma}(t) = \sum_k p_k(t)|k\rangle\langle k| \quad \text{for } t \geq t_0\,. \tag{7.123}$$

In the long-time limit, the atom will reach a steady state, which can be found by setting $\dot{p}_n(t \to \infty) = 0$ in (7.122) above. The combining the relation (7.29) for the thermal photon number with the definition (7.121) of the transition rates, we find

$$\Gamma_{nk} = \mathrm{e}^{\hbar\tilde{\omega}_{nk}/(k_\mathrm{B}T)}\Gamma_{kn}\,, \tag{7.124}$$

so that

$$0 = -\Gamma_n p_n + \sum_k \Gamma_{kn} p_k = -\sum_k \Gamma_{nk} p_n + \sum_k \Gamma_{nk} e^{-\hbar\tilde{\omega}_{nk}/(k_\mathrm{B}T)} p_k\,. \tag{7.125}$$

It follows that

$$p_n = \mathrm{e}^{-\hbar\tilde{\omega}_{nk}/(k_\mathrm{B}T)} p_k\,. \tag{7.126}$$

After normalisation, the steady-state probabilities hence read

$$p_n(t \to \infty) = \frac{\mathrm{e}^{-\tilde{E}_n/(k_\mathrm{B}T)}}{\sum_k \mathrm{e}^{-\tilde{E}_k/(k_\mathrm{B}T)}}\,, \tag{7.127}$$

with

$$\tilde{E}_n = \tilde{E}_n(\mathbf{r}_A) = E_n + \hbar\delta\omega_n \tag{7.128}$$

denoting the shifted atomic eigenenergies. In the long-time limit, the atom hence reaches thermal equilibrium with its environment, and its internal state is a thermal state

$$\hat{\sigma}(t \to \infty) = \hat{\sigma}_T = \frac{\mathrm{e}^{-\hat{\tilde{H}}_A/(k_B T)}}{\mathrm{tr}\left[\mathrm{e}^{-\hat{\tilde{H}}_A/(k_B T)}\right]}, \tag{7.129}$$

where

$$\hat{\tilde{H}}_A = \sum_n \tilde{E}_n |n\rangle\langle n| . \tag{7.130}$$

Note that the steady state is independent of the initial state preparation. It only depends on the environment temperature T and the internal level structure of the atom.

The transition rates determine how fast an atom reaches thermal equilibrium with its environment. In free space, we use (A.26) to find the transition rates

$$\Gamma_{nk} = \begin{cases} \dfrac{\tilde{\omega}_{nk}^3 |\mathbf{d}_{nk}|^2}{3\pi\varepsilon_0\hbar c^3}\,[n(\tilde{\omega}_{nk}) + 1] & \text{for } k < n , \\ \dfrac{\tilde{\omega}_{kn}^3 |\mathbf{d}_{nk}|^2}{3\pi\varepsilon_0\hbar c^3}\,n(\tilde{\omega}_{kn}) & \text{for } k > n . \end{cases} \tag{7.131}$$

By splitting off the energy density of the thermal radiation

$$\rho_T(\omega) = \frac{2\hbar}{\pi c^2}\, n(\omega)\omega^2 \mathrm{tr}\,\mathrm{Im}\,\boldsymbol{G}^{(0)}(\mathbf{r}, \mathbf{r}, \omega) = \frac{\hbar\omega^3}{\pi^2 c^3}\, n(\omega) , \tag{7.132}$$

the rates can be expressed in terms of the Einstein A- and B-coefficients [34],

$$\Gamma_{nk} = \begin{cases} A_{nk} + B_{nk}\rho_T(\tilde{\omega}_{nk}) & \text{for } k < n , \\ B_{nk}\rho_T(\tilde{\omega}_{kn}) & \text{for } k > n . \end{cases} \tag{7.133}$$

By comparing (7.131) and (7.133), we find [35]

$$A_{nk} = \frac{\tilde{\omega}_{nk}^3 |\mathbf{d}_{nk}|^2}{3\pi\varepsilon_0\hbar c^3} \quad (k < n) , \tag{7.134}$$

$$B_{nk} = \frac{\pi |\mathbf{d}_{nk}|^2}{3\varepsilon_0\hbar^2} . \tag{7.135}$$

The Einstein coefficients are important atomic parameters. The A-coefficients describe the ability of an atom to undergo spontaneous emission, while the B-coefficients represent its susceptibility to stimulated photon emission ($k < n$) or absorption ($k > n$).

The life time $T_1 = 1/\Gamma_n$ of a pure state $|n\rangle$ against equilibration is governed by the total transition rate (7.119). In particular, the heating rate of a ground-state atom (initial temperature $T = 0\,\mathrm{K}$) in free space reads

$$\Gamma_0 = \sum_k B_{0k}\rho_T(\tilde{\omega}_{k0}) = \frac{1}{3\pi\varepsilon_0\hbar c^3}\sum_k \tilde{\omega}_{k0}^3|\boldsymbol{d}_{0k}|^2\, n(\tilde{\omega}_{k0})\,. \tag{7.136}$$

Finally, our dynamical analysis has also revealed that the atomic energy levels (7.128) and transition frequencies (5.65) acquire shifts due to the interaction of the atom with the electromagnetic field. These shifts are given by (7.118) together with (7.120). As in the zero-temperature case, we have assumed that the infinite zero-temperature shift associated with the free-space Green's tensor is already included in the bare transition frequencies ω_{mn}. Its contribution has thus been removed from (7.120) by making the replacement $\boldsymbol{G} \mapsto \boldsymbol{G}^{(1)}$. On the contrary, the thermal free-space shift is finite and has been retained. By using (A.26) for the free-space Green's tensor and employing the integration contour from Fig. 7.1, equation (7.120) for the frequency shifts can be brought into the more explicit form

$$\begin{aligned}\delta\omega_{nk} = &-\frac{\mu_0}{\hbar}\{\Theta(\tilde{\omega}_{nk})[n(\tilde{\omega}_{nk})+1] - \Theta(\tilde{\omega}_{kn})n(\tilde{\omega}_{kn})\}\tilde{\omega}_{nk}^2\\ &\times \boldsymbol{d}_{nk}\cdot\mathrm{Re}\,\boldsymbol{G}^{(1)}(\boldsymbol{r}_A,\boldsymbol{r}_A,\tilde{\omega}_{nk})\cdot\boldsymbol{d}_{kn}\\ &+\frac{2\mu_0 k_B T}{\hbar^2}\sum_{j=0}^{\infty}{}'\,\tilde{\omega}_{kn}\xi_j^2\,\frac{\boldsymbol{d}_{nk}\cdot\boldsymbol{G}^{(1)}(\boldsymbol{r}_A,\boldsymbol{r}_A,\xi_j)\cdot\boldsymbol{d}_{kn}}{\tilde{\omega}_{kn}^2+\xi_j^2}\\ &+\frac{\mu_0|\boldsymbol{d}_{nk}|^2}{6\pi^2 c\hbar}\mathcal{P}\int_0^{\infty}\mathrm{d}\omega\,\omega^3\left[\frac{n(\omega)}{\tilde{\omega}_{nk}-\omega}+\frac{n(\omega)}{\tilde{\omega}_{nk}+\omega}\right].\end{aligned} \tag{7.137}$$

The first term represents the absorption, spontaneous or stimulated emission of real photons which are reflected off the surfaces of present bodies. The second term is due to reflected virtual photons. The third, position-independent contribution is due to real, thermal photons which propagate without reflection.

As in the zero-temperature case (5.78), the shifts have to be obtained from a self-consistent solution of the equations (7.137), since the shifted frequencies appear on the right hand side. his complication does not arise in the perturbative limit $\delta\omega_m, \delta\omega_n \ll \omega_{mn}$ where we have

$$\delta\omega_{nk} = -\frac{\mu_0}{\hbar}\{\Theta(\omega_{nk})[n(\omega_{nk})+1] - \Theta(\omega_{kn})n(\omega_{kn})\}\omega_{nk}^2$$
$$\times\, \boldsymbol{d}_{nk}\cdot\mathrm{Re}\boldsymbol{G}^{(1)}(\boldsymbol{r}_A,\boldsymbol{r}_A,\omega_{nk})\cdot\boldsymbol{d}_{kn}$$
$$+\frac{2\mu_0 k_{\mathrm{B}}T}{\hbar^2}\sum_{j=0}^{\infty}{}'\omega_{kn}\xi_j^2\,\frac{\boldsymbol{d}_{nk}\cdot\boldsymbol{G}^{(1)}(\boldsymbol{r}_A,\boldsymbol{r}_A,\xi_j)\cdot\boldsymbol{d}_{kn}}{\omega_{kn}^2+\xi_j^2}$$
$$+\frac{\mu_0|\boldsymbol{d}_{nk}|^2}{6\pi^2 c\hbar}\mathcal{P}\int_0^\infty \mathrm{d}\omega\,\omega^3\left[\frac{n(\omega)}{\omega_{nk}-\omega}+\frac{n(\omega)}{\omega_{nk}+\omega}\right]. \tag{7.138}$$

when neglecting the position-independent last term, the perturbative energy shift $\Delta E_n = \hbar\delta\omega_n = \sum_k \hbar\delta\omega_{nk}$ coincides with the thermal CP potential (7.17)–(7.19) as derived in Sect. 7.1.

7.3.2 Casimir–Polder Force

Following the method outlined in Chap. 5, the thermal CP force can be found by evaluating the average Lorentz force acting on the atom. Note that we again neglect the effect of atomic motion [$\boldsymbol{r}_A(t) \equiv \boldsymbol{r}_A$]. Our result (5.94) for the average force on a non-magnetic atom is valid for arbitrary initial preparations of the atom and the electromagnetic field. It contains a contribution from the electric field and one from the magnetic field, where the latter involves a total time derivative. For zero temperature, we have seen that this latter term is only relevant for atoms in coherent superpositions of energy eigenstates, cf. (5.119) and (5.120). For simplicity, we assume the atom to be initially prepared in an incoherent superposition of energy eigenstates. Our results (7.116) and (7.117) from the previous section show that it will then remain in an incoherent superposition at all times. The second term in (5.94) can then be ignored and the average Lorentz force is given by

$$\boldsymbol{F}(t) = \sum_{m,n}\int_0^\infty \mathrm{d}\omega\big[\nabla\langle\hat{A}_{mn}(t)\boldsymbol{d}_{mn}\cdot\hat{\underline{\boldsymbol{E}}}(\boldsymbol{r},\omega,t)\rangle$$
$$+\nabla\langle\hat{\underline{\boldsymbol{E}}}^\dagger(\boldsymbol{r},\omega,t)\cdot\boldsymbol{d}_{mn}\hat{A}_{mn}(t)\rangle\big]_{\boldsymbol{r}=\boldsymbol{r}_A}. \tag{7.139}$$

Substituting the formal solution (5.44) for the time-dependent electric field, we find

$$\boldsymbol{F}(t) = \sum_{m,n}\int_0^\infty \mathrm{d}\omega\,\mathrm{e}^{-\mathrm{i}\omega(t-t_0)}\big[\nabla\langle\hat{A}_{mn}(t)\boldsymbol{d}_{mn}\cdot\hat{\underline{\boldsymbol{E}}}(\boldsymbol{r},\omega)\rangle\big]_{\boldsymbol{r}=\boldsymbol{r}_A}$$
$$+\frac{\mathrm{i}\mu_0}{\pi}\sum_{m,n,k,l}\int_0^\infty \mathrm{d}\omega\,\omega^2\nabla\boldsymbol{d}_{mn}\cdot\mathrm{Im}\,\boldsymbol{G}(\boldsymbol{r}_A,\boldsymbol{r}_A,\omega)\cdot\boldsymbol{d}_{kl}$$

$$\times \int_{t_0}^{t} \mathrm{d}t' \mathrm{e}^{-\mathrm{i}\omega(t-t')} \langle \hat{A}_{mn}(t)\hat{A}_{kl}(t')\rangle + \mathrm{C.c.} \tag{7.140}$$

To account for the influence of the thermal electromagnetic fields, we require an expression that is quadratic in the free fields. To that and, we use (7.117) and improve our approximate solution (7.111) for the atomic flip operators,

$$\begin{aligned}\hat{A}_{mn}(t) = {} & \mathrm{e}^{[\mathrm{i}\tilde{\omega}_{mn}-(\Gamma_m+\Gamma_n)/2](t-t_0)}\hat{A}_{mn} \\ & + \frac{\mathrm{i}}{\hbar}\sum_k \int_0^\infty \mathrm{d}\omega \int_{t_0}^{t} \mathrm{d}t'\, \mathrm{e}^{[\mathrm{i}\tilde{\omega}_{mn}-(\Gamma_m+\Gamma_n)/2](t-t')} \\ & \times \left\{ \mathrm{e}^{-\mathrm{i}\omega(t'-t_0)} \left[\hat{A}_{mk}(t')\boldsymbol{d}_{nk} - \hat{A}_{kn}(t')\boldsymbol{d}_{km}\right]\cdot \underline{\hat{\boldsymbol{E}}}(\boldsymbol{r}_A,\omega) \right. \\ & \left. + \mathrm{e}^{\mathrm{i}\omega(t'-t_0)} \underline{\hat{\boldsymbol{E}}}^\dagger(\boldsymbol{r}_A,\omega)\cdot\left[\boldsymbol{d}_{nk}\hat{A}_{mk}(t') - \boldsymbol{d}_{km}\hat{A}_{kn}(t')\right]\right\} \quad (m\neq n)\,.\end{aligned} \tag{7.141}$$

We substitute it into the first term (7.140) in above. Terms linear in $\underline{\hat{\boldsymbol{E}}}$, $\underline{\hat{\boldsymbol{E}}}^\dagger$ do not contribute for a thermal electromagnetic field, so we find

$$\begin{aligned}\boldsymbol{F}(t) = {} & \frac{\mathrm{i}}{\hbar}\sum_{m,n,k}\int_0^\infty \mathrm{d}\omega \int_0^\infty \mathrm{d}\omega' \int_{t_0}^{t} \mathrm{d}t'\, \mathrm{e}^{-\mathrm{i}\omega(t-t_0)+\mathrm{i}\omega'(t'-t_0)} \mathrm{e}^{[\mathrm{i}\tilde{\omega}_{mn}-(\Gamma_m+\Gamma_n)/2](t-t')} \\ & \times \left[\nabla\left\langle\left[\boldsymbol{d}_{nk}\hat{A}_{mk}(t') - \boldsymbol{d}_{km}\hat{A}_{kn}(t')\right]\cdot\underline{\hat{\boldsymbol{E}}}^\dagger(\boldsymbol{r}_A,\omega)\boldsymbol{d}_{mn}\cdot\underline{\hat{\boldsymbol{E}}}(\boldsymbol{r},\omega)\right\rangle\right]_{\boldsymbol{r}=\boldsymbol{r}_A} \\ & + \frac{\mathrm{i}\mu_0}{\pi}\sum_{m,n,k,l}\int_0^\infty \mathrm{d}\omega\, \omega^2 \nabla \boldsymbol{d}_{mn}\cdot \mathrm{Im}\,\boldsymbol{G}(\boldsymbol{r}_A,\boldsymbol{r}_A,\omega)\cdot\boldsymbol{d}_{kl} \\ & \times \int_{t_0}^{t} \mathrm{d}t' \mathrm{e}^{-\mathrm{i}\omega(t-t')} \langle \hat{A}_{mn}(t)\hat{A}_{kl}(t')\rangle + \mathrm{C.c.}\end{aligned} \tag{7.142}$$

Next, we need to evaluate the expectation values. The thermal averages of the electric field featuring in the first term above are given in (7.113). To find the atomic two-time correlation functions governing in the second term, we note that the equations of motion (7.117) for the off-diagonal flip operators have exactly the same form (5.64) as in the zero-temperature case. Using the quantum regression theorem (5.69) it follows that their correlation functions are given by (5.70). With these results, the average Lorentz force reads

$$
\begin{aligned}
\boldsymbol{F}(\boldsymbol{r}_A, t) = \frac{\mathrm{i}\mu_0}{\pi} \sum_{n,k} \int_0^\infty \mathrm{d}\omega\, \omega^2 \nabla \boldsymbol{d}_{nk} \cdot \operatorname{Im} \boldsymbol{G}^{(1)}(\boldsymbol{r}_A, \boldsymbol{r}_A, \omega) \cdot \boldsymbol{d}_{kn} \\
\times \int_{t_0}^{t} \mathrm{d}t' \left\{ [n(\omega) + 1] \mathrm{e}^{[-\mathrm{i}(\omega - \tilde{\omega}_{nk}) - (\Gamma_n + \Gamma_k)/2](t - t')} \right. \\
\left. + n(\omega) \mathrm{e}^{[\mathrm{i}(\omega + \tilde{\omega}_{nk}) - (\Gamma_n + \Gamma_k)/2](t - t')} \right\} \langle \hat{A}_{nn}(t') \rangle + \text{C.c.}
\end{aligned}
\tag{7.143}
$$

The time integral can be evaluated via the Markov approximation (5.99). Using $\langle \hat{A}_{nn}(t) \rangle = p_n(t)$, the thermal CP force reads [2, 4, 36]

$$
\boldsymbol{F}(\boldsymbol{r}_A, t) = \sum_n p_n(t) \boldsymbol{F}_n(\boldsymbol{r}_A) \tag{7.144}
$$

with

$$
\begin{aligned}
\boldsymbol{F}_n(\boldsymbol{r}_A) = \frac{\mu_0}{\pi} \sum_k \int_0^\infty \mathrm{d}\omega\, \omega^2\, \nabla \boldsymbol{d}_{nk} \cdot \operatorname{Im} \boldsymbol{G}^{(1)}(\boldsymbol{r}_A, \boldsymbol{r}_A, \omega) \cdot \boldsymbol{d}_{kn} \\
\times \left[\frac{n(\omega) + 1}{\omega - \tilde{\omega}_{nk} - \frac{\mathrm{i}}{2}(\Gamma_n + \Gamma_k)} - \frac{n(\omega)}{\omega - \tilde{\omega}_{kn} + \frac{\mathrm{i}}{2}(\Gamma_n + \Gamma_k)} \right] + \text{C.c.}
\end{aligned}
\tag{7.145}
$$

As in the zero-temperature case, the force on an atom in an incoherent superposition state is a sum over force components, weighted by the populations of the respective energy eigenstates. Note that we have discarded self-forces by making the replacement $\boldsymbol{G} \mapsto \boldsymbol{G}^{(1)}$.

The force components can be cast into a more explicit form by using contour-integral techniques. Writing $\operatorname{Im} \boldsymbol{G} = (\boldsymbol{G} - \boldsymbol{G}^*)/(2\mathrm{i})$ and employing the Schwarz reflection principle (A.3), we apply the integration contour depicted in Fig. 7.8. It transforms integrals along the real frequency axis into Matsubara sums plus contributions from the atomic resonances. The force components can then be written as a sum (5.103) of non-resonant contributions

$$
\begin{aligned}
\boldsymbol{F}_n^{\mathrm{nres}}(\boldsymbol{r}_A) = -\frac{\mu_0 k_{\mathrm{B}} T}{2\pi} \sum_{j=0}^{\infty}{}' \xi_j^2 \\
\times \nabla \operatorname{tr}\left\{ [\boldsymbol{\alpha}_n(\mathrm{i}\xi_j) + \boldsymbol{\alpha}_n(-\mathrm{i}\xi_j)] \cdot \boldsymbol{G}^{(1)}(\boldsymbol{r}, \boldsymbol{r}, \mathrm{i}\xi_j) \right\} \Big|_{\boldsymbol{r} = \boldsymbol{r}_A}
\end{aligned}
\tag{7.146}
$$

due to virtual photons and resonant contributions

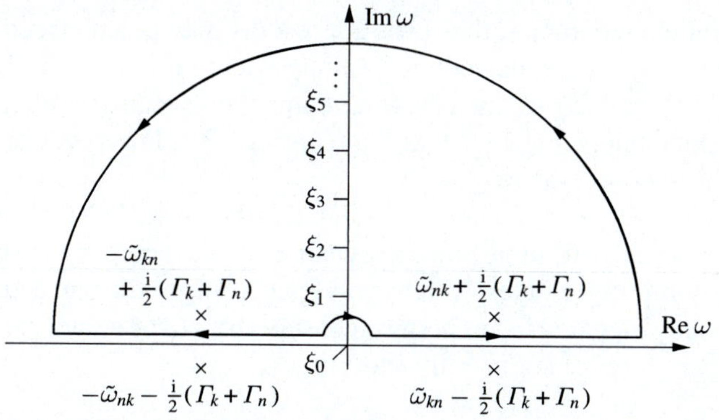

Fig. 7.8 Integration contour used for transforming the real-frequency CP integral into a Matsubara sum plus contributions from the atomic poles

$$\boldsymbol{F}_n^{\mathrm{res}}(\boldsymbol{r}_A) = \frac{\mu_0}{2}\sum_{k<n}[n(\Omega_{nk})+1]\Omega_{nk}^2 \nabla \boldsymbol{d}_{nk}\cdot \boldsymbol{G}^{(1)}(\boldsymbol{r},\boldsymbol{r},\Omega_{nk})\cdot \boldsymbol{d}_{kn}\big|_{\boldsymbol{r}=\boldsymbol{r}_A} - \frac{\mu_0}{2}\sum_{k>n} n(\Omega_{kn})\Omega_{kn}^2 \nabla \boldsymbol{d}_{nk}\cdot \boldsymbol{G}^{(1)}(\boldsymbol{r},\boldsymbol{r},\Omega_{kn})\cdot \boldsymbol{d}_{kn}\big|_{\boldsymbol{r}=\boldsymbol{r}_A} + \mathrm{C.c.} \tag{7.147}$$

from real photons. Here, we have used (5.107) to introduce a total derivative, assuming the atom to be time-reversal invariant with real dipole-matrix elements, $\boldsymbol{d}_{kn} = \boldsymbol{d}_{nk}$. The definitions (5.91) and (5.106) of atomic polarisability and complex transition frequencies are now given in terms of the thermal frequency shifts (7.137) and widths (7.121).

Comparing the thermal CP force with the respective zero-temperature results, we see that the integral in the non-resonant force (5.108) has simply been replaced by a Matsubara sum (7.20) at finite temperature in accordance with the replacement rule (7.24). For the resonant forces, contributions (5.109) due to the spontaneous emission of real photons have been augmented by stimulated emission. In addition, a contribution due to the absorption of thermal photons has appeared which carries an opposite sign. These changes due to the presence of thermal photons may be summarised in a replacement rule

$$\sum_{k<n} f(\Omega_{nk}) \mapsto \sum_{k<n}[n(\Omega_{nk})+1]f(\Omega_{nk}) - \sum_{k>n} n(\Omega_{kn})f(\Omega_{kn})\,. \tag{7.148}$$

As in the zero-temperature case, the thermal CP force is seen to depend on the shifted and broadened atomic transition frequencies. These effects have an impact as outlined in Sect. 5.5 for zero temperature. The shifts affect both non-resonant and resonant force components whereas the widths only have an influence on the resonant forces. At finite temperature, the shifts and widths may be enhanced. In particular, even the atomic ground state acquires a finite width at $T \neq 0$. In the perturbative limit $|\delta\omega_n|$, $|\delta\omega_k|$, Γ_n, $\Gamma_k \ll \omega_{nk}$, the force components reduce to the gradients of the thermal CP potential (7.17)–(7.19) as found in Sect. 7.1. In accordance with (1.119), the force is then conservative.

Most importantly, our dynamical calculation has revealed that the thermal CP force is time-dependent, in general. It dynamics is governed by the populations of the internal atomic eigenstates, which evolve according to the rate equations (7.122). An atom initially prepared in an energy eigenstate $|n\rangle$ will be subject to the respective, state-dependent force at sufficiently short times,

$$\boldsymbol{F}(\boldsymbol{r}_A, t) \simeq \boldsymbol{F}(\boldsymbol{r}_A, t_0) = \boldsymbol{F}_n(\boldsymbol{r}_A) \quad \text{for } (t - t_0)\Gamma_n \ll 1\,. \tag{7.149}$$

Due to photon emission and absorption, both higher and lower energy levels become populated, leading to a superposition force

$$\boldsymbol{F}(\boldsymbol{r}_A, t) = \sum_k p_k(t) \boldsymbol{F}_k(\boldsymbol{r}_A) \tag{7.150}$$

at intermediate times. In the long-time limit, the atom will reach thermal equilibrium with its environment such that its populations are given by (7.59). The thermal CP force is then purely non-resonant

$$\boldsymbol{F}(\boldsymbol{r}_A, t \to \infty) = -\frac{\mu_0 k_\mathrm{B} T}{\pi} \sum_{j=0}^{\infty}{}' \xi_j^2 \nabla_A \mathrm{tr}\big[\boldsymbol{\alpha}_T(\mathrm{i}\xi_j) \cdot \boldsymbol{G}^{(1)}(\boldsymbol{r}_A, \boldsymbol{r}_A, \mathrm{i}\xi_j)\big]\,. \tag{7.151}$$

7.3.3 Molecule in Front of a Plate

To illustrate the dynamics of the thermal CP force, let us once more consider an molecule in front of a semi-infinite electric half space. We are going to neglect the influence of the frequency shifts and broadenings on the force, so that our perturbative results (7.81), (7.93) and (7.97) from Sect. 7.2.2 remain valid.

The transition rates (7.121) governing the internal dynamics can be found by using the free-space Green's tensor (A.26) and the scattering Green's tensor (4.58). They read

$$\Gamma_{nk}(z_A) = \Gamma_{nk}^{(0)} + \Gamma_{nk}^{(1)}(z_A)\,, \tag{7.152}$$

where the free-space rates $\Gamma_{nk}^{(0)}$ are given by (7.131) and the plate-induced rates read

$$\begin{aligned}\Gamma_{nk}^{(1)}(z_A) = \frac{\mu_0}{4\pi\hbar}[n(\omega_{nk})+1]\omega_{nk}^2\Bigg(\int_0^{\omega_{nk}/c} \mathrm{d}k^\perp \bigg\{|\boldsymbol{d}_{nk}^\parallel|^2 \mathrm{Re}\bigg[\mathrm{e}^{2\mathrm{i}k^\perp z_A}\frac{k^\perp - k_1^\perp}{k^\perp + k_1^\perp}\bigg] \\ - \bigg[\frac{k^{\perp 2}c^2}{\omega_{nk}^2}|\boldsymbol{d}_{nk}^\parallel|^2 + \bigg(2\frac{k^{\perp 2}c^2}{\omega_{nk}^2} - 2\bigg)|\boldsymbol{d}_{nk}^\perp|^2\bigg] \\ \times \mathrm{Re}\bigg[\mathrm{e}^{2\mathrm{i}k^\perp z_A}\frac{\varepsilon(\omega_{nk})k^\perp - k_1^\perp}{\varepsilon(\omega_{nk})k^\perp + k_1^\perp}\bigg]\bigg\} \\ + \int_0^\infty \mathrm{d}\kappa^\perp\, \mathrm{e}^{-2\kappa^\perp z_A}\bigg\{|\boldsymbol{d}_{nk}^\parallel|^2 \mathrm{Im}\bigg[\frac{\kappa^\perp - \kappa_1^\perp}{\kappa^\perp + \kappa_1^\perp}\bigg] + \bigg[\frac{\kappa^{\perp 2}c^2}{\omega_{nk}^2}|\boldsymbol{d}_{nk}^\parallel|^2 \\ + \bigg(2\frac{\kappa^{\perp 2}c^2}{\omega_{nk}^2} + 2\bigg)|\boldsymbol{d}_{nk}^\perp|^2\bigg]\mathrm{Im}\bigg[\frac{\varepsilon(\omega_{nk})\kappa^\perp - \kappa_1^\perp}{\varepsilon(\omega_{nk})\kappa^\perp + \kappa_1^\perp}\bigg]\bigg\}\end{aligned} \tag{7.153}$$

for $k < n$ and

$$\begin{aligned}\Gamma_{nk}^{(1)}(z_A) = \frac{\mu_0}{4\pi\hbar}n(\omega_{kn})\omega_{kn}^2\Bigg(\int_0^{\omega_{kn}/c} \mathrm{d}k^\perp \bigg\{|\boldsymbol{d}_{nk}^\parallel|^2 \mathrm{Re}\bigg[\mathrm{e}^{2\mathrm{i}k^\perp z_A}\frac{k^\perp - k_1^\perp}{k^\perp + k_1^\perp}\bigg] \\ - \bigg[\frac{k^{\perp 2}c^2}{\omega_{kn}^2}|\boldsymbol{d}_{nk}^\parallel|^2 + \bigg(2\frac{k^{\perp 2}c^2}{\omega_{kn}^2} - 2\bigg)|\boldsymbol{d}_{nk}^\perp|^2\bigg] \\ \times \mathrm{Re}\bigg[\mathrm{e}^{2\mathrm{i}k^\perp z_A}\frac{\varepsilon(\omega_{kn})k^\perp - k_1^\perp}{\varepsilon(\omega_{kn})k^\perp + k_1^\perp}\bigg]\bigg\} \\ + \int_0^\infty \mathrm{d}\kappa^\perp\, \mathrm{e}^{-2\kappa^\perp z_A}\bigg\{|\boldsymbol{d}_{nk}^\parallel|^2 \mathrm{Im}\bigg[\frac{\kappa^\perp - \kappa_1^\perp}{\kappa^\perp + \kappa_1^\perp}\bigg] + \bigg[\frac{\kappa^{\perp 2}c^2}{\omega_{kn}^2}|\boldsymbol{d}_{nk}^\parallel|^2 \\ + \bigg(2\frac{\kappa^{\perp 2}c^2}{\omega_{kn}^2} + 2\bigg)|\boldsymbol{d}_{nk}^\perp|^2\bigg]\mathrm{Im}\bigg[\frac{\varepsilon(\omega_{kn})\kappa^\perp - \kappa_1^\perp}{\varepsilon(\omega_{kn})\kappa^\perp + \kappa_1^\perp}\bigg]\bigg\}\end{aligned} \tag{7.154}$$

for $k > n$, recall (7.94) and (7.98). The two integrals represent contributions from propagating and evanescent waves, respectively.

At retarded distances $z_A \gg c/|\omega_{nk}|$, propagating waves dominate, and the approximations $k^\perp \simeq |\omega_{nk}|/c$ and $k_1^\perp \simeq \sqrt{\varepsilon(|\omega_{nk}|)}|\omega_{nk}|/c$ lead to

$$\Gamma_{nk}^{(1)}(z_A) = -\frac{\mu_0}{4\pi\hbar z_A}[n(\omega_{nk})+1]\,\omega_{nk}^2|\boldsymbol{d}_{nk}^\parallel|^2 \mathrm{Im}\bigg[\mathrm{e}^{2\mathrm{i}|\omega_{nk}|z_A/c}\frac{\sqrt{\varepsilon(\omega_{nk})}-1}{\sqrt{\varepsilon(\omega_{nk})}+1}\bigg] \tag{7.155}$$

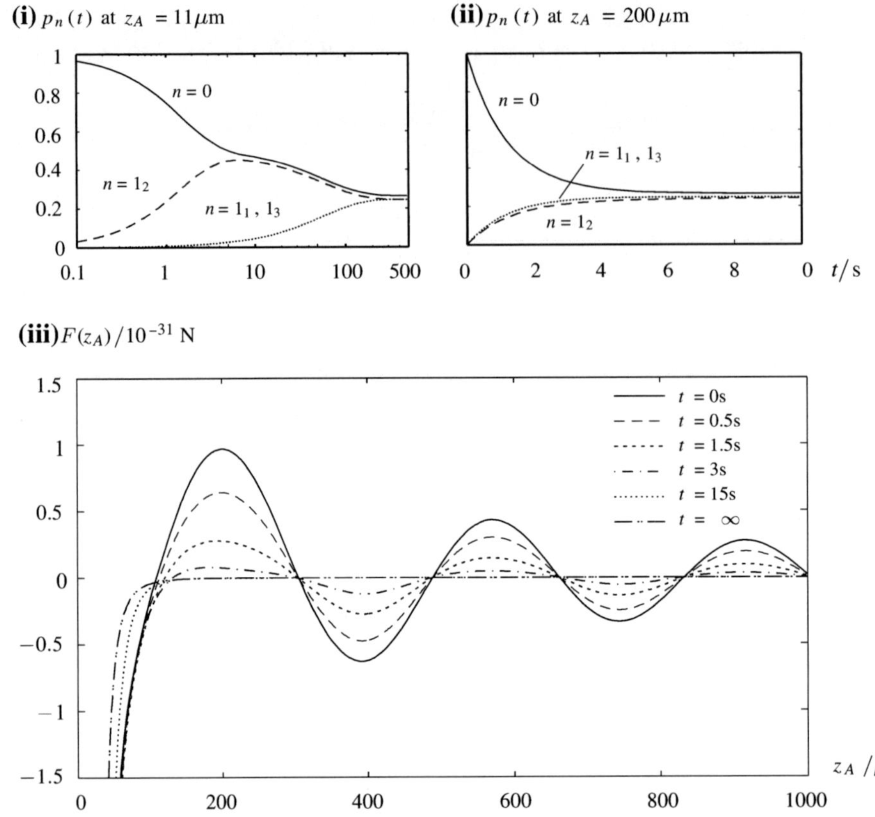

Fig. 7.9 (**i, ii**) Time-dependent internal-state populations of a LiH molecule at fixed distances from an Au half space at $T = 300\,\mathrm{K}$. (**iii**) Dynamics of the thermal CP force on a LiH molecule in front of an Au half space

for $k < n$ and

$$\Gamma_{nk}^{(1)}(z_A) = \frac{\mu_0}{4\pi\hbar z_A}\, n(\omega_{kn})\, \omega_{kn}^2 |\boldsymbol{d}_{nk}^{\parallel}|^2 \mathrm{Im}\left[\mathrm{e}^{2\mathrm{i}|\omega_{kn}|z_A/c}\, \frac{\sqrt{\varepsilon(\omega_{kn})}-1}{\sqrt{\varepsilon(\omega_{kn})}+1}\right]. \tag{7.156}$$

for $k > n$. In this limit, the plate-induced rate is a spatially oscillating correction to the free-space rate. At nonretarded distances $z_A \ll c/|\omega_{nk}|$, the estimates $\mathrm{e}^{2\mathrm{i}k^{\perp}z_A} \simeq 1$, $k_1^{\perp} \simeq k^{\perp}$ and $\kappa_1^{\perp} \simeq \kappa^{\perp}$ result in

$$\Gamma_{nk}^{(1)}(z_A) = \frac{1}{8\pi\hbar\varepsilon_0 z_A^3}\, [n(\omega_{nk})+1]\big(|\boldsymbol{d}_{nk}^{\parallel}|^2 + 2|\boldsymbol{d}_{nk}^{\perp}|^2\big)\frac{\mathrm{Im}\,\varepsilon(\omega_{nk})}{|\varepsilon(\omega_{nk})+1|^2} \tag{7.157}$$

for $k < n$ and

$$\Gamma_{nk}^{(1)}(z_A) = \frac{1}{8\pi\hbar\varepsilon_0 z_A^3} n(\omega_{kn})\big(|d_{nk}^{\parallel}|^2 + 2|d_{nk}^{\perp}|^2\big)\frac{\mathrm{Im}\,\varepsilon(\omega_{kn})}{|\varepsilon(\omega_{kn})+1|^2} \tag{7.158}$$

for $k > n$. They hence dominate the internal dynamics of the molecule at short distances.

As an example, let us once more consider the thermal CP force on a LiH molecule in front of a room-temperature Au surface ($T = 300$ K). Using the parameters of Au and LiH as listed below (7.103), we can evaluate the transition rates (7.152) together with (7.131) and (7.153). Assuming the molecule to be prepared in its ground state $|0\rangle$ at initial time, we can then solve the rate equations (7.122) to find the time-dependent state populations $p_n(t)$ for given atom–surface separations. The results are shown in Fig. 7.9(i, ii). At $z_A = 200\,\mu$m, the dynamics is dominated by the free-space transition rates. As seen from Fig. 7.9(ii), the molecule reaches thermal equilibrium on a time-scale of about 10 s, where the three degenerate excited states have become equally populated. The situation is different at $z_A = 11\,\mu$m. At this distance, the rates $\Gamma_{01_1}(z_A) = \Gamma_{01_3}(z_A)$ are strongly reduced due to a negative contribution from the spatially oscillating plate-induced rate. On the contrary, the $\Gamma_{01_2}(z_A)$ is not affected by spatial oscillations and takes a larger value. As we can see from Fig. 7.9(i), the population of the state $|1_2\rangle$ hence reaches thermal equilibrium with the ground-state population on a time scale of 10 s, while the thermalisation of the other two excited states $|1_1\rangle$, $|1_1\rangle$ sets in much later after about 500 s. As a result of the dynamics in an anisotropic environment, the molecule is hence in an anisotropic state at intermediate times.

The resulting dynamics of the thermal CP force is shown in Fig. 7.9(iii) as obtained from (7.144) together with (7.81), (7.93) and (7.97). At initial time, the force exhibits spatial oscillations due to the absorption of propagating thermal photons. These oscillations die out as the molecule reaches thermal equilibrium with its environment, leaving a purely attractive equilibrium force.

References

1. T. Matsubara, Prog. Theor. Phys. **14**(4), 351 (1955)
2. S.Y. Buhmann, S. Scheel, Phys. Rev. Lett. **100**(25), 253201 (2008)
3. S. Scheel, S.Y. Buhmann, Acta Phys. Slovaca **58**(5), 675 (2008)
4. S.Y. Buhmann, S. Scheel, Phys. Scripta T **135**, 014013 (2008)
5. M.P. Gorza, M. Ducloy, Eur. Phys. J. D **40**(3), 343 (2006)
6. A.D. McLachlan, Proc. R. Soc. Lond. Ser. A **274**(1356), 80 (1963)
7. C. Henkel, K. Joulain, J.P. Mulet, J.J. Greffet, J. Opt. A: Pure Appl. Opt. **4**(5), S109 (2002)
8. G.H. Hardy, *Divergent Series* (Oxford University Press, London, 1963)
9. J.P. Dowling, Math. Mag. **62**(5), 324 (1989)
10. B.W. Ninham, J. Daicic, Phys. Rev. A **57**(3), 1870 (1998)
11. H. Wennerström, J. Daicic, B.W. Ninham, Phys. Rev. A **60**(3), 2581 (1999)
12. C. Eberlein, R. Zietal, Phys. Rev. A **75**(3), 032516 (2007)

13. V.M. Nabutovskiĭ, V.R. Belosludov, A.M. Korotkikh, Sov. Phys. JETP **50**, 352 (1979)
14. E.V. Blagov, G.L. Klimchitskaya, V.M. Mostepanenko, Phys. Rev. B **71**(23), 235401 (2005)
15. G.L. Klimchitskaya, E.V. Blagov, V.M. Mostepanenko, J. Phys. A: Math. Gen. 39(21), 6481 (2006)
16. E.V. Blagov, G.L. Klimchitskaya, V.M. Mostepanenko, Phys. Rev. B **75**(23), 235413 (2007)
17. T. Nakajima, P. Lambropoulos, H. Walther, Phys. Rev. A **56**(6), 5100 (1997)
18. S.Å. Ellingsen, S.Y. Buhmann, S. Scheel, Phys. Rev. A **80**(2), 022901 (2009)
19. S.Å. Ellingsen, Y. Sherkunov, S.Y. Buhmann, S. Scheel, in *Proceedings of the Ninth Conference on Quantum Field Theory Under the Influence of External Conditions (QFEXT09)*, ed. by K.A. Milton, M. Bordag (World Scientific, New Jersey, 2010), pp. 168–177
20. S.Å. Ellingsen, S.Y. Buhmann, S. Scheel, Phys. Rev. A **82**(3), 032516 (2010)
21. S.Å. Ellingsen, S.Y. Buhmann, S. Scheel, Phys. Rev. Lett. **104**(22), 223003 (2010)
22. T.H. Boyer, Phys. Rev. A **11**(5), 1650 (1975)
23. M. Boström, B.E. Sernelius, Phys. Rev. A **61**(5), 052703 (2000)
24. M. Bordag, B. Geyer, G.L. Klimchitskaya, V.M. Mostepanenko, Phys. Rev. B **74**(20), 205431 (2006)
25. M. Boström, B.E. Sernelius, Phys. Rev. Lett. **84**(20), 4757 (2000)
26. E.M. Lifshitz, Sov. Phys. JETP **2**(1), 73 (1956)
27. I.E. Dzyaloshinskii, E.M. Lifshitz, L.P. Pitaevskii, Adv. Phys. **10**(38), 165 (1961)
28. M.S. Tomaš, Phys. Rev. A **72**(3), 034104 (2005)
29. S.Å. Ellingsen, S.Y. Buhmann, S. Scheel, Phys. Rev. A **79**(5), 052903 (2009)
30. S.T. Wu, C. Eberlein, Proc. R. Soc. Lond. Ser. A **456**(2000), 1931 (2000)
31. S.Y. Buhmann, M.R. Tarbutt, S. Scheel, E.A. Hinds, Phys. Rev. A **78**(5), 052901 (2008)
32. A. Lambrecht, S. Reynaud, Eur. Phys. J. D **8**(3), 309 (2000)
33. C.E. Moore, *Atomic Energy Levels: As Derived from the Analyses of Optical Spectra* (U.S. Government Printing Office, Washington, 1971)
34. A. Einstein, Phys. Z. **18**, 121 (1917)
35. R.C. Hilborn, Am. J. Phys. **50**(11), 982 (1982)
36. S.Y. Buhmann, S. Scheel, Phys. Scripta T **135**, 014013 (2008)

Chapter 8
Casimir–Polder Forces on Moving Atoms

So far, we have completely ignored the impact of motion on dispersion forces. Our description of the medium-assisted electromagnetic field is only valid for bodies at rest; the velocity-dependent Röntgen interaction has been discarded from atom–field couplings; and the distance travelled by an atom between photon emission and reabsorption has been disregarded. Strictly speaking, all of our results for the Casimir, CP and vdW forces are thus only valid for objects at rest.

In reality, the validity limits imposed by the disregard of atomic and body motion are a lot less severe. In an order-of-magnitude estimate, velocity-dependent dispersion forces should be smaller than the corresponding velocity-independent ones by a factor v/c, with v denoting the relative velocity of the interacting objects. Unless the objects move with relativistic speed, we may expect the velocity-independent results to be an excellent approximation.

A fully relativistic analysis is beyond the scope of this book. In particular, the constitutive relations between the electromagnetic fields and excitations as presented in Sect. 1.1.1 as well as the employed Coulomb gauge are not Lorentz-invariant. In addition, our description of an atom and its coupling to the electromagnetic field as given in Sect. 1.1.2 is valid only in a non-relativistic approximation. In a much more modest approach, we will instead develop leading-order non-relativistic account of velocity-dependent dispersion forces. This can be achieved most easily for the CP force on a single moving atom which interacts with bodies at rest. Our investigation of this model system help us identify scenarios where velocity-dependent forces might play a role and show in which cases they can be safely neglected.

To study the impact of motion on the CP force, we extend the dynamical approach developed in Chap. 5. We will begin by solving the atom–field dynamics for a slowly moving atom and proceed by calculating the CP force as the average Lorentz force. As an example, we study the quantum friction force on an atom moving parallel to a plate.

S. Y. Buhmann, *Dispersion Forces II*, Springer Tracts in Modern Physics 248,
DOI: 10.1007/978-3-642-32466-6_8, © Springer-Verlag Berlin Heidelberg 2012

8.1 Internal Atomic Dynamics

The impact of atomic motion on the atom–field interaction is most explicit in the multipolar coupling scheme where it is manifest in the Röntgen interaction. We will employ this scheme throughout this chapter, discarding the primes identifying multipolar variables. The multipolar interaction in long-wavelength approximation is given by (1.96) where the centre-of-mass momentum appears in the Röntgen interaction. It is related to the atomic velocity according via (5.17) which reduces to

$$\hat{\boldsymbol{p}}_A = m_A\dot{\hat{\boldsymbol{r}}}_A - \hat{\boldsymbol{d}}\times\hat{\boldsymbol{B}}(\hat{\boldsymbol{r}}_A) \tag{8.1}$$

in long-wavelength approximation. Substitution into (1.96) yields

$$\hat{H}_{\mathrm{AF}} = -\,\hat{\boldsymbol{d}}\cdot\hat{\boldsymbol{E}}(\hat{\boldsymbol{r}}_A) - \hat{\boldsymbol{m}}\cdot\hat{\boldsymbol{B}}(\hat{\boldsymbol{r}}_A) + \sum_{\alpha\in A}\frac{q_\alpha^2}{8m_\alpha}\big[\hat{\overline{\boldsymbol{r}}}_\alpha\times\hat{\boldsymbol{B}}(\hat{\boldsymbol{r}}_A)\big]^2 - \frac{5}{8m_A}\big[\hat{\boldsymbol{d}}\times\hat{\boldsymbol{B}}(\hat{\boldsymbol{r}}_A)\big]^2$$
$$+\,\dot{\hat{\boldsymbol{r}}}_A\cdot\hat{\boldsymbol{d}}\times\hat{\boldsymbol{B}}(\hat{\boldsymbol{r}}_A)\,. \tag{8.2}$$

For a non-magnetic atom, we may discard the magnetic dipole and diamagnetic interactions to obtain the electric-dipole Hamiltonian

$$\hat{H}_{\mathrm{AF}} = -\hat{\boldsymbol{d}}\cdot\hat{\boldsymbol{E}}(\hat{\boldsymbol{r}}_A) + \dot{\hat{\boldsymbol{r}}}_A\cdot\hat{\boldsymbol{d}}\times\hat{\boldsymbol{B}}(\hat{\boldsymbol{r}}_A)\,. \tag{8.3}$$

It now explicitly depends on the atomic velocity via the Röntgen term. We make use of the Born–Oppenheimer approximation by solving the atom–field dynamics for given centre-of-mass position $\boldsymbol{r}_A$ and velocity $\boldsymbol{v}$. Using the expansion (5.38) of the electric-dipole operator and introducing the atomic flip operators, the multipolar coupling Hamiltonian can be given in the form

$$\hat{H}_{\mathrm{AF}} = -\sum_{m,n}\hat{A}_{mn}\boldsymbol{d}_{mn}\cdot\hat{\boldsymbol{E}}(\boldsymbol{r}_A) + \sum_{m,n}\hat{A}_{mn}\boldsymbol{v}\cdot\boldsymbol{d}_{mn}\times\hat{\boldsymbol{B}}(\boldsymbol{r}_A)\,. \tag{8.4}$$

In close analogy to the treatment in Sect. 5.2, we can now derive and solve the equations governing the atom–field dynamics. Using the Hamiltonians (5.37) and (8.4) and the commutation relation (5.36), the Heisenberg equations for the atomic flip operators read

$$\dot{\hat{A}}_{mn} = \frac{1}{\mathrm{i}\hbar}\big[\hat{A}_{mn},\hat{H}\big]$$
$$= \mathrm{i}\omega_{mn}\hat{A}_{mn} + \frac{\mathrm{i}}{\hbar}\sum_k\big(\hat{A}_{mk}\boldsymbol{d}_{nk} - \hat{A}_{kn}\boldsymbol{d}_{km}\big)\cdot\big[\hat{\boldsymbol{E}}(\boldsymbol{r}_A) + \boldsymbol{v}\times\hat{\boldsymbol{B}}(\boldsymbol{r}_A)\big]\,. \tag{8.5}$$

The equation of motion for the fundamental fields follow from the Hamiltonians (1.93) and (8.4) by recalling the field expansions (1.22) and (1.26) and invoking the bosonic commutation relations (1.17) and (1.18),

$$\dot{\hat{\boldsymbol{f}}}_\lambda(\boldsymbol{r},\omega)=\frac{1}{\mathrm{i}\hbar}\big[\hat{\boldsymbol{f}}_\lambda(\boldsymbol{r},\omega),\hat{H}\big]=-\mathrm{i}\omega\hat{\boldsymbol{f}}_\lambda(\boldsymbol{r},\omega)+\frac{\mathrm{i}}{\hbar}\sum_{m,n}\boldsymbol{G}_\lambda^{*\mathrm{T}}(\boldsymbol{r}_A,\boldsymbol{r},\omega)\cdot\boldsymbol{d}_{mn}\hat{A}_{mn}$$
$$-\frac{1}{\hbar\omega}\sum_{m,n}\big[\boldsymbol{G}_\lambda^{*\mathrm{T}}(\boldsymbol{r},\boldsymbol{r}_A,\omega)\times\overleftarrow{\nabla}'\big]\times\boldsymbol{d}_{mn}\cdot\boldsymbol{v}\hat{A}_{mn} \tag{8.6}$$

where $\overleftarrow{\nabla}'$ is acting on the second position argument of the Green's tensor. This equation is solved by

$$\hat{\boldsymbol{f}}_\lambda(\boldsymbol{r},\omega,t)=\mathrm{e}^{-\mathrm{i}\omega(t-t_0)}\hat{\boldsymbol{f}}_\lambda(\boldsymbol{r},\omega)$$
$$+\frac{\mathrm{i}}{\hbar}\sum_{m,n}\int_{t_0}^{t}\mathrm{d}t'\,\mathrm{e}^{-\mathrm{i}\omega(t-t')}\boldsymbol{G}_\lambda^{*\mathrm{T}}[\boldsymbol{r}_A(t'),\boldsymbol{r},\omega]\cdot\hat{A}_{mn}(t')\boldsymbol{d}_{mn}$$
$$+\frac{1}{\hbar\omega}\sum_{m,n}\int_{t_0}^{t}\mathrm{d}t'\,\mathrm{e}^{-\mathrm{i}\omega(t-t')}\big\{\boldsymbol{G}_\lambda^{*\mathrm{T}}[\boldsymbol{r},\boldsymbol{r}_A(t'),\omega]\times\overleftarrow{\nabla}'\big\}\times\boldsymbol{d}_{mn}\cdot\boldsymbol{v}\hat{A}_{mn}(t')\,. \tag{8.7}$$

Assuming that the atom moves with non-relativistic speed, $v\ll c$, we are seeking a solution to the atom–field dynamics that is correct within linear order of v/c. According to the Born–Oppenheimer approximation, we may assume the centre-of-mass velocity to remain constant on the time-scale relevant for the internal atom–field dynamics, so that

$$\boldsymbol{r}_A(t')=\boldsymbol{r}_A(t)+(t'-t)\boldsymbol{v}\,. \tag{8.8}$$

Substituting this relation into the solution for the fundamental fields and discarding quadratic terms in v/c, we find $[\boldsymbol{r}_A(t)\equiv\boldsymbol{r}_A]$

$$\hat{\boldsymbol{f}}_\lambda(\boldsymbol{r},\omega,t)=\mathrm{e}^{-\mathrm{i}\omega(t-t_0)}\hat{\boldsymbol{f}}_\lambda(\boldsymbol{r},\omega)$$
$$+\frac{\mathrm{i}}{\hbar}\sum_{m,n}\int_{t_0}^{t}\mathrm{d}t'\,\mathrm{e}^{-\mathrm{i}\omega(t-t')}\boldsymbol{G}_\lambda^{*\mathrm{T}}(\boldsymbol{r}_A,\boldsymbol{r},\omega)\cdot\boldsymbol{d}_{mn}\hat{A}_{mn}(t')$$
$$-\frac{\mathrm{i}}{\hbar}\sum_{m,n}\int_{t_0}^{t}\mathrm{d}t'\,(t-t')\mathrm{e}^{-\mathrm{i}\omega(t-t')}\boldsymbol{G}_\lambda^{*\mathrm{T}}(\boldsymbol{r}_A,\boldsymbol{r},\omega)\cdot\boldsymbol{d}_{mn}\big(\overleftarrow{\nabla}'\cdot\boldsymbol{v}\big)\hat{A}_{mn}(t')$$

$$+\frac{1}{\hbar\omega}\sum_{m,n}\int_{t_0}^{t}\mathrm{d}t'\,\mathrm{e}^{-\mathrm{i}\omega(t-t')}\big[\boldsymbol{G}_\lambda^{*\mathrm{T}}(\boldsymbol{r},\boldsymbol{r}_A,\omega)\times\overleftarrow{\nabla}'\big]\times\boldsymbol{d}_{mn}\cdot\boldsymbol{v}\hat{A}_{mn}(t')\,. \tag{8.9}$$

Using the integral relation (1.25) for the Green's tensor, the electric field (1.22) is hence given by

$$\begin{aligned}\underline{\hat{\boldsymbol{E}}}(\boldsymbol{r},\omega,t) &= \mathrm{e}^{-\mathrm{i}\omega(t-t_0)}\underline{\hat{\boldsymbol{E}}}(\boldsymbol{r},\omega)\\ &+\frac{\mathrm{i}\mu_0}{\pi}\sum_{m,n}\omega^2\int_{t_0}^{t}\mathrm{d}t'\,\mathrm{e}^{-\mathrm{i}\omega(t-t')}\mathrm{Im}\,\boldsymbol{G}(\boldsymbol{r},\boldsymbol{r}_A,\omega)\cdot\boldsymbol{d}_{mn}\hat{A}_{mn}(t')\\ &-\frac{\mathrm{i}\mu_0}{\pi}\sum_{m,n}\omega^2\int_{t_0}^{t}\mathrm{d}t'\,(t-t')\mathrm{e}^{-\mathrm{i}\omega(t-t')}\mathrm{Im}\,\boldsymbol{G}(\boldsymbol{r},\boldsymbol{r}_A,\omega)\cdot\boldsymbol{d}_{mn}\big(\overleftarrow{\nabla}'\cdot\boldsymbol{v}\big)\hat{A}_{mn}(t')\\ &+\frac{\mu_0}{\pi}\sum_{m,n}\omega\int_{t_0}^{t}\mathrm{d}t'\,\mathrm{e}^{-\mathrm{i}\omega(t-t')}\big[\mathrm{Im}\,\boldsymbol{G}(\boldsymbol{r},\boldsymbol{r}_A,\omega)\times\overleftarrow{\nabla}'\big]\times\boldsymbol{d}_{mn}\cdot\boldsymbol{v}\hat{A}_{mn}(t')\,.\end{aligned} \tag{8.10}$$

In order to obtain a closed equation for the atomic flip operators, we also require an expression for the time-dependent magnetic field. However, since the magnetic field only appears in conjunction with a factor $\boldsymbol{v}$, the zero-order approximation in v/c is sufficient. Substituting the field operators (8.9) into the expansion (1.26), discarding all terms linear in v/c and making use of the integral relation (1.25), we find

$$\begin{aligned}\underline{\hat{\boldsymbol{B}}}(\boldsymbol{r},\omega,t) &= \mathrm{e}^{-\mathrm{i}\omega(t-t_0)}\underline{\hat{\boldsymbol{B}}}(\boldsymbol{r},\omega)\\ &+\frac{\mu_0}{\pi}\sum_{m,n}\omega\int_{t_0}^{t}\mathrm{d}t'\,\mathrm{e}^{-\mathrm{i}\omega(t-t')}\nabla\times\mathrm{Im}\,\boldsymbol{G}(\boldsymbol{r},\boldsymbol{r}_A,\omega)\cdot\boldsymbol{d}_{mn}\hat{A}_{mn}(t')\,.\end{aligned} \tag{8.11}$$

For weak atom–field coupling, the time integrals may be evaluated by means of the Markov approximation. We recall the result (5.47) from Sect. 5.2 and take its derivative with respect to ω to find

$$\int_{t_0}^{t}\mathrm{d}t'\,(t-t')\mathrm{e}^{-\mathrm{i}\omega(t-t')}\hat{A}_{mn}(t')\simeq\frac{\mathrm{d}}{\mathrm{d}\omega}\bigg[\frac{\mathcal{P}}{\omega-\tilde{\omega}_{nm}}+\mathrm{i}\pi\delta(\omega-\tilde{\omega}_{nm})\bigg]\hat{A}_{mn}(t)\,. \tag{8.12}$$

In Markov approximation, the electric and magnetic fields hence read

$$
\begin{aligned}
\hat{\underline{\boldsymbol{E}}}(\boldsymbol{r},\omega,t) &= \mathrm{e}^{-\mathrm{i}\omega(t-t_0)}\hat{\underline{\boldsymbol{E}}}(\boldsymbol{r},\omega) \\
&+ \mathrm{i}\mu_0\sum_{m,n}\left[\delta(\omega-\tilde{\omega}_{nm}) - \frac{\mathrm{i}}{\pi}\frac{\mathcal{P}}{\omega-\tilde{\omega}_{nm}}\right]\omega^2\mathrm{Im}\boldsymbol{G}(\boldsymbol{r},\boldsymbol{r}_A,\omega)\cdot\boldsymbol{d}_{mn}\hat{A}_{mn}(t) \\
&- \mathrm{i}\mu_0\sum_{m,n}\omega^2\frac{\mathrm{d}}{\mathrm{d}\omega}\left[\frac{1}{\pi}\frac{\mathcal{P}}{\omega-\tilde{\omega}_{nm}} + \mathrm{i}\delta(\omega-\tilde{\omega}_{nm})\right]\mathrm{Im}\boldsymbol{G}(\boldsymbol{r},\boldsymbol{r}_A,\omega)\cdot\boldsymbol{d}_{mn}\left(\overleftarrow{\nabla}'\cdot\boldsymbol{v}\right)\hat{A}_{mn}(t) \\
&+ \mu_0\sum_{m,n}\omega\left[\delta(\omega-\tilde{\omega}_{nm}) - \frac{\mathrm{i}}{\pi}\frac{\mathcal{P}}{\omega-\tilde{\omega}_{nm}}\right]\left[\mathrm{Im}\boldsymbol{G}(\boldsymbol{r},\boldsymbol{r}_A,\omega)\times\overleftarrow{\nabla}'\right]\times\boldsymbol{d}_{mn}\cdot\boldsymbol{v}\hat{A}_{mn}(t)
\end{aligned}
\tag{8.13}
$$

and

$$
\begin{aligned}
\hat{\underline{\boldsymbol{B}}}(\boldsymbol{r},\omega,t) &= \mathrm{e}^{-\mathrm{i}\omega(t-t_0)}\hat{\underline{\boldsymbol{B}}}(\boldsymbol{r},\omega) \\
&+ \mu_0\sum_{m,n}\omega\left[\delta(\omega-\tilde{\omega}_{nm}) - \frac{\mathrm{i}}{\pi}\frac{\mathcal{P}}{\omega-\tilde{\omega}_{nm}}\right]\nabla\times\mathrm{Im}\,\boldsymbol{G}(\boldsymbol{r},\boldsymbol{r}_A,\omega)\cdot\boldsymbol{d}_{mn}\hat{A}_{mn}(t)\,.
\end{aligned}
\tag{8.14}
$$

We arrange the equation of motion (8.5) for the atomic flip operators in normal ordering (note that the time-dependence is not shown for brevity)

$$
\begin{aligned}
\dot{\hat{A}}_{mn} &= \mathrm{i}\omega_{mn}\hat{A}_{mn} \\
&+ \frac{\mathrm{i}}{\hbar}\sum_k\int_0^\infty \mathrm{d}\omega\Big[\left(\hat{A}_{mk}\boldsymbol{d}_{nk} - \hat{A}_{kn}\boldsymbol{d}_{km}\right)\cdot\left[\hat{\underline{\boldsymbol{E}}}(\boldsymbol{r}_A,\omega) + \boldsymbol{v}\times\hat{\underline{\boldsymbol{B}}}(\boldsymbol{r}_A,\omega)\right] \\
&+ \left[\hat{\underline{\boldsymbol{E}}}^\dagger(\boldsymbol{r}_A,\omega) + \boldsymbol{v}\times\hat{\underline{\boldsymbol{B}}}^\dagger(\boldsymbol{r}_A,\omega)\right]\cdot\left(\boldsymbol{d}_{nk}\hat{A}_{mk} - \boldsymbol{d}_{km}\hat{A}_{kn}\right)\Big]\,,
\end{aligned}
\tag{8.15}
$$

substitute our solutions for the time-dependent electromagnetic field and evaluate operator products in accordance with (5.50). Taking expectation values and assuming the field to be prepared in its ground state at initial time, $\hat{\rho} = \hat{\rho}(t_0) = |\{0\}\rangle\langle\{0\}|$, the free fields do not contribute according to (5.53) and (5.96). We are left with an equation of the form (5.54), but the coefficients

$$
\boldsymbol{C}_{mn} = \boldsymbol{C}_{mn}(\boldsymbol{r}_A) + \boldsymbol{C}_{mn}(\boldsymbol{r}_A,\boldsymbol{v}) \tag{8.16}
$$

now contain a purely position-dependent component (5.52) as well as a position- and velocity-dependent correction

$$
\begin{aligned}
\boldsymbol{C}_{mn}(\boldsymbol{r}_A,\boldsymbol{v}) &= \frac{\mu_0}{\pi\hbar}\mathcal{P}\int_0^\infty\frac{\mathrm{d}\omega}{\omega-\tilde{\omega}_{nm}}\left[\omega^2\mathrm{Im}\,\boldsymbol{G}(\boldsymbol{r}_A,\boldsymbol{r}_A,\omega)\right]'\cdot\boldsymbol{d}_{mn}\left(\overleftarrow{\nabla}'\cdot\boldsymbol{v}\right) \\
&+ \frac{\mathrm{i}\mu_0}{\hbar}\Theta(\tilde{\omega}_{nm})\left[\tilde{\omega}_{nm}^2\mathrm{Im}\,\boldsymbol{G}(\boldsymbol{r}_A,\boldsymbol{r}_A,\tilde{\omega}_{nm})\right]'\cdot\boldsymbol{d}_{mn}\left(\overleftarrow{\nabla}'\cdot\boldsymbol{v}\right)
\end{aligned}
$$

$$
\begin{aligned}
&-\frac{\mu_0}{\pi\hbar}\mathcal{P}\int_0^\infty \frac{\mathrm{d}\omega}{\omega-\tilde{\omega}_{nm}}\,\omega\big[\mathrm{Im}\,\boldsymbol{G}(\boldsymbol{r}_A,\boldsymbol{r}_A,\omega)\times\overleftarrow{\nabla}'\big]\times\boldsymbol{d}_{mn}\cdot\boldsymbol{v}\\
&-\frac{\mathrm{i}\mu_0}{\hbar}\Theta(\tilde{\omega}_{nm})\tilde{\omega}_{nm}\big[\mathrm{Im}\,\boldsymbol{G}(\boldsymbol{r}_A,\boldsymbol{r}_A,\tilde{\omega}_{nm})\times\overleftarrow{\nabla}'\big]\times\boldsymbol{d}_{mn}\cdot\boldsymbol{v}\\
&-\frac{\mu_0}{\pi\hbar}\mathcal{P}\int_0^\infty \frac{\mathrm{d}\omega}{\omega-\tilde{\omega}_{nm}}\,\omega\boldsymbol{v}\times\big[\nabla\times\mathrm{Im}\,\boldsymbol{G}(\boldsymbol{r}_A,\boldsymbol{r}_A,\omega)\big]\cdot\boldsymbol{d}_{mn}\\
&-\frac{\mathrm{i}\mu_0}{\hbar}\Theta(\tilde{\omega}_{nm})\tilde{\omega}_{nm}\boldsymbol{v}\times\big[\nabla\times\mathrm{Im}\,\boldsymbol{G}(\boldsymbol{r}_A,\boldsymbol{r}_A,\tilde{\omega}_{nm})\big]\cdot\boldsymbol{d}_{mn}\,.
\end{aligned}
\tag{8.17}
$$

Note that the primes in the first and second terms denote derivatives with respect to the respective frequency arguments.

We take real and imaginary parts of the coefficients according to (5.57) and (5.58). For an atom with time-reversal invariant internal Hamiltonian and hence real dipole-matrix elements, the last four terms in (8.17) due not contribute due to pairwise cancellations. For an atom that is free of quasi-degenerate transitions, off-diagonal atomic flip operators decouple from the diagonal ones as well as from each other and the internal dynamics of the atom is governed by the same Eqs. (5.71) and (5.72) that are valid for an atom at rest. However, the frequency shifts and transition rates [1]

$$
\begin{aligned}
\delta\omega_n &= \delta\omega_n(\boldsymbol{r}_A)+\delta\omega_n(\boldsymbol{r}_A,\boldsymbol{v})\\
&=\sum_k\delta\omega_{nk}=\sum_k\delta\omega_{nk}(\boldsymbol{r}_A)+\sum_k\delta\omega_{nk}(\boldsymbol{r}_A,\boldsymbol{v})\,,
\end{aligned}
\tag{8.18}
$$

$$
\begin{aligned}
\Gamma_n &= \Gamma_n(\boldsymbol{r}_A)+\Gamma_n(\boldsymbol{r}_A,\boldsymbol{v})\\
&=\sum_k\Gamma_{nk}=\sum_{k<n}\Gamma_{nk}(\boldsymbol{r}_A)+\sum_{k<n}\Gamma_{nk}(\boldsymbol{r}_A,\boldsymbol{v})
\end{aligned}
\tag{8.19}
$$

now contain motion-induced corrections

$$
\delta\omega_{nk}(\boldsymbol{r}_A,\boldsymbol{v})=\frac{\mu_0}{2\hbar}(\boldsymbol{v}\cdot\nabla_A)\big[\tilde{\omega}_{nk}^2\boldsymbol{d}_{nk}\cdot\mathrm{Im}\,\boldsymbol{G}^{(1)}(\boldsymbol{r}_A,\boldsymbol{r}_A,\tilde{\omega}_{nk})\cdot\boldsymbol{d}_{kn}\big]'\,,
\tag{8.20}
$$

$$
\Gamma_{nk}(\boldsymbol{r}_A,\boldsymbol{v})=\frac{\mu_0}{\pi\hbar}(\boldsymbol{v}\cdot\nabla_A)\mathcal{P}\int_0^\infty\frac{\mathrm{d}\omega}{\omega-\tilde{\omega}_{nk}}\big[\omega^2\boldsymbol{d}_{nk}\cdot\mathrm{Im}\,\boldsymbol{G}^{(1)}(\boldsymbol{r}_A,\boldsymbol{r}_A,\omega)\cdot\boldsymbol{d}_{kn}\big]'
\tag{8.21}
$$

in addition to the purely position-dependent contributions (5.61) and (5.62). The above results have been obtained by exploiting the Onsager reciprocity (A.4) of the Green's tensor in the form $\nabla'\boldsymbol{G}^{(1)}(\boldsymbol{r}_A,\boldsymbol{r}_A,\omega)\mapsto\frac{1}{2}\nabla_A\boldsymbol{G}^{(1)}(\boldsymbol{r}_A,\boldsymbol{r}_A,\omega)$ and making use of the fact that the free-space Green's tensor does not contribute.

The centre-of-mass motion hence affects the internal dynamics of an atom by inducing velocity-dependent frequency shifts and decay rates. These shifts and rates are due to the time delay between photon emission and reabsorption and they may be identified as a vacuum Doppler effect. The vacuum Doppler shifts and rates depend on the gradient of the Green's tensor in the direction of motion. As a consequence, the decay rates and frequency shifts are unaffected by atomic motion if the environment is translationally invariant along the respective direction. This is the case, e.g., for an atom moving parallel to a plate or a cylinder. In close analogy, the transverse Doppler effect induced by an external electromagnetic field vanishes within linear order in v/c. In particular, the shifts and rates are unaffected by motion in free space. This is required by the fact that the electromagnetic vacuum in free space does not define a specific reference frame; it is Lorentz invariant and hence also Galilean invariant. Note that our results are only valid in a non-relativistic approximation within leading, linear order in v/c.

8.2 Casimir–Polder Force

Using our solution for the atom–field dynamics, we can proceed by calculating the CP force on the atom where we proceed in close analogy to Sect. 5.4. The Lorentz force on a moving non-magnetic atom is given by (5.29). To find the CP force, we have to calculate its quantum average

$$\boldsymbol{F} = \nabla\big\langle\hat{\boldsymbol{d}}\cdot\big[\hat{\boldsymbol{E}}(\boldsymbol{r}) + \boldsymbol{v}\times\hat{\boldsymbol{B}}(\boldsymbol{r})\big]\big\rangle\big|_{\boldsymbol{r}=\boldsymbol{r}_A}\,. \tag{8.22}$$

Note that we restrict our attention to an atom prepared in an incoherent superposition of energy eigenstates, so that the term involving a total time-derivative does not contribute. Expanding the dipole operator according to (5.38) and employing normal ordering, we have

$$\begin{aligned}\boldsymbol{F}(t) = \int_0^\infty \mathrm{d}\omega \sum_{m,n}\Big\{&\nabla\big\langle\hat{A}_{mn}(t)\boldsymbol{d}_{mn}\cdot\big[\hat{\underline{\boldsymbol{E}}}(\boldsymbol{r},\omega,t) + \boldsymbol{v}\times\hat{\underline{\boldsymbol{B}}}(\boldsymbol{r},\omega,t)\big]\big\rangle\\ &+ \nabla\big\langle\big[\hat{\underline{\boldsymbol{E}}}^\dagger(\boldsymbol{r},\omega,t) + \boldsymbol{v}\times\hat{\underline{\boldsymbol{B}}}^\dagger(\boldsymbol{r},\omega,t)\big]\cdot\boldsymbol{d}_{mn}\hat{A}_{mn}(t)\big\rangle\Big\}_{\boldsymbol{r}=\boldsymbol{r}_A}\,.\end{aligned} \tag{8.23}$$

We make use of the solutions (8.10) and (8.11) for the time-dependent electromagnetic field as found in the previous section. Assuming the electromagnetic field to be prepared in its ground state at initial time, $\hat{\rho} = \hat{\rho}(t_0) = |\{0\}\rangle\langle\{0\}|$, the free fields do not contribute and we are left with

$$\begin{aligned}
\boldsymbol{F}(t) = {} & \frac{\mathrm{i}\mu_0}{\pi} \sum_{m,n,k,l} \int_0^\infty \mathrm{d}\omega\, \omega^2 \nabla \boldsymbol{d}_{mn} \cdot \operatorname{Im} \boldsymbol{G}(\boldsymbol{r}_A, \boldsymbol{r}_A, \omega) \cdot \boldsymbol{d}_{kl} \\
& \times \int_{t_0}^{t} \mathrm{d}t' \mathrm{e}^{-\mathrm{i}\omega(t-t')} \langle \hat{A}_{mn}(t) \hat{A}_{kl}(t') \rangle \\
& - \frac{\mathrm{i}\mu_0}{\pi} \sum_{m,n,k,l} \int_0^\infty \mathrm{d}\omega\, \omega^2 \nabla \boldsymbol{d}_{mn} \cdot \operatorname{Im} \boldsymbol{G}(\boldsymbol{r}_A, \boldsymbol{r}_A, \omega) \cdot \boldsymbol{d}_{kl} \big(\overleftarrow{\nabla}' \cdot \boldsymbol{v} \big) \\
& \times \int_{t_0}^{t} \mathrm{d}t' (t-t') \mathrm{e}^{-\mathrm{i}\omega(t-t')} \langle \hat{A}_{mn}(t) \hat{A}_{kl}(t') \rangle \\
& + \frac{\mu_0}{\pi} \sum_{m,n,k,l} \int_0^\infty \mathrm{d}\omega\, \omega \nabla \boldsymbol{d}_{mn} \cdot \big[\operatorname{Im} \boldsymbol{G}(\boldsymbol{r}_A, \boldsymbol{r}_A, \omega) \times \overleftarrow{\nabla}' \big] \times \boldsymbol{d}_{kl} \cdot \boldsymbol{v} \\
& \times \int_{t_0}^{t} \mathrm{d}t' \mathrm{e}^{-\mathrm{i}\omega(t-t')} \langle \hat{A}_{mn}(t) \hat{A}_{kl}(t') \rangle \\
& + \frac{\mu_0}{\pi} \sum_{m,n,k,l} \int_0^\infty \mathrm{d}\omega\, \omega \nabla \boldsymbol{d}_{mn} \cdot \boldsymbol{v} \times \big[\nabla \times \operatorname{Im} \boldsymbol{G}(\boldsymbol{r}_A, \boldsymbol{r}_A, \omega) \big] \cdot \boldsymbol{d}_{kl} \\
& \times \int_{t_0}^{t} \mathrm{d}t' \mathrm{e}^{-\mathrm{i}\omega(t-t')} \langle \hat{A}_{mn}(t) \hat{A}_{kl}(t') \rangle + \mathrm{C.c}\,.
\end{aligned} \tag{8.24}$$

With the internal atomic dynamics being governed by (5.66), correlation functions of the off-diagonal flip operators can be determined by means of the quantum regression theorem (5.70). The time integrals can subsequently be carried out using the Markov approximation where we have (5.99) and similarly

$$\begin{aligned}
& \int_{t_0}^{t} \mathrm{d}t'\, (t-t') \mathrm{e}^{[\mathrm{i}(\tilde{\omega}_{mk}-\omega)-(\Gamma_m+\Gamma_k)/2](t-t')} \langle \hat{A}_{mn}(t') \rangle \\
& \simeq \langle \hat{A}_{mn}(t) \rangle \int_{-\infty}^{t} \mathrm{d}t'\, \mathrm{e}^{[\mathrm{i}(\tilde{\omega}_{nk}-\omega)-(\Gamma_m+\Gamma_k)/2](t-t')} = -\frac{\langle \hat{A}_{mn}(t) \rangle}{[\omega-\tilde{\omega}_{nk}+\frac{\mathrm{i}}{2}(\Gamma_m+\Gamma_k)]^2}\,.
\end{aligned} \tag{8.25}$$

After these steps, the CP force acting on a moving atom is found to be

$$\boldsymbol{F}(t) = \sum_n p_n(t)\boldsymbol{F}_n \tag{8.26}$$

with

$$\begin{aligned}
\boldsymbol{F}_n = {} & \frac{\mu_0}{\pi}\sum_k \int_0^\infty \mathrm{d}\omega\,\omega^2 \frac{\nabla \boldsymbol{d}_{nk}\cdot \mathrm{Im}\,\boldsymbol{G}^{(1)}(\boldsymbol{r}_A,\boldsymbol{r}_A,\omega)\cdot \boldsymbol{d}_{kn}}{\omega - \tilde{\omega}_{nk} - \frac{\mathrm{i}}{2}(\Gamma_n + \Gamma_k)} \\
& + \frac{\mathrm{i}\mu_0}{\pi}\sum_k \int_0^\infty \mathrm{d}\omega\,\omega^2 \frac{(\boldsymbol{v}\cdot\nabla')\nabla \boldsymbol{d}_{nk}\cdot \mathrm{Im}\,\boldsymbol{G}^{(1)}(\boldsymbol{r}_A,\boldsymbol{r}_A,\omega)\cdot \boldsymbol{d}_{kn}}{\left[\omega - \tilde{\omega}_{nk} - \frac{\mathrm{i}}{2}(\Gamma_n + \Gamma_k)\right]^2} \\
& - \frac{\mathrm{i}\mu_0}{\pi}\sum_k \int_0^\infty \mathrm{d}\omega\,\omega \frac{(\nabla' - \nabla)\boldsymbol{d}_{nk}\cdot \boldsymbol{v}\times\left[\nabla\times \mathrm{Im}\,\boldsymbol{G}^{(1)}(\boldsymbol{r}_A,\boldsymbol{r}_A,\omega)\right]\cdot \boldsymbol{d}_{kn}}{\omega - \tilde{\omega}_{nk} - \frac{\mathrm{i}}{2}(\Gamma_n + \Gamma_k)} \\
& + \mathrm{C.c.}
\end{aligned} \tag{8.27}$$

Note that the last two contributions in (8.24) have been collected in a single term by making use of the Onsager reciprocity (A.4) of the Green's tensor. In addition, the vanishing contributions from the free-space Green's tensor have been discarded.

To identify the velocity-dependent part of the CP force, we recall that the frequency shifts (8.19) and widths (8.18) may be separated into purely position-dependent quantities plus velocity-dependent corrections. Using this decomposition, retaining only leading-order terms in v/c and introducing the abbreviating notation $\delta\omega_n(\boldsymbol{r}_A) \equiv \delta\omega_n$, $\delta\omega_n(\boldsymbol{r}_A, \boldsymbol{v}) \equiv \delta\omega_n(\boldsymbol{v})$, $\Gamma_n(\boldsymbol{r}_A) \equiv \Gamma_n$, $\Gamma_n(\boldsymbol{r}_A, \boldsymbol{v}) \equiv \Gamma_n(\boldsymbol{v})$, the CP force on a moving atom reads

$$\boldsymbol{F}(t) = \boldsymbol{F}(\boldsymbol{r}_A, t) + \boldsymbol{F}(\boldsymbol{r}_A, \boldsymbol{v}, t) \tag{8.28}$$

where $\boldsymbol{F}(\boldsymbol{r}_A, t)$ is the velocity-independent part of the force as given by (5.101) with (5.103), (5.108) and (5.109); and

$$\boldsymbol{F}(\boldsymbol{r}_A, \boldsymbol{v}, t) = \sum_n p_n(t)\boldsymbol{F}_n(\boldsymbol{r}_A, \boldsymbol{v}) \tag{8.29}$$

with

$$\begin{aligned}
\boldsymbol{F}_n(\boldsymbol{r}_A, \boldsymbol{v}) = {} & \frac{\mu_0}{\pi}\sum_k \int_0^\infty \mathrm{d}\omega\,\omega^2 \left\{\delta\omega_n(\boldsymbol{v}) - \delta\omega_k(\boldsymbol{v}) + \tfrac{\mathrm{i}}{2}[\Gamma_n(\boldsymbol{v}) + \Gamma_k(\boldsymbol{v})]\right\} \\
& \times \frac{\nabla \boldsymbol{d}_{nk}\cdot \mathrm{Im}\,\boldsymbol{G}^{(1)}(\boldsymbol{r}_A,\boldsymbol{r}_A,\omega)\cdot \boldsymbol{d}_{kn}}{\left[\omega - \tilde{\omega}_{nk} - \frac{\mathrm{i}}{2}(\Gamma_n + \Gamma_k)\right]^2}
\end{aligned}$$

$$
\begin{aligned}
&+\frac{\mathrm{i}\mu_0}{\pi}\sum_k\int_0^\infty \mathrm{d}\omega\,\omega^2\,\frac{\nabla(\boldsymbol{v}\cdot\nabla')\boldsymbol{d}_{nk}\cdot\mathrm{Im}\,\boldsymbol{G}^{(1)}(\boldsymbol{r}_A,\boldsymbol{r}_A,\omega)\cdot\boldsymbol{d}_{kn}}{\left[\omega-\tilde{\omega}_{nk}-\frac{\mathrm{i}}{2}(\Gamma_n+\Gamma_k)\right]^2}\\
&+\frac{\mathrm{i}\mu_0}{\pi}\sum_k\int_0^\infty \mathrm{d}\omega\,\omega\,\frac{(\nabla'-\nabla)\boldsymbol{d}_{nk}\cdot\boldsymbol{v}\times\left[\nabla\times\mathrm{Im}\,\boldsymbol{G}^{(1)}(\boldsymbol{r}_A,\boldsymbol{r}_A,\omega)\right]\cdot\boldsymbol{d}_{kn}}{\omega-\tilde{\omega}_{nk}-\frac{\mathrm{i}}{2}(\Gamma_n+\Gamma_k)}\\
&+\mathrm{C.c.}
\end{aligned}
\tag{8.30}
$$

is the motion-induced CP force.

The velocity-independent force has already been discussed in Sects. 5.4 and 5.5. In the following, we concentrate on the motion-induced force alone. Just like the force on an atom at rest, it is given by a weighted average over force components associated with the internal atomic eigenstates. Let us separate the force components into non-resonant contributions from virtual photons and off-resonant contributions due to real photons. To that end we write $\mathrm{Im}\,\boldsymbol{G} = (\boldsymbol{G}-\boldsymbol{G}^*)/(2\mathrm{i})$, make use of the Schwarz reflection principle (A.3), and apply the integration contour depicted in Fig. 5.1. Using Cauchy's residue theorem, we find [1]

$$
\boldsymbol{F}_n(\boldsymbol{r}_A,\boldsymbol{v}) = \boldsymbol{F}_n^{\mathrm{nres}}(\boldsymbol{r}_A,\boldsymbol{v}) + \boldsymbol{F}_n^{\mathrm{res}}(\boldsymbol{r}_A,\boldsymbol{v})
\tag{8.31}
$$

with non-resonant forces

$$
\begin{aligned}
\boldsymbol{F}_n^{\mathrm{nres}}(\boldsymbol{r}_A,\boldsymbol{v}) = &-\frac{\hbar\mu_0}{2\pi}\int_0^\infty \mathrm{d}\xi\,\xi^2\nabla\mathrm{tr}\left\{[\boldsymbol{\alpha}_n(\boldsymbol{v},\mathrm{i}\xi)+\boldsymbol{\alpha}_n(\boldsymbol{v},-\mathrm{i}\xi)]\cdot\boldsymbol{G}^{(1)}(\boldsymbol{r}_A,\boldsymbol{r}_A,\mathrm{i}\xi)\right\}\\
&-\frac{\mathrm{i}\hbar\mu_0}{2\pi}\int_0^\infty \mathrm{d}\xi\,\xi^2(\boldsymbol{v}\cdot\nabla')\nabla\mathrm{tr}\left\{[\boldsymbol{\alpha}'_n(\mathrm{i}\xi)+\boldsymbol{\alpha}'_n(-\mathrm{i}\xi)]\cdot\boldsymbol{G}^{(1)}(\boldsymbol{r}_A,\boldsymbol{r}_A,\mathrm{i}\xi)\right\}\\
&+\frac{\hbar\mu_0}{2\pi}\int_0^\infty \mathrm{d}\xi\,\xi(\nabla'-\nabla)\mathrm{tr}\left\{[\boldsymbol{\alpha}_n(\mathrm{i}\xi)-\boldsymbol{\alpha}_n(-\mathrm{i}\xi)]\cdot\boldsymbol{v}\right.\\
&\left.\times\left[\nabla\times\boldsymbol{G}^{(1)}(\boldsymbol{r}_A,\boldsymbol{r}_A,\mathrm{i}\xi)\right]\right\}
\end{aligned}
\tag{8.32}
$$

and resonant forces

$$
\begin{aligned}
\boldsymbol{F}_n^{\mathrm{res}}(\boldsymbol{r}_A,\boldsymbol{v}) = \mu_0&\sum_{k<n}\Omega_{nk}(\boldsymbol{v})\nabla\left[\Omega_{nk}^2\boldsymbol{d}_{nk}\cdot\boldsymbol{G}^{(1)}(\boldsymbol{r}_A,\boldsymbol{r}_A,\Omega_{nk})\cdot\boldsymbol{d}_{kn}\right]'\\
&+\mathrm{i}\mu_0\sum_{k<n}(\boldsymbol{v}\cdot\nabla')\nabla\left[\Omega_{nk}^2\,\boldsymbol{d}_{nk}\cdot\boldsymbol{G}^{(1)}(\boldsymbol{r}_A,\boldsymbol{r}_A,\Omega_{nk})\cdot\boldsymbol{d}_{kn}\right]'\\
&+\mathrm{i}\mu_0\sum_{k<n}\Omega_{nk}(\nabla'-\nabla)\boldsymbol{d}_{nk}\cdot\boldsymbol{v}\times\left[\nabla\times\boldsymbol{G}^{(1)}(\boldsymbol{r}_A,\boldsymbol{r}_A,\Omega_{nk})\right]\cdot\boldsymbol{d}_{kn}\\
&+\mathrm{C.c.}
\end{aligned}
\tag{8.33}
$$

Here,

$$\boldsymbol{\alpha}_n(\boldsymbol{v},\omega)=\frac{1}{\hbar}\sum_k\left(\frac{\{\delta\omega_n(\boldsymbol{v})-\delta\omega_k(\boldsymbol{v})-\frac{\mathrm{i}}{2}[\Gamma_n(\boldsymbol{v})+\Gamma_k(\boldsymbol{v})]\}\boldsymbol{d}_{nk}\boldsymbol{d}_{kn}}{[\tilde{\omega}_{kn}-\omega-\frac{\mathrm{i}}{2}(\Gamma_n+\Gamma_k)]^2}\right.$$
$$\left.+\frac{\{\delta\omega_n(\boldsymbol{v})-\delta\omega_k(\boldsymbol{v})+\frac{\mathrm{i}}{2}[\Gamma_n(\boldsymbol{v})+\Gamma_k(\boldsymbol{v})]\}\boldsymbol{d}_{kn}\boldsymbol{d}_{nk}}{[\tilde{\omega}_{kn}+\omega+\frac{\mathrm{i}}{2}(\Gamma_n+\Gamma_k)]^2}\right) \quad (8.34)$$

and

$$\Omega_{nk}(\boldsymbol{v})=\delta\omega_n(\boldsymbol{v})-\delta\omega_k(\boldsymbol{v})+\tfrac{\mathrm{i}}{2}[\Gamma_n(\boldsymbol{v})+\Gamma_k(\boldsymbol{v})] \quad (8.35)$$

are the motion-induced corrections to the atomic polarisability (5.91) and the complex transitions frequencies (5.106) within linear order in v/c. Both resonant and non-resonant forces on the moving atom have three contributions. Firstly, the atomic motion leads to a vacuum Doppler shifting and broadening of the atomic transitions. This induces a correction to the CP force as given by the first terms in (8.32) and (8.33) above. The second term is due to the time delay between photon emission and reabsorption resulting from the atomic motion. Finally, the Röntgen interaction between the magnetic field and the current associated with the moving atomic dipole moment leads to a third term.

In the perturbative limit $|\delta\omega_n|$, $|\delta\omega_k|$, Γ_n, $\Gamma_k\ll\omega_{mn}$, the resonant force simplifies to

$$\begin{aligned}\boldsymbol{F}_n^{\mathrm{res}}(\boldsymbol{r}_A,\boldsymbol{v})&=2\mu_0\sum_{k<n}[\delta\omega_n(\boldsymbol{v})-\delta\omega_k(\boldsymbol{v})]\nabla\big[\omega_{nk}^2\boldsymbol{d}_{nk}\cdot\mathrm{Re}\,\boldsymbol{G}^{(1)}(\boldsymbol{r}_A,\boldsymbol{r}_A,\omega_{nk})\cdot\boldsymbol{d}_{kn}\big]'\\&-\mu_0\sum_{k<n}[\Gamma_n(\boldsymbol{v})+\Gamma_k(\boldsymbol{v})]\nabla\big[\omega_{nk}^2\boldsymbol{d}_{nk}\cdot\mathrm{Im}\,\boldsymbol{G}^{(1)}(\boldsymbol{r}_A,\boldsymbol{r}_A,\omega_{nk})\cdot\boldsymbol{d}_{kn}\big]'\\&-2\mu_0\sum_{k<n}(\boldsymbol{v}\cdot\nabla')\nabla\big[\omega_{nk}^2\boldsymbol{d}_{nk}\cdot\mathrm{Im}\,\boldsymbol{G}^{(1)}(\boldsymbol{r}_A,\boldsymbol{r}_A,\omega_{nk})\cdot\boldsymbol{d}_{kn}\big]'\\&-2\mu_0\sum_{k<n}\omega_{nk}(\nabla'-\nabla)\boldsymbol{d}_{nk}\cdot\boldsymbol{v}\times\big[\nabla\times\mathrm{Im}\,\boldsymbol{G}^{(1)}(\boldsymbol{r}_A,\boldsymbol{r}_A,\omega_{nk})\big]\cdot\boldsymbol{d}_{kn}\end{aligned} \quad (8.36)$$

where we have set $\delta\omega_n$, $\delta\omega_k$, Γ_n, $\Gamma_k\simeq0$, but retained the velocity-dependent vacuum Doppler shifts and broadenings. The non-resonant force has to be treated with more care, because the delay and Röntgen terms vanish when setting the level widths to zero. Retaining linear terms in Γ_n, Γ_k, we find

$$\begin{aligned}
\boldsymbol{F}_n^{\mathrm{nres}}(\boldsymbol{r}_A, \boldsymbol{v}) = & -\frac{2\mu_0}{\pi} \sum_k [\delta\omega_n(\boldsymbol{v}) - \delta\omega_k(\boldsymbol{v})] \int_0^\infty \mathrm{d}\xi\, \xi^2 \frac{\omega_{kn}^2 - \xi^2}{(\omega_{kn}^2 + \xi^2)^2} \\
& \times \nabla \boldsymbol{d}_{nk} \cdot \boldsymbol{G}^{(1)}(\boldsymbol{r}_A, \boldsymbol{r}_A, \mathrm{i}\xi) \cdot \boldsymbol{d}_{kn} \\
& + \frac{2\mu_0}{\pi} \sum_k \omega_{kn} (\Gamma_n + \Gamma_k) \int_0^\infty \mathrm{d}\xi\, \xi^2 \frac{\omega_{kn}^2 - 3\xi^2}{(\omega_{kn}^2 + \xi^2)^3} \\
& \times (\boldsymbol{v} \cdot \nabla') \nabla \boldsymbol{d}_{nk} \cdot \boldsymbol{G}^{(1)}(\boldsymbol{r}_A, \boldsymbol{r}_A, \mathrm{i}\xi) \cdot \boldsymbol{d}_{kn} \\
& + \frac{2\mu_0}{\pi} \sum_k \omega_{kn} (\Gamma_n + \Gamma_k) \int_0^\infty \mathrm{d}\xi \frac{\xi^2}{(\omega_{kn}^2 + \xi^2)^2} \\
& \times (\nabla' - \nabla) \boldsymbol{d}_{nk} \cdot \boldsymbol{v} \times \left[\nabla \times \boldsymbol{G}^{(1)}(\boldsymbol{r}_A, \boldsymbol{r}_A, \mathrm{i}\xi)\right] \cdot \boldsymbol{d}_{kn} \,. \qquad (8.37)
\end{aligned}$$

The CP force on a moving atom thus depends crucially on the atomic level shifts and widths. Recall from Sect. 5.2 that they may induce an anisotropy of the atomic polarisability. In particular, the explicit dependence of the force on the Doppler shifts $\delta\omega_n(\boldsymbol{v})$, $\delta\omega_k(\boldsymbol{v})$ and widths $\Gamma_n(\boldsymbol{v})$, $\Gamma_k(\boldsymbol{v})$ as well as the level widths Γ_n, Γ_k will in general prohibit a simplification of the result for an atom in an isotropic state. Two conditions must be fulfilled for the force to be expressible in terms of an isotropic polarisability. Firstly, the atom must be moving in a direction along which the environment is translationally invariant, $\delta\omega_n(\boldsymbol{v}), \delta\omega_k(\boldsymbol{v}), \Gamma_n(\boldsymbol{v}), \Gamma_k(\boldsymbol{v}) = 0$. Secondly, the atom must be sufficiently far from any body, so that its level widths are well approximated by their isotropic free-space values, $\Gamma_{k'} = \Gamma_{k''}$ whenever $k', k'' \in \{\overline{k}\}$. In this case, (4.16) can be used to simplify the force components to

$$\begin{aligned}
\boldsymbol{F}_n^{\mathrm{res}}(\boldsymbol{r}_A, \boldsymbol{v}) = & -\frac{2\mu_0}{3} \sum_{k<n} |\boldsymbol{d}_{nk}|^2 (\boldsymbol{v} \cdot \nabla') \nabla \left\{\omega_{nk}^2 \operatorname{tr}\left[\operatorname{Im} \boldsymbol{G}^{(1)}(\boldsymbol{r}_A, \boldsymbol{r}_A, \omega_{nk})\right]\right\}' \\
& - \frac{2\mu_0}{3} \sum_{k<n} \omega_{nk} |\boldsymbol{d}_{nk}|^2 (\nabla' - \nabla) \operatorname{tr}\left\{\boldsymbol{v} \times \left[\nabla \times \operatorname{Im} \boldsymbol{G}^{(1)}(\boldsymbol{r}_A, \boldsymbol{r}_A, \omega_{nk})\right]\right\}
\end{aligned} \qquad (8.38)$$

and

$$\begin{aligned}
\boldsymbol{F}_n^{\mathrm{nres}}(\boldsymbol{r}_A, \boldsymbol{v}) = & \frac{2\mu_0}{3\pi} \sum_k \omega_{kn} (\Gamma_n + \Gamma_k) |\boldsymbol{d}_{nk}|^2 \int_0^\infty \mathrm{d}\xi\, \xi^2 \frac{\omega_{kn}^2 - 3\xi^2}{(\omega_{kn}^2 + \xi^2)^3} \\
& \times (\boldsymbol{v} \cdot \nabla') \nabla \operatorname{tr} \boldsymbol{G}^{(1)}(\boldsymbol{r}_A, \boldsymbol{r}_A, \mathrm{i}\xi) \\
& + \frac{2\mu_0}{3\pi} \sum_k \omega_{kn} (\Gamma_n + \Gamma_k) |\boldsymbol{d}_{nk}|^2 \int_0^\infty \mathrm{d}\xi \frac{\xi^2}{(\omega_{kn}^2 + \xi^2)^2}
\end{aligned}$$

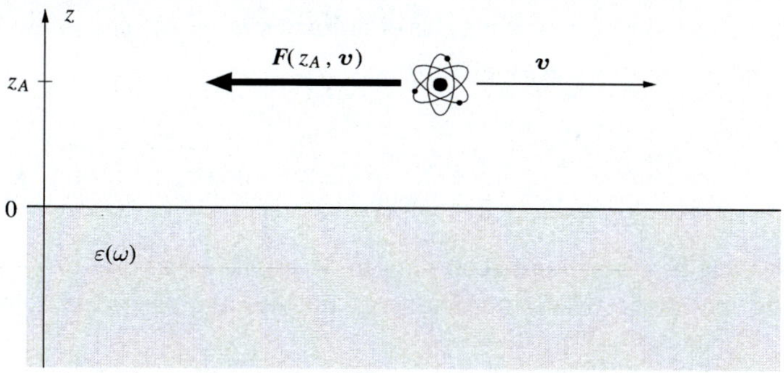

Fig. 8.1 Quantum friction experienced by an atom moving parallel to plate

$$\times\,(\nabla' - \nabla)\mathrm{tr}\big\{v\times\big[\nabla\times \boldsymbol{G}^{(1)}(r_A, r_A, \mathrm{i}\xi)\big]\big\} \tag{8.39}$$

for an atom in an isotropic state.

8.3 Quantum Friction

To illustrate the general results obtained in the previous section, let us apply them to the standard quantum friction scenario as depicted in Fig. 8.1. We consider an atom moving parallel ($v\cdot e_z = 0$) to an infinitely thick plate of permittivity $\varepsilon(\omega)$ and determine the force induced by the atomic motion. Recall that the velocity-dependent force is generally a small effect in the non-relativistic limit considered. In order to achieve a measurable effect, we hence concentrate on the nonretarded regime of small atom–plate separations $z_A \ll c/(\omega_{+}n)$ where the force is expected to be largest. For simplicity, we further employ the perturbative limit.

With the half space being invariant along the direction of motion, the vacuum Doppler shifts (8.20) and widths (8.21) vanish, $\delta\omega_n(r_A, v) = \Gamma_n(r_A, v) = 0$. The Doppler contribution is hence absent from the motion-induced forces (8.36) and (8.37). To calculate the remaining delay and Röntgen contributions, we require the Green's tensor of the half space. According to (A.35) in App. A.3.2, it is given by

$$\boldsymbol{G}^{(1)}(r, r', \omega) = \frac{\mathrm{i}}{8\pi^2}\int \frac{\mathrm{d}^2k^{\perp}}{k^{\perp}}\,\mathrm{e}^{\mathrm{i}k^{\parallel}\cdot(r-r')+\mathrm{i}k^{\perp}(z+z')}\sum_{\sigma=s,p} r_\sigma e_{\sigma+}e_{\sigma-} \tag{8.40}$$

with

$$k^{\perp} = \sqrt{\frac{\omega^2}{c^2} - k^{\perp 2}}\,,\quad \mathrm{Im}\,k^{\perp} > 0\,. \tag{8.41}$$

Recall from (4.51) in Sect. 4.2 in polar coordinates $\boldsymbol{k}^{\|} = (k^{\|}\cos\phi, k^{\|}\sin\phi, 0)$, the polarisation unit vectors can be given as

$$\boldsymbol{e}_{s\pm} = (\sin\phi, -\cos\phi, 0)\,, \quad \boldsymbol{e}_{p\pm} = \frac{c}{\omega} = (\mp k^{\perp}\cos\phi, \mp k^{\perp}\sin\phi, k^{\|})\,. \tag{8.42}$$

Derivatives can be carried out according to $\boldsymbol{\nabla} \mapsto \mathrm{i}\boldsymbol{k}^{\|} + \mathrm{i}k^{\perp}\boldsymbol{e}_z$, $\boldsymbol{v}\cdot\boldsymbol{\nabla}' \mapsto -\mathrm{i}\boldsymbol{v}\cdot\boldsymbol{k}^{\|}$. Combining the above results and carrying out the angular integral, we calculate ($\int \mathrm{d}^2k^{\|} = \int_0^\infty k^{\|}\mathrm{d}k^{\|} \int_0^{2\pi} \mathrm{d}\phi$)

$$\begin{aligned}&(\boldsymbol{v}\cdot\boldsymbol{\nabla}')\boldsymbol{\nabla}\boldsymbol{d}\cdot\boldsymbol{G}^{(1)}(\boldsymbol{r}_A,\boldsymbol{r}_A,\omega)\cdot\boldsymbol{d}\\ &\quad= \frac{\mathrm{i}}{32\pi}\int_0^\infty \mathrm{d}k^{\|}\,\frac{k^{\|3}}{k^{\perp}}\,\mathrm{e}^{2\mathrm{i}k^{\perp}z_A}\Bigg(\big[3\boldsymbol{d}^{\|2}\boldsymbol{v} - 2\boldsymbol{d}^{\|}(\boldsymbol{d}^{\|}\cdot\boldsymbol{v})\big]r_s\\ &\qquad+\left\{4\,\frac{k^{\|2}c^2}{\omega^2}\,\boldsymbol{d}^{\perp 2}\boldsymbol{v} - \frac{k^{\perp 2}c^2}{\omega^2}\big[\boldsymbol{d}^{\|2}\boldsymbol{v} + 2\boldsymbol{d}^{\|}(\boldsymbol{d}^{\|}\cdot\boldsymbol{v})\big]\right\}r_p\Bigg)\end{aligned} \tag{8.43}$$

with $\boldsymbol{d}^{\|} = d_x\boldsymbol{e}_x + d_y\boldsymbol{e}_y$ and $\boldsymbol{d}^{\perp} = d_z\boldsymbol{e}_z$ denoting the components of $\boldsymbol{d}$ parallel and perpendicular to the plate surface. Note that the dipole matrix elements have been assumed to be real in accordance with our assumption from Sect. 8.1. In a similar way, we find

$$\begin{aligned}&(\boldsymbol{\nabla}' - \boldsymbol{\nabla})\boldsymbol{d}\cdot\boldsymbol{v}\times\big[\boldsymbol{\nabla}\times\boldsymbol{G}^{(1)}(\boldsymbol{r}_A,\boldsymbol{r}_A,\omega)\big]\cdot\boldsymbol{d}\\ &\quad= \frac{\mathrm{i}}{8\pi}\int_0^\infty \mathrm{d}k^{\|}\,\frac{k^{\|3}}{k^{\perp}}\,\mathrm{e}^{2\mathrm{i}k^{\perp}z_A}\big\{\big[2\boldsymbol{d}^{\|2}\boldsymbol{v} - \boldsymbol{d}^{\|}(\boldsymbol{d}^{\|}\cdot\boldsymbol{v})\big]r_s + 2\boldsymbol{d}^{\perp 2}\boldsymbol{v}r_p\big\}\,.\end{aligned} \tag{8.44}$$

In the nonretarded limit $z_A \ll c/(\omega_+ n)$, the integral is dominated by evanescent waves ($k^{\|} > \omega/c$) with large imaginary wave vectors $k^{\perp} \simeq \mathrm{i}k^{\|}$. With this approximation, the reflection coefficients (A.41) and (A.42) of the plate read

$$r_s \simeq 0\,, \quad r_p \simeq \frac{\varepsilon(\omega) - 1}{\varepsilon(\omega) + 1}\,. \tag{8.45}$$

The $k^{\|}$-integrals can then be performed to give

$$
\begin{aligned}
&(\boldsymbol{v}\cdot\nabla')\nabla\boldsymbol{d}\cdot\boldsymbol{G}^{(1)}(\boldsymbol{r}_A,\boldsymbol{r}_A,\omega)\cdot\boldsymbol{d}\\
&=\frac{c^2}{32\pi\omega^2}\int_0^\infty \mathrm{d}k^\parallel\, k^{\parallel 4}\,\mathrm{e}^{-2k^\parallel z_A}\big[\big(4\boldsymbol{d}^{\perp 2}+\boldsymbol{d}^{\parallel 2}\big)\boldsymbol{v}+2\boldsymbol{d}^\parallel\big(\boldsymbol{d}^\parallel\cdot\boldsymbol{v}\big)\big]\frac{\varepsilon(\omega)-1}{\varepsilon(\omega)+1}\\
&=\frac{3c^2\big[\big(4\boldsymbol{d}^{\perp 2}+\boldsymbol{d}^{\parallel 2}\big)\boldsymbol{v}+2\boldsymbol{d}^\parallel\big(\boldsymbol{d}^\parallel\cdot\boldsymbol{v}\big)\big]}{128\pi\omega^2 z_A^5}\,\frac{\varepsilon(\omega)-1}{\varepsilon(\omega)+1}
\end{aligned}
\tag{8.46}
$$

and similarly

$$
(\nabla'-\nabla)\boldsymbol{d}\cdot\boldsymbol{v}\times\big[\nabla\times\boldsymbol{G}^{(1)}(\boldsymbol{r}_A,\boldsymbol{r}_A,\omega)\big]\cdot\boldsymbol{d}=\frac{\boldsymbol{d}^{\perp 2}\boldsymbol{v}}{16\pi z_A^3}\frac{\varepsilon(\omega)-1}{\varepsilon(\omega)+1}\,. \tag{8.47}
$$

A comparison of these two results reveals that the delay contribution to the force with its $1/z_A^5$ asymptote dominates over the Röntgen term with its $1/z_A^3$ power law in the nonretarded limit. This can be understood from the fact that the Röntgen interaction couples the atom to the magnetic field which is weaker than the electric field responsible for the delay term. We have encountered an analogous difference in power laws for the CP force on a stationary atom in Table 3.1 of Sect. 3.1: Here, the force between a magnetic atom and an electric plate is proportional to $1/z_A^2$ and hence weaker then $1/z_A^4$ the force between an electric atom and the plate.

For parallel motion and nonretarded distances, the velocity-dependent CP force is hence entirely due to the time delay between photon emission and reabsorption. Note that the Röntgen contribution becomes relevant for larger atom–surface separations while the vacuum Doppler shifts and widths contribute for motion towards or away from the plate. Using the Green's tensor, the non-resonant force (8.37) reads

$$
\begin{aligned}
\boldsymbol{F}_n^{\mathrm{nres}}(z_A,\boldsymbol{v})=&-\frac{3}{64\pi^2\varepsilon_0 z_A^5}\sum_k\omega_{kn}(\Gamma_n+\Gamma_k)\big[\big(4\boldsymbol{d}_{nk}^{\perp 2}+\boldsymbol{d}_{nk}^{\parallel 2}\big)\boldsymbol{v}+2\boldsymbol{d}_{nk}^{\parallel}\big(\boldsymbol{d}_{nk}^{\parallel}\cdot\boldsymbol{v}\big)\big]\\
&\times\int_0^\infty\mathrm{d}\xi\,\frac{\omega_{kn}^2-3\xi^2}{(\omega_{kn}^2+\xi^2)^3}\,\frac{\varepsilon(\mathrm{i}\xi)-1}{\varepsilon(\mathrm{i}\xi)+1}\,.
\end{aligned}
\tag{8.48}
$$

The relevant decay rates (5.60) with (5.62) can be found by using the nonretarded Green's tensor (5.123) and recalling the relation (4.98),

$$
\Gamma_n(z_A)=\frac{1}{8\pi\varepsilon_0\hbar z_A^3}\sum_{k<n}\big(2\boldsymbol{d}_{nk}^{\perp 2}+\boldsymbol{d}_{nk}^{\parallel 2}\big)\frac{\operatorname{Im}\varepsilon(\omega_{nk})}{|\varepsilon(\omega_{nk})+1|^2}\,. \tag{8.49}
$$

For an atom in an isotropic state, the configuration exhibits an axial symmetry around the z-axis, so that

$$
\sum_k\Gamma_k\boldsymbol{d}_{nk}^{\parallel}\boldsymbol{d}_{nk}^{\parallel}=\tfrac{1}{2}\sum_k\Gamma_k\boldsymbol{d}_{nk}^{\parallel 2}\boldsymbol{I}\,. \tag{8.50}
$$

The non-resonant force then simplifies to

$$F_n^{\mathrm{nres}}(z_A, \boldsymbol{v}) = -\frac{3\boldsymbol{v}}{32\pi^2\varepsilon_0 z_A^5}\sum_k \omega_{kn}(\Gamma_n+\Gamma_k)\big(2d_{nk}^{\perp 2}+d_{nk}^{\parallel 2}\big) \times \int_0^\infty \mathrm{d}\xi\, \frac{\omega_{kn}^2-3\xi^2}{(\omega_{kn}^2+\xi^2)^3}\,\frac{\varepsilon(\mathrm{i}\xi)-1}{\varepsilon(\mathrm{i}\xi)+1}\,. \tag{8.51}$$

The motion-induced force on a ground-state atom is purely non-resonant,

$$F_0(z_A, \boldsymbol{v}) = -\frac{3\boldsymbol{v}}{32\pi^2\varepsilon_0 z_A^5}\sum_k \omega_{k0}\Gamma_k\big(2d_{0k}^{\perp 2}+d_{0k}^{\parallel 2}\big) \times \int_0^\infty \mathrm{d}\xi\, \frac{\omega_{k0}^2-3\xi^2}{(\omega_{k0}^2+\xi^2)^3}\,\frac{\varepsilon(\mathrm{i}\xi)-1}{\varepsilon(\mathrm{i}\xi)+1}\,. \tag{8.52}$$

To determine its direction, we note that the factor $(\omega_{k0}^2-3\xi^2)/(\omega_{k0}^2+\xi^2)^3$ is positive for small ξ and negative for large ξ. The contributions from these regions exactly balance each other,

$$\int_0^\infty \mathrm{d}\xi\, \frac{\omega_{k0}^2-3\xi^2}{(\omega_{k0}^2+\xi^2)^3} = 0\,. \tag{8.53}$$

With $\varepsilon(\omega)$ being a causal response function, $\varepsilon(\mathrm{i}\xi)$ and hence $[\varepsilon(\mathrm{i}\xi)-1]/[\varepsilon(\mathrm{i}\xi)+1]$ are monotonously decreasing functions of ξ. As a result, the positive contributions from small ξ dominate the above integral, so that

$$\int_0^\infty \mathrm{d}\xi\, \frac{\omega_{k0}^2-3\xi^2}{(\omega_{k0}^2+\xi^2)^3}\,\frac{\varepsilon(\mathrm{i}\xi)-1}{\varepsilon(\mathrm{i}\xi)+1} > 0\,. \tag{8.54}$$

The motion-induced CP force on a ground-state atom is hence a genuine quantum friction which points in the direction opposite to the velocity and decelerates the atom.

As a simple example, consider an atom in an isotropic state moving parallel to a metal plate whose permittivity can be described by the Drude model (4.93). Carrying out the ξ-integral and exploiting the fact that $\gamma \ll \omega_\mathrm{P}$ for most metals, the quantum friction force reads

$$F_0(z_A, \boldsymbol{v}) = -\frac{3\omega_\mathrm{S}\boldsymbol{v}}{64\pi\varepsilon_0 z_A^5}\sum_k \frac{\Gamma_k\big(2d_{0k}^{\perp 2}+d_{0k}^{\parallel 2}\big)}{(\omega_{k0}+\omega_\mathrm{S})^3}\,, \tag{8.55}$$

with $\omega_S = \omega_P/\sqrt{2}$ being the surface-plasmon frequency. The decay rates are found to be

$$\Gamma_n(z_A) = \frac{\gamma \omega_S^2}{16\pi\varepsilon_0 \hbar z_A^3} \sum_{k<n} \frac{\omega_{nk}\left(2\boldsymbol{d}_{nk}^{\perp 2} + \boldsymbol{d}_{nk}^{\parallel 2}\right)}{(\omega_{nk}^2 - \omega_S^2)^2}\,, \tag{8.56}$$

We note that quantum friction is proportional to the atomic damping parameters Γ_k which in turn are proportional to the metallic damping parameter γ. As expected for a dissipative force, quantum friction vanishes in the absence of damping. The dependence on the γ indicates that the kinetic energy lost by the moving atom leads to an Ohmic heating of the plate.

The simplest example of an isotropic ground state $|0\rangle$ is an S state. The first excited manifold is then triply degenerate and it consists of three P states which we shall label as $|1_1\rangle, |1_2\rangle, |1_3\rangle$. As seen from (B.19) and (B.20) in App B, the relevant dipole matrix elements are

$$\boldsymbol{d}_{01_1}\boldsymbol{d}_{1_10} = \frac{d_{10}^2}{6}\begin{pmatrix} 1 & \mathrm{i} & 0 \\ -\mathrm{i} & 1 & 0 \\ 0 & 0 & 0 \end{pmatrix}, \tag{8.57}$$

$$\boldsymbol{d}_{01_2}\boldsymbol{d}_{1_20} = \frac{d_{10}^2}{3}\begin{pmatrix} 0 & 0 & 0 \\ 0 & 0 & 0 \\ 0 & 0 & 1 \end{pmatrix}, \tag{8.58}$$

$$\boldsymbol{d}_{01_3}\boldsymbol{d}_{1_30} = \frac{d_{10}^2}{6}\begin{pmatrix} 1 & -\mathrm{i} & 0 \\ \mathrm{i} & 1 & 0 \\ 0 & 0 & 0 \end{pmatrix} \tag{8.59}$$

with d_{10} denoting the reduced matrix element. The force thus reads

$$\boldsymbol{F}_0(z_A, \boldsymbol{v}) = -\frac{d_{01}^2 \omega_S (\Gamma_{1_1} + 2\Gamma_{1_2} + \Gamma_{1_3})}{64\pi\varepsilon_0(\omega_{10} + \omega_S)^3} \frac{\boldsymbol{v}}{z_A^5}. \tag{8.60}$$

Substituting the damping parameters

$$\Gamma_{1_k}(z_A) = \frac{\omega_{10} d_{01}^2}{48\pi\varepsilon_0 \hbar z_A^3} \frac{\gamma\omega_S^2}{(\omega_{10}^2 - \omega_S^2)^2} \times \begin{cases} 1 & \text{for } k = 1, 3\,, \\ 2 & \text{for } k = 2\,, \end{cases} \tag{8.61}$$

we find

$$\boldsymbol{F}_0(z_A, \boldsymbol{v}) = -\frac{\omega_{10} d_{01}^4}{512\pi^2\varepsilon_0^2\hbar} \frac{\gamma\omega_S^3}{(\omega_{10}+\omega_S)^3(\omega_{10}^2-\omega_S^2)^2} \frac{\boldsymbol{v}}{z_A^8}\,. \tag{8.62}$$

The nonretarded quantum friction on an isotropic two-level atom moving parallel to a metal plate has recently been calculated from an alternative point of view [2].

Considering the energy deposited into metal by the moving atom in the long-time limit, a power

$$P = \frac{\omega_{10} d_{01}^4}{512\pi^2 \varepsilon_0^2 \hbar} \frac{\gamma \omega_S^3}{(\omega_{10}+\omega_S)^3 (\omega_{10}^2-\omega_S^2)^2} \frac{v^2}{z_A^8} \tag{8.63}$$

was found. This is in exact agreement with the result $P = -\boldsymbol{F}_0(z_A, \boldsymbol{v})\cdot\boldsymbol{v}$ implied by (8.62).

To determine the order of magnitude of quantum friction, let us consider the example of a ^{87}Rb atom moving parallel to an Au plate. The ground state $5^2S_{1/2}$ of ^{87}Rb is doubly degenerate, see App. B. To evaluate the quantum friction force, we only consider the dominant transition $5^2S_{1/2} \to 5^2P_{3/2}$ to the excited states of the $5^2P_{3/2}$ manifold, which is fourfold degenerate. The associated atomic spectral line is known as the D_2 line. We label the ground states as $|0_1\rangle$, $|0_2\rangle$ and the excited states as $|1_1\rangle \ldots |1_4\rangle$. The frequency for transitions between the ground-state and excited manifolds is $\omega_{10} = 2.41 \times 10^{15}$ rad/s [3].

Due to the degeneracy of the $5^2S_{1/2}$ state, the ground-state atom will be in an incoherent, equal-weight superposition of the states $|0_1\rangle$, $|0_2\rangle$, so that its internal density matrix reads $\hat{\sigma}_0 = \frac{1}{2}|0_1\rangle\langle 0_1| + \frac{1}{2}|0_2\rangle\langle 0_2|$. The friction force (8.29) is thus given by

$$\boldsymbol{F}_0(z_A, \boldsymbol{v}) = \tfrac{1}{2}[\boldsymbol{F}_{0_1}(z_A, \boldsymbol{v}) + \boldsymbol{F}_{0_2}(z_A, \boldsymbol{v})] \tag{8.64}$$

with

$$\boldsymbol{F}_{0_i}(z_A, \boldsymbol{v}) = -\frac{3\omega_S \boldsymbol{v}}{64\pi\varepsilon_0(\omega_{10}+\omega_S)^3 z_A^5} \sum_{1_k} \Gamma_k \left(2\boldsymbol{d}_{0_i 1_k}^{\perp 2} + \boldsymbol{d}_{0_i 1_k}^{\parallel 2}\right) . \tag{8.65}$$

As seen from (B.12) and (B.13), the dipole-matrix elements for the relevant transitions read

$$\boldsymbol{d}_{0_1 1_1}\boldsymbol{d}_{1_1 0_1} + \boldsymbol{d}_{0_2 1_1}\boldsymbol{d}_{1_1 0_2} = \frac{d_{10}^2}{8}\begin{pmatrix} 1 & \mathrm{i} & 0 \\ -\mathrm{i} & 1 & 0 \\ 0 & 0 & 0 \end{pmatrix}, \tag{8.66}$$

$$\boldsymbol{d}_{0_1 1_2}\boldsymbol{d}_{1_2 0_1} + \boldsymbol{d}_{0_2 1_2}\boldsymbol{d}_{1_2 0_2} = \frac{d_{10}^2}{24}\begin{pmatrix} 1 & \mathrm{i} & 0 \\ -\mathrm{i} & 1 & 0 \\ 0 & 0 & 4 \end{pmatrix}, \tag{8.67}$$

$$\boldsymbol{d}_{0_1 1_3}\boldsymbol{d}_{1_3 0_1} + \boldsymbol{d}_{0_2 1_3}\boldsymbol{d}_{1_3 0_2} = \frac{d_{10}^2}{24}\begin{pmatrix} 1 & -\mathrm{i} & 0 \\ \mathrm{i} & 1 & 0 \\ 0 & 0 & 4 \end{pmatrix}, \tag{8.68}$$

$$\boldsymbol{d}_{0_1 1_4}\boldsymbol{d}_{1_4 0_1} + \boldsymbol{d}_{0_2 1_4}\boldsymbol{d}_{1_4 0_2} = \frac{d_{10}^2}{8}\begin{pmatrix} 1 & -\mathrm{i} & 0 \\ \mathrm{i} & 1 & 0 \\ 0 & 0 & 0 \end{pmatrix} \tag{8.69}$$

where $d_{10} = 3.58 \times 10^{-29}$ Cm is the reduced matrix element for the D_2 transition [4]. With these results, the force (8.64) with (8.55) reads

$$\boldsymbol{F}_0(z_A, \boldsymbol{v}) = -\frac{d_{10}^2 \omega_S(\Gamma_{1_1} + 5\Gamma_{1_2} + 5\Gamma_{1_3} + 3\Gamma_{1_4})}{512\pi\varepsilon_0(\omega_{10} + \omega_S)^3} \frac{\boldsymbol{v}}{z_A^5} \tag{8.70}$$

with the decay rates (8.56) being given by

$$\Gamma_{1_k}(z_A) = \frac{\omega_{10} d_{10}^2}{192\pi\varepsilon_0 \hbar z_A^3} \frac{\gamma\omega_S^2}{(\omega_{10}^2 - \omega_S^2)^2} \times \begin{cases} 3 & \text{for } k = 1, 4\,, \\ 5 & \text{for } k = 2, 3\,. \end{cases} \tag{8.71}$$

Substituting the rates into the force, we have

$$\boldsymbol{F}_0(z_A, \boldsymbol{v}) = -\frac{17\omega_{10} d_{10}^4}{24576\pi^2\varepsilon_0^2\hbar} \frac{\gamma\omega_S^3}{(\omega_{10}+\omega_S)^3(\omega_{10}^2-\omega_S^2)^2} \frac{\boldsymbol{v}}{z_A^8}\,. \tag{8.72}$$

Using the atomic parameters as given above together with $\omega_P = 1.37 \times 10^{16}$ rad/s and $\gamma = 4.12 \times 10^{13}$ rad/s for Au [5] and noting that $m_{^{87}\mathrm{Rb}} = 1.44 \times 10^{-25}$ kg, we find that the deceleration due to quantum friction is

$$\boldsymbol{a} = -\boldsymbol{v}\big(0.64\,\mathrm{s}^{-1}\big)\left[\frac{1\,\mathrm{nm}}{z_A}\right]^8. \tag{8.73}$$

This shows that ground-state quantum friction extremely short-ranged, making it very difficult to observe experimentally.

Let us next turn our attention to the velocity-dependent force acting on an excited atom, which is dominated by the resonant component (8.36). As for the non-resonant force, the vacuum Doppler shift does not contribute for parallel motion. A comparison of the Green's tensors (8.46) and (8.47) shows that the delay term dominates over the Röntgen interaction in the nonretarded limit and we find

$$\boldsymbol{F}_n^{\mathrm{res}}(z_A, \boldsymbol{v}) = -\frac{3}{32\pi\varepsilon_0 z_A^5} \sum_{k<n} \big[\big(4\boldsymbol{d}_{nk}^{\perp 2} + \boldsymbol{d}_{nk}^{\parallel 2}\big)\boldsymbol{v} + 2\boldsymbol{d}_{nk}^{\parallel}\big(\boldsymbol{d}_{nk}^{\parallel}\cdot\boldsymbol{v}\big)\big] \times \left[\frac{\operatorname{Im}\varepsilon(\omega_{nk})}{|\varepsilon(\omega_{nk}) + 1|^2}\right]'. \tag{8.74}$$

For an atom in an isotropic state with (4.16), the force simplifies to

$$\boldsymbol{F}_n^{\mathrm{res}}(z_A, \boldsymbol{v}) = -\frac{\boldsymbol{v}}{4\pi\varepsilon_0 z_A^5} \sum_{k<n} \boldsymbol{d}_{nk}^2 \left[\frac{\operatorname{Im}\varepsilon(\omega_{nk})}{|\varepsilon(\omega_{nk}) + 1|^2}\right]'. \tag{8.75}$$

The resonant force is proportional to $1/z_A^5$ and falls off much less rapidly than the $1/z_A^8$ non-resonant force (8.51) with (8.49). Neglecting the latter, the force on an excited atom hence reads

$$F_n(z_A, v) = -\frac{3}{32\pi\varepsilon_0 z_A^5} \sum_{k<n} \left[\left(4d_{nk}^{\perp 2} + d_{nk}^{\parallel 2}\right)v + 2d_{nk}^{\parallel}\left(d_{nk}^{\parallel}\cdot v\right)\right] \times \left[\frac{\operatorname{Im}\varepsilon(\omega_{nk})}{|\varepsilon(\omega_{nk})+1|^2}\right]' . \tag{8.76}$$

To determine the sign of the force, let us again consider an atom moving parallel to a metal plate. Using the Drude model (4.93) with $\gamma \ll \omega_P$, we find

$$F_n(z_A, v) = -\frac{3\gamma\omega_S^2}{64\pi\varepsilon_0 z_A^5} \sum_{k<n} \frac{\omega_S^2 + 3\omega_{nk}^2}{(\omega_S^2 - \omega_{nk}^2)^3} \left[\left(4d_{nk}^{\perp 2} + d_{nk}^{\parallel 2}\right)v + 2d_{nk}^{\parallel}\left(d_{nk}^{\parallel}\cdot v\right)\right] . \tag{8.77}$$

The surface-plasmon frequencies are typically much larger than the atomic transition frequencies. In this case, the motion-induced force on an excited atom is a genuine quantum friction with

$$F_n(z_A, v)\cdot v < 0 . \tag{8.78}$$

The force is again proportional to the metallic damping parameter γ, indicating that the energy loss by the moving atom is balanced by Ohmic heating of the metal.

As an example, we again consider a ^{87}Rb atom moving parallel to a gold plate. Let us consider the case where the excited atom is prepared in the anisotropic excited state $\hat{\sigma}_1 = \frac{1}{2}|1_2\rangle\langle 1_2| + \frac{1}{2}|1_3\rangle\langle 1_3|$. Using the dipole matrix elements (8.67) and (8.68), we then have

$$F_1(z_A, v) = -\frac{5d_{01}^2}{128\pi\varepsilon_0} \frac{\gamma\omega_S^2(\omega_S^2 + 3\omega_{10}^2)}{(\omega_S^2 - \omega_{10}^2)^3} \frac{v}{z_A^5} . \tag{8.79}$$

With the numerical values for the Rb and gold parameters given above, quantum friction leads to a deceleration

$$a = -v\left(7.9 \times 10^3\,\mathrm{s}^{-1}\right)\left[\frac{1\,\mathrm{nm}}{z_A}\right]^5 \tag{8.80}$$

of the excited atom. In comparison to the ground-state force, excited-state quantum friction is strongly enhanced and has a much longer range. For instance, for an atomic velocity of $v = 100\,\mathrm{m/s}$, we find a deceleration $a = 8\,\mathrm{m/s}^2$ at distance $z_A = 10\,\mathrm{nm}$.

Recall that the excited-state force only acts for a short time due to the surface-enhanced spontaneous decay. In order to maintain the enhanced quantum friction, the atom must be continuously repumped to its excited state by a laser. The relative velocity reduction per photon can be estimated in a straightforward way if the excited state is an S state. In this case, the quantum friction (8.77) simplifies to

$$\boldsymbol{F}_1(z_A, \boldsymbol{v}) = -\frac{d_{01}^2}{8\pi\varepsilon_0} \frac{\gamma\omega_S^2(\omega_S^2 + 3\omega_{10}^2)}{(\omega_S^2 - \omega_{10}^2)^3} \frac{\boldsymbol{v}}{z_A^5} . \tag{8.81}$$

Without repumping, it acts during a time interval $\Delta t \simeq 1/\Gamma_1$, where the decay rate (8.56) reads

$$\Gamma_1(z_A) = \frac{\omega_{10} d_{01}^2}{12\pi\varepsilon_0\hbar} \frac{\gamma\omega_S^2}{(\omega_S^2 - \omega_{10}^2)^2} \frac{1}{z_A^3} . \tag{8.82}$$

The relative change in velocity induced by a single excitation quantum is hence

$$\frac{\Delta v}{v} \simeq -\frac{F_1(z_A, \boldsymbol{v})}{\Gamma_1(z_A) m_A v} \simeq -\frac{3\hbar}{m_A z_A^2} \frac{1}{2\omega_{10}} \tag{8.83}$$

where we have exploited the fact that $\omega_{10} \ll \omega_S$ typically holds for metals. The velocity reduction per photon for a metal hence only depends on the atomic mass and transition frequency and neither on its dipole moment nor on the parameters of the metal.

The quantum friction force on an atom moving parallel to a metal surface is far off-resonant, $\omega_{10} \ll \omega_S$. Our result (8.77) shows that the effect can be strongly enhanced near resonance, $\omega_{10} \simeq \omega_S$. This can be achieved for dielectric plates whose surface-plasmon frequencies are generally lower than those for metals. As an example, consider a sapphire plate whose surface-plasmon resonance is $\omega_S = 1.54 \times 10^{14}$ rad/s [6]. In the vicinity of this resonance, the permittivity of sapphire can be given as

$$\varepsilon(\omega) = \eta + \frac{\omega_P^2}{\omega_T^2 - \omega^2 - i\gamma\omega} \tag{8.84}$$

with parameters $\eta = 2.71$, $\omega_P = 2.12 \times 10^{14}$ rad/s, $\omega_T = 1.08 \times 10^{14}$ rad/s and $\gamma = 2.31 \times 10^{12}$ rad/s. With this model, the excited-state force (8.76) becomes ($\gamma \ll \omega_P$)

$$\boldsymbol{F}_n(z_A, \boldsymbol{v}) = -\frac{3\gamma\omega_P^2}{32\pi\varepsilon_0(\eta+1)^2 z_A^5} \sum_{k<n} \frac{(\omega_S^2 - \omega_{nk}^2)(\omega_S^2 + 3\omega_{nk}^2)}{[(\omega_S^2 - \omega_{nk}^2)^2 + \gamma^2\omega_{nk}^2]^2} \times \left[\left(4d_{nk}^{\perp 2} + \boldsymbol{d}_{nk}^{\parallel 2}\right)\boldsymbol{v} + 2\boldsymbol{d}_{nk}^{\parallel}\left(\boldsymbol{d}_{nk}^{\parallel} \cdot \boldsymbol{v}\right) \right] \tag{8.85}$$

with $\omega_S = \sqrt{\omega_T^2 + \omega_P/(\eta+1)}$.

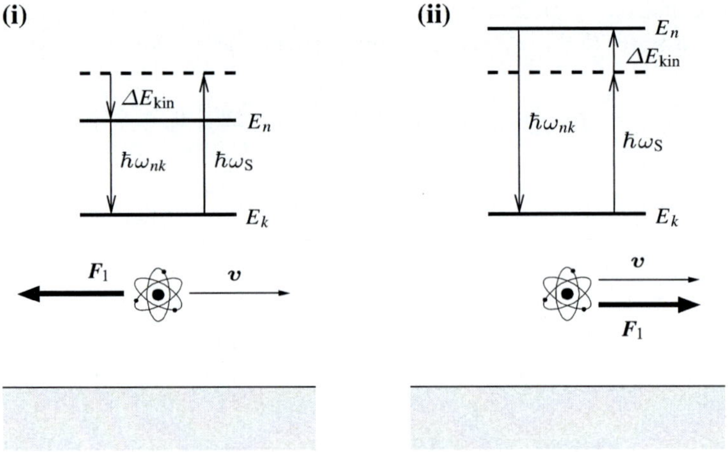

Fig. 8.2 (**i**) Quantum friction vs (**ii**) quantum acceleration experienced by an excited atom moving parallel to plate

As expected, the force is strongly enhanced in the vicinity of the surface-plasmon resonance $\omega_{nk} \simeq \omega_S$ where it can be written in the simpler form

$$\boldsymbol{F}_n(z_A, \boldsymbol{v}) = -\frac{3\gamma\omega_P^2}{64\pi\varepsilon_0(\eta+1)^2\omega_S z_A^5}\sum_{k<n}\frac{\omega_S - \omega_{nk}}{[(\omega_S-\omega_{nk})^2+\frac{1}{4}\gamma^2]^2}$$
$$\times\left[(4d_{nk}^{\perp 2} + \boldsymbol{d}_{nk}^{\parallel 2})\boldsymbol{v} + 2\boldsymbol{d}_{nk}^{\parallel}(\boldsymbol{d}_{nk}^{\parallel}\cdot\boldsymbol{v})\right]. \tag{8.86}$$

In addition, we note that the velocity-dependent CP force on the excited atom changes sign when the atomic transition frequency becomes larger than the surface-plasmon frequency: A dissipative quantum friction for $\omega_{nk} < \omega_S$ changes to a quantum acceleration force for $\omega_{nk} > \omega_S$. This behaviour can be easily understood from an energy conservation argument. The excited atom emits a photon of energy $\hbar\omega_{nk}$ whereas the dielectric surface can only absorb photons of the energy $\hbar\omega_S$, which excite surface plasmons. In the case $\omega_{nk} < \omega_S$, the atom does not have sufficient internal energy and it must loose kinetic energy ΔE_{kin} to excite the surface plasmon. The situation is depicted in Fig. 8.2(i). For $\hbar\omega_{nk} > \omega_S$, the atomic internal energy is greater than the required surface-plasmon frequency. The excess is converted into kinetic energy, accelerating the atom, cf. Fig. 8.2(ii).

As an example, let us consider ^{133}Cs whose $6^2D_{3/2} \to 7^2P_{1/2}$ transition has a frequency $\omega_{10} = 1.55 \times 10^{14}$ rad/s [6], very close to the sapphire resonance. The lower manifold $7^2P_{1/2}$ is doubly degenerate, we label its states as $|0_1\rangle$, $|0_2\rangle$, see App. B. Similarly, the states of the upper, fourfold degenerate $6^2D_{3/2}$ manifold are denoted by $|1_1\rangle \ldots |1_4\rangle$. We assume the atom to be prepared in the exited state $\hat{\sigma}_1 = \frac{1}{2}|1_1\rangle\langle 1_1| + \frac{1}{2}|1_4\rangle\langle 1_4|$. According to (B.14), the relevant dipole matrix elements are

$$\boldsymbol{d}_{1_10_1}\boldsymbol{d}_{0_11_1} + \boldsymbol{d}_{1_40_1}\boldsymbol{d}_{0_11_4} = \frac{d_{01}^2}{8}\begin{pmatrix} 1 & -\mathrm{i} & 0 \\ \mathrm{i} & 1 & 0 \\ 0 & 0 & 0 \end{pmatrix}, \tag{8.87}$$

$$\boldsymbol{d}_{1_10_2}\boldsymbol{d}_{0_21_1} + \boldsymbol{d}_{1_40_2}\boldsymbol{d}_{0_21_4} = \frac{d_{01}^2}{8}\begin{pmatrix} 1 & \mathrm{i} & 0 \\ -\mathrm{i} & 1 & 0 \\ 0 & 0 & 0 \end{pmatrix} \tag{8.88}$$

with a reduced matrix element $d_{01} = 1.17 \times 10^{-28}$ Cm [7]. The $\boldsymbol{v}$-dependent force on the excited Cs atom hence reads

$$\boldsymbol{F}_1(z_A, \boldsymbol{v}) = -\frac{3d_{01}^2}{128\pi\varepsilon_0(\eta+1)^2}\frac{\gamma\omega_\mathrm{P}^2(\omega_\mathrm{S}-\omega_{10})}{\omega_\mathrm{S}[(\omega_\mathrm{S}-\omega_{10})^2+\frac{1}{4}\gamma^2]^2}\frac{\boldsymbol{v}}{z_A^5}. \tag{8.89}$$

Using the numerical values for the sapphire and Cs parameters, we find a quantum acceleration ($m_{^{133}\mathrm{Cs}} = 2.21 \times 10^{-25}$ kg)

$$\boldsymbol{a} = +\boldsymbol{v}\big(5.3 \times 10^{11}\text{-1}\big)\left[\frac{1\mathrm{nm}}{z_A}\right]^5. \tag{8.90}$$

We note that the CP force is parallel to the velocity. This is due to the atomic transition frequency being slightly larger than the surface-plasmon frequency, recall Fig. 8.2(ii). We note that the force is much larger than the off-resonant quantum friction (8.80) near a gold surface. Even at a distance of $z_A = 100$ nm the acceleration of an atom of velocity $v = 100$ m/s is as large as $a = 5300\,\mathrm{m/s^2}$.

Again, the force on the excited atom only acts for a very short amount of time unless it is continuously repumped. Let us estimate the relative velocity boost per photon for an atom in an S state. In this case, the excited-state force (8.86) simplifies to

$$\boldsymbol{F}_1(z_A, \boldsymbol{v}) = -\frac{d_{01}^2}{8\pi\varepsilon_0(\eta+1)^2}\frac{\gamma\omega_\mathrm{P}^2(\omega_\mathrm{S}-\omega_{10})}{\omega_\mathrm{S}[(\omega_\mathrm{S}-\omega_{10})^2+\frac{1}{4}\gamma^2]^2}\frac{\boldsymbol{v}}{z_A^5}. \tag{8.91}$$

With the sapphire permittivity (8.84), the decay rate (8.49) reads

$$\Gamma_1(z_A) = \frac{d_{01}^2}{24\pi\varepsilon_0\hbar(\eta+1)^2}\frac{\gamma\omega_\mathrm{P}^2}{\omega_\mathrm{S}[(\omega_\mathrm{S}-\omega_{10})^2+\frac{1}{4}\gamma^2]}\frac{1}{z_A^3} \tag{8.92}$$

for an isotropic state. The relative velocity boost per photon is thus

$$\frac{\Delta v}{v} \simeq -\frac{F_1(z_A, \boldsymbol{v})}{\Gamma_1(z_A)m_A v} \simeq -\frac{3\hbar}{m_A z_A^2}\frac{\omega_\mathrm{S}-\omega_{10}}{(\omega_\mathrm{S}-\omega_{10})^2+\frac{1}{4}\gamma^2}. \tag{8.93}$$

It is strongly enhanced in comparison with the far off-resonant result (8.83) for a metal plate.

References

1. S. Scheel, S.Y. Buhmann, Phy. Rev. A **80**(4), 042902 (2009)
2. G. Barton, New J. Phys. **12**, 113045 (2010)
3. C.E. Moore, *Atomic Energy Levels: As Derived from the Analyses of Optical Spectra* (U.S. Government Printing Office, Washington, D.C., 1971)
4. D.A. Steck, Rubidium 87 D line data (2009), http://steck.us/alkalidata
5. E.D. Palik (ed.), *Handbook of Optical Constants of Solids II* (Academic Press, New York, 1991)
6. M.P. Gorza, M. Ducloy, Eur. Phys. J. D **40**(3), 343 (2006)
7. A. Lindgård, S.E. Nielsen, At. Data Nucl. Data Tables **19**(6), 533 (1977)

Appendix A
The Green's Tensor

In this appendix, we review some properties of the classical Green's tensor for the electromagnetic field as laid out in more detail in App. B of Vol. I. After defining it and summarising some general properties, we explicitly present the Green's tensors for a bulk medium and a planar multilayer system. Born-series expansions and scaling laws are also given.

A.1 Definition and General Properties

The classical Green's tensor for the electromagnetic field is uniquely defined by the inhomogeneous Helmholtz equation

$$\left[\nabla\times\frac{1}{\mu(\boldsymbol{r},\omega)}\nabla\times-\frac{\omega^2}{c^2}\,\varepsilon(\boldsymbol{r},\omega)\right]\boldsymbol{G}(\boldsymbol{r},\boldsymbol{r}',\omega)=\boldsymbol{\delta}(\boldsymbol{r}-\boldsymbol{r}') \tag{A.1}$$

together with the boundary condition

$$\boldsymbol{G}(\boldsymbol{r},\boldsymbol{r}',\omega)\to\boldsymbol{0}\quad\text{for}\quad|\boldsymbol{r}-\boldsymbol{r}'|\to\infty\,. \tag{A.2}$$

Here, the local and isotropic relative electric permittivity $\varepsilon(\boldsymbol{r},\omega)$ and magnetic permeability $\mu(\boldsymbol{r},\omega)$ characterise a given arrangement of absorbing and dispersing magnetoelectric bodies and/or media with $\mathrm{Im}\,\varepsilon(\boldsymbol{r},\omega)$, $\mathrm{Im}\,\mu(\boldsymbol{r},\omega)>0$.

As shown in App. B.1 of Vol. I, the Green's tensor has a number of useful general properties. It is an analytic function of frequency on the upper half of the complex plane. Furthermore, it satisfies the Schwarz reflection principle

S. Y. Buhmann, *Dispersion Forces II*, Springer Tracts in Modern Physics 248,
DOI: 10.1007/978-3-642-32466-6, © Springer-Verlag Berlin Heidelberg 2012

$$\boldsymbol{G}^*(\boldsymbol{r},\boldsymbol{r}',\omega)=\boldsymbol{G}(\boldsymbol{r},\boldsymbol{r}',-\omega^*) \tag{A.3}$$

as well as Onsager reciprocity

$$\boldsymbol{G}^{\mathrm{T}}(\boldsymbol{r},\boldsymbol{r}',\omega)=\boldsymbol{G}(\boldsymbol{r}',\boldsymbol{r},\omega) \tag{A.4}$$

and the integral relation

$$\int \mathrm{d}^3s\,\biggl\{-\frac{\operatorname{Im}\mu(\boldsymbol{s},\omega)}{|\mu(\boldsymbol{s},\omega)|^2}\Bigl[\boldsymbol{G}(\boldsymbol{r},\boldsymbol{s},\omega)\times\overleftarrow{\nabla}_s\Bigr]\cdot\bigl[\nabla_s\times\boldsymbol{G}^*(\boldsymbol{s},\boldsymbol{r}',\omega)\bigr] \\ +\frac{\omega^2}{c^2}\operatorname{Im}\varepsilon(\boldsymbol{s},\omega)\,\boldsymbol{G}(\boldsymbol{r},\boldsymbol{s},\omega)\cdot\boldsymbol{G}^*(\boldsymbol{s},\boldsymbol{r}',\omega)\biggr\}=\operatorname{Im}\boldsymbol{G}(\boldsymbol{r},\boldsymbol{r}',\omega) \tag{A.5}$$

holds [1, 2]. Large- and small-frequency limits of the Green's tensor are

$$\lim_{|\omega|\to\infty}\frac{\omega^2}{c^2}\,\boldsymbol{G}(\boldsymbol{r},\boldsymbol{r}',\omega)=-\boldsymbol{\delta}(\boldsymbol{r}-\boldsymbol{r}') \tag{A.6}$$

and

$$\lim_{|\omega|\to 0}\frac{\omega^2}{c^2}\,\boldsymbol{G}(\boldsymbol{r},\boldsymbol{r}',\omega)=-{}^{\|}\bigl[({}^{\|}\varepsilon^{\|})^{-1}\bigr]^{\|}(\boldsymbol{r},\boldsymbol{r}')\,, \tag{A.7}$$

respectively. Here, ${}^{\|}\bigl[({}^{\|}\varepsilon^{\|})^{-1}\bigr]^{\|}$ denotes the inverse of the operator with components $\varepsilon(\boldsymbol{r},\boldsymbol{r}')=\varepsilon(\boldsymbol{r},\omega{=}0)\boldsymbol{\delta}(\boldsymbol{r}-\boldsymbol{r}')$ in the space of longitudinal functions, so that

$$\lim_{|\omega|\to 0}\frac{\omega^2}{c^2}\,\boldsymbol{G}^{\perp}(\boldsymbol{r},\boldsymbol{r}',\omega)=\lim_{|\omega|\to 0}\frac{\omega^2}{c^2}\,{}^{\perp}\boldsymbol{G}(\boldsymbol{r},\boldsymbol{r}',\omega)=\boldsymbol{0}\,. \tag{A.8}$$

The electric–magnetic dual Green's tensor $\boldsymbol{G}^{\circledast}$ is the solution to the Helmholtz equation with a global exchange of electric and magnetic properties,

$$\left[\nabla\times\frac{1}{\varepsilon(\boldsymbol{r},\omega)}\nabla\times-\frac{\omega^2}{c^2}\,\mu(\boldsymbol{r},\omega)\right]\boldsymbol{G}^{\circledast}(\boldsymbol{r},\boldsymbol{r}',\omega)=\boldsymbol{\delta}(\boldsymbol{r}-\boldsymbol{r}')\,. \tag{A.9}$$

By introducing the tensors

$$\boldsymbol{G}_{ee}(\boldsymbol{r},\boldsymbol{r}',\omega)=\frac{\mathrm{i}\omega}{c}\,\boldsymbol{G}(\boldsymbol{r},\boldsymbol{r}',\omega)\,\frac{\mathrm{i}\omega}{c}\,, \tag{A.10}$$

$$\boldsymbol{G}_{mm}(\boldsymbol{r},\boldsymbol{r}',\omega)=\nabla\times\boldsymbol{G}(\boldsymbol{r},\boldsymbol{r}',\omega)\times\overleftarrow{\nabla}'\,, \tag{A.11}$$

$$\boldsymbol{G}_{em}(\boldsymbol{r},\boldsymbol{r}',\omega)=\frac{\mathrm{i}\omega}{c}\,\boldsymbol{G}(\boldsymbol{r},\boldsymbol{r}',\omega)\times\overleftarrow{\nabla}'\,,\tag{A.12}$$

$$\boldsymbol{G}_{me}(\boldsymbol{r},\boldsymbol{r}',\omega)=\nabla\times\boldsymbol{G}(\boldsymbol{r},\boldsymbol{r}',\omega)\,\frac{\mathrm{i}\omega}{c}\,,\tag{A.13}$$

one can show that the dual Green's tensor is related to the original one according to [3, 4]

$$\begin{aligned}\boldsymbol{G}^{\circledast}_{ee}(\boldsymbol{r},\boldsymbol{r}',\omega)&=\frac{1}{\mu(\boldsymbol{r},\omega)}\boldsymbol{G}_{mm}(\boldsymbol{r},\boldsymbol{r}',\omega)\,\frac{1}{\mu(\boldsymbol{r}',\omega)}\\&\quad+\frac{1}{\mu(\boldsymbol{r},\omega)}\boldsymbol{\delta}(\boldsymbol{r}-\boldsymbol{r}')\,,\end{aligned}\tag{A.14}$$

$$\begin{aligned}\boldsymbol{G}^{\circledast}_{mm}(\boldsymbol{r},\boldsymbol{r}',\omega)&=\varepsilon(\boldsymbol{r},\omega)\boldsymbol{G}_{ee}(\boldsymbol{r},\boldsymbol{r}',\omega)\varepsilon(\boldsymbol{r}',\omega)\\&\quad-\varepsilon(\boldsymbol{r},\omega)\boldsymbol{\delta}(\boldsymbol{r}-\boldsymbol{r}')\,,\end{aligned}\tag{A.15}$$

$$\boldsymbol{G}^{\circledast}_{em}(\boldsymbol{r},\boldsymbol{r}',\omega)=-\frac{1}{\mu(\boldsymbol{r},\omega)}\,\boldsymbol{G}_{me}(\boldsymbol{r},\boldsymbol{r}',\omega)\varepsilon(\boldsymbol{r}',\omega)\,,\tag{A.16}$$

$$\boldsymbol{G}^{\circledast}_{me}(\boldsymbol{r},\boldsymbol{r}',\omega)=-\varepsilon(\boldsymbol{r},\omega)\boldsymbol{G}_{em}(\boldsymbol{r},\boldsymbol{r}',\omega)\,\frac{1}{\mu(\boldsymbol{r}',\omega)}\,.\tag{A.17}$$

A.2 Bulk Green's Tensor

For an infinitely extended homogeneous bulk medium of permittivity $\varepsilon(\omega)$ and permeability $\mu(\omega)$, the Green's tensor takes the form (cf. App. B.2 of Vol. I) [1, 2]

$$\begin{aligned}\boldsymbol{G}^{(0)}(\boldsymbol{r},\boldsymbol{r}',\omega)&=-\frac{\mu(\omega)}{3k^2}\,\boldsymbol{\delta}(\boldsymbol{\rho})-\frac{\mu(\omega)\mathrm{e}^{\mathrm{i}k\rho}}{4\pi k^2\rho^3}\big\{\big[1-\mathrm{i}k\rho-(k\rho)^2\big]\boldsymbol{I}\\&\quad-\big[3-3\mathrm{i}k\rho-(k\rho)^2\big]\boldsymbol{e}_\rho\boldsymbol{e}_\rho\big\}\end{aligned}\tag{A.18}$$

($\boldsymbol{\rho}=\boldsymbol{r}-\boldsymbol{r}'$; $\rho=|\boldsymbol{\rho}|$; $\boldsymbol{e}_\rho=\boldsymbol{\rho}/\rho$) with $k=\sqrt{\varepsilon(\omega)\mu(\omega)}\,\omega/c$ denoting the wave number. It has to satisfy $\mathrm{Im}\,k>0$ for an absorbing medium. The large- and small-frequencies of this bulk Green's tensor read

$$\lim_{|\omega|\to\infty}\frac{\omega^2}{c^2}\,\boldsymbol{G}^{(0)}(\boldsymbol{r},\boldsymbol{r}',\omega)=-\boldsymbol{\delta}(\boldsymbol{r}-\boldsymbol{r}')\,,\tag{A.19}$$

$$\lim_{|\omega|\to 0}\frac{\omega^2}{c^2}\,\boldsymbol{G}^{(0)}(\boldsymbol{r},\boldsymbol{r}',\omega)=-\frac{1}{\varepsilon(0)}\,\boldsymbol{\delta}^{\parallel}(\boldsymbol{r}-\boldsymbol{r}')\,,\tag{A.20}$$

respectively. One can explicitly verify that the dual of the bulk Green's tensor is given by (A.14)–(A.17).

An important special case is free-space Green's tensor

$$\begin{aligned}\boldsymbol{G}^{(0)}(\boldsymbol{r},\boldsymbol{r}',\omega) = -\frac{c^2}{3\omega^2}\,\delta(\boldsymbol{\rho}) - \frac{c^2 e^{i\omega\rho/c}}{4\pi\omega^2\rho^3}\bigg\{\bigg[1 - i\frac{\omega\rho}{c} - \Big(\frac{\omega\rho}{c}\Big)^2\bigg]\boldsymbol{I} \\ -\bigg[3 - 3i\frac{\omega\rho}{c} - \Big(\frac{\omega\rho}{c}\Big)^2\bigg]\boldsymbol{e}_\rho\boldsymbol{e}_\rho\bigg\}\,. \qquad (A.21)\end{aligned}$$

It can be decomposed into transverse and longitudinal components

$$\boldsymbol{G}^{(0)\parallel}(\boldsymbol{r},\boldsymbol{r}',\omega) = -\frac{c^2}{3\omega^2}\,\delta(\boldsymbol{\rho}) - \frac{c^2}{4\pi\omega^2\rho^3}[\boldsymbol{I} - 3\boldsymbol{e}_\rho\boldsymbol{e}_\rho]\,, \qquad (A.22)$$

$$\begin{aligned}\boldsymbol{G}^{(0)\perp}(\boldsymbol{r},\boldsymbol{r}',\omega) = \frac{c^2}{4\pi\omega^2\rho^3}\bigg([\boldsymbol{I} - 3\boldsymbol{e}_\rho\boldsymbol{e}_\rho] - \bigg\{\bigg[1 - i\frac{\omega\rho}{c} - \Big(\frac{\omega\rho}{c}\Big)^2\bigg]\boldsymbol{I} \\ -\bigg[3 - 3i\frac{\omega\rho}{c} - \Big(\frac{\omega\rho}{c}\Big)^2\bigg]\boldsymbol{e}_\rho\boldsymbol{e}_\rho\bigg\}e^{i\omega\rho/c}\bigg) \qquad (A.23)\end{aligned}$$

where the displayed right-longitudinal and -transverse components are also left-longitudinal and -transverse components. Note that the free-space Green's tensor becomes purely transverse in the retarded, long-distance limit $\omega\rho/c \gg 1$,

$$\boldsymbol{G}^{(0)}(\boldsymbol{r},\boldsymbol{r}',\omega) = \boldsymbol{G}^{(0)\perp}(\boldsymbol{r},\boldsymbol{r}',\omega) = \frac{e^{i\omega\rho/c}}{4\pi\rho}\,(\boldsymbol{I} - \boldsymbol{e}_\rho\boldsymbol{e}_\rho)\,, \qquad (A.24)$$

while being purely longitudinal in the nonretarded, short-distance limit $\omega\rho/c \ll 1$,

$$\boldsymbol{G}^{(0)}(\boldsymbol{r},\boldsymbol{r}',\omega) = \boldsymbol{G}^{(0)\parallel}(\boldsymbol{r},\boldsymbol{r}',\omega) = -\frac{c^2}{3\omega^2}\,\delta(\boldsymbol{\rho}) - \frac{c^2}{4\pi\omega^2\rho^3}(\boldsymbol{I} - 3\boldsymbol{e}_\rho\boldsymbol{e}_\rho)\,. \quad (A.25)$$

Using a Born series (see App. A.4), one can show that even in the presence of magnetoelectric bodies, the Green's tensor has purely transverse/longitudinal asymptotes in the retarded/nonretarded limits.

Taking imaginary parts of (A.21) above, one can further show that the imaginary part of the free-space Green's tensor takes the finite value

$$\mathrm{Im}\,\boldsymbol{G}^{(0)}(\boldsymbol{r},\boldsymbol{r},\omega) = \frac{\omega}{6\pi c}\,\boldsymbol{I} \qquad (A.26)$$

in the coincidence limit.

A.3 Scattering Green's Tensor

Whenever the source and field points $\mathbf{r}'$ and $\mathbf{r}$ are situated in a common connected region of uniform magnetoelectric properties, the Green's tensor can be separated into bulk and scattering parts according to

$$\boldsymbol{G}(\mathbf{r},\mathbf{r}',\omega) = \boldsymbol{G}^{(0)}(\mathbf{r},\mathbf{r}',\omega) + \boldsymbol{G}^{(1)}(\mathbf{r},\mathbf{r}',\omega)\,. \tag{A.27}$$

The bulk part of the Green's tensor is simply given by (A.18) while the scattering part depends on the particular environment. Note that for source and field points in regions of different magnetoelectric properties, the notion of a bulk part becomes meaningless and the Green's tensor coincides with its scattering part.

A.3.1 General Properties

The scattering Green's tensor is purely transverse,

$$\boldsymbol{G}^{(1)}(\mathbf{r},\mathbf{r}',\omega) = {}^{\perp}\boldsymbol{G}^{(1)}(\mathbf{r},\mathbf{r}',\omega) = \boldsymbol{G}^{(1)\perp}(\mathbf{r},\mathbf{r}',\omega)\,, \tag{A.28}$$

and its large- and small-frequency limits are given by

$$\lim_{|\omega|\to\infty} \frac{\omega^2}{c^2}\,\boldsymbol{G}^{(1)}(\mathbf{r},\mathbf{r}',\omega) = \mathbf{0}\,, \tag{A.29}$$

$$\lim_{|\omega|\to 0} \frac{\omega^2}{c^2}\,\boldsymbol{G}^{(1)}(\mathbf{r},\mathbf{r}',\omega) = \mathbf{0}\,. \tag{A.30}$$

The dual of the scattering Green's tensor is given by [3, 4]

$$\boldsymbol{G}^{(1)\circledast}_{ee}(\mathbf{r},\mathbf{r}',\omega) = \frac{1}{\mu(\mathbf{r},\omega)}\,\boldsymbol{G}^{(1)}_{mm}(\mathbf{r},\mathbf{r}',\omega)\,\frac{1}{\mu(\mathbf{r},\omega)}\,, \tag{A.31}$$

$$\boldsymbol{G}^{(1)\circledast}_{mm}(\mathbf{r},\mathbf{r}',\omega) = \varepsilon(\mathbf{r},\omega)\boldsymbol{G}^{(1)}_{ee}(\mathbf{r},\mathbf{r}',\omega)\varepsilon(\mathbf{r},\omega)\,, \tag{A.32}$$

$$\boldsymbol{G}^{(1)\circledast}_{em}(\mathbf{r},\mathbf{r}',\omega) = -\frac{1}{\mu(\mathbf{r},\omega)}\,\boldsymbol{G}^{(1)}_{me}(\mathbf{r},\mathbf{r}',\omega)\varepsilon(\mathbf{r},\omega)\,, \tag{A.33}$$

$$\boldsymbol{G}^{(1)\circledast}_{me}(\mathbf{r},\mathbf{r}',\omega) = -\varepsilon(\mathbf{r},\omega)\boldsymbol{G}^{(1)}_{em}(\mathbf{r},\mathbf{r}',\omega)\,\frac{1}{\mu(\mathbf{r},\omega)}\,. \tag{A.34}$$

As shown in Appendix B of Vol. I, scattering Green's tensors for piecewise constant media in highly symmetric configurations can be determined analytically by invoking conditions of continuity.

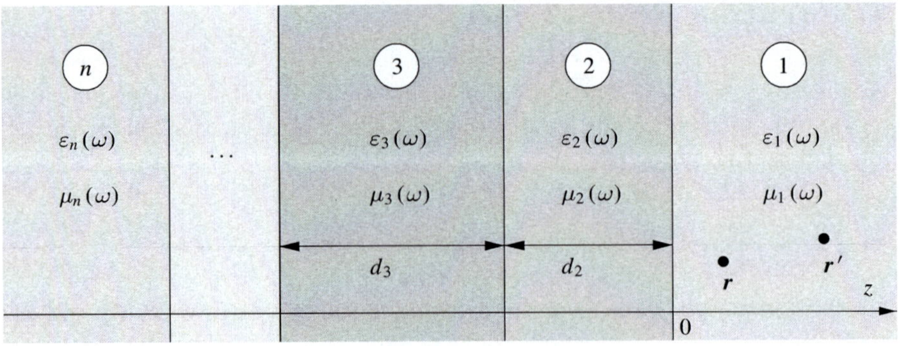

Fig. A.1 Planar n-layer system with source and field points in an outer layer

A.3.2 Planar Multilayer System

As illustrated in Fig. A.1, a planar multilayer system consists of n homogeneous layers of permittivities $\varepsilon_j(\omega)$ and permeabilities $\mu_j(\omega)$ $(j = 1\ldots n)$ where the intermediate layers have thickness d_j $(j = 2\ldots n-1)$. When both source and field points are situated in the outer layer 1, then the scattering Green's tensor of the system can be given as [5, 6]

$$\mathbf{G}^{(1)}(\boldsymbol{r},\boldsymbol{r}',\omega) = \frac{\mathrm{i}\mu_1(\omega)}{8\pi^2}\int\frac{\mathrm{d}^2k^{\|}}{k_1^{\perp}}\,\mathrm{e}^{\mathrm{i}\boldsymbol{k}^{\|}\cdot(\boldsymbol{r}-\boldsymbol{r}')+\mathrm{i}k_1^{\perp}(z+z')}\sum_{\sigma=s,p} r_\sigma^1 \boldsymbol{e}_{\sigma+}\boldsymbol{e}_{\sigma-} \quad \text{for } \boldsymbol{r},\boldsymbol{r}'\in V_1\,. \tag{A.35}$$

cf. App. B.3.2 of Vol. I. Here, $\boldsymbol{k}^{\|}\perp\boldsymbol{e}_z$ and $\pm k_j^{\perp}$ with

$$k_j^{\perp} = k_j^{\perp}(k^{\|},\omega) = \sqrt{\varepsilon_j(\omega)\mu_j(\omega)\frac{\omega^2}{c^2} - k^{\|2}}\,,\quad \mathrm{Im}k_j^{\perp} > 0 \tag{A.36}$$

are the components of the wave vector parallel and perpendicular to the interfaces. The polarisation unit vectors for s- and p-polarised waves in layer 1 read

$$\boldsymbol{e}_{s\pm} = \boldsymbol{e}_{s\pm}(\boldsymbol{k}^{\|},\omega) = \boldsymbol{e}_{k^{\|}}\times\boldsymbol{e}_z\,, \tag{A.37}$$

$$\boldsymbol{e}_{p\pm} = \boldsymbol{e}_{p\pm}(\boldsymbol{k}^{\|},\omega) = \frac{1}{k_1}\left(k^{\|}\boldsymbol{e}_z \mp k_1^{\perp}\boldsymbol{e}_{k^{\|}}\right) \tag{A.38}$$

with $k_1 = \sqrt{\varepsilon_1(\omega)\mu_1(\omega)}\,\omega/c$, $\mathrm{Im}\,k_1 > 0$; they are perpendicular (German: "*s*enkrecht") and parallel (German: "*p*arallel") to the plane of incidence, respectively.

The generalised Fresnel reflection coefficients can be found from the recursive relations

$$
\begin{aligned}
r_s^j &= r_s^j(k^\parallel,\omega)\\
&= \frac{(\mu_{j+1}k_j^\perp - \mu_j k_{j+1}^\perp) + (\mu_{j+1}k_j^\perp + \mu_j k_{j+1}^\perp)\mathrm{e}^{2\mathrm{i}k_{j+1}^\perp d_{j+1}} r_s^{j+1}}{(\mu_{j+1}k_j^\perp + \mu_j k_{j+1}^\perp) + (\mu_{j+1}k_j^\perp - \mu_j k_{j+1}^\perp)\mathrm{e}^{2\mathrm{i}k_{j+1}^\perp d_{j+1}} r_s^{j+1}}, \qquad \text{(A.39)}\\
r_p^j &= r_p^j(k^\parallel,\omega)\\
&= \frac{(\varepsilon_{j+1}k_j^\perp - \varepsilon_j k_{j+1}^\perp) + (\varepsilon_{j+1}k_j^\perp + \varepsilon_j k_{j+1}^\perp)\mathrm{e}^{2\mathrm{i}k_{j+1}^\perp d_{j+1}} r_s^{j+1}}{(\varepsilon_{j+1}k_j^\perp + \varepsilon_j k_{j+1}^\perp) + (\varepsilon_{j+1}k_j^\perp - \varepsilon_j k_{j+1}^\perp)\mathrm{e}^{2\mathrm{i}k_{j+1}^\perp d_{j+1}} r_p^{j+1}} \qquad \text{(A.40)}
\end{aligned}
$$

for $j = 1 \ldots n-1$ with $\varepsilon_j = \varepsilon_j(\omega)$, $\mu_j = \mu_j(\omega)$ and termination condition $r_\sigma^n = 0$. For a two-layer system ($n = 2$), they reduce to the ordinary Fresnel coefficients

$$
r_s^1 = r_s^1(k^\parallel,\omega) = \frac{\mu_2(\omega)k_1^\perp - \mu_1(\omega)k_2^\perp}{\mu_2(\omega)k_1^\perp + \mu_1(\omega)k_2^\perp}, \qquad \text{(A.41)}
$$

$$
r_p^1 = r_p^1(k^\parallel,\omega) = \frac{\varepsilon_2(\omega)k_1^\perp - \varepsilon_1(\omega)k_2^\perp}{\varepsilon_2(\omega)k_1^\perp + \varepsilon_1(\omega)k_2^\perp}. \qquad \text{(A.42)}
$$

In the context of CP forces, the equal-position Green's tensor is of particular interest. For $\boldsymbol{r} = \boldsymbol{r}'$ above a semi-infinite half space of permittivity $\varepsilon(\omega)$ and permeability $\mu(\omega)$, it is explicitly given by

$$
\begin{aligned}
\boldsymbol{G}^{(1)}(\boldsymbol{r},\boldsymbol{r},\omega) &= \frac{\mathrm{i}}{8\pi}\int_0^\infty \mathrm{d}k^\parallel \, \frac{k^\parallel}{k^\perp}\,\mathrm{e}^{2\mathrm{i}k^\perp z}\\
&\times \left[\begin{pmatrix} 1 & 0 & 0\\ 0 & 1 & 0\\ 0 & 0 & 0\end{pmatrix} r_s + \frac{c^2}{\omega^2}\begin{pmatrix} -k^{\perp 2} & 0 & 0\\ 0 & -k^{\perp 2} & 0\\ 0 & 0 & 2k^{\parallel 2}\end{pmatrix} r_p\right], \qquad \text{(A.43)}
\end{aligned}
$$

see (4.56) in Sect. 4.2. Here, the reflection coefficients (A.41) and (A.42) take the simple form

$$
r_s = \frac{\mu(\omega)k^\perp - k_1^\perp}{\mu(\omega)k_1^\perp + k_1^\perp}, \quad r_p = \frac{\varepsilon(\omega)k^\perp - k_1^\perp}{\varepsilon(\omega)k^\perp + k_1^\perp} \qquad \text{(A.44)}
$$

with

$$k^{\perp} = \sqrt{\frac{\omega^2}{c^2} - k^{\parallel 2}}\,, \quad \operatorname{Im} k^{\perp} > 0\,, \tag{A.45}$$

$$k_1^{\perp} = \sqrt{\varepsilon(\omega)\mu(\omega)\,\frac{\omega^2}{c^2} - k^{\parallel 2}}\,, \quad \operatorname{Im} k_1^{\perp} > 0\,. \tag{A.46}$$

In the retarded limit $z \gg c/\omega$, the stationary-phase point $k^{\parallel} = 0$ with $\mathrm{d}k^{\perp}/\mathrm{d}k^{\parallel} = 0$ gives the main contribution to the oscillatory $k^{\parallel}$-integral. We may hence approximate $k^{\perp} \simeq \omega/c$ and $k_1^{\perp} \simeq \sqrt{\varepsilon(\omega)\mu(\omega)}\,\omega/c$ in the reflection coefficients. The integral can then be solved, and to leading order in $c/(\omega z)$, one finds the retarded Green's tensor

$$\boldsymbol{G}^{(1)}(\boldsymbol{r}, \boldsymbol{r}, \omega) = \frac{e^{2\mathrm{i}z\omega/c}}{8\pi z}\,\frac{\sqrt{\mu(\omega)} - \sqrt{\varepsilon(\omega)}}{\sqrt{\mu(\omega)} + \sqrt{\varepsilon(\omega)}} \begin{pmatrix} 1 & 0 & 0 \\ 0 & 1 & 0 \\ 0 & 0 & 0 \end{pmatrix}. \tag{A.47}$$

The roots have to be taken such that $\operatorname{Im}\sqrt{\varepsilon}\,, \operatorname{Im}\sqrt{\mu} > 0$. In the opposite nonretarded limit $z \ll c/[\sqrt{\varepsilon(\omega)\mu(\omega)}\,\omega]$, the main contribution to the integral is due to regions of large $k^{\parallel}$ where $k^{\perp} \simeq k_1^{\perp} \simeq \mathrm{i}k^{\parallel}$. With these approximations, the integral can again be carried out. Retaining only the leading-order contribution in $z\omega/c$, one finds the nonretarded Green's tensor

$$\boldsymbol{G}^{(1)}(\boldsymbol{r}, \boldsymbol{r}, \omega) = \frac{c^2}{32\pi\omega^2 z^3}\,\frac{\varepsilon(\omega) - 1}{\varepsilon(\omega) + 1} \begin{pmatrix} 1 & 0 & 0 \\ 0 & 1 & 0 \\ 0 & 0 & 2 \end{pmatrix}. \tag{A.48}$$

A.4 Born Expansion

As shown in Sect. 2.1, the Green's tensor can be approximated in a systematic way whenever the system's permittivity and permeability deviate only slightly from a background permittivity $\overline{\varepsilon}(\boldsymbol{r}, \omega)$ and permeability $\overline{\mu}(\boldsymbol{r}, \omega)$ whose Green's tensor

$$\left[\boldsymbol{\nabla} \times \frac{1}{\overline{\mu}(\boldsymbol{r}, \omega)}\,\boldsymbol{\nabla} \times -\frac{\omega^2}{c^2}\,\overline{\varepsilon}(\boldsymbol{r}, \omega)\right] \overline{\boldsymbol{G}}(\boldsymbol{r}, \boldsymbol{r}', \omega) = \boldsymbol{\delta}(\boldsymbol{r} - \boldsymbol{r}') \tag{A.49}$$

is known. For a purely electric correction $\chi(\boldsymbol{r}, \omega)$, the total permittivity can be decomposed as

$$\varepsilon(\boldsymbol{r}, \omega) = \overline{\varepsilon}(\boldsymbol{r}, \omega) + \chi(\boldsymbol{r}, \omega)\,. \tag{A.50}$$

The full Green's tensor is then the solution to the Dyson equation

$$\boldsymbol{G}(\boldsymbol{r},\boldsymbol{r}',\omega)=\overline{\boldsymbol{G}}(\boldsymbol{r},\boldsymbol{r}',\omega)+\frac{\omega^2}{c^2}\int \mathrm{d}^3s\,\chi(\boldsymbol{s},\omega)\overline{\boldsymbol{G}}(\boldsymbol{r},\boldsymbol{s},\omega)\cdot\boldsymbol{G}(\boldsymbol{s},\boldsymbol{r}',\omega)\,. \quad \text{(A.51)}$$

The Born expansion is obtained as an iterative solution to the Dyson equation in powers of χ. One finds [7]

$$\boldsymbol{G}(\boldsymbol{r},\boldsymbol{r}',\omega)=\overline{\boldsymbol{G}}(\boldsymbol{r},\boldsymbol{r}',\omega)+\sum_{K=1}^{\infty}\Delta_K\boldsymbol{G}(\boldsymbol{r},\boldsymbol{r}',\omega) \quad \text{(A.52)}$$

with

$$\Delta_K\boldsymbol{G}(\boldsymbol{r},\boldsymbol{r}',\omega)=\frac{\omega^{2K}}{c^{2K}}\int \mathrm{d}^3s_1\,\chi(s_1,\omega)\cdots\int \mathrm{d}^3s_K\,\chi(s_K,\omega)$$
$$\times\,\overline{\boldsymbol{G}}(\boldsymbol{r},s_1,\omega)\cdot\overline{\boldsymbol{G}}(s_1,s_2,\omega)\cdots\overline{\boldsymbol{G}}(s_K,\boldsymbol{r}',\omega)\,. \quad \text{(A.53)}$$

For a purely magnetic correction $\zeta(\boldsymbol{r},\omega)$, the decomposition

$$\frac{1}{\mu(\boldsymbol{r},\omega)}=\frac{1}{\overline{\mu}(\boldsymbol{r},\omega)}-\zeta(\boldsymbol{r},\omega)\,. \quad \text{(A.54)}$$

leads to the Dyson equation

$$\boldsymbol{G}(\boldsymbol{r},\boldsymbol{r}',\omega)=\overline{\boldsymbol{G}}(\boldsymbol{r},\boldsymbol{r}',\omega)$$
$$-\int \mathrm{d}^3s\,\zeta(\boldsymbol{s},\omega)\Big[\overline{\boldsymbol{G}}(\boldsymbol{r},\boldsymbol{s},\omega)\times\overleftarrow{\nabla}_{\boldsymbol{s}}\Big]\cdot\Big[\nabla_{\boldsymbol{s}}\times\boldsymbol{G}(\boldsymbol{s},\boldsymbol{r}',\omega)\Big]. \quad \text{(A.55)}$$

In this case, the terms of the Born expansion (A.52) are found to be

$$\Delta_K\boldsymbol{G}(\boldsymbol{r},\boldsymbol{r}',\omega)=(-1)^K\int \mathrm{d}^3s_1\,\zeta(s_1,\omega)\cdots\int \mathrm{d}^3s_K\,\zeta(s_K,\omega)$$
$$\times\Big[\overline{\boldsymbol{G}}(\boldsymbol{r},s_1,\omega)\times\overleftarrow{\nabla}_{s_1}\Big]\cdot\Big[\nabla_{s_1}\times\overline{\boldsymbol{G}}(s_1,s_2,\omega)\times\overleftarrow{\nabla}_{s_2}\Big]$$
$$\cdots\Big[\nabla_{s_K}\times\overline{\boldsymbol{G}}(s_K,\boldsymbol{r}',\omega)\Big]. \quad \text{(A.56)}$$

When the correction is genuinely magnetodielectric, so that (A.50) and (A.54) hold, then the Dyson equation can be given in compact notation as (λ, $\lambda'=e,m$)

$$\boldsymbol{G}_{\lambda\lambda'}(\boldsymbol{r},\boldsymbol{r}',\omega)=\overline{\boldsymbol{G}}_{\lambda\lambda'}(\boldsymbol{r},\boldsymbol{r}',\omega)$$
$$-\sum_{\lambda''=e,m}\int \mathrm{d}^3s\,\chi_{\lambda''}(\boldsymbol{s},\omega)\overline{\boldsymbol{G}}_{\lambda\lambda''}(\boldsymbol{r},\boldsymbol{s},\omega)\cdot\boldsymbol{G}_{\lambda''\lambda'}(\boldsymbol{s},\boldsymbol{r}',\omega)$$
$$\text{(A.57)}$$

where $\chi_e = \chi$, $\chi_m = \zeta$. The Born expansion thus takes the form

$$\boldsymbol{G}_{\lambda\lambda'}(\boldsymbol{r},\boldsymbol{r}',\omega) = \overline{\boldsymbol{G}}_{\lambda\lambda'}(\boldsymbol{r},\boldsymbol{r}',\omega) + \sum_{K=1}^{\infty} \Delta_K \boldsymbol{G}_{\lambda\lambda'}(\boldsymbol{r},\boldsymbol{r}',\omega) \tag{A.58}$$

with

$$\begin{aligned} &\Delta_K \boldsymbol{G}_{\lambda\lambda'}(\boldsymbol{r},\boldsymbol{r}',\omega) \\ &= (-1)^K \sum_{\lambda_1=e,m} \int \mathrm{d}^3 s_1\, \chi_{\lambda_1}(\boldsymbol{s}_1,\omega) \cdots \sum_{\lambda_K=e,m} \int \mathrm{d}^3 s_K\, \chi_{\lambda_K}(\boldsymbol{s}_K,\omega) \\ &\quad \times \overline{\boldsymbol{G}}_{\lambda\lambda_1}(\boldsymbol{r},\boldsymbol{s}_1,\omega)\cdot \overline{\boldsymbol{G}}_{\lambda_1\lambda_2}(\boldsymbol{s}_1,\boldsymbol{s}_2,\omega) \cdots \overline{\boldsymbol{G}}_{\lambda_K\lambda'}(\boldsymbol{s}_K,\boldsymbol{r}',\omega)\,. \end{aligned} \tag{A.59}$$

Note that by using different expansion parameters, one can obtain alternative Born series. As shown in Sect. 2.1, the terms of the alternative series are given by (2.15), (2.28) and (2.44) in place of (A.53), (A.56) and (A.59), respectively.

A.5 Scaling Behaviour

The Green's tensor has unique scaling properties. A scaling transformation is a global stretching or shrinking of an arrangement of bodies by a factor $a > 0$, so that the new permittivity and permeability can be given as

$$\tilde{\varepsilon}(a\boldsymbol{r},\omega) = \varepsilon(\boldsymbol{r},\omega)\,, \quad \tilde{\mu}(a\boldsymbol{r},\omega) = \mu(\boldsymbol{r},\omega) \tag{A.60}$$

For frequency-independent $\varepsilon(\boldsymbol{r},\omega) \equiv \varepsilon(\boldsymbol{r})$ and $\mu(\boldsymbol{r},\omega) \equiv \mu(\boldsymbol{r})$, we have shown in Sect. 3.2.1 that the Green's tensor scales according to [8]

$$\tilde{\boldsymbol{G}}(a\boldsymbol{r},a\boldsymbol{r}',\omega/a) = \frac{1}{a}\,\boldsymbol{G}(\boldsymbol{r},\boldsymbol{r}',\omega) \tag{A.61}$$

under such a transformation. The same scaling behaviour holds for the tensors $\boldsymbol{G}_{\lambda\lambda'}$ (λ, $\lambda' = e, m$):

$$\tilde{\boldsymbol{G}}_{\lambda\lambda'}(a\boldsymbol{r},a\boldsymbol{r}',\omega/a) = \frac{1}{a}\,\boldsymbol{G}_{\lambda\lambda'}(\boldsymbol{r},\boldsymbol{r}',\omega)\,. \tag{A.62}$$

For frequency-dependent permittivities and permeabilities, general scaling laws can only be formulated in the nonretarded limit and when distinguishing the cases of purely electric vs purely magnetic bodies, cf. Sect. 3.2.2. For purely electric bodies, one has [8]

$$\tilde{\boldsymbol{G}}(a\boldsymbol{r}, a\boldsymbol{r}', \omega) = \frac{1}{a^3}\,\boldsymbol{G}(\boldsymbol{r}, \boldsymbol{r}', \omega)\,. \tag{A.63}$$

For purely magnetic bodies, the bulk part of the nonretarded Green's tensor still scales as above, while the scattering Green's tensor behaves as [8]

$$\tilde{\boldsymbol{G}}^{(1)}(a\boldsymbol{r}, a\boldsymbol{r}', \omega) = \frac{1}{a}\,\boldsymbol{G}^{(1)}(\boldsymbol{r}, \boldsymbol{r}', \omega)\,. \tag{A.64}$$

More generally, we have

$$\tilde{\boldsymbol{G}}^{(1)}_{\lambda\lambda}(a\boldsymbol{r}, a\boldsymbol{r}', \tilde{\omega}) = \begin{cases} \dfrac{1}{a^3}\,\boldsymbol{G}^{(1)}_{\lambda\lambda}(\boldsymbol{r}, \boldsymbol{r}', \omega) \text{ for } \lambda = e\,, \\ \dfrac{1}{a}\,\boldsymbol{G}^{(1)}_{\lambda\lambda}(\boldsymbol{r}, \boldsymbol{r}', \omega) \quad \text{for } \lambda = m\,, \end{cases} \tag{A.65}$$

for purely electric bodies and

$$\tilde{\boldsymbol{G}}^{(1)}_{\lambda\lambda}(a\boldsymbol{r}, a\boldsymbol{r}', \tilde{\omega}) = \begin{cases} \dfrac{1}{a}\,\boldsymbol{G}^{(1)}_{\lambda\lambda}(\boldsymbol{r}, \boldsymbol{r}', \omega) \quad \text{for } \lambda = e\,, \\ \dfrac{1}{a^3}\,\boldsymbol{G}^{(1)}_{\lambda\lambda}(\boldsymbol{r}, \boldsymbol{r}', \omega) \quad \text{for } \lambda = m\,. \end{cases} \tag{A.66}$$

for magnetic bodies. The mixed tensors scale as

$$\tilde{\boldsymbol{G}}^{(1)}_{\lambda\lambda'}(a\boldsymbol{r}, a\boldsymbol{r}', \tilde{\omega}) = \frac{1}{a^2}\,\boldsymbol{G}^{(1)}_{\lambda\lambda'}(\boldsymbol{r}, \boldsymbol{r}', \omega) \quad (\lambda \neq \lambda') \tag{A.67}$$

in both cases.

References

1. L. Knöll, S. Scheel, D.G. Welsch, in *Coherence and Statistics of Photons and Atoms*, ed. by J. Perina (Wiley, New York, 2001), p. 1
2. D.T. Ho, S.Y. Buhmann, L. Knöll, D.G. Welsch, S. Scheel, J. Kästel, Phys. Rev. A **68**(4), 043816 (2003)
3. S.Y. Buhmann, S. Scheel, H. Safari, D.G. Welsch, Int. J. Mod. Phys. A **24**(8–9), 1796 (2009)
4. S.Y. Buhmann, S. Scheel, Phys. Rev. Lett. **102**(14), 140404 (2009)
5. L.W. Li, P.S. Kooi, M.S. Leong, T.S. Yeo, J. Electromagn. Waves Appl. **8**(6), 663 (1994)
6. M.S. Tomaš, Phys. Rev. A **51**(3), 2545 (1995)
7. S.Y. Buhmann, D.G. Welsch, Appl. Phys. B **82**(2), 189 (2006)
8. S.Y. Buhmann, S. Scheel, J. Babington, Phys. Rev. Lett. **104**(7), 070404 (2010)

Appendix B
Atomic Physics

In the following, we review some basic concepts from atomic physics which are needed for the examples studied in Chaps. 7 and 8.

The energy eigenstates $|\{n_i l_i\} LSJM\rangle$ of a multi-electron atom can be parametrised by the principal quantum numbers n_i and the quantum numbers for the orbital angular momenta $\hat{\boldsymbol{l}}_i$ ($l_i = 0, 1, \ldots, n_i - 1$) of each electron together with the quantum numbers for the total orbital angular momentum $\hat{\boldsymbol{L}} = \sum_i \hat{\boldsymbol{l}}_i$ (L), spin $\hat{\boldsymbol{S}}$ (S), angular momentum $\hat{\boldsymbol{J}} = \hat{\boldsymbol{L}} + \hat{\boldsymbol{S}}$ ($J = |L - S| \ldots L + S$) and its z-component $\hat{J}_z$ ($M = -J \ldots J$). By convention, they are denoted by giving electronic configuration as a list of occupied orbitals $n_i l_i^{k_i}$ (with $l = 0, 1, 2, 3, 4, \ldots$ being represented by the letters $s, p, d, f, g, \ldots$ and k_i denoting the number of electrons in the respective orbital) and specifying the remaining quantum numbers via a term symbol ${}^{2S+1}L_J$ (with $L = 0, 1, 2, 3, 4, \ldots$ being represented by the letters $S, P, D, F, G, \ldots$). For instance, the ground state of Li is $1s^2\, 2s^1\, {}^2S_{1/2}$, meaning that two electrons occupy the orbitals $n = 1, l = 0$, one occupies the orbital $n = 2, l = 0$ and the remaining quantum numbers are $L = 0$, $S = \frac{1}{2}$, $J = \frac{1}{2}$ (and $M = \pm\frac{1}{2}$). For atoms with a single valence electron, one commonly only displays its principal quantum number together with the term symbol, e.g., $2^2S_{1/2}$ for the Li ground-state.

The associated eigenenergies depend on the quantum numbers n_i, l_i, L, S, J. The dependence on S is due to the electron–electron Coulomb interaction together with the Pauli exclusion principle, while the dependences on L, J are due to the electronic spin–orbit interaction. They are is commonly referred to as the fine structure. We ignore the atomic hyperfine structure which is due to the interaction of the electrons with the multipole moments of the nucleus.

To determine dipole-matrix elements between energy eigenstates, we represent the dipole operator $\hat{\boldsymbol{d}}$ in terms of its spherical components $q = 0, \pm 1$

$$\hat{d}_{\pm 1} = \mp \frac{1}{\sqrt{2}} \left(\hat{d}_x \pm \mathrm{i} \hat{d}_y \right), \quad \hat{d}_0 = \hat{d}_z \tag{B.1}$$

S. Y. Buhmann, *Dispersion Forces II*, Springer Tracts in Modern Physics 248,
DOI: 10.1007/978-3-642-32466-6, © Springer-Verlag Berlin Heidelberg 2012

by writing

$$\hat{d}_x = \frac{1}{\sqrt{2}}\left(\hat{d}_{-1} - \hat{d}_{+1}\right) , \qquad \hat{d}_x = \frac{\mathrm{i}}{\sqrt{2}}\left(\hat{d}_{+1} + \hat{d}_{-1}\right) , \qquad \hat{d}_z = \hat{d}_0 . \tag{B.2}$$

According to the Wigner–Eckhart theorem [1–3], matrix elements of the spherical vector components can be factorised as

$$\langle \alpha' J' M' | \hat{d}_q | \alpha J M \rangle = (-1)^{J'-M'} \begin{pmatrix} J' & 1 & J \\ -M' & q & M \end{pmatrix} \langle \alpha' J' \| \hat{\boldsymbol{d}} \| \alpha J \rangle . \tag{B.3}$$

The first factor is the Wigner 3-j symbol; it depends only on the spherical tensor component q as well as the quantum numbers for the angular momenta J, J' and their z-components M, M'. The second factor is the reduced matrix element $\langle \alpha J \| \hat{d} \| \alpha' J' \rangle$ is independent of q, M and M'. It depends on J, J' and all other quantum numbers needed to characterise the respective states, represented in shorthand notation by α and α'.

Let us demonstrate the general case for the D_2 transitions of ^{87}Rb from the doubly degenerate ground state $5^2S_{1/2}$ ($n = 5$, $L = 0$, $S = \frac{1}{2}$, $J = \frac{1}{2}$, $M = \pm\frac{1}{2}$) to the excited states of the $5^2P_{3/2}$ manifold ($n' = 5$, $L' = 1$, $S' = \frac{1}{2}$, $J' = \frac{3}{2}$, $M' = -\frac{3}{2} \dots \frac{3}{2}$) which is fourfold degenerate. Using the Wigner–Eckart theorem (B.3) together with the spherical tensor components (B.2) and evaluating the 3-j symbols, we find

$$\left\langle J' = \tfrac{3}{2}, M' = \tfrac{3}{2} \middle| \hat{\boldsymbol{d}} \middle| J = \tfrac{1}{2}, M = \tfrac{1}{2} \right\rangle = \frac{\langle J' = \frac{3}{2} \| \hat{\boldsymbol{d}} \| J = \frac{1}{2} \rangle}{2\sqrt{2}} \begin{pmatrix} -1 \\ \mathrm{i} \\ 0 \end{pmatrix} , \tag{B.4}$$

$$\left\langle J' = \tfrac{3}{2}, M' = \tfrac{1}{2} \middle| \hat{\boldsymbol{d}} \middle| J = \tfrac{1}{2}, M = \tfrac{1}{2} \right\rangle = \frac{\langle J' = \frac{3}{2} \| \hat{\boldsymbol{d}} \| J = \frac{1}{2} \rangle}{\sqrt{6}} \begin{pmatrix} 0 \\ 0 \\ 1 \end{pmatrix} , \tag{B.5}$$

$$\left\langle J' = \tfrac{3}{2}, M' = -\tfrac{1}{2} \middle| \hat{\boldsymbol{d}} \middle| J = \tfrac{1}{2}, M = \tfrac{1}{2} \right\rangle = \frac{\langle J' = \frac{3}{2} \| \hat{\boldsymbol{d}} \| J = \frac{1}{2} \rangle}{2\sqrt{6}} \begin{pmatrix} 1 \\ \mathrm{i} \\ 0 \end{pmatrix} , \tag{B.6}$$

$$\left\langle J' = \tfrac{3}{2}, M' = -\tfrac{3}{2} \middle| \hat{\boldsymbol{d}} \middle| J = \tfrac{1}{2}, M = \tfrac{1}{2} \right\rangle = 0 , \tag{B.7}$$

$$\left\langle J' = \tfrac{3}{2}, M' = \tfrac{3}{2} \middle| \hat{\boldsymbol{d}} \middle| J = \tfrac{1}{2}, M = -\tfrac{1}{2} \right\rangle = 0 , \tag{B.8}$$

$$\left\langle J' = \tfrac{3}{2}, M' = \tfrac{1}{2} \middle| \hat{\boldsymbol{d}} \middle| J = \tfrac{1}{2}, M = -\tfrac{1}{2} \right\rangle = \frac{\langle J' = \frac{3}{2} \| \hat{\boldsymbol{d}} \| J = \frac{1}{2} \rangle}{2\sqrt{6}} \begin{pmatrix} -1 \\ \mathrm{i} \\ 0 \end{pmatrix} , \tag{B.9}$$

$$\left\langle J' = \tfrac{3}{2}, M' = -\tfrac{1}{2} \middle| \hat{\boldsymbol{d}} \middle| J = \tfrac{1}{2}, M = -\tfrac{1}{2} \right\rangle = \frac{\langle J' = \frac{3}{2} \| \hat{\boldsymbol{d}} \| J = \frac{1}{2} \rangle}{\sqrt{6}} \begin{pmatrix} 0 \\ 0 \\ 1 \end{pmatrix} , \tag{B.10}$$

$$\left\langle J'=\tfrac{3}{2}, M'=-\tfrac{3}{2}\middle|\hat{\boldsymbol{d}}\middle|J=\tfrac{1}{2}, M=-\tfrac{3}{2}\right\rangle = \frac{\langle J'=\frac{3}{2}\|\hat{\boldsymbol{d}}\|J=\frac{1}{2}\rangle}{2\sqrt{2}}\begin{pmatrix}1\\ \mathrm{i}\\ 0\end{pmatrix}. \tag{B.11}$$

where for convenience, we have suppressed the arguments $n = n' = 5$, $L = 0$, $L = 1$ and $S = S' = 1/2$. These results can be combined to give

$$\sum_{M=\pm 1/2}\left\langle J=\tfrac{1}{2}, M\middle|\hat{\boldsymbol{d}}\middle| J'=\tfrac{3}{2}, M'=\pm\tfrac{3}{2}\right\rangle\left\langle J'=\tfrac{3}{2}, M'=\pm\tfrac{3}{2}\middle|\hat{\boldsymbol{d}}\middle| J=\tfrac{1}{2}, M\right\rangle$$
$$= \frac{\left|\left\langle J'=\frac{3}{2}\middle\|\hat{\boldsymbol{d}}\middle\| J=\frac{1}{2}\right\rangle\right|^2}{8}\begin{pmatrix}1 & \mp\mathrm{i} & 0\\ \pm\mathrm{i} & 1 & 0\\ 0 & 0 & 0\end{pmatrix} \tag{B.12}$$

and

$$\sum_{M=\pm 1/2}\left\langle J=\tfrac{1}{2}, M\middle|\hat{\boldsymbol{d}}\middle| J'=\tfrac{3}{2}, M'=\pm\tfrac{1}{2}\right\rangle\left\langle J'=\tfrac{3}{2}, M'=\pm\tfrac{1}{2}\middle|\hat{\boldsymbol{d}}\middle| J=\tfrac{1}{2}, M\right\rangle$$
$$= \frac{\left|\left\langle J'=\frac{3}{2}\middle\|\hat{\boldsymbol{d}}\middle\| J=\frac{1}{2}\right\rangle\right|^2}{24}\begin{pmatrix}1 & \mp\mathrm{i} & 0\\ \pm\mathrm{i} & 1 & 0\\ 0 & 0 & 4\end{pmatrix}. \tag{B.13}$$

By introducing the abbreviating notations $|J=\frac{1}{2}, M=-\frac{1}{2}, \frac{1}{2}\rangle \equiv |0_1\rangle, |0_2\rangle$ for the $5^2S_{1/2}$ manifold and $|J'=\frac{3}{2}, M' = -\frac{3}{2}\ldots\frac{3}{2}\rangle \equiv |1_1\rangle\ldots|1_4\rangle$ for the $5^2P_{3/2}$ manifold as well as $\langle J'=\frac{3}{2}\|\hat{\boldsymbol{d}}\|J=\frac{1}{2}\rangle \equiv d_{10}$, these results assume the form of (8.66)–(8.69).

As a second example, we study ^{133}Cs, considering transitions from the doubly degenerate $7^2P_{1/2}$ manifold ($n = 7$, $L = 1$, $S = \frac{1}{2}$, $J = \frac{1}{2}$, $M = \pm\frac{1}{2}$) to the four higher-lying $6^2D_{3/2}$ states ($n' = 6$, $L' = 2$, $S' = \frac{1}{2}$, $J' = \frac{3}{2}$, $M' = -\frac{3}{2}\ldots\frac{3}{2}$). The dipole matrix elements between these transitions are given by (B.4)–(B.11), as before, but with different values of n, n', L and L'. They result in products

$$\sum_{M=\pm 3/2}\left\langle J=\tfrac{3}{2}, M\middle|\hat{\boldsymbol{d}}\middle| J'=\tfrac{1}{2}, M'=\pm\tfrac{1}{2}\right\rangle\left\langle J'=\tfrac{1}{2}, M'=\pm\tfrac{1}{2}\middle|\hat{\boldsymbol{d}}\middle| J=\tfrac{3}{2}, M\right\rangle$$
$$= \frac{\left|\left\langle J'=\frac{1}{2}\middle\|\hat{\boldsymbol{d}}\middle\| J=\frac{3}{2}\right\rangle\right|^2}{8}\begin{pmatrix}1 & \pm\mathrm{i} & 0\\ \mp\mathrm{i} & 1 & 0\\ 0 & 0 & 0\end{pmatrix} \tag{B.14}$$

Writing $|J=\frac{1}{2}, M=-\frac{1}{2}, \frac{1}{2}\rangle \equiv |0_1\rangle, |0_2\rangle$ and $|J=\frac{3}{2}, M=-\frac{3}{2}\ldots\frac{3}{2}\rangle \equiv |1_1\rangle\ldots|1_4\rangle$ for the states and $\langle J'=\frac{1}{2}\|\hat{d}\|J=\frac{3}{2}\rangle \equiv d_{01}$ for the reduced matrix element, this is equivalent to (8.87) and (8.88).

The analysis of electric-dipole transitions simplifies when neglecting the fine structure. In this case, the atomic energy eigenstates can alternatively be characterised by means of the quantum numbers of total orbital angular momentum $\hat{\boldsymbol{L}}$ (L) and spin $\hat{\boldsymbol{S}}$ (S) together with the their z-components $\hat{L}_z$ ($M_L = -L \dots L$) and $\hat{S}_z$ ($M_S = -S \dots S$): $|\{n_i l_i\} LSJM\rangle \mapsto |\{n_i l_i\} L M_L S M_S\rangle$. The spin degrees of freedom can be completely ignored, as the electric dipole operator only acts on the orbital angular momentum. The matrix elements of the latter follow from the Wigner–Eckart theorem for L, M_L,

$$\langle \alpha' L' M_L' | \hat{d}_q | \alpha L M_L \rangle = (-1)^{L'-M_L'} \begin{pmatrix} L' & 1 & L \\ -M_L' & q & M_L \end{pmatrix} \langle \alpha' L' \| \hat{\boldsymbol{d}} \| \alpha L \rangle \,, \tag{B.15}$$

together with the spherical tensor components (B.2). In particular, for transitions from an S state ($L = 0$, $M = 0$) to a P manifold ($L' = 1$, $M' = -1 \dots 1$), one finds

$$\big\langle L' = 1, M_L' = 1 \big| \hat{\boldsymbol{d}} \big| L = 0, M_L = 0 \big\rangle = \frac{\langle L' = 1 \| \hat{\boldsymbol{d}} \| L = 0 \rangle}{\sqrt{6}} \begin{pmatrix} -1 \\ \mathrm{i} \\ 0 \end{pmatrix} , \tag{B.16}$$

$$\big\langle L' = 1, M_L' = 0 \big| \hat{\boldsymbol{d}} \big| L = 0, M_L = 0 \big\rangle = \frac{\langle L' = 1 \| \hat{\boldsymbol{d}} \| L = 0 \rangle}{\sqrt{3}} \begin{pmatrix} 0 \\ 0 \\ 1 \end{pmatrix} , \tag{B.17}$$

$$\big\langle L' = 1, M_L' = -1 \big| \hat{\boldsymbol{d}} \big| L = 0, M_L = 0 \big\rangle = \frac{\langle L' = 1 \| \hat{\boldsymbol{d}} \| L = 0 \rangle}{\sqrt{6}} \begin{pmatrix} 1 \\ \mathrm{i} \\ 0 \end{pmatrix} . \tag{B.18}$$

After introducing the short-hand notations $|L = 0, M_L = 0\rangle \equiv |0\rangle$ for the S state, $|L' = 1, M_L' = -1, 0, 1\rangle \equiv |1_1\rangle, |1_2\rangle, |1_3\rangle$ for the P manifold as well as a reduced matrix element $\langle L = 1 \| \hat{d} \| L' = 0 \rangle \equiv d_{10}$, these results assume the form of (7.103). They immediately lead to

$$\begin{aligned} &\big\langle L = 0, M_L = 0 \big| \hat{\boldsymbol{d}} \big| L' = 1, M_L' = \pm 1 \big\rangle \big\langle L' = 1, M_L' = \pm 1 \big| \hat{\boldsymbol{d}} \big| L = 0, M_L = 0 \big\rangle \\ &\quad = \frac{|\langle L = 0 \| \hat{\boldsymbol{d}} \| L' = 1 \rangle|^2}{6} \begin{pmatrix} 1 & \mp\mathrm{i} & 0 \\ \pm\mathrm{i} & 1 & 0 \\ 0 & 0 & 0 \end{pmatrix} \end{aligned} \tag{B.19}$$

and

$$\begin{aligned} &\big\langle L = 0, M_L = 0 \big| \hat{\boldsymbol{d}} \big| L' = 1, M_L' = 0 \big\rangle \big\langle L' = 1, M_L' = 0 \big| \hat{\boldsymbol{d}} \big| L = 0, M_L = 0 \big\rangle \\ &\quad = \frac{|\langle L' = 1 \| \hat{\boldsymbol{d}} \| L = 0 \rangle|^2}{3} \begin{pmatrix} 0 & 0 & 0 \\ 0 & 0 & 0 \\ 0 & 0 & 1 \end{pmatrix} , \end{aligned} \tag{B.20}$$

as stated in (8.57)–(8.59). The total dipole transition tensor is isotropic, as required for transitions from an S state,

$$\sum_{M_L'=-1,0,1} \langle L=0,\, M_L=0|\hat{\boldsymbol{d}}|L'=1,\, M_L'\rangle\langle L'=1,\, M_L'|\hat{\boldsymbol{d}}|L=0,\, M_L=0\rangle$$

$$= \frac{|\langle L'=1\|\hat{\boldsymbol{d}}\|L=0\rangle|^2}{3}\,\boldsymbol{I}\,. \tag{B.21}$$

The consistency of this result with the more general description in terms of J and M can easily be established. Combining (B.4)–(B.11), we find

$$\sum_{J'=1/2,3/2}\ \sum_{M'=-J\ldots J} \langle J=\tfrac{1}{2},\, M=\pm\tfrac{1}{2}|\hat{\boldsymbol{d}}|J'M'\rangle\langle J',\, M'|\hat{\boldsymbol{d}}|J=\tfrac{1}{2},\, M\rangle$$

$$= |\langle J'=\tfrac{3}{2}\|\hat{\boldsymbol{d}}\|J=\tfrac{1}{2}\rangle|^2\left[\tfrac{1}{8}\begin{pmatrix}1 & \mp\mathrm{i} & 0\\ \pm\mathrm{i} & 1 & 0\\ 0 & 0 & 0\end{pmatrix} + \tfrac{1}{6}\begin{pmatrix}0&0&0\\0&0&0\\0&0&1\end{pmatrix} + \tfrac{1}{24}\begin{pmatrix}1 & \pm\mathrm{i} & 0\\ \mp\mathrm{i} & 1 & 0\\ 0 & 0 & 0\end{pmatrix}\right]$$

$$+ |\langle J'=\tfrac{1}{2}\|\hat{\boldsymbol{d}}\|J=\tfrac{1}{2}\rangle|^2\left[\tfrac{1}{6}\begin{pmatrix}1 & \pm\mathrm{i} & 0\\ \mp\mathrm{i} & 1 & 0\\ 0 & 0 & 0\end{pmatrix} + \tfrac{1}{6}\begin{pmatrix}0&0&0\\0&0&0\\0&0&1\end{pmatrix}\right]. \tag{B.22}$$

The reduced matrix elements $\langle J'=\frac{3}{2}\|\hat{\boldsymbol{d}}\|J=\frac{1}{2}\rangle$ and $\langle J'=\frac{1}{2}\|\hat{\boldsymbol{d}}\|J=\frac{1}{2}\rangle$ of the J, M scheme can expressed in terms of the reduced matrix element $\langle L'=1\|\hat{\boldsymbol{d}}\|L=0\rangle$ of the L, M_L scheme by means of the general relation [3]

$$\langle\alpha' L' S J'\|\hat{\boldsymbol{d}}\|\alpha L S J\rangle = (-1)^{L'+S+J+1}\sqrt{(2J+1)(2J'+1)}$$

$$\times\begin{Bmatrix}L' & J' & S\\ J & L & 1\end{Bmatrix}\langle\alpha' L'\|\hat{\boldsymbol{d}}\|\alpha L\rangle\,. \tag{B.23}$$

The Wigner 6-j symbol in curly brackets can be evaluated to give

$$\langle J'=\tfrac{1}{2}\|\hat{\boldsymbol{d}}\|J=\tfrac{1}{2}\rangle = -\sqrt{\tfrac{2}{3}}\,\langle\alpha' L'=1\|\hat{\boldsymbol{d}}\|L=0\rangle\,, \tag{B.24}$$

$$\langle J'=\tfrac{3}{2}\|\hat{\boldsymbol{d}}\|J=\tfrac{1}{2}\rangle = \tfrac{2}{\sqrt{3}}\,\langle\alpha' L'=1\|\hat{\boldsymbol{d}}\|L=0\rangle\,. \tag{B.25}$$

Substituting these results into (B.22), we obtain

$$\sum_{J'=1/2,3/2} \sum_{M'=-J\ldots J} \langle J=\tfrac{1}{2}, M|\hat{\boldsymbol{d}}|J'M'\rangle\langle J', M'|\hat{\boldsymbol{d}}|J=\tfrac{1}{2}, M\rangle$$
$$= \frac{|\langle L'=1\|\hat{\boldsymbol{d}}\|L=0\rangle|^2}{3}\,\boldsymbol{I}, \tag{B.26}$$

in agreement with the L, M_L result (B.21).

References

1. E. Wigner, Z. Phys. **43**(9–10), 624 (1927)
2. C. Eckart, Rev. Mod. Phys. **2**(3), 305 (1930)
3. A.R. Edmods, *Angular Momentum in Quantum Mechanics*, 4th edn. (Princeton University Press, Princeton, 1996)

Index

DOI: 10.1007/978-3-642-32466-6, © Springer-Verlag Berlin Heidelberg 2012